KT-131-354

THE UNIVERSITY
ANATOMY
DEPARTMENT
ST. ANDREWS

THE OVARY

SECOND EDITION

Volume I

General Aspects

Contributors

T. G. Baker
L. Bjersing
J. M. Dodd
Katy Haffen
R. J. Harrison
I. W. Rowlands
R. V. Short
H. G. Vevers
Barbara J. Weir
E. Wolff
S. Zuckerman

THE OVARY

SECOND EDITION

Volume I

General Aspects

EDITED BY **Professor Lord Zuckerman**

Zoological Society of London
London, England
and
University of East Anglia
Norwich, Norfolk
England

AND **Barbara J. Weir**

Wellcome Institute of Comparative Physiology
Zoological Society of London
London, England
and
Journal of Reproduction and Fertility,
Cambridge, England

ACADEMIC PRESS *New York San Francisco London 1977*

A Subsidiary of Harcourt Brace Jovanovich, Publishers

ACADEMIC PRESS, INC.
111 Fifth Avenue, New York, New York 10003

United Kingdom Edition published by
ACADEMIC PRESS, INC. (LONDON) LTD.
24/28 Oval Road, London NW1

Library of Congress Cataloging in Publication Data

Zuckerman, Professor Lord, Date ed.
The ovary.

Includes bibliographies and index.
1. Ovaries. I. Title. [DNLM: 1. Ovary. QL876
096]
QP261.Z8 1976 599′.01′66 76-13955
ISBN 0–12–782601–7

PRINTED IN THE UNITED STATES OF AMERICA

Contents

4 *Structure of the Mammalian Ovary*

R. J. HARRISON and BARBARA J. WEIR

5 *The Structure of the Ovary of Nonmammalian Vertebrates*

J. M. DODD

6 *Ovulation and Atresia*

BARBARA J. WEIR and I. W. ROWLANDS

7 *Ovarian Histochemistry*

L. BJERSING

List of Contributors

Numbers in parentheses indicate the pages on which the authors' contributions begin.

T. G. BAKER (41), Department of Obstetrics and Gynaecology, Hormone Laboratory, University of Edinburgh, Edinburgh, Scotland

L. BJERSING (303), Institute of Pathology, University of Umeå, Umeå, Sweden

J. M. DODD (219), Department of Zoology, University College of North Wales, Bangor, Gwynedd, Wales

KATY HAFFEN (69, 393),* Institute de Embryologie et Tèratologie Expèrimentale, Collège de France, Nagent-sur-Marne, France

R. J. HARRISON (113), Anatomy School, University of Cambridge, Cambridge, England

I. W. ROWLANDS (265),† Wellcome Institute of Comparative Physiology, Zoological Society of London, London, England

R. V. SHORT (1), Medical Research Council Unit of Reproductive Biology, Edinburgh, Scotland

H. G. VEVERS (447), Zoological Society of London, London, England

BARBARA J. WEIR (113, 265),‡ Wellcome Institute of Comparative Physiology, Zoological Society of London, London, England

E. WOLFF (393), Laboratoire d'Embryologie Experimentale, Collège de France, Paris, France

S. ZUCKERMAN (41), Zoological Society of London, and University of East Anglia, Norwich, Norfolk, England

* Present address: Unite de Recherche 61 de l'INSERM, Strasbourg-Hautespierre, France

† Present address: Department of Anatomy, University of Cambridge, Cambridge, England.

‡ Present address: Journal of Reproduction and Fertility, 7 Downing Place, Cambridge, England.

Preface

Insofar as its speed is necessarily that of its slowest member, the completion of a book by several hands is somewhat like the voyage of a convoy of ships. This does not, however, make the enterprise any the less valuable. The first edition of "The Ovary" appeared over fifteen years ago, and it is time the reviews it incorporated were brought up-to-date. The slowness with which some chapters arrived was not, however, the only reason for the delay in the appearance of this new edition. The editors could perhaps have tried to be more demanding of their contributors than they were. As it turned out, however, illness held up some chapters, and one, which was all but completed, had to be restarted when its main author was killed in an air crash. In 1961 I found myself constrained to apologize for the tardy appearance of the original edition, and I therefore do so again for this new one, not only to those authors who were first with their contributions and who have had to wait longest to see them in print, but also to our patient publishers, Academic Press, and to the scientific public for whom the work is designed.

Like any book of reviews, it is obvious that the chapters of this new edition will fail to mention some papers that may have appeared in the past three to four years. This, however, detracts little from their value. Immediately before the texts were sent to the publisher every author was invited to update, given she or he so wished, what had been written. No doubt there is a lesson to be drawn from the fact that few felt this necessary. A review of a subject does not gain in value if it merely catalogues the names of authors who have written about it, with a brief reference to the summaries of their published papers. For a review is nothing if it is not critical, and has little merit if its purpose is not to focus on such generalizations as are justified by the pieces of information on which it is based. In the ideal, the updating of a review should be an exercise in which the validity of any general proposition that has already been defined is examined in the light of new experimental data and in which new hypotheses are formulated where they are called for by new findings. This, of course, is the personal view of one editor who knows full well that it is not necessarily accepted universally.

I have heard it said that with the enormous growth of the world's scientific effort over the past two to three decades budding scientists are often advised that when they survey the literature which bears on the problems they are investigating there is little point in going back more than ten years, so rapidly is what is already known overlaid by new observation. This is something which I am sure many who belong to an older generation of scientists greatly regret. Sometimes it results in what is already established being "rediscovered." Often new findings are treated out of all proportion to the major generalizations to which they relate. What is more—and this applies not only to the fields of science with which "The Ovary" deals—the vast expansion of scientific activity over recent years has inevitably resulted in resources becoming available not only for what is called "big science" but for a wider range of enquiry in "small science" than would ever have been possible in more penurious times. When editing a work such as the present one, it is difficult to avoid the impression that on occasion an experiment or set of experiments has been undertaken merely because a professor or supervisor has had to provide a theme for a postgraduate's thesis. One also finds that new techniques have defined topics for experiment without the kind of critical preliminary evaluation of the limitations of new methods of enquiry in relation to the central questions which it is hoped they will help elucidate. With the vast growth in experimental work—and inevitably, therefore, the occasional dilution of its quality—"controls" also sometimes appear inadequate. In spite of these general observations, I have, however, no hesitation in saying that there has been a considerable increase in our knowledge of the ovary over the period since the appearance of the first edition.

Dr. Anita Mandl and Professor Peter Eckstein collaborated with me in the editing of the first edition but were unable to help in this, Dr. Mandl because she had retired from academic life and Professor Eckstein because of the heavy load of other work which he has since assumed. Fortunately I was able to recruit as coeditor Dr. Barbara Weir, to whom my thanks, as well as that of the contributors, are due, as they also are to Academic Press.

S. Zuckerman

Acknowledgments

Professor Short wishes to acknowledge his gratitude to Sir Geoffrey Keynes for his first introduction to the fascinations of the history of science, to Mr. W. E. Le Fanu, Mr. R. C. Martineau, and Professor U. Limentani for their help in translating original works from the Latin, Greek, and Italian. Miss Jessie Dobson, Curator of the Hunterian Museum of the Royal College of Surgeons, guided Professor Short through the treasure trove of Hunterian specimens and enabled him to reproduce original drawings of some of them. Professor W. J. Hamilton kindly provided a photograph of an early human embryo for Chapter 1, and Professor E. C. Amoroso, FRS, and Professor Sir Alan Parkes, FRS, made many helpful suggestions for this chapter.

The research activities presented by Dr. Bjersing in his chapter were supported in part by the Swedish Medical Research Council (Project Nos. 13X-78-01, 12X-78-02, 12X-78-03, B69-12X-78-04A, B70-12X-78-05B, and B71-12X-78-06C).

The help of the librarians and their staff of the Royal Society of Medicine (Mr. R. Wade), Wellcome History of Science Library (Mr. L. Symons) and of the Zoological Society of London (Mr. R. A. Fish) in checking references is greatly appreciated.

Preface to the First Edition

To the best of my knowledge no book on the normal ovary has appeared in English since the publication, in 1929, of Professor A. S. Parkes' monograph entitled "The Internal Secretions of the Ovary". The greater length of the present work, the object of which is to provide a detailed account of the principal aspects of ovarian development, structure and function as understood today, reflects the vigour with which researches on these subjects have been pursued over the past thirty years.

An almost unlimited number of topics could have been regarded as falling within the scope of the review. Since there was a necessary limit to size, the two volumes it constitutes cannot, accordingly, be claimed to exhaust the subject they were designed to cover. To whatever extent arbitrariness marks the fields dealt with, the treatise also partakes of a characteristic common to all scientific reviews, and one which reflects the fact that the content, pattern and emphasis of different fields of knowledge are always in a state of change.

The original intention was to publish the work in a single volume. When it became necessary to allocate the material to two, some rearrangement of chapters was called for, and the original sequence of topics which had been planned was changed. In the main, those chapters which relate to what might be called the natural history of the ovary are now included in Volume I, while information derived from more experimental and chemical studies is assembled in Volume II. The two volumes overlap to some extent, as do certain topics, but so far as possible this has been dealt with by means of cross-references.

I am deeply grateful to the many contributors for their generous assent to my invitation to participate in what has proved a lengthier and more arduous task than I, and perhaps they, anticipated at the outset. The authors of the various chapters are of course individually responsible for the content and bibliographic references as well as the style and accuracy of their contributions.

When manuscripts started to arrive, I had to turn to two of my colleagues, Dr. Anita Mandl and Dr. Peter Eckstein, for assistance in the work of editing, and of arranging for the translations of those chapters which were submitted in French. I am deeply grateful for their help, as I am also to the Academic Press for its tolerance during the long period in which this review has been in train. My thanks are also due to Miss Heather Paterson for her able help in preparing manuscripts and checking proofs, to Mr. L. T. Morton for compiling the Subject Index, and to the Academic Press for constructing the Author Index.

The delays which are inevitably associated with the production of a lengthy treatise have meant that a number of contributions appear less up-to-date in print than they did in typescript. Even though references to papers published in last year's scientific journals may be lacking, I nonetheless believe that at the moment the two volumes provide a more comprehensive picture of the whole subject than can be found in any other single work.

August, 1961 *S. Zuckerman*

Contents of Other Volumes

VOLUME II: PHYSIOLOGY

VOLUME III: REGULATION OF OOGENESIS AND STEROIDOGENESIS

THE OVARY

SECOND EDITION

Volume I

General Aspects

1

The Discovery of the Ovaries

R. V. Short

Who first discovered the ovaries? A simple enough question to pose, but one that is impossible to answer with any degree of certainty. As is so often the case, there was no sudden revelation, but rather a gradual process of realization that took place over the last 2000 years.

The earliest reference point we have is in the writings of Aristotle (384–322 B.C.), and although he provided a detailed description of the uterus, he did not recognize the existence of the ovaries. Nevertheless, in "Historia Animalium" he gives a detailed account of the spaying or gelding of sows, which was evidently a common agricultural practice at the time of his writing, and he also refers to the spaying of camels:

> The ovaries of sows are excised with the view of quenching in them sexual appetites and of stimulating growth in size and fatness. The sow has first to be kept two days without food, and, after being hung up by the hind legs, is operated on; they cut the lower belly about the place where the boars have their testicles, for it is there that the ovary grows, adhering to the two divisions (or horns) of the womb; they cut off a little piece and stitch up the incision. Female camels are mutilated when they are wanted for war purposes, and are mutilated to prevent their being got with young. (Thompson, 1910.)

The use of the word "ovary" here results from a free translation of the original Greek word *καπρια*; its literal meaning is a matter of speculation. The closely related word *κάπραω* meant the state of estrus in the sow (Liddell and Scott, 1940).

On reading Aristotle's remarkable account, two questions immediately come to mind: where was the operation first developed, and how did agriculturalists come to discover it almost 2000 years before the first scientists realized that ovaries play a part in reproduction? Aristotle states that sows were spayed in order to make them fatten better, and this may provide a clue. The fattening of livestock is a highly developed form of agricultural practice that would only be carried out by a stable, settled community. Since the pig has always been regarded as unclean by Jews and Moslems, neither of these peoples would have developed the operation. There is also no record of spaying ever having been carried out in ancient China (Needham and Gwei-Djen, 1966). The Egyptians are known to have kept pigs in large numbers since about 3000 B.C. (Zeuner, 1963) and the heiroglyph for womb, ♈ (Dawson, 1929; Gardiner, 1964), presumably depicts the bicornuate uterus of a pig, indicating that they had some knowledge of the anatomy of the female reproductive organs. However, the Egyptians did not fatten pigs for eating, but they used them as biological ploughs, for treading in the seed after rain or floods (Zeuner, 1963). Pork itself was an unpopular dish; most people would not eat it, and in any case its consumption was only permitted on certain specified days of the year. To many, the very touch of the animal was regarded as pollution, and the souls of the wicked were thought to migrate into pigs. None of these facts suggests that the Egyptians would have had the motivation to discover the technique of spaying, and there is certainly no mention of the operation in the Ebers or Kahun medical or veterinary papyri. On the other hand, we do know that pork was held in high esteem in Greece and Rome, where the fattening of pigs for the table became an important agricultural pursuit. So maybe it was in these countries that the operation was first discovered. But whatever could have been the motivation for spaying camels?

Aristotle tells us that camels were spayed when they were wanted for war purposes. We know that they were mainly used as pack animals, although some armies also used them as mounts for their cavalry; in either case, a heavily pregnant animal would be at a severe disadvantage, so it would have been desirable to stop them becoming pregnant. But perhaps the chief reason for spaying a camel was to prevent it coming into estrus. As Aristotle correctly states:

> Camels copulate with the female in a sitting posture, and the male straddles over and covers her . . . and they pass the whole day long in the operation. (Thompson, 1910.)

We now know that camels are induced ovulators, and if mating does not take place, the female will remain in estrus for up to 15 days (Nawito *et al.*, 1967; Novoa, 1970), ready to sit down at the sight of a male. Camels were domesticated in the Arabian peninsula at least as early as the eighteenth century B.C. (Zeuner, 1963) but were unknown in early Egypt. The spaying operation is unknown in Arabia today (W. Thesiger, personal communication), and so there is no clue to tell us where this might have originated. But there can be no doubt that the procedure would have conferred a distinct military advantage on the operated animal. Even today, the Touareg Arabs in North Africa prevent their transport camels from becoming pregnant by inserting a small pebble into the uterus, and this technique has probably been in use for centuries (Guttmacher, 1965).

The principal impediment to an earlier understanding of the role of the ovaries in reproduction was undoubtedly the all-pervading influence of the Aristotelean "seed and soil" concept of reproduction. This was to dominate scientific thought for almost 2000 years and confuse even such great men as William Harvey (see pp. 12–15). Aristotle believed that the male was the giver of "seed" and the female played a passive role by providing the "soil" in which the "seed" could grow. Although he associated the semen of the male in some way with the seed, he was of course completely unaware of the existence of spermatozoa. Semen was regarded as a secretion of the male duct system, and the testes themselves were only remotely concerned in its formation, acting as weights to keep the ducts straight, but not in themselves necessary for fertility. This was an understandable mistake, because Aristotle correctly observed that a bull could still get a cow in calf if allowed to serve her soon after he had been castrated.

The "soil" of the female was thought to be the catamenia, or menstrual coagulum. Seed and soil united to produce an egg, which was therefore a *product* of conception; the egg then went on to develop into an embryo. In Aristotle's own words,

> After the seed reaches the womb and remains there for a while, a membrane forms around it; for when it happens to escape before it is distinctly formed, it looks like an egg enveloped in its membranes after removal of the eggshell; and the membrane is full of veins. (Thompson, 1910.)

If one looks at an early human embryo within its membranes (see Fig. 1), it is easy to see how Aristotle thought it developed initially as an egg. But the chief objection to the idea that the egg could only arise as a product of conception was the domestic hen, which continues to lay eggs even in the absence of a cockerel. This perplexed Aristotle, who was forced to conclude that such eggs were formed by accident when a strong breeze blew into the vagina, and he called them "wind eggs."

Even if Aristotle failed to recognize the physiological significance of the ovaries, he inadvertently described the hypertrophied fetal gonads of the horse. Describing the pregnant mare, he says:

> Her foal, if dissected, is found to have other kidney-shaped substances around its kidneys, presenting the appearance of having four kidneys. (Thompson, 1910.)

As can be seen from Fig. 2, the fetal ovaries of the horse do indeed look remarkably like kidneys.

Aristotle was also fascinated by mules (donkey ♂ × horse ♀) and hinnies (horse ♂ × donkey ♀) and their apparent sterility. The intermediate appearance of the mule and hinny is the most dramatic evidence of an active maternal involvement in the makeup of the embryo, although Aristotle failed to appreciate this point. Recognizing that female mules can

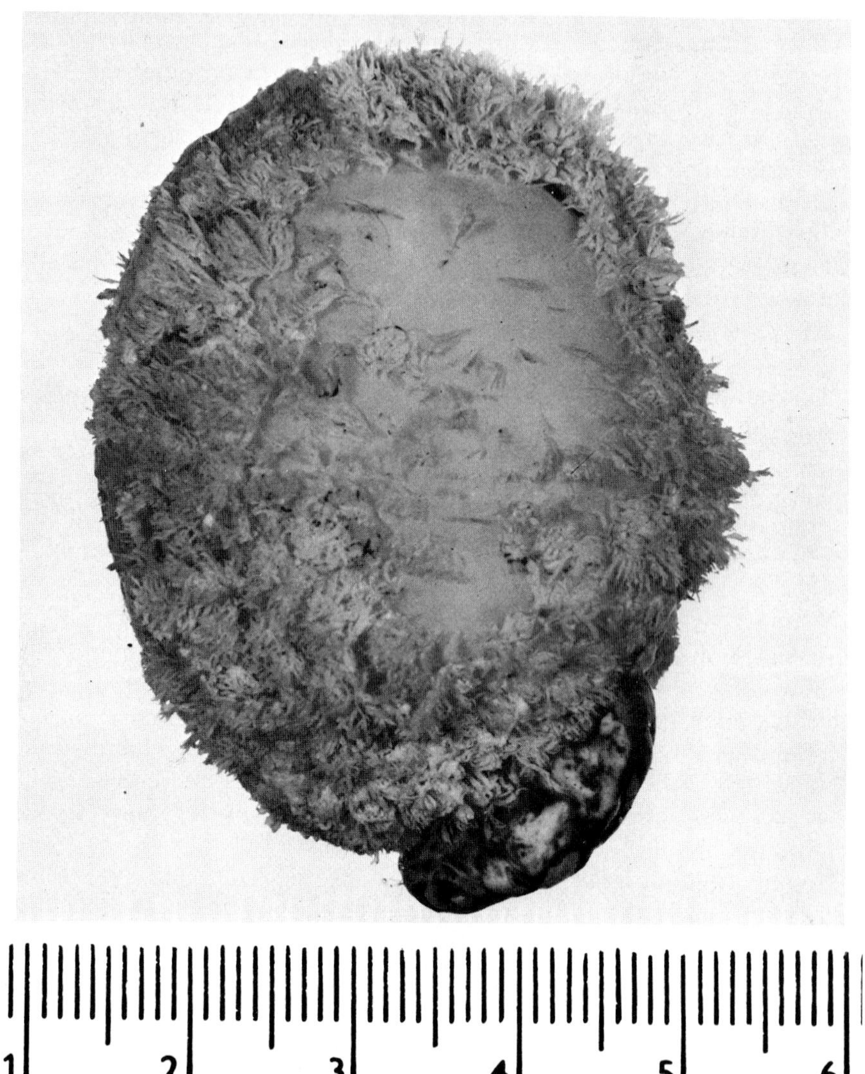

Fig. 1. Early human conceptus (10 mm crown–rump embryo) with membranes intact. It is easy to see how Aristotle mistook it for an egg. Scale in cm.

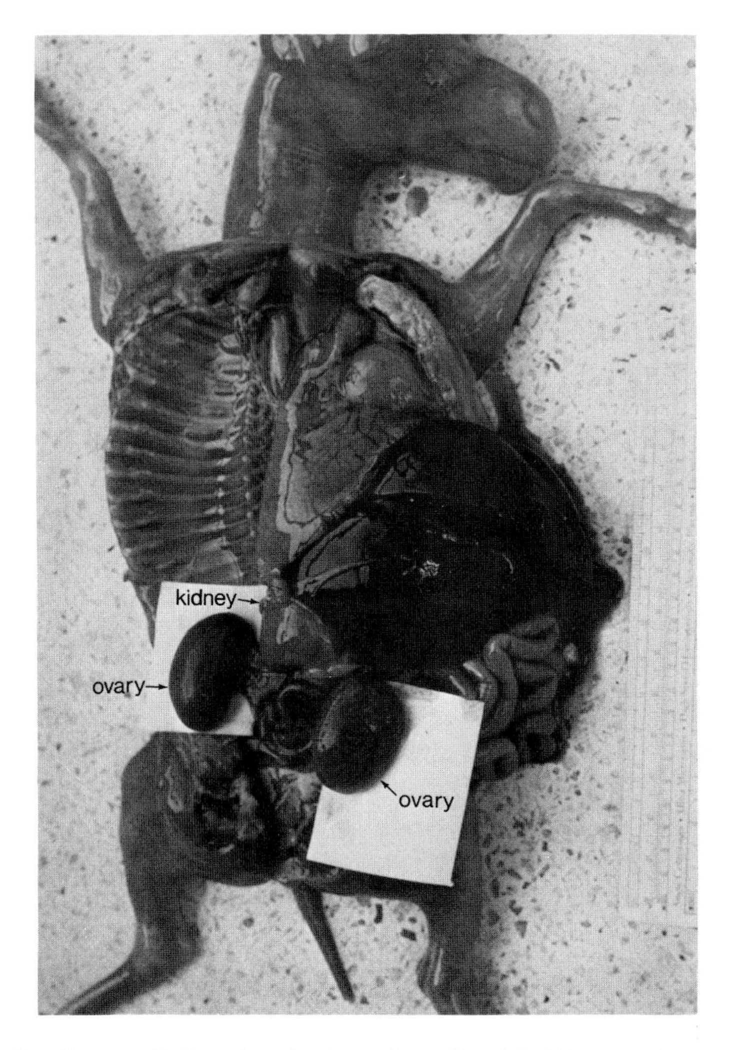

Fig. 2. Hypertrophied ovaries of a horse fetus. The right kidney can be seen as a paler structure lying between the gonads and the liver.

come into estrus and be mated but do not conceive, he attributed their infertility to a lack of catamenia.

Herophilus of Alexandria (*ca.* 300 B.C.) was probably the first person to recognize ovaries as anatomical entities, likening them to the testes of the male. However, he did not appreciate either their structure or function. It was Soranus of Ephesus, born in about 50 A.D., who gave the first detailed description of the ovaries, which he referred to as didymi (paired organs):

Furthermore, the didymi are attached to the outside of the uterus, near its isthmus, one on each side. They are of loose texture, and like glands are covered by a particular membrane. Their shape is not longish as in the males; rather they are slightly flattened, rounded and a little broadened at the base. The seminal duct runs from the uterus through each didymus and extending along the side of the uterus as far as the bladder, is implanted in its neck. Therefore, the female seed seems not be be drawn upon in generation since it is excreted externally. (Temkin, 1956.)

Galen (129–199 A.D.) added very little to Aristotle's accounts of reproduction, although he gave a somewhat more detailed description of how to spay sows, and he considered that the ovaries manufactured a kind of "sperm" that was important in reproduction. Following the death of Galen, the spirit of enquiry was extinguished as the Western world entered the Dark Ages; no discoveries relevant to reproduction were to be made for more than 1000 years.

A renewed interest in anatomy and a reawakening of independent scientific thought began to develop in the fourteenth century instigated by men such as Henri de Mondeville, who wrote a book on surgery in 1314 that contained a drawing of a woman with her abdomen opened, in which the uterus and ovaries are just visible (Singer, 1957). Leonardo da Vinci (1452–1519) provided a good drawing of the anatomical relationships between the human uterus and ovaries (see Fig. 3); unfortunately, however, he drew the cow's cotyledonary placenta attaching the fetus to the uterus. Berengario da Carpi (*ca.* 1460–1530) added a little to the description of the female testicles, as they were than called, in his book "Isagogae Breves":

Near these horns (of the uterus) on both sides is one testicle, harder and smaller than the testicle in man, not perfectly round but compressed like an almond. In them is generated a sperm, not thick as in the male nor warm, but watery, thin and cold. (Lind, 1959.)

Thus, people were beginning to appreciate the structural differences between testis and ovary, and the nature of follicular fluid.

The first person who really seems to have described the follicles, and probably the corpus luteum, was that great anatomist, Andreas Vesalius of Brussels (1514–1564). In the 1555 edition of his "Fabrica," he gives the following account of the human ovaries:

The testes of women contain, besides blood vessles, some sinuses full of a thin watery fluid which, if the testis has not been previously damaged, but is squeezed and makes a noise like an inflated bladder, will spurt out like a fountain to a great height during the dissection. As this fluid is white and like a milky serum in healthy women, so I have found it to be a wonderful saffran yellow colour and a little thicker in two well-bred girls who were troubled before death with strangulation of the womb; the testis of one of the girls, or at any rate one of the sinuses in it, protruded like a

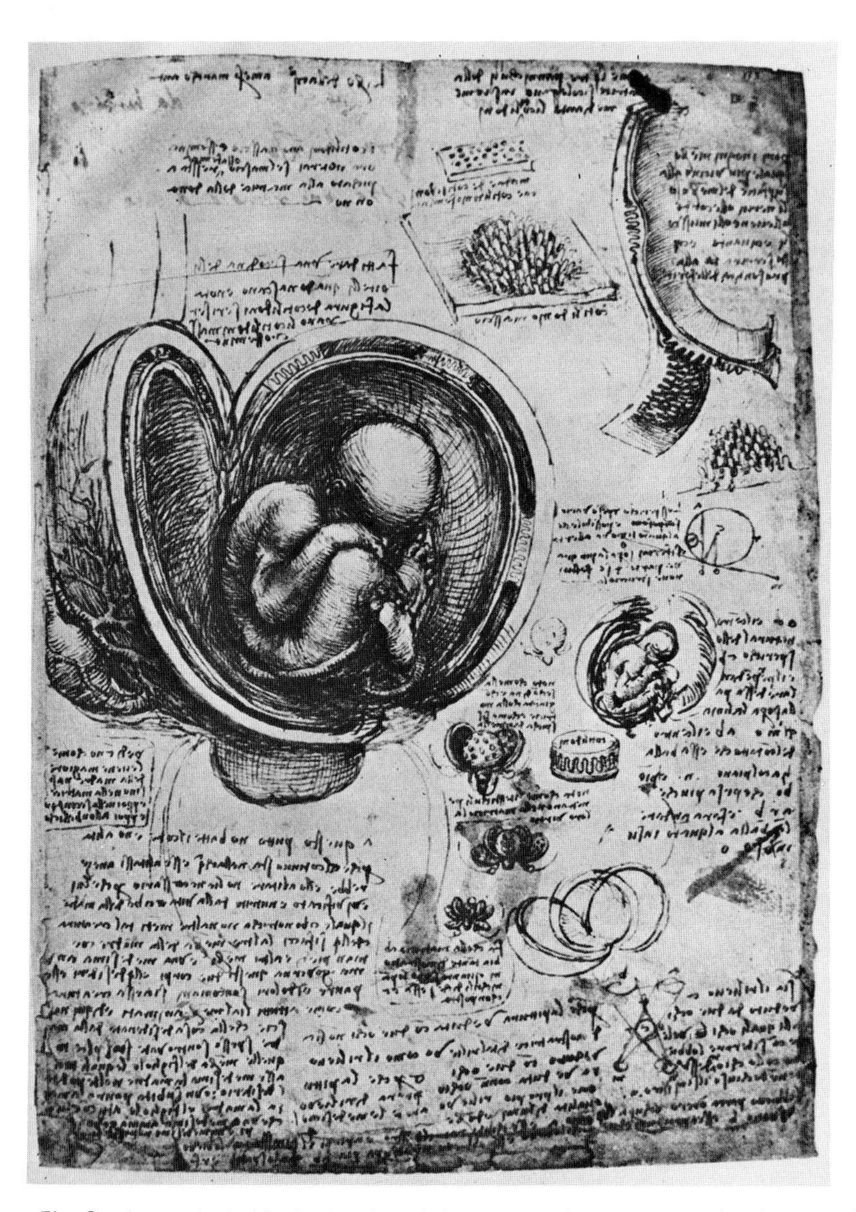

Fig. 3. Leonardo da Vinci's drawing of the pregnant human uterus, showing one of the ovaries and the cotyledonary placenta of a cow.

> rather large pea full of a yellow fluid, colouring the adjacent tissues just as in man the colon is coloured yellow where it passes beneath the liver, by the gall bladder. This colour rarely occurs in the tissues or the fluids; it also smelt very bad and had something poisonous and foul in it.

There can be no doubt from this passage that Vesalius appreciated the existence of fluid-filled vesicles or follicles within the ovary. It is interesting to note that in his early illustrations of the female reproductive tract in the "Tabulae Sex" of 1538, the ovaries are shown as having smooth, rounded contours (see Fig. 4), whereas in the "Fabrica" of 1543 they clearly appear

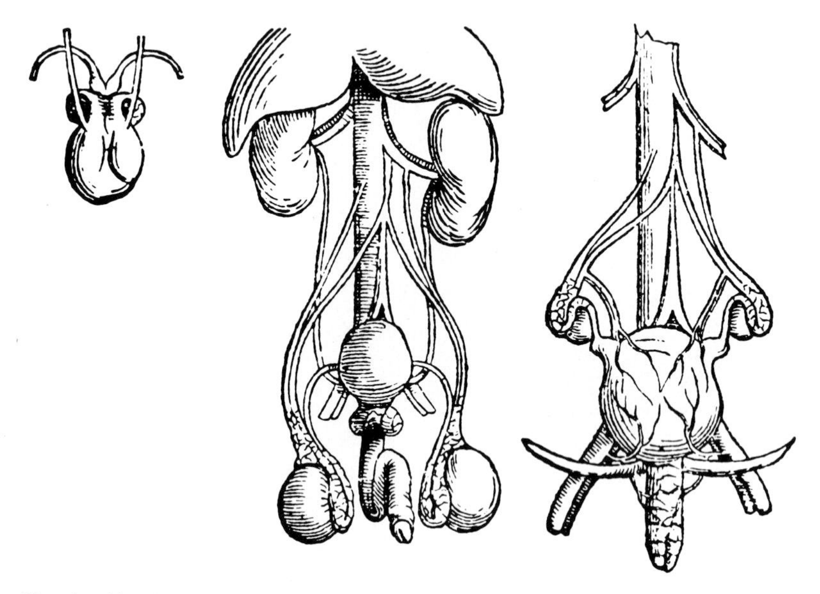

Fig. 4. Vesalius' crude drawing of the human female reproductive tract (right) in his "Tabulae Anatomicae Sex" (1538). No structures can be seen within the ovaries, which are surrounded by "epididymi" that appear to be confluent with the Fallopian tubes. Also seen is the human male reproductive system.

to be full of follicles (see Fig. 5). The "rather large pea full of a yellow fluid" which protruded from the surface of the ovary was possibly the ovulation stigma of a corpus luteum, and the "poisonous and foul" contents probably refer to the coagulated blood and serum found in a recently formed corpus luteum. The coloring of the adjacent tissues could have been the luteal tissue itself.

Vesalius obviously had great difficulty in obtaining corpses, and it was rare indeed to get hold of the body of a woman in her reproductive years when follicles and corpora lutea would be apparent. It is a credit to his powers of observation that he was able to make such good use of the very limited amount of material that did come his way.

The 25 years following the birth of Vesalius were to see the birth of a number of other famous anatomists. Ulysses Aldrovandrus (1522–1605) wrote a fanciful book on monsters (1642), devoting several pages to the subject of castration. In addition to accounts of several people who had successfully ovariectomized women, he tells of one Ioannis of Essen, a mare gelder by profession, who was so infuriated by his daughter perpetually run-

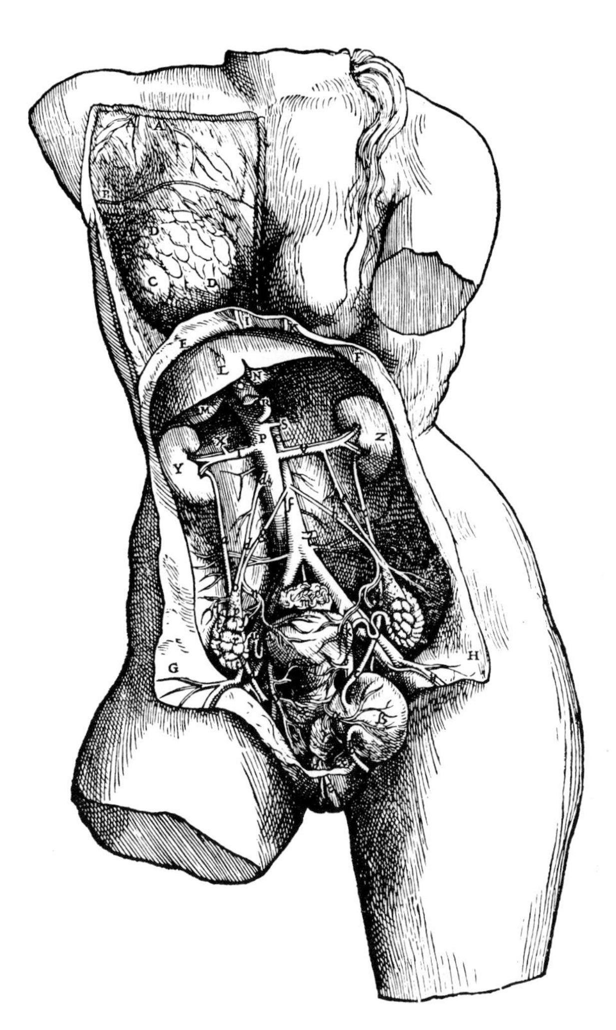

Fig. 5. Female reproductive tract illustrated in the first edition of Vesalius' "Fabrica" (1543). Some attempt has been made to depict the follicular nature of the ovaries, and the Fallopian tubes are still clearly evident.

ning around with the stable lads that he ended up by gelding her as well; she made an uneventful recovery.

Fallopius (1534–1562), who had been a pupil of Vesalius, is of course chiefly remembered for his discovery of the Fallopian tubes, but in his "Anatomical Observations" (1562) he also has something to say about the female "testicles":

> All anatomists assert with one voice that seed is made in the testicles of females and that they may be found full of seed, which I have never been able to see though I have used no light effort to perceive it. I have indeed seen in them some vesicles, as it were, turgid with water or a watery humour, in some cases muddy, in others clear. But I have never seen seed, except in the spermatic vessels themselves or those called deferent.

Fallopius was a successful and popular teacher of anatomy at Padua, and during his short lifetime he produced an outstanding pupil in Fabricius (1533–1619). Another great anatomist of those days was Volcher Coiter (1534–1576), a pupil of Aldrovandrus, and a person who is often regarded as the "Father of Embryology" in recognition of his detailed studies of the development of the chick embryo. Coiter died at a relatively early age from leprosy, and he deserves some mention in this account because others have wrongly concluded that he was the first to describe the corpus luteum (Solomons and Gatenby, 1924; Harrison, 1948). However, there is no foundation for this claim (Adelmann, 1933), as the following extract from his book, "Externarum et Internarum Principalium Humani Corporis" (1573), shows:

> The uteri of Ruminants are bicornuate, as are those of bitches, cats, pigs, and many other animals. The testes of the female are distended by many small glands and vesicles, mixed up and joined to one another. Some vesicles contain clear water, some a yellow fluid.

Fabricius was a contemporary of Coiter, and succeeded Fallopius to the chair of anatomy at Padua, where at his own expense he built the anatomy lecture theatre that is still standing. Fabricius made an enormous contribution to embryology (Adelmann, 1942), and he was probably the first person to give an accurate account of the role of the ovary in the formation of the hen's egg, for he appreciated that the yolk itself was shed from the ovary, and that the white, the shell membrane, and the shell were all formed during the passage of the egg down the oviduct. This key observation was the first indication that the egg might be produced directly by the female, rather than as a consequence of the union of male "seed" and female "soil," as envisaged by Aristotle. In his book, "De Formato Foetu" (1604), Fabricius also provided one of the first illustrated accounts of the comparative anatomy of the female reproductive tract. The quality of the illustrations is superb, and in one drawing (see Fig. 6) he depicts the uterus and ovaries of a sow in the early stages of gestation, and in the legend he refers to the corpora lutea as "glandulae multae simul junctae in utraque uteri parte" (numerous conjoined little glands). This was the first time that the corpora lutea were actually illustrated.

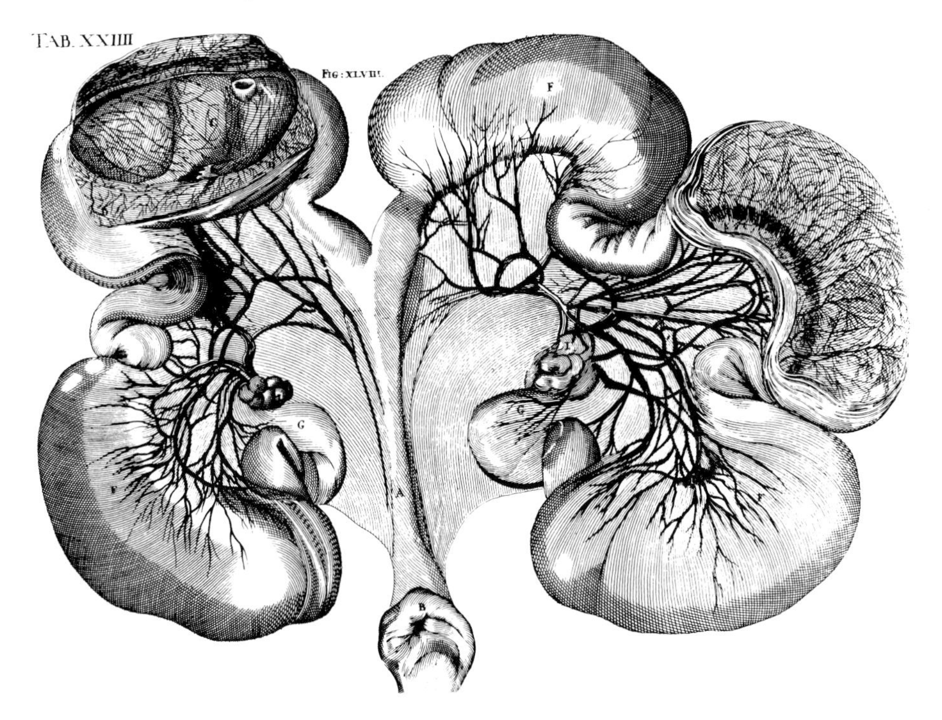

Fig. 6. Fabricius' magnificent drawing of the uterus and ovaries of a pregnant sow, in "De Formato Foetu" (1604). Corpora lutea are seen in the ovaries; this was the first time that they had been illustrated.

Although a lawyer rather than an anatomist by profession, we must also mention the remarkable contributions of Carlo Ruini, who in 1598 published his "Anatomia del Cavallo." In a series of exquisite illustrations of the uterus of pregnant and nonpregnant mares, the ovaries are clearly shown, and in one drawing (see Fig. 7) he actually depicts the hypertrophied ovaries of a horse fetus, first remarked upon by Aristotle (see Fig. 2). Ruini was intrigued by the fact that during the later months of gestation the ovaries of the fetus were up to four times as big as those of the mother,

> . . . which monstrous size could perhaps result either from the excess humidity of the foetus itself, or from the foetal testes having to contain within themselves all the seed and the blood and the spirit which in pregnant mares is dispersed through the foetus, the body and horns of the

> uterus, and all the other parts, which is why the mare's testicles, being so empty, are so small.

Although Ruini's explanation may not seem very plausible today, it is equally true to say that we still have no idea of the cause or the significance of this remarkable hypertrophy of the fetal gonads in the horse.

Summarizing the events of the sixteenth century, we may conclude that it was an era in which the ovaries received recognition as structures, even though their function was not appreciated. But the seventeenth century was to be the period of discovery, and the time when the ovaries first acquired their name.

The first seventeenth century scientist to contribute to our knowledge of the ovaries was William Harvey (1578–1657), trained by Fabricius in Padua

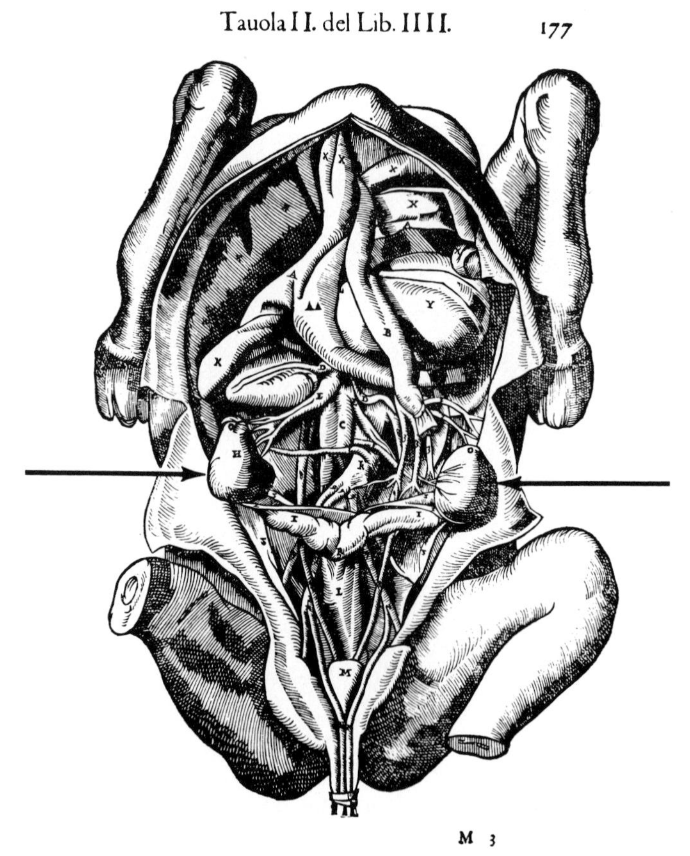

Fig. 7. Ruini's remarkable drawing of a female horse fetus, showing the hypertrophied fetal gonads (H; arrows) ("Anatomia del Cavallo," 1598). Compare this drawing with the photograph in Fig. 2.

before returning to London to practise medicine. Harvey is famed for his discovery of the circulation of the blood in 1628, and he was one of the first anatomists to bring the experimental approach to bear on a subject that had hitherto been the domain of the descriptive morphologist. Unfortunately, his work on reproduction, "De Generatione Animalium," published in 1651 toward the end of his life, was destined to confuse rather than to clarify scientific thinking. While Harvey's experimental observations were perfectly valid, he erred in their interpretation as he was thoroughly indoctrinated with the Aristotelean belief that the egg was a product of conception. In the frontispiece of his book (see Fig. 8) the seated Jove is seen holding apart the two halves of an egg inscribed "ex ovo omnia," from which all manner of creatures are emerging.

Harvey used to accompany King Charles, whose custom it was to hunt red and fallow deer in the royal parks and chases almost every week. This gave Harvey a unique opportunity to examine the anatomy of the reproductive organs throughout the year, and since deer are seasonal breeders with a restricted period of rut in the autumn, Harvey was particularly interested in the animals at this time. From his Aristotelean upbringing, he naturally expected to find a large, spherical egg in the lumen of the uterus after mating. It was common knowledge that red deer stags started to rut and to collect their harems of hinds in mid-September; therefore Harvey (1653) was baffled to find no visible products of conception present in the uterus until November. In exasperation, he concluded:

> In so much that I am very well ascertained that in Viviparous, (as well as in Oviparous creatures) the Foetus doth neither proceed from the Seed of the Male and Female emitted in Coition, nor yet from any conmixture of that seed (as the Physitians will have it) nor yet out of the Menstrous blood, as Aristotle conceits.

He also paid close attention to the ovaries, and concluded the following:

> The Testicles, as they call them, (like things utterly unconcerned in the matter of Generation) do neither swell, nor differ any way from the constitution they were of (either before, or after coition) affording no testimony at all of their use either in respect of coition or generation.

It is only now that we can begin to appreciate how Harvey was led so sadly astray by his observations. Because the stags started to rut in the middle of September, he naturally assumed that this was when the first conceptions occurred. But we now know that stags in fact start to rut a full month before the hinds first begin to ovulate (Guinness *et al.*, 1971). Therefore, Harvey could not have seen any changes in the ovaries until October, and the developing embryo would not be visible to the naked eye until the end of the month. Even then, the elongated chorioallantois of the deer embryo is so

Fig. 8. Frontispiece of William Harvey's "De Generatione Animalium" (1651), showing Jove seated on a throne and holding an egg in his hands, inscribed with the words *"ex ovo omnia."*

unlike the spherical, egg-shaped human or rabbit conceptus (see Fig. 1) that it is no wonder Harvey failed to recognize it, commenting on the fact that in the uterus at the end of October

> a certain purulent matter doth stick to the sides, (in manner of sweat) such as is visible in wounds, and ulcers, when they are said to be concocted, and cast forth a white, smooth and equall matter. When I first discovered this kind of substance, I was in suspence, whether I should conceit it to be the seed of the Male, or some concocted substance arising from it. But because I did observe this matter but seldom, and in few onely, and also seeing twenty days were now past, since any commerce with the male had been celebrated, and likewise for as much as this substance was not thick, clammy, or froathy (as seed is) but more friable and purulent, inclining to yellow, I concluded that it arrived thither casually rather, or else proceeded from over much sweat.

The practice of spaying mares, sows, and bitches was widespread in Harvey's time, and the procedure was well described in at least four contemporary publications, namely "The Noble Art of Venerie or Hunting" (Turbervile, 1576), "Foure Bookes of Husbandry" (Heresbach, 1577), "Five Hundred Pointes of Good Husbandrie" (Tusser, 1580), and "The History of Four-Footed Beasts and Serpents" (Topsell, 1607). Furthermore, we know that on one occasion Harvey entered into a long conversation with a sowgelder, who made the shrewd comment: "From the meanest person in some way or other the learnedst man may learn something" (Keynes, 1966). If only Harvey had paused to consider the physiological significance of the sowgelder's surgical skills, he might have appreciated the true significance of the ovaries, a discovery that would have been at least of comparable magnitude to his description of the circulation of the blood. As it was, Harvey's scientific reputation lent weight to his erroneous views on reproduction, and probably impeded progress in the field.

Harvey's mistaken views about the ovaries were corrected by an unlikely person, Bishop Niels Stensen of Denmark (1638–1686). In the year in which he was consecrated a Bishop, he published a work entitled "Myologiae Specimen" (1667), and in the second supplement, which described the dissection of the dogfish, he put forward for the very first time the view that the female testes of mammals contained eggs, and should be regarded as analogous to the well-recognized ovaries of oviparous species: "Inde vero, cum veridum, viviparorum testes ova in se continere . . . quin mulierum testes ovario analogi sint."

In 1675, Stensen published "Ova Viviparorum Spectantes Observationes," dealing specifically with the ovaries, although curiously he reverted to calling them "testicles." In addition to descriptions of the ovaries of wolves, red deer, fallow deer, and donkeys, he gives a most interesting account of the ovaries of two mules. In one, no follicles were visible at all, whereas the other animal had larger ovaries which contained a number of follicles and a corpus luteum. Commenting on these findings, he said that it might be possible for a mule to give birth to young on rare occasions since its ovaries were not entirely devoid of eggs; however, absence of ova was likely to be the principal cause of the mule's proverbial infertility. We have been unable to improve upon this conclusion, even 300 years later (Taylor and Short, 1973).

Mention should also be made of the work of Theodore Kerckring (1640–1693), who dissected the follicles out of human and bovine ovaries, believing them to be the eggs themselves; he communicated these results to the Royal Society in 1672. But of much greater importance was the publication by the young Dutchman, Regnier de Graaf (1641–1673), of his outstanding work on female reproduction, "De Mulierum Organis Generationi Inservientibus" (1672). Chapter 12 of de Graaf's work entitled "on the female testes or ovaries" was first translated into English by Corner (1943), and Jocelyn and Setchell (1972) have translated the complete work. On reading it, one wonders why de Graaf is chiefly remembered today for the discovery of the follicle; as de Graaf himself points out, this had been well described in the preceding century by men such as Vesalius, Fallopius, and Coiter. De Graaf, like Stensen and Van Horne before him, thought that the whole follicle was an egg, and he was encouraged in this belief by the similarity of size between the large, yolky eggs in the hen's ovary and the follicles in mammalian ovaries (see Figs. 9 and 10). He even produced some biochemical evidence to support the similarity:

> That albumen is actually contained in the ova of women will be beautifully demonstrated if they are boiled, for the liquor contained in the ova of the testicles acquires upon cooking the same colour, the same taste and consistence as the albumen contained in the eggs of birds.

It must have required some courage, even in those days, to eat the boiled ovaries of a rotting human cadaver in order to see what they tasted like.

Undoubtedly de Graaf's major contribution was the first accurate and detailed description of the corpora lutea. His experimental observations on rabbits taught him that ovulation could be induced by coitus, and he realized that the number of "globular bodies" (corpora lutea) that formed in place of the "ova" (follicles), gave an indication of the number of embryos present:

> We assert that these globules do not exist at all times in the testicles of females; on the contrary, they are only detected in them after coitus, being one or more in number, according as the animal brings forth one or more foetuses from that congress. Nor are these always of the same nature in all

> animals, or in the same kind of animal; for in cows they exhibit a yellow colour, in sheep red, in others ashen. . . . When the foetus is delivered these globular bodies again diminish and finally disappear.

De Graaf's drawings of the corpus luteum (see Fig. 11) were a vast improvement on those of his predecessors, and it was undoubtedly his appreciation of the diversity of colors of the gland in various species that prevented him giving it the totally illogical name of "corpus luteum."

By the end of the seventeenth century, a lively controversy had developed on the subject of generation. There were those like Harvey who still adhered to the old ideas of Aristotle, believing that the ovaries were quite unimportant for reproduction. Then there were those like Wharton (1656), Descartes (1664), and Le Grand (1672) who thought that the female "testes" produced their own "semen," the follicular fluid, which had to mix with the semen of the male in order to give rise to an embryo. And then

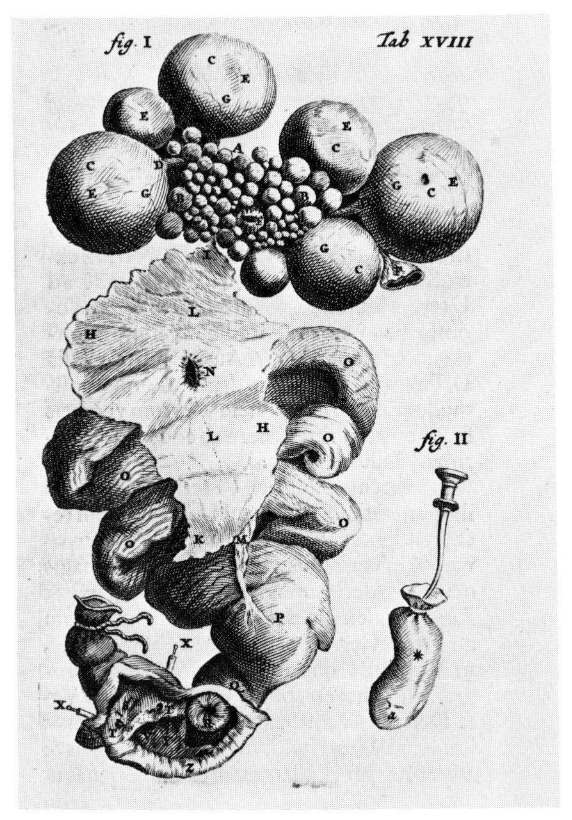

Fig. 9. De Graaf's drawings of the ovaries and uterus of a hen ("De Mulierum Organis Generationi Inservientibus," 1672).

there was the new "ovist" school of Stensen, Kerckring, and de Graaf, subsequently joined by Langly (1674) and Bartholinus (1678); they believed that the whole follicle was an egg, and that fertilization occurred within the ovary itself, only fertilized eggs being capable of escaping from the ovary. Then van Leeuwenhoek (1678) of Delft communicated his discovery of the mammalian spermatozoon in a letter to the Fellows of the Royal Society in November 1677. His observation was the basis of the Animalculist school, whose most fanciful member was probably Hartsoeker (1694); he imagined that he could see in the head of the spermatozoon a complete, preformed fetus. Van Leeuwenhoek (1683) suggested that de Graaf's egg had to be impregnated by one of these animalcules at a particular point on its circumference for fertilization to occur. This, he reasoned, made it essential for many thousands of spermatozoa to be ejaculated, if one of them was to

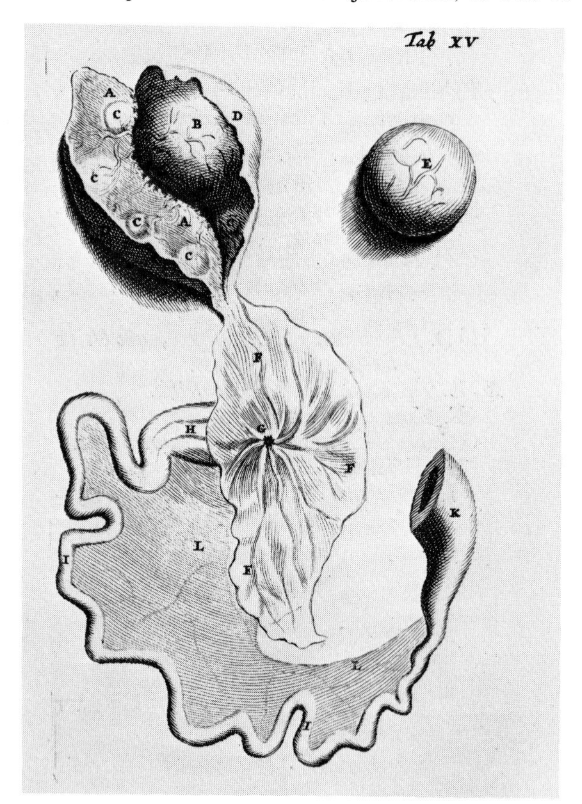

Fig. 10. De Graaf's drawing of the ovary, fimbria, and Fallopian tube of a cow with an intact follicle dissected out. He thought these follicles were the eggs themselves ("De Mulierum Organis Generationi Inservientibus," 1672).

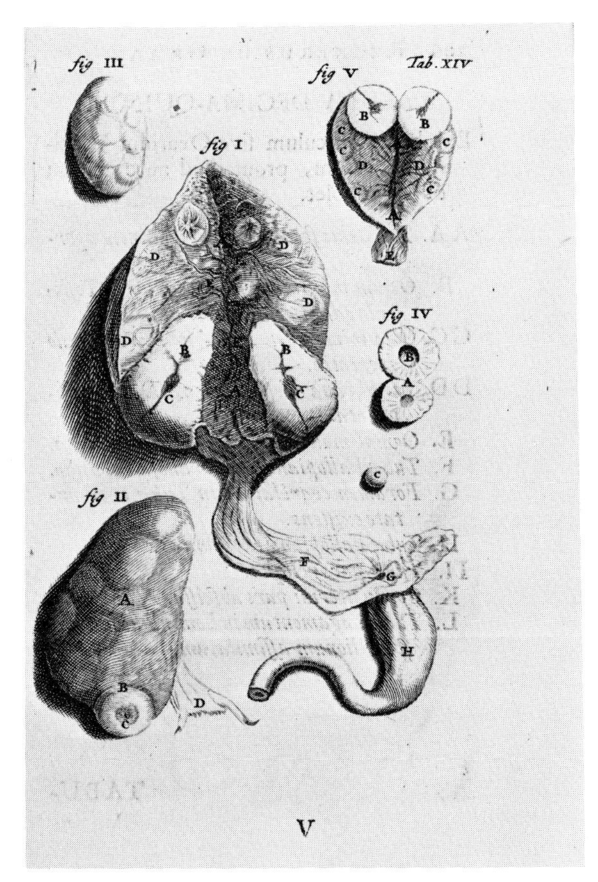

Fig. 11. De Graaf's drawing of the sheep ovary, showing various forms of the corpus luteum. This was regarded as scar tissue that developed after the "ovum" (follicle) had been shed ("De Mulierum Organis Generationi Inservientibus," 1672).

have a reasonable chance of reaching its target. Drelincourt (1685) even suggested that the true egg lay within de Graaf's follicle, but the truth of this suggestion was not to be revealed for almost another 150 years.

The coexistence of so many sharply conflicting views about the origins of life stimulated people's powers of observation and experimentation. It was this growing spirit of enquiry that led Nuck (1650–1692) to carry out an interesting experiment on a bitch in 1691. He opened the abdomen 3 days after mating and noted the presence of two "ova" (follicles) in the left ovary. He then placed a ligature around the middle of the left uterine horn, and closed the abdomen. The bitch was killed 20 days later. Two fetuses were found in the left uterine horn above the ligature, and none below.

Nuck concluded that this proved the correctness of the ovist school, since the fetuses could only have come from "fecundated" ova, and could not have been formed from the seed of the male.

The last seventeenth century anatomist who deserves some mention in this account is Malpighi (1628–1694). In his works, which were published posthumously (1697), he correctly surmised that the Graafian follicle never left the ovary, but served instead as a protection for the true ovum that must lie within it. In a few instances he claimed to have seen an object the size of a millet seed within the follicle; this could conceivably have been the ovum surrounded by cumulus, although the ovum itself, which is only about 0.1 mm in diameter in mammals, would probably escape detection by the naked eye. It was also Malpighi who invented the name "corpus luteum"; he was working with cow's ovaries, in which the corpus luteum really is yellow, whereas this is not a true description of the color of the gland in the dog, cat, sheep, pig, horse, or human. Although Malpighi referred to the corpus luteum as being glandular in nature, he did not really appreciate that it was formed from the lining of the ruptured follicle:

> My observations of different states seem to demonstrate as probable not only that Nature produced this yellow body (corpus luteum) to provide for the protection and ejection of the ovule but also that the yellow body perhaps contributes to the formation of the ovum and hence should be regarded as glandular rather than muscular; for instead of being fibrous or fleshy in structure it very closely resembles the suprarenal gland. (Adelmann, 1966.)

Another writer in the seventeenth century with views on the ovary was Snape, the Royal Farrier; he was one of the first to try and relate the widespread practice of spaying to the new theories on reproduction:

> And whereas it has been thought a strong argument for the Female's having true Seed, and that these Testicles make it, in that when they are cut out of the body in Bitches, Hogs, or any other Creature, such Creatures are always barren afterwards; this new Opinion shows that there is no strength or certainty at all in that argument. For granting, what is most certainly true, that Females that are gelt or spay'd, have never any Young after; yet it does not at all follow, that therefore their Stones make and conserve Seed; but onely that they contain something that is absolutely necessary for generation and conception; but whether that be seed or somewhat else, is indifferent. (Snape, 1687.)

Looking back on the seventeenth century, we can see that anatomists were entirely concerned with descriptions of ovarian structure, and speculation about whether or not the ovary produced eggs. Some of this controversy was to carry on well into the 18th century. Although Boerhaave (cited in Haller, 1744) outlined the theory of how the ovum must escape from the ovary, leaving a scar or corpus luteum in the process, and then pass down the Fallopian tube and be fertilized by a spermatozoon before entering the uterus, one of Boerhaave's pupils, Haller, flatly contradicted his teacher with the words:

> We may conclude from all this that the ovarian vesicles are not eggs and that they do not contain the rudiments of the animal. (See Needham and Hughes, 1959.)

The 18th century was to see an awakening interest in ovarian pathology and also in the concept of the control of ovarian function. But curiously, nobody had yet begun to pay any attention to Aristotle's observation that spayed animals ceased to show estrous behavior.

Historians of science have given undue prominence to Cruikshank (1797), whose studies of the reproductive anatomy of rabbits persuaded him that eggs were only released from the ovary after fertilization. But, in the same 1797 volume of the Proceedings of the Royal Society, Haighton made some far more significant observations on rabbits. He carried out a series of ligations of one Fallopian tube at various times after mating, and subsequently recorded whether or not embryos developed in the uterine horns. His experiments clearly demonstrated that unilateral tubal ligation 6, 12, and 24 hours after mating prevented the development of embryos in the adjacent uterine horn, even though corpora lutea were formed in the ovary. But if ligation was delayed until 2½ days after mating, then embryos would develop normally in the uterine horn on the ligated side. Haighton concluded:

> . . . that the ovaries can be affected by the stimulus of impregnation, without the contact either of palpable semen, or of the *aura seminalis*.

He therefore deserves the credit for being the first person to describe the phenomenon of induced ovulation.

Toward the end of the eighteenth century, two Scottish anatomists, the brothers William and John Hunter, were also investigating a number of aspects of reproduction in man and animals. William dissected about 400 gravid human uteri, still probably one of the largest series ever studied, and in 1774 he published a beautifully illustrated folio, "The Anatomy of the Human Gravid Uterus Exhibited in Figures." The text to this volume only appeared in 1794, after William's death, but he had some interesting comments to make on the corpus luteum during pregnancy:

> When there is only one child, there is only one corpus luteum; and two in case of twins. I have had opportunities of examining the ovaria with care in several cases of twins, and always found two copora lutea. In some of these cases there were two distinct corpora lutea in one ovarium.

William Hunter's young brother, John, undoubtedly made by far the greater scientific contribution of the two. In 1779, he published what must have been the first scientific study of the bovine freemartin. The animals were probably located for him by his friend Dr. Jenner, of smallpox vaccine fame; Hunter dissected the reproductive tracts, and preserved them in his museum, where they can still be seen. The freemartin condition occurs in the majority of heifers born cotwin to bulls, and has been recognized since Roman times (Forbes, 1946). Although Hunter did not appreciate the significance of the placental vascular anastomosis in the etiology of the condition, he noticed that the ovaries of freemartins were abnormal, since testicular tissue was present. He also noted that the reproductive tract consisted of some male duct derivatives, such as vasa deferentia and seminal vesicles, in addition to hypoplastic female duct derivatives such as uterus and vagina. One of the animals that Hunter dissected, "Mr. Wright's Free-Martin," is of particular interest, since it was almost certainly not a freemartin, but the first recorded case of testicular feminization (Short, 1969). This is a sex-linked genetic defect in which all the target organs for testosterone are unresponsive to the hormone. The animal, whose reproductive tract is still available for inspection in the Hunterian Museum of the Royal College of Surgeons, shows the large intraabdominal testes, absence of epididymis and vas deferens, uterine and vaginal aplasia, and female external genitalia typical of testicular feminization (see Fig. 12).

Probably Hunter's most important contribution to our understanding of ovarian function was a paper entitled "An experiment to determine the effect of extirpating one ovarium upon the number of young produced," which was published in 1787. In Hunter's own words:

> As the female, in most classes of animals, has two ovaria, I imagined that by removing one it might be possible to determine how far their actions were reciprocally influenced by each other. . . . There are two views in which this subject may be considered. The first, that the ovaria, when properly employed, may be bodies determined and unalterable respecting the number of young to be produced. . . . The second view of the subject is, by supposing that there is not originally any fixed number which the ovarium must produce, but that the number is increased or diminished according to circumstances; that it is rather the constitution at large that determines the number; and that if one ovarium is removed, the other will be called upon by the constitution to perform the operations of both, by which means the animal should produce with one ovarium the same number of young as would have been produced if both had remained.

For his actual experiments, Hunter took two littermate sows born in 1777, and removed one ovary from one of them. He also kept a boar from the same litter to run with the sows; the animals were served as soon as they came into estrus, and the experiment was only terminated when they had

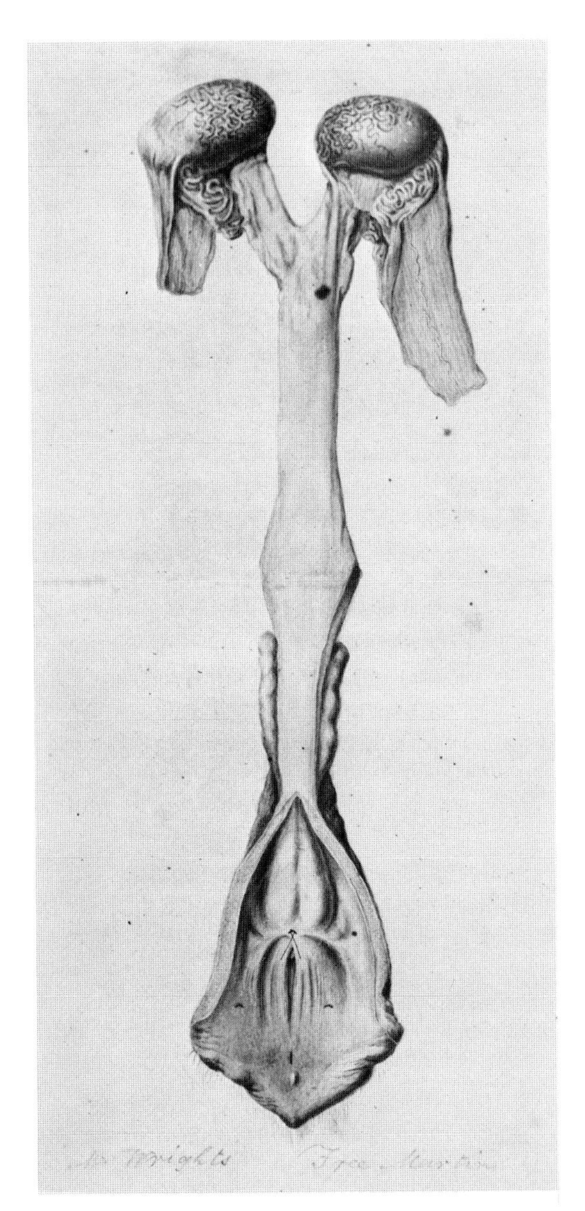

Fig. 12. The dissected reproductive tract of "Mr. Wright's Free-martin" (Hunter, 1779). The specimen is certainly not that of a freemartin, and it is almost certain that this is the first recorded case of testicular feminization.

TABLE I

The Results of John Hunter's Experiments on Two Sows

Litter number	Hemi-spayed sow		Intact sow	
	No. of young born	Date of birth	No. of young born	Date of birth
1	6	Dec. 1779	9	—
2	8	July 1780	6	—
3	6	Jan. 1781	8	—
4	10	Aug. 1781	13	Dec. 1781
5	10	Mar. 1782	10	June 1782
6	9	Sept. 1782	16	Dec. 1782
7	14	May 1783	13	June 1783
8	13	Sept. 1783	12	Oct. 1783
	76		87	

ceased to breed. The results are summarized in Table I. After farrowing for the eighth time, the hemi-spayed animal returned to estrus in November, 1783 and April, 1784, but did not conceive; Hunter therefore had her killed in November, 1784, when, it seems, the remaining ovary had become cystic.* The behavior of this animal, which was sterile but nevertheless continued to return to estrus, aroused Hunter's curiosity:

> From a circumstance mentioned in the course of this experiment it appears that the desire for the male continues after the power of breeding is exhausted in the female; and therefore does not altogether depend on the powers of the ovaria to propagate, although it may probably be influenced by the existence of such parts.

It was most unfortunate that the hemi-spayed animal should have developed a cystic ovary and stopped breeding prematurely, because her littermate control went on to have five more litters before she, too, stopped breeding in December, 1785. By this time she had produced 162 piglets, in contrast to the 76 of the hemi-spayed animal, so Hunter, ignoring the similarity in the number of piglets produced by each sow in her first eight litters, concluded:

> From this experiment it seems most probable that the ovaria are from the beginning destined to produce a fixed number, beyond which they cannot go, although circumstances may tend to diminish that number; but that the constitution at large has no power of giving to one ovarium the power of propagating equal to both; for in the present experiment the animal with one ovarium produced ten pigs less than half the number brought forth by the sow with both ovaria.

* This ovary is almost certainly specimen No. 3545 in his museum.

The "law of follicular constancy" was eventually formulated by Lipschütz (1928).

A contemporary of John Hunter, the great surgeon Percival Pott of St. Bartholomew's Hospital, London, made an important observation on the function of the ovaries (Pott, 1775), and Corner (1950) has related how this was really the first intimation that menstruation was under ovarian control. In his "Chirurgical Observations," Pott recounts the following case:

> Case XXIV An Ovarian Hernia
> A healthy young woman about 23, was taken into St. Bartholomew's hospital on account of two small swellings, one in each groin, which for some months had been so painful, that she could not do her work as a servant.
> The woman was in full health, large breasted, stout, and menstruated regularly, had no obstruction to the discharge per anum, nor any complaint but what arose from the uneasiness these tumours gave her, when she stooped or moved so as to press them.

Pott operated upon her, and removed both the herniated ovaries; as a result of the operation, he reported that:

> She has enjoyed good health ever since, but is become thinner and more apparently muscular; her breasts, which were large, are gone; *nor has she ever menstruated since the operation*, which is now some years.

This was the first accurate clinical description of the consequences of ovariectomy in a woman and the first clue to the fact that menstruation is under ovarian control. Unfortunately Pott did not appreciate the significance of his observations, which passed unnoticed by the scientific community.

Clinical interest in the ovaries started to develop in the nineteenth century. Surgeons like Sir Thomas Spencer Wells and Charles Clay were becoming bolder and began to develop techniques for the removal of diseased or neoplastic ovaries from women. Initially, the mortality from the operation was terrifyingly high, but as it began to fall, ovariectomy became the operation of choice for a variety of ill-diagnosed and half-imagined neuroses and psychiatric disorders. Fortunately, Battey's operation, as it was called, did not endure for long (Graham, 1960). By the end of the century, clinical opinion had swung to the opposite extreme, and endocrine replacement therapy began to gain favor.

Corner (1950) has given a most interesting account of how the relationship between the ovaries and menstruation was first discovered following Percival Pott's unrecognized observations at the end of the eighteenth century. It was generally believed that a woman's body periodically became replete with blood, and that menstruation was Nature's way of getting rid of the excess. John Davidge (1814), a young American who had obtained his medical degree in Scotland, demonstrated the futility of this explanation by pointing out that only about 8 oz of blood were lost in a typical menstrual flow, whereas removal of much larger quantities by venesection did not interfere with menstruation at all. He concluded that menstruation

> is attributable to a peculiar condition of the ovaries serving as a source of excitement to the vessels of the womb, rather than to the doctrine of repletion of the body,

but no notice was taken of his observations.

It was the misinterpreted diagnoses of a young French doctor, Roberts, that reawakened medical interest in the mechanism of menstruation. Roberts had been sent on a mission into Central Asia, and in giving a public account of his travels in 1843 he related how, on the way from Delhi to Bombay, he came across a group of dancing girls whom he concluded had been castrated for social or religious purposes. These women were referred to as "Hedjera," an Urdu word meaning eunuch or hermaphrodite. They had poorly developed breasts, did not menstruate, and had no libido. Roberts examined and questioned three of them, but was unable to elicit any history of ovariectomy and could see no surgical scars. Nevertheless, he concluded that these women had been forcibly ovariectomized, and public and medical opinion of the day was incensed. This was taken as proof that removal of the ovaries suppressed menstruation, and, in his textbook "On Diseases of Menstruation," John Tilt of London (1850) stated:

> The ovaries are also the organs of menstruation, for if they have not existed, though the uterus may be present, it cannot secrete the menstrual fluid.
>
> These *"testes muliebrum"* have evidently the same influence over the development of woman as the testes have over that of man, and their absence or destruction by disease, or by artificial means, to serve the licentious propensities of Eastern despots of antiquity, or of the present day, is followed by the arrest of that characteristic luxuriance of form which we admire in women, and by their assuming the drier texture, the harder outline, and the angular harshness of men.

But the story has an ironical twist, for it now seems certain that the "Hedjera" examined by Roberts were not castrates at all. The castration of women has never been practiced in India (H. Heilig, personal communication); however, even to this day there are wandering bands of Hijras, most of whom are male eunuchs, who turn up at any house where a boy has recently been born, since their appearance at such an event is a good omen, and an occasion for feasting and dancing. A few infertile women may appear in such a group; apparently news soon gets around about the existence of any masculinized, hermaphroditic girls, and they are recruited by persuasion, financial reward, or force. Roberts had thus examined three

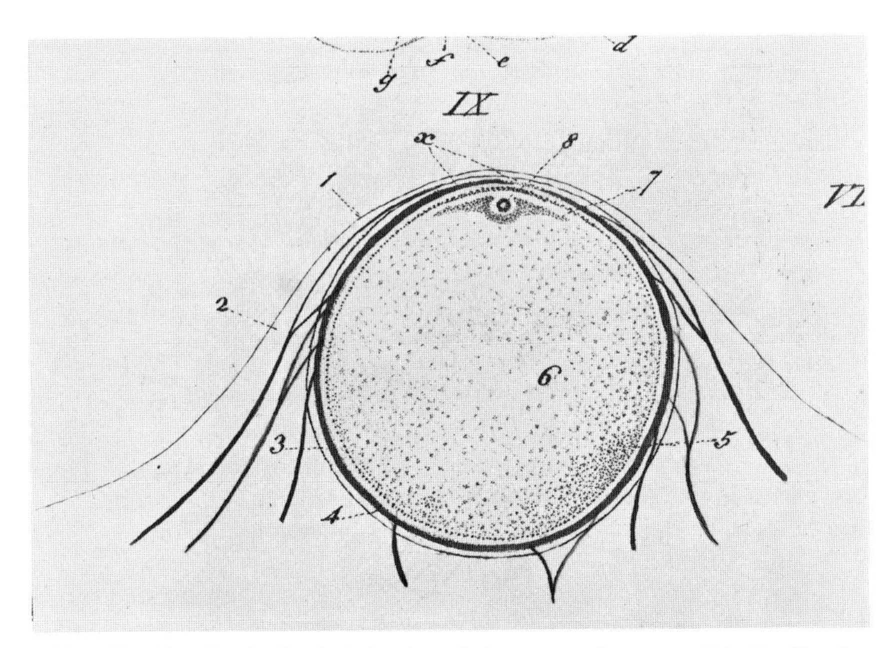

Fig. 13. Von Baer's classical drawing of the mammalian egg within the Graafian follicle ("De Ovi Mammalium et Hominis Genesi," 1827). The legend is as follows. IX, Vesicula Graafiana (mediae magnitudinis) scrofae, decies aucta ad axin dissecta. (1) epithelium peritoneale; (2) tela formativa (stroma); (3) stratum externum (thecae), (4) stratum internum, (thecae), (x) stigma; (5) membrana granulosa (nuclei), (6) fluidum contentum, (7) discus proligerus (nuclei); (8) ovulum (nuclei).

such cases of primary amenorrhoea. Nevertheless his observations implied, at least to some, that there might be a functional relationship between the ovaries and menstruation.

Other discoveries relevant to reproduction were also proceeding apace at the beginning of the nineteenth century. Two Frenchmen, Prévost and Dumas (1824), gave an excellent account of ovulation and the formation of the corpus luteum in the bitch. They realized that fertilization must occur after the egg had left the ovary, and they came to the conclusion that the follicles described by de Graaf were probably not the ova themselves but were likely to contain the ova, and that the follicular fluid was probably necessary to transport the minute eggs into the uterus.

Following this line of reasoning, von Baer, in 1827, first described and illustrated the mammalian egg within the follicle (see Fig. 13). In a series of hand-painted drawings, the early development of the dog embryo and the unovulated egg within the Graafian follicle of a sow were depicted. Von Baer's conclusion

> Omne animal quod coitu maris et feminae gignitur, ex ovo evolvitur. [Every animal which is generated by coitus of male and female is evolved from an egg.]

was not far removed from Harvey's "Ex Ovo Omnia," two centuries earlier. But the intervening years had seen the mammalian egg redefined as a component rather than a product of conception. Soon after von Baer's discovery, Newport (1853, 1854) went on to describe in detail the actual process of fertilization and early cleavage. So the great controversy about the mechanism of conception was finally laid to rest. The era of gross descriptive anatomy of the ovary was drawing to a close and that of physiology and endocrinology was about to begin.

The idea that the ovary might be a gland of internal secretion had lain dormant since the time of Aristotle, who had described how the spaying of sows "quenched their sexual appetites." Fabricius, Malpighi, and de Graaf had all referred to the corpora lutea as glands, and yet, in his classical work on the histological appearance of secreting glands throughout the body, Müller (1830) made no mention of the ovaries. Bischoff, the great German anatomist, gave detailed descriptions of the corpora lutea in many species but made no suggestions as to their possible functional significance. The general opinion among the anatomists of the day was that the corpora lutea were just a peculiar form of scar tissue formed after ovulation.

A popular misconception, presumably originating from the work of Haighton in the preceding century, was that ovulation was always induced by the act of coitus. This view was first contradicted by Pouchet (1842, 1847), the Professor of Zoology at Rouen. He believed that *all* mammals were spontaneous ovulators, and he was one of the first people to speculate about the time during the menstrual cycle when women ovulated. He erroneously concluded that:

> Menstruation in women corresponds to the period of sexual excitement in other female mammals.

However, he did lay the basis for a "rhythm" method of contraception:

> Fertilization bears a constant relationship to the time of menstruation; thus in women it is possible to draw a clear distinction between the times when conception is a physical impossibility, and the times when it is likely to occur.

Unfortunately, he thought that pregnancy could only occur if intercourse took place during or immediately after menstruation; thus, his "safe period" actually coincided with the time of ovulation. The first person to appreciate when a woman was most likely to ovulate and conceive was Raciborski (1843) who noted that girls who were married soon after a menstrual period usually conceived at once without menstruating again.

If the wedding occurred 12 or more days after menstruation, however, then the girl almost invariably menstruated once more before becoming pregnant. But even though people slowly began to appreciate the temporal relationship of menstruation and ovulation, it was not until the 1930's that it was realized that menstruation marked the end of a cycle, rather than the premonitory signs of the next impending ovulation.

In 1849, Berthold showed experimentally that the atrophy of a castrated cockerel's comb, so clearly described by Aristotle, could be prevented by grafting a piece of testis into the bird's abdomen. But Berthold's classical experiment passed unnoticed, and it required the tragicomic claims of Brown-Séquard (1889), an aging Professor at the Collège de France and President of the Société de Biologie, to focus scientific interest on the subject of endocrinology. Brown-Séquard, at the age of 72, was worried by his diminishing vigor and so gave himself a course of subcutaneous injections of testicular extracts prepared from young, vigorous guinea pigs and dogs. He imagined that these injections had a spectacular effect on him; he felt that his intellectual capacity was enhanced, his defecation improved, and his force of urination markedly increased. He even advised two middle-aged colleagues to get themselves into a state of extreme sexual excitement in preparation for a particular intellectual effort, and they reported that the results of this therapy were entirely favorable. These spurious claims, although having no real foundation, were so dramatic and came from such an eminent source that they probably stimulated the development of endocrinology. In 1890, Brown-Séquard reported on the effects of injecting guinea pig and rabbit ovarian extracts into ovariectomized or hysterical women, again with spectacular results.

Puech, a French physician, had already pointed out in 1873 that the spaying of young sows inhibited subsequent uterine growth, and this was confirmed by Hegar (1878). The observation was extended to the rabbit by Kehrer (1879). But the first direct proof that the ovary was indeed a gland of internal secretion came from experiments in which the ovaries were grafted into the abdominal cavity of spayed bitches, rabbits, and guinea pigs (Knauer, 1896); this was shown to prevent castration atrophy of the uterus.

In 1897, John Beard, the Edinburgh embryologist, published an important book entitled "The Span of Gestation and the Cause of Birth," in which he ascribed a specific function to the corpus luteum:

> In this connection the corpus luteum and its history during Mammalian gestation afford interesting material for reflection. The structure appears to be entirely absent below the mammalia, and its increase in size during gestation would appear to be a contrivance for diminishing the nutrition of the ovary and ova, and, thus, of preventing a normal ovulation.
>
> The commencing degeneration of this structure some little time before the end of the gestation (like its rapid atrophy, where fertilization has not

> taken place), allows of preparation being made for a new ovulation. As already indicated, the approach of this ovulation is, in a reflex manner, the direct cause of birth.
> In mammals the uterine development from beginning to end, gestation, birth, ovulation, and, in many cases at least lactation, all obey one rhythm, whose basis is in the ovary.

One year later, Prenant (1898) first put forward the view that the corpus luteum was a gland of internal secretion, releasing its products directly into the circulation:

> Quant au rôle physiologique du corps jaune, on ne peut douter, a la suite de l'examen histologique de cet organe, qu'il s'agisse d'une glande, d'une glande a sécrétion interne, déversant un ou plusieurs produits dans le liquidé sanguin. [As for the physiological role of the corpus luteum, there can be no doubt from a study of its histological appearance that it acts as a gland, and as a gland of internal secretion, releasing one or more products into the bloodstream.]

He suggested that this idea could be put to the test by extracting corpora lutea and isolating the active principle. The challenge was taken up at the beginning of the twentieth century by Ludwig Fraenkel (Fraenkel and Cohn, 1901; Fraenkel, 1903) and Vilhelm Magnus (1901), both of whom had been stimulated by the teachings of Gustav Born, the embryologist from Breslau. They were able to show, in separate experiments, that spaying pregnant rabbits or removing their corpora lutea by thermocautery resulted in abortion or resorption of the embryos. Magnus also showed that unilateral ovariectomy of the pregnant doe did not interrupt the pregnancy, hence nervous reflexes could be ruled out as a cause of the abortion. He concluded that the ovaries had two functions—a maternal effect, maintaining uterine size, and a fetal effect, allowing the maintenance of pregnancy. Magnus also proposed that one might investigate the biological activity of luteal extracts by studying their ability to maintain pregnancy in ovariectomized rabbits, but it was several years before this suggestion was followed up.

Of the two obvious anatomical structures in the ovary, the Graafian follicle and the corpus luteum, the latter was the more attractive experimental proposition. It was easily removed surgically, so that the biological effects of ablation could be recorded, and great quantities of luteal tissue could be collected from slaughterhouses for the preparation of chemical extracts. But before any progress could be made in the isolation of a luteal hormone, it was first necessary to characterize the biological effects of the corpus luteum, so that a bioassay could be devised.

Although pregnancy maintenance was one obvious function of the corpora lutea, other activities were soon discovered. Leo Loeb noted that if the endometrium of a nonpregnant guinea pig or rabbit was subjected to trauma when corpora lutea were present, a maternal placenta or deciduoma was formed, whereas no such structure developed in the absence of luteal tissue (Loeb, 1907, 1908). Loeb also showed that enucleation of corpora lutea from the ovaries of guinea pigs hastened the onset of the next estrus (Loeb, 1911), in confirmation of Beard's earlier suggestion that the corpus luteum controlled the time of ovulation.

The isolation and identification of estrogens from the Graafian follicle was also dependent on the discovery of a suitable bioassay. The initial discovery was made by Lataste (1886, 1887), in Bordeaux, who demonstrated cyclical changes in the vaginal epithelium of rodents that were correlated with the stage of follicular development in the ovaries. Lataste's work proved a great stimulus to Walter Heape, working in the Zoology Department in Cambridge, who in 1900 published his classical paper entitled "The 'Sexual Season' of Mammals and the relation of the 'pro-oestrum' to menstruation." In this paper, Heape clearly defined the terms estrus, proestrus, diestrus, metestrus and anestrus, and we are still using his terminology today. Stockard and Papanicolaou (1917) showed that it was possible to tell the stage of the estrous cycle of the guinea pig by studying the cytology of the vaginal smear, and this discovery enabled Long and Evans (1922) in California to give a detailed description of the vaginal cytology of the rat throughout reproductive life. Vaginal cornification was a simple, sensitive and specific bioassay for estrogenic activity that was soon taken up by Edgar Allen and Doisy at Washington University, St. Louis. They showed that the follicular fluid from sows could produce vaginal cornification when injected into rats and mice (Allen and Doisy, 1923). Doisy, the chemist, then went on to purify the active principle; he found it to be lipid soluble (Doisy *et al.*, 1924), and named it "theelin," from the Greek *thelys* meaning female. Robert Courrier in France proposed the alternative name of "folliculine." But this had to be abandoned when Parkes (1926), working at the Physiology Department in University College, London, showed that X-irradiation of 3-week-old mice destroyed all the oocytes and follicular tissue in the ovaries, and yet the majority of animals reached puberty at the normal time and for several months thereafter had estrous cycles of normal length. Parkes concluded:

> that the oestrus-producing hormone is elaborated by the inter-follicular tissue, although also possibly by the follicle under normal conditions.

Parkes and Bellerby (1926) proposed an alternative name, "oestrin," which found favor for a time. Meanwhile, the search for estrogens had switched from follicular fluid to human pregnancy urine, following the discovery by Aschheim and Zondek that this was a particularly rich source (see reviews by Parkes, 1966; Marrian, 1966). It was not long before Doisy in St. Louis, Butenandt in Gottingen, Laqueur in Amsterdam, and Marrian in London had succeeded in isolating and characterizing estrone and estriol, names which were given to the hydroxyketone and triol in 1933. The search for estradiol-17β turned once more to follicular fluid, and in 1936 Doisy and his colleagues reported the isolation of a few milligrams of an estrogenic diol from follicular fluid obtained from 4 tons of sows' ovaries (MacCorquodale *et al.*, 1936), which proved to be identical with the estradiol-17β previously isolated from the urine of pregnant mares.

Meanwhile Bouin and Ancel in France (1909, 1910) made the important discovery that the proliferation of the endometrial glands in the rabbit's uterus was dependent on the corpus luteum. This observation was to form the basis of the bioassay that enabled chemists to isolate progesterone from the corpus luteum several years later.

By the beginning of the First World War, many investigators had begun to make crude extracts of whole ovaries, which they then injected into a variety of test animals. Their results were confused because the extracts invariably contained both estrogenic and progestational activity, and the greater biological activity of estrogens meant that these effects usually masked those of progesterone. However, Edmund Herrmann (1915), an Austrian obstetrician, showed that a lipid extract of pig's corpora lutea produced a fully progestational endometrium when injected into immature rabbits. The war put a stop to this promising line of investigation, and it was only taken up again by George Corner in the United States in 1928. Having confirmed the observations of Fraenkel (Magnus' work was unknown to him), he enlisted the help of a young medical student at Rochester University, Willard Allen, and between them they prepared an alcoholic extract of pig corpora lutea which gave a progestational proliferation of the endometrium in the rabbit and could also maintain pregnancy for a time in ovariectomized animals (Corner and Allen, 1929). Allen and Corner (1930) successfully maintained pregnancy up to term in rabbits ovariectomized only 18 hours after mating, and Hisaw *et al.* (1930) at the University of Wisconsin showed that luteal extracts could also produce a full progestational proliferation of the endometrium in rhesus monkeys.

These biological experiments enabled chemists to isolate progesterone in crystalline form from luteal tissue and culminated in the elucidation of the structure of the hormone in 1934 by four independent groups, working in the United States (Allen and Wintersteiner, 1934), Breslau (Slotta *et al.*, 1934), Danzig (Butenandt *et al.*, 1934), and Switzerland (Hartmann and Wettstein, 1934). Some workers preferred to call the new hormone "luteosterone," a word that described both its origin and its chemical structure, whereas others preferred the name "progestin," which described its biological activity. Parkes (1962) has explained how the composite name "progesterone" was agreed.

By the late 1930's, the principal steroids secreted by the ovary, estradiol-17β and progesterone, had been isolated, characterized, and synthesized. The increasing availability of these hormones stimulated more research into their biological effects in man and animals.

John Beard (1897) had first postulated that the function of the corpus luteum was to prevent ovulation during pregnancy; Leo Loeb (1911) showed that removal of the corpora lutea hastened the onset of the next ovulation; and Makepeace *et al.* (1937) showed that the administration of progesterone to rabbits could inhibit ovulation. Clinical exploitation of this biological action was not possible until a whole range of orally active steroids had been synthesized. In particular it was the work of Pincus and Chang (1953) that made the "pill" a reality. No other scientific discovery in the field of reproduction has had greater impact on the general well-being of mankind.

Another great problem of endocrinology that was only solved comparatively late in the twentieth century was the hormonal control of menstruation. It was generally believed that without ovulation there could be no menstruation, but in 1923 Corner showed that menstruation could occur in the rhesus monkey in the absence of any corpus luteum. This was explained when Edgar Allen (1927) found that menstruation could be induced in castrated rhesus monkeys following a period of estrogen treatment. Numerous investigators were able to show that progesterone could prevent this estrogen-withdrawal bleeding (Smith and Engle, 1932; Engle *et al.*, 1935; Hisaw, 1935), and these studies, together with subsequent work by Zuckerman (1937) and Hisaw and Greep (1938), finally provided convincing proof that menstruation was normally the consequence of regression of the corpus luteum. However, if there was no corpus luteum present, menstruation could still occur in response to a declining secretion of estrogen.

Another important discovery made at about this time was Loeb's observation (1923, 1927) that hysterectomy in the guinea pig would prolong the life of the corpora lutea. This led him to postulate that

> an internal secretion of the uterine mucosa might have a specific abbreviating effect upon the corpus luteum.

The truth of this suggestion did not become apparent until the 1970's when prostaglandin $F_{2\alpha}$ was identified as the uterine luteolytic agent (see Volume III, Chapter 3).

The elucidation of the pituitary tropic control of the ovaries is still an active area of research. But the real breakthrough occurred in 1927 when P. E. Smith at Columbia University, New York, showed that it was possible to hypophysectomize rats. Animals treated in this way became the standard preparations for biologists and chemists doing work designed to determine whether there were one or more pituitary gonadotropins and how they acted on the ovaries (see review by Greep, 1967; Volume III, Chapter 2).

This brief account of the history of our knowledge of the ovaries outlines the value of a broad comparative approach to the subject. By our increasing tendency to specialize we have lost the breadth of interest which our predecessors had. Many of their observations were close to what we now believe to be the truth, but they could not be developed in the prevailing scientific climate. A similar problem exists today, for with the enormous volume of research findings now being published, a new concept is only "on top of the pile" for a short time. A concept may remain unrecognized or be swamped by newer material, and it may actually need to be rediscovered at the "right" time to be influential in scientific thinking.

Although much is known, much still remains to be discovered; and much of what has already been discovered needs to be uncovered.

REFERENCES

Adelmann, H. B. (1933). The "De ovorum gallinaceorum generationis primo exordio progressusque, et pulli gallinacei creationis ordine" of Volcher Coiter. *Ann. Med. Hist.* **5,** 327–341.

Adelmann, H. B. (1942). "The Embryological Treatises of Hieronymus Fabricius of Aquapendente. Cornell Univ. Press, Ithaca, New York.

Adelmann, H. B. (1966). "Marcello Malpighi and the Evolution of Embryology," 5 vols. Cornell Univ. Press, Ithaca, New York.

Aldrovandrus, U. (1642). "Monstrorum Historia." Bononiae.

Allen, E. (1927). The menstrual cycle of the monkey, Macacus Rhesus: observations on normal animals, the effects of removal of the ovaries and the effects of injections of ovarian and placental extracts into the spayed animals. *Contrib. Embryol. Carnegie Inst.* **19,** 1–44.

Allen, E., and Doisy, E. A. (1923). An ovarian hormone; preliminary report on its localization, extraction and partial purification, and action in test animals. *J. Am. Med. Assoc.* **81,** 819–821.

Allen, W. M., and Corner, G. W. (1930). Physiology of corpus luteum; maintenance of pregnancy in rabbit after very early castration by corpus luteum extracts. *Proc. Soc. Exp. Biol. Med.* **27,** 403–405.

Allen, W. M., and Wintersteiner, O. (1934). Crystalline progestin. *Science* **80,** 190–191.

Bartholinus, C. (1678). "De Ovariis Mulierum et Generationis Historia Epistola Anatomica." Westen, Amsterdam.

Beard, J. (1897). "The Span of Gestation and the Cause of Birth." Fischer, Jena.

Berthold, A. A. (1849). Transplantation der Hoden. *Arch. Anat. Physiol. Wiss. Med.* **16,** 42–46.

Bouin, P., and Ancel, P. (1909). Sur la fonction du corps jaune. Action du corps jaune vrai sur l'utérus. *C. R. Soc. Biol.* **66,** 505–507.

Bouin, P., and Ancel, P. (1910). Récherches sur les fonctions du corps jaune gestatif. I. Sur le déterminisme de la préparation de l'utérus à la fixation de l'oeuf. *J. Physiol. Pathol. Gen.* **12,** 1–16.

Brown-Séquard, C. É. (1889). Des effets produits chez l'homme par des injections sous-cutanées d'un liquide retiré des testicules frais de cobaye et de chien. *C. R. Soc. Biol.* **1,** 415–419.

Brown-Séquard, C. É. (1890). Remarques sur les effets produits sur la femme par des injections sous-cutanées d'un liquide retiré d'ovaires d'animaux. *Arch. Physiol. Norm. Pathol.* **2,** 456–457.

Butenandt, A., Westphal, U., and Cobler, H. (1934). Über einen Abbau des Stigmasterins zu corpus-luteum-worksamen Stoffen; ein Beitrag zur Konstitution des Corpus-luteum-Hormons (Vorläuf Mitteil). *Ber. Dtsch. Chem. Ges. B* **67,** 1611–1616.

Coiter, V. (1573). "Externarum et Internarum Principalium Humani Corporis Partium Tabulae." Noribergae.

Corner, G. W. (1923). Ovulation and menstruation in *Macacus rhesus. Contrib. Embryol. Carnegie Inst.* **15,** 73–102.

Corner, G. W. (1928). Physiology of corpus luteum; effect of very early ablation of corpus luteum upon embryos and uterus. *Am. J. Physiol.* **86,** 74–81.

Corner, G. W. (1943). On the female testes or ovaries. *In* "Essays in Honor of H. M. Evans" (transl. of Regner de Graaf), pp. 122–137. Univ. of California Press, Berkeley and Los Angeles.

Corner, G. W. (1950). The relation of the ovary to the menstrual cycle. Notes on the history of a belated discovery. *An. Fac. Med., Univ. Repub., Montevideo* **35,** 758–766.

Corner, G. W., and Allen, W. M. (1929). Physiology of corpus luteum; production of special uterine reaction (progestational proliferation) by extracts of corpus luteum. *Am. J. Physiol.* **88,** 326–339.

Cruikshank, W. (1797). 9. Experiments in which, on the third day after impregnation, the ova of rabbits were found in the fallopian tubes; and on the fourth day after impregnation in the uterus itself; with the first appearances of the foetus. *Philos. Trans. R. Soc. London* **87,** 197–214.

Davidge, J. B. (1814). "Physical Sketches, or Outlines of Correctives, Applied to Certain Modern Errours in Physick." Baltimore, Maryland.

Dawson, W. R. (1929). "Magician and Leach." Methuen, London.

de Graaf, R. (1672). "De Mulierum Organis Generationi Inservientibus." Leyden.

Descartes, R. (1664). "L'Homme, et un traitté de la formation du foetus du mesme autheur." C. Angot, Paris.

Doisy, E. A., Rolls, J. O., Allen, E., and Johnston, C. G. (1924). The extraction and some properties of an ovarian hormone. *J. Biol. Chem.* **61,** 711–727.

Drelincourt, C. (1685). "De conceptione adversaria." Lugduni Batavorum.

Engle, E. T., Smith, P. E., and Shelesnyak, M. C. (1935). Rôle of estrin and progestin in experimental menstruation, with especial reference to complete ovulatory cycle in monkeys and human beings. *Am. J. Obstet. Gynecol.* **29,** 787–797.

Fabricius, H. (1604). "De Formato Foetu." Patavii.

Fallopius, G. (1562). "Observationes Anatomicae." Coloniae.

Forbes, T. R. (1946). Origin of "freemartin." *Bull. Hist. Med.* **20,** 461–466.
Fraenkel, L. (1903). Die Function des Corpus luteum. *Arch. Gynaekol.* **68,** 438–545.
Fraenkel, L., and Cohn, F. (1901). Experimentelle Untersuchungen über den Einfluss des Corpus luteum auf die Insertion des Eies. *Anat. Anz.* **20,** 294–300.
Gardiner, A. (1964). "Egyptian Grammar," 3rd ed. Oxford Univ. Press, London and New York.
Graham, H. (1960). "Eternal Eve." Hutchinson, London.
Greep, R. O. (1967). The saga and the science of the gonadotrophins. *J. Endocrinol.* **39,** ii–ix.
Guinness, F., Lincoln, G. A., and Short, R. V. (1971). The reproductive cycle of female red deer, *Cervus elaphus* L. *J. Reprod. Fertil.* **27,** 427–438.
Guttmacher, A. (1965). Intra-uterine contraceptive devices. *Fam. Planning* **13,** 91–96.
Haighton, J. (1797). 8. An experimental enquiry concerning animal impregnation. *Philos. Trans. R. Soc. London* **87,** 159–196.
Haller, A., ed. (1744). "Praelectiones Academicae in Proprias Institutiones Rei Medicae," Vol. 5. Gottingae.
Harrison, R. J. (1948). The development and fate of the corpus luteum in the vertebrate series. *Biol. Rev. Cambridge Philos. Soc.* **23,** 296–331.
Hartmann, M., and Wettstein, A. (1934). Ein krystallisiertes Hormon aus Corpus luteum. *Helv. Chim. Acta* **17,** 878–882.
Hartsoeker, N. (1694). "Essay de Dioptrique." Paris.
Harvey, W. (1628). "De Motu Cordis et Sanguinis in Animalibus." Frankfurt.
Harvey, W. (1651). "Exercitationes de Generatione Animalium. Quibus Accedunt Quaedam de Partu: de Membranis ac Humoribus Uteri: & de Conceptione." Londini.
Harvey, W. (1653). "Anatomical Exercitations, Concerning the Generation of Living Creatures: To Which are Added Particular Discourses, of Births, and of Conceptions, &c." J. Young for O. Pulleyn, London.
Heape, W. (1900). The "sexual season" of mammals and the relation of the "Pro-oestrum" to menstruation. *Q. J. Microsc. Sci.* [N.S.] **44,** 1–70.
Hegar, A. (1878). Die Castration der Frauen. *Sammlung klin. Vortr. Volkmann Gynaekol.* **42,** 925–1068.
Heresbach, C. (1577). "Rei Rustici Libri Quatuor" (transl. by B. Googe as "Foure Bookes of Husbandry"). Londini.
Herrmann, E. (1915). Ueber eine wirksame Substanz im Eierstocke und in der Placenta. *Monatsschr. Geburtshilfe Gynaekol.* **41,** 1–50.
Hisaw, F. L. (1935). Physiology of menstruation in *Macacus rhesus* monkeys; influence of follicular and corpus luteum hormones; effects of anterior pituitary extracts. *Am. J. Obstet. Gynecol.* **29,** 638–659.
Hisaw, F. L., and Greep, R. O. (1938). Inhibition of uterine bleeding with estradiol and progesterone and associated endometrial modifications. *Endocrinology* **23,** 1–14.
Hisaw, F. L., Fevold, H. L., and Meyer, R. K. (1930). The corpus luteum hormone. II. Methods of extraction. *Physiol. Zool.* **3,** 135–144.
Hunter, J. (1779). Account of the free martin. *Philos. Trans. R. Soc. London* **69,** 279–293.
Hunter, J. (1787). 21. An experiment to determine the effect of extirpating one ovarium upon the number of young produced. *Philos. Trans. R. Soc. London* **77,** 233–239.

Hunter, W. (1774). "The Anatomy of the Human Gravid Uterus Exhibited in Figures." Birmingham.

Hunter, W. (1794). "An Anatomical Description of the Human Gravid Uterus and Its Contents." London.

Jocelyn, H. D., and Setchell, B. P. (1972). Regnier de Graaf on the human reproductive organs. *J. Reprod. Fertil., Suppl.* **17.**

Kehrer, F. A. (1879). "Beiträge zur klinischen und experimentelle Geburtskunden und Gynakologie." Giessen.

Kerckring, T. (1672). An account of what hath been of late observed by Dr. Kerkringius concerning Eggs to be found in all sorts of Females. *Philos. Trans. R. Soc. London* **7,** 4018–4026.

Keynes, G. (1966). "The Life of William Harvey." Oxford Univ. Press (Clarendon), London and New York.

Knauer, E. (1896). II. Einige Versuche uber Ovarientransplantation bei Kaninchen. *Zentralbl. Gynaekol.* **20,** 524–528.

Langly, W. (1674). De generatione animalium observationes quaedam. In Schrader, J. "Observationes et Historiae Omnes et Singulae." Wolfgang, Amstelodami.

Lataste, F. (1886). Notes prises au jour le jour sur différentes espèces de l'ordre des Rongeurs observées en captivité. *Actes Soc. Linn. Bordeaux* **40,** 293–466.

Lataste, F. (1887). Notes prises au jour le jour sur différentes espèces de l'ordre des Rongeurs observées en captivité. *Actes. Soc. Linn. Bordeaux* **41,** 201–536.

Le Grand, A. (1672). "Institutio Philosophiae, Secundum Principia Renati Descartes." Londini.

Liddell, H. G., and Scott, R. (1940). "A Greek-English Lexicon." Oxford Univ. Press (Clarendon), London and New York.

Lind, L. R. (1959). "Jacopo Berengario da Carpi—A Short Introduction to Anatomy (*Isagogae breves*)." Univ. of Chicago Press, Chicago, Illinois.

Lipschütz, A. (1928). New developments in ovarian dynamics and the law of follicular constancy. *Br. J. Exp. Biol.* **5,** 283–291.

Loeb, L. (1907). Ueber die experimentelle Erzeugung von Knoten von Deciduagewebe in dem Uterus des Meerschweinchens nach stattgefundener Copulation. *Zentralbl. Allg. Pathol. Pathol. Anat.* **18,** 563–565.

Loeb, L. (1908). The experimental production of the maternal part of the placenta in the rabbit. *Proc. Soc. Exp. Biol. Med.* **5,** 102–104.

Loeb, L. (1911). Ueber die Bedertung des Corpus luteum für die Periodizitat des sexuellen Zyklus beim weiblichen Saugetierorganismus. *Dtsch. Med. Wochenschr.* **37,** 17–21.

Loeb, L. (1923). The effect of extirpation of the uterus on the life and function of the corpus luteum in the guinea pig. *Proc. Soc. Exp. Biol. Med.* **20,** 441–443.

Loeb, L. (1927). Effects of hysterectomy on system of sex organs and on periodicity of sexual cycle in guinea pig. *Am. J. Physiol.* **83,** 202–224.

Long, J. A., and Evans, H. M. (1922). The oestrous cycle in the rat and its associated phenomena. "Memoirs of University of California," No 6. University of California Press, Berkeley.

MacCorquodale, D. W., Thayer, S. A., and Doisy, E. A. (1936). Isolation of principal estrogenic substance of *liquor folliculi*. *J. Biol. Chem.* **115,** 435–448.

Magnus, V. (1901). Ovariets betydning for svangerskabet med saerligt hensyn til corpus luteum. *Nor. Mag. Laegevidensk.* **62,** 1138–1145.

Makepeace, A. W., Weinstein, G. L., and Friedman, M. H. (1937). Effect of progestin and progesterone on ovulation in rabbit. *Am. J. Physiol.* **119,** 512–516.

Malpighi, M. (1697). "Opera Posthuma." Londini.

Marrian, G. F. (1966). Early work on the chemistry of pregnanediol and the oestrogenic hormones. *J. Endocrinol.* **35,** vi–xvi.

Müller, J. (1830). "De Glandularum Secrenentium Structura Penitiori." Lipsia

Nawito, M. F., Shalash, M. R., Hoppe, R., and Rakha, A. M. (1967). "Reproduction in Female Camel," Bull. No. 2. Anim. Sci. Res. Inst., Nat. Res. Cent., Cairo, U.A.R.

Needham, J., and Gwei-Djen, L. (1966). Proto-Endocrinology in Mediaeval China. *Jpn. Stud. Hist. Sci.* **5,** 150–171.

Needham, J., and Hughes, A. (1959). "A History of Embryology," 2nd ed. Cambridge Univ. Press, London and New York.

Newport, G. (1853). 10. On the impregnation of the ovum in the Amphibia. And on the direct agency of the spermatozoon. *Philos. Trans. R. Soc. London* **143,** 233–290.

Newport, G. (1854). 8. Researches on the impregnation of the ovum in the amphibia; and on the early stages of development of the embryo. *Philos. Trans. R. Soc. London* **144,** 229–244.

Novoa, C. J. (1970). Reproduction in Camelidae. *J. Reprod. Fertil.* **22,** 3–20.

Nuck, A. (1691). "Adenographia Curiosa et Uteri Foeminei Anatome Nova." Leiden.

Parkes, A. S. (1926). On the occurrence of the oestrous cycle after x-ray sterilisation. Part 1. Irradiation of mice at three weeks old. *Proc. R. Soc. London, Ser, B* **100,** 172–199.

Parkes, A. S. (1962). Prospect and retrospect in the physiology of reproduction. *Br. Med. J.* **2,** 71–75.

Parkes, A. S. (1966). The rise of reproductive endocrinology, 1926–1940. *J. Endocrinol.* **34,** xx–xxxii.

Parkes, A. S., and Bellerby, C. W. (1926). Studies on the internal secretions of the ovary. 1. The distribution in the ovary of the oestrus-producing hormone. *J. Physiol.* (*London*) **61,** 562–575.

Pincus, G., and Chang, M. C. (1953). Effects of progesterone and related compounds on ovulation and early development in rabbit. *Acta Physiol. Lat. Am.* **3,** 177–183.

Pott, P. (1775). "Chirurgical Observations." London.

Pouchet, F.-A. (1842). "Théorie Positive de la Fécondation des Mammifères." Baillière et Fils, Paris.

Pouchet, F.-A. (1847). "Théorie Positive de l'Ovulation Spontanée et de la Fécondation des Mammifères et de l'Espèce Humaine." Baillière et Fils, Paris.

Prenant, A. (1898). La valeur morphologique du corps jaune. Son action physiologique et therapéutique possible. *Rev. Gen. Sci. Pure Appl.* **9,** 646–650.

Prévost, J. L., and Dumas, J. A. B. (1824). De la génération dans les Mammifères, et des premiers indices du développement de l'Embryon. *Ann. Sci. Nat.* **3,** 113–138.

Puech, A. (1873). "Les Ovaires, de Leur Anomalies." Paris.

Raciborski, A. (1843). "De la Puberté et de l'Age Critique Chez la Femme." Paris.

Roberts, G. (1843). Fragment d'un voyage dans les provinces interieures de l'Inde en 1841; publié par la Societé Orientale, Paris. *Experience* **11,** 99.

Ruini, C. (1598). "Anatomia del Cavallo, Infermità et Suoi Rimedii." Venetia.

Short, R. V. (1969). An introduction to some of the problems of intersexuality. *J. Reprod. Fertil., Suppl.* **7,** 1–8.

Singer, C. (1957). "A Short History of Anatomy and Physiology from the Greeks to Harvey." Dover, New York.

Slotta, K. H., Ruschig, H., and Fels, E. (1934). Reindarstellung der Hormone aus dem Corpus luteum (II. Mitteil). *Ber. Dtsch. Chem. Ges. A* **67,** 1624–1626.

Smith, P. E. (1927). Disabilities caused by hypophysectomy and their repair; tuberal (hypothalamic) syndrome in rat. *J. Amer. Med. Assoc.* **88,** 158–161.

Smith, P. E., and Engle, E. T. (1932). Prevention of experimental uterine bleeding in macacus monkey by corpus luteum extract (progestin). *Proc. Soc. Exp. Biol. Med.* **29,** 1225–1227.

Snape, A. (1687). "The Anatomy of an Horse." M. Flesher, London.

Solomons, B., and Gatenby, J. W. B. (1924). Formation, structure and physiology of corpus luteum of man, pig, and duck-billed platypus. *J. Obstet. Gynaecol. Br. Emp.* **31,** 580–594.

Stensen, N. (1667). "Elementorum Myologiae Specimen, Seu Musculi Descriptio Geometrica. Cui Accedunt Canis Carchariae Dissectum Caput, et Dissectus Piscis ex Canum Genere." Florence.

Stensen, N. (1675). Ova viviparorum spectantes observationes. *In* "Thomae Bartholini Acta Medica et Philosophica Hafniensia," vol. II, pp. 219–232. Hafniae.

Stockard, C. R., and Papanicolaou, G. N. (1917). The existence of a typical oestrous cycle in the guinea-pig—with a study of its histological and physiological changes. *Am. J. Anat.* **22,** 225–283.

Taylor, M. J., and Short, R. V. (1973). The development of the germ cells in the ovary of the mule and hinny. *J. Reprod. Fertil.* **32,** 441–445.

Temkin, O. (1956). "Soranus' Gynecology." Johns Hopkins Press, Baltimore, Maryland.

Thompson, D'A. W. (1910). "The Works of Aristotle Translated into English. Vol. IV. *Historia Animalium*" (J. A. Smith and W. D. Ross, eds). Oxford Univ. Press (Clarendon), London and New York.

Tilt, E. J. (1850). "On Diseases of Menstruation and Ovarian Inflammation, in Connexion with Sterility, Pelvic Tumours, and Affections of the Womb." London.

Topsell, E. (1607). "The History of Four-Footed Beasts and Serpents." London.

Turbervile, G. (1576). "The Noble Art of Venerie or Hunting." London.

Tusser, T. (1580). "Five Hundred Pointes of Good Husbandrie." London.

van Leeuwenhoek, A. (1678). Observationes D. Anthonii Leeuwenhoek, de Natis e semine genitali Animalculis. Letter dated November 1677. *Philos. Trans. R. Soc. London* **12,** 1040.

van Leeuwenhoek, A. (1683). An Abstract of a Letter from Mr. Anthony Leeuwenhoek of Delft about Generation by an Animacule of the Male seed. Animals in the seed of a frog. Some other Observables in the parts of a Frog. Digestion, and the motion of the blood in a Feavor. *Philos. Trans. R. Soc. London* **13,** 347–355.

Vesalius, A. (1538). "Tabulae Anatomicae Sex." Venetiis.

Vesalius, A. (1543). "De Humani Corporis Fabrica Libri Septem." Basileae.

Vesalius, A. (1555). "De Humani Corporis Fabrica Libri Septem." Basileae.

von Baer, C. E. (1827). "De Ovi Mammalium et Hominis Genesi." Lipsiae.

Wharton, T. (1656). "Adenographia." London.

Zeuner, F. E. (1963). "A History of Domesticated Animals." Hutchinson, London.

Zuckerman, S. (1937). Menstrual cycle of Primates; part played by oestrogenic hormone in menstrual cycle. *Proc. R. Soc. London, Ser. B* **123,** 457–471.

2

The Development of the Ovary and the Process of Oogenesis

S. Zuckerman and T. G. Baker

The chapter which appeared under this title in the 1962 edition of "The Ovary"* dealt, first, with the origin, development, and differentiation of the individual constituents of the ovary in different classes of vertebrate. It then reviewed such evidence as was available on the nature of the process of oogenesis, which was strictly defined as the transformation of oogonia, which multiply in the ovary by mitotic division into oocytes.

These topics then commanded considerable intellectual attention, with the main interest focusing on the source of the primordial germ cells (see, e.g., Waldeyer, 1870, 1906; Maréchal, 1907; Heys, 1931; Bounoure, 1939; Brambell, 1956), about which two main views were then current. The first was Waldeyer's (1870) thesis that germ cells arise during embryonic development through the proliferation of the so-called germinal epithelium that covers the presumptive gonad. The contrary view, which had gained considerable support from Weismann's (1885) theory of the continuity of

* Dr. Anita Mandl and Dr. L. L. Franchi collaborated with the senior author in its preparation.

germ plasm, was that the germ cells become segregated before the formation of the organ systems of the embryo (e.g., Goette, 1875; Balfour, 1878; Nussbaum, 1880).

I. INTRODUCTION—PHASES IN OVARIAN DEVELOPMENT

According to the view most widely accepted in the early 1960s, the development of the ovary of vertebrates could be divided into four major phases. During the first, *primordial germ cells* (i.e., undifferentiated germ cells or "Urkeimzellen"), which become segregated very early in development, migrate from their sites of origin and finally settle in bilateral thickenings of the coelomic epithelium ventral to the developing mesonephroi. This thickened epithelium is known as the *genital ridge.* Although adult females of some species have only one ovary (e.g., cyclostomes), the genital ridges always appear bilaterally during early development.

The second phase, which occurs after the arrival of the primordial germ cells at the genital ridges, consists in the proliferation of both nongerminal and germinal cells. This leads to the formation of distinct gonadal primordia which are identical in both sexes (*the indifferent gonad stage*).

During the third phase, the gonads become divided into a peripheral *cortex* and a central *medulla,* separated by a primary *tunica albuginea* (the medulla is absent in some lower vertebrates). The germ cells are at first mainly peripheral in position.

The fourth and final phase, characteristic of sex differentiation, consists in the development of the cortex and involution of the medulla in the female and proliferation of the medulla and involution of the cortex in the male. While the germ cells in the male are drawn into the medulla, those in the female remain in the cortex, and in higher vertebrates they become organized into cords (so-called Pflüger's cords). At this stage, the germ cells in female embryos of some species may enter meiotic prophase and therefore become morphologically distinguishable from corresponding cells in the male. (The term germ cell is nonspecific and is applicable equally to oogonia, oocytes, and ova, and to spermatogonia, spermatocytes, spermatids, and spermatozoa.) Cells derived from the surface of the gonad invest the female germ cells with a follicular envelope corresponding to the *granulosa cells* of the adult. The germ cells of the male become associated with corresponding cells of epithelial origin ("supporting" cells or future *Sertoli cells*) and become arranged in *spermatic cords* (future seminiferous tubules). It needs to be emphasized that the adjective "germinal" in the term germinal (coelomic) epithelium is retained only because it has been in common use for many decades, and even though it is now clear that the epithelial cells which cover the ovary lack any ability to give rise to germ cells either during development or after birth.

II. MIGRATION OF PRIMORDIAL GERM CELLS

Numerous experimental and histological studies of invertebrates and vertebrates have indicated that certain specialized cytoplasmic components (see Hegner, 1914; Bounoure, 1939), called Keimbahn determinants, can be traced from the unsegmented egg through successive developmental stages into the gonads of the sexually mature animal (for reviews, see also Hegner, 1914; Bounoure, 1939; Nieuwkoop, 1949). It had been fully established before 1960 that Keimbahn determinants are present in the vegetative pole of the amphibian egg shortly after fertilization and before the first cleavage (Bounoure, 1931a,b, 1935a,b, 1939; Bounoure *et al.*, 1954).

More recent observations on the subject add little of significance to what was known before. The Keimbahn determinants have been shown to be rich in ribonucleoprotein, and it has been confirmed by Blackler (1966) and Czolowska (1969) that if they are destroyed by irradiation or microsurgery, the frog or toad which develops from the egg will be devoid of primordial germ cells and will be permanently sterile (see review by Smith and Williams, 1975).

As has long been known, the primordial germ cells in mammals are first detected in the wall of the yolk sac from which they migrate via the gut mesentery to the gonadal *anlagen*. Hardisty (1967) has compared the number of primordial germ cells in a range of vertebrate species and has assessed the role of germ cell mitosis during migration.

In all vertebrates that have been investigated, the primordial germ cells originate extragonadally and migrate to the genital ridges. The paths by which they migrate from their sites of origin to their final position in the genital ridges have been studied by standard histological as well as histochemical methods and by a variety of experimental techniques.

Four modes of migration of primordial germ cells have been postulated: active, by ameboid movements of the primordial germ cells themselves; passive, by differential growth of surrounding tissues; passive, by transport via the blood stream; and chemotactic, under the influence of some local inductors diffusing from the presumptive gonadal area.

A. Active Movements

The view that the primordial germ cells migrate actively by an ameboid form of movement is usually associated with the observation that they possess blunt pseudopodia-like processes. Zamboni and Merchant (1973), using the electron microscope, have confirmed the presence of pseudopodia on mouse primordial germ cells undergoing migration. There is, however, little evidence that the pseudopodia are constantly present, regardless of the method of fixation, although Muratori (1937) and Blandau *et al.* (1963) have observed them in the living germ cells of the chick and mouse *in vitro*. A combination of active and passive (see below) migration is claimed to occur in a number of animals, including cyclostomes, fishes, and some reptiles (Dodds, 1910; Okkelberg, 1921), but no experimental proof is provided.

It is generally held that mammalian primordial germ cells migrate largely by their own active ameboid movement (see, e.g., Everett, 1943; Chiquoine, 1954), possibly aided by histiolytic action (Witschi, 1948). The only experimental evidence supporting this view derives from Mintz's (1959) study in which primordial germ cells were held up in their movement toward the genital ridges even though organogenesis appeared to be proceeding normally.

B. Passive Movements by Differential Growth of Tissues

Since differential growth of the germ layers occurs consistently in all vertebrates during organogenesis, it is clear that the primordial germ cells could be drawn passively from an extra- to an intraembryonic position. Most of the evidence for amphibians indicates that germ cells reach the genital ridges solely by this means (e.g., Humphrey, 1925; Blackler, 1958).

C. Passive Movements by Vascular Transport

Stanley and Witschi (1940) consider that the presence of germ cells in the blood stream in birds is accidental, and that any such cells are destined to degenerate. Other workers (e.g., Firket, 1913; Matsumoto, 1932; Witschi, 1935) believe that the avian germ cells migrate actively, and Berenberg-Gossler (1913) suggests that they migrate by differential tissue growth. Experimental work on the chick embryo (Simon, 1957; see also Willier, 1937) largely disposes of these earlier claims, in that it clearly demonstrates that, in birds at least, the vascular path is responsible for the transport of the primordial germ cells. Simon (1957) bisected the chick embryo at the level of the twentieth presumptive somite and found that the primordial germ cells situated in the germinal crescent reached the severed caudal region of the embryo, given that the vascular path remained intact. In a second experiment, he grafted the caudal half of another embryo on to the blastoderm, and found that, given that the graft became vascularized, primordial germ cells originating from the cranial region of the host embryo were also able to reach the graft. In his final experiment, Simon united a chick embryo, whose germinal crescent had been removed, in parabiosis with a normal chick embryo. If a circulation became established between the two parabionts, the sterilized embryo's genital ridges became populated by some primordial germ cells. The implication of this experiment is that the cells were transported there via the blood stream from the nonsterilized partner. The view that the primordial germ cells in birds are carried to their definitive site of development by means of the blood stream is also supported by the repeated finding that the primordial germ cells are initially widely scattered in the embryo, particularly in the head (e.g., Swift, 1914), and that relatively large numbers become trapped in ectopic positions within fine capillaries. Such primordial germ cells as fail to reach the genital ridges degenerate.

The hypothesis that in certain teleosts the primordial germ cells also migrate via the blood stream (Stolk, 1958) requires substantiation.

D. Chemotaxis

The suggestion that the migration of the primordial germ cells to the presumptive gonads is aided by the emission of chemotactic or inductor substances from the gonad-forming areas has been put forward for fishes (Woods, 1902), birds (Firket, 1913, 1920; Willier, 1939), and mammals (Witschi, 1948, 1951a). It would appear probable, however, that although such specific influences could be responsible for directing the primordial germ cells at later stages of migration, when they are already in the vicinity of the future gonads, their initial movements are more likely due to the differential growth of the germ layers in which the cells reside. The latter possibility is supported by the observation that anomalies arising in the course of grafting experiments, such as the formation of sterile gonads from grafts either containing or lacking the germinal crescent, are due, in birds at least, to the absence of a suitable transporting mechanism (in this instance, the blood stream: Willier, 1937; Simon, 1957). On the other hand, the final distribution of the primordial germ cells in the right and left presumptive ovaries in birds may well be mediated by chemotactic substances. An alternative explanation for the uneven distribution of primordial germ cells between the two avian ovaries is that the right ovary, which is destined to be nonfunctional, and which receives fewer primordial germ cells than does the left, is poorly vascularized at very early stages of development (Dantschakoff, 1931).

The few enquiries on the manner of migration of the primordial germ cells that have been undertaken since the first edition of this book appear to favor the "active-movement hypothesis" (Blandau, 1969; Blandau *et al.*,

1963; also Peters, 1970). When the ovaries of human or mouse embryos are disintegrated by means of enzyme action, and the dissociated cells are then placed in culture, it can be seen that the primordial germ cells and oogonia possess pseudopodia and are able to undergo ameboid movement, whereas the somatic cells remain stationary. When a piece of gonadal ridge is placed at the edge of the Petri dish, the germ cells move in an attempt to colonize the strip of tissue (Blandau *et al.*, 1963; Tarkowski, 1970).

III. THE SEXUALLY INDIFFERENT PHASE OF GONADAL DEVELOPMENT

It is widely taught that two proliferations of the surface epithelium of the indifferent gonad invade the underlying mesenchyme. If the genetic sex of the embryo is male, the initial proliferation of the coelomic epithelium results in the production of sex cords, which then differentiate into the definitive seminiferous tubules. In the female, the cells of the first proliferation give rise to sex cords which appear identical with those of the male. These do not branch, but elongate, and become separated from the cortical region of the gonad by a transient connective tissue layer or primary tunica albuginea. As a result, the medulla of the ovary becomes organized into radially arranged cords which come into contact with the distal ends of cords of cells that arise from the mesonephric rudiment, and which are themselves the rudiments of the rete system (or "urinogenital connection"; see Brambell, 1956).

The development of the medulla which, like the early genital ridge but unlike the cortex, is largely independent of the presence of germ cells is, thus, at first the same in both sexes and constitutes the main feature of the indifferent stage of gonad development.

The second proliferation of the surface epithelium is generally held to give rise to the ovarian cortex in amphibians and amniotes. In males, the second, cortical, proliferation is either transitory or absent.

This conventional picture of the course of gonadal development has been strongly challenged by Witschi (1914, 1951a, 1956) who argued that throughout the vertebrates, including the mammals, the medulla is formed not from the first proliferation of the germinal epithelium but from extragonadal undifferentiated cells related to the mesonephric rudiment (mesonephric "blastema"), which is separated from the germinal epithelium by a previously formed core of mesenchyme cells known as the primary gonad cavity or albuginea. The presumptive rete cords also arise from this undifferentiated mesonephric "blastema," the connection between the medulla and the rete apparatus being present from the earliest stages of development. In Witschi's words: "The misconception that the medulla might arise as a 'first proliferation' from the 'germinal epithelium' could only arise from a study of mammalian materials, where the gonad cavity is reduced to a very narrow space or albuginea." It is worth noting, however, that from an independent examination of Witschi's (1948) material, Gillman (1948) concluded that the medullary cords arise from the coelomic epithelium. At the same time, Gillman noted that the structure was looser in the area "just below the epithelium" (i.e., the region designated by Witschi as the albuginea) than elsewhere. This difference in interpretation indicates the great difficulty in reaching final conclusions about detailed embryological processes from histological study alone.

Toward the end of the indifferent stage, the gonad still retains a relatively broad connection with the mesonephros. Expansion of the interior of the gonad, and involution of the mesonephric kidney, later cause the "nipping off" of the ovary so that it becomes suspended at the future hilar region by a narrow peduncle. The end of the indifferent stage of gonad development is also marked in the male by the further expansion of connective tissue cells in a layer, under the germinal epithelium, which is destined to become the definitive fibrous tunica albuginea of the testis.

In the earlier part of the indifferent phase, some of the germ cells which are scattered in and under the germinal epithelium become incorporated into the medullary cords, and thus pass down into the substance of the gonad. This process differs quantitatively in the two sexes and heralds the onset of sexual differentiation.

IV. THE SEXUAL DIFFERENTIATION OF THE GONAD

Before sex differentiation begins, the indifferent gonad contains all the cell types it needs to render it capable of developing into either a testis or an ovary (Gillman, 1948). The direction of further development depends upon both the genetic sex of the individual (see, e.g., Dodd, 1960; Beatty, 1960) and on internal and external environmental factors.

Experimental evidence indicating that the direction of development is largely predetermined genetically derives from Humphrey's (1928a,b) observation that, after transplantation, the gonad-forming area develops into a testis or an ovary according to the sex of the donor, and regardless of that of the host. The frequency with which such grafts develop into a gonad of specific sex increases the later the transplantations are performed during the stages of development represented by the genital ridge and the indifferent gonad (see also Willier, 1939; Burns, 1955).

The fact that environmental factors, particularly internal ones, can also play an essential role in gonad development is demonstrated by the ease with which the normal pattern of development for either sex in lower vertebrates can be modified by hormonal or other physiological disturbances (Witschi, 1939, 1951b; Burns, 1955; Dodd, 1960).

The process of sex differentiation consists essentially of the proliferation of the cortex and regression of the medulla in the female and the reverse process in the male. The recessive component (cortex in the male; medulla in the female) rarely differentiates further than the indifferent or bisexual stage. The germ cells become preferentially associated with the dominant component, frequently before sex differentiation has progressed to any extent. The distinction between the indifferent gonad stage and the onset of sex differentiation is therefore not clear-cut (see, e.g., Allen, 1906; Swift, 1915, 1916; Bascom, 1923; Brambell, 1927), except perhaps in elasmobranchs (Chieffi, 1949, 1959) and amphibians (Witschi, 1914, 1929, 1956; Christensen, 1930; Jurand, 1957).

In these groups of vertebrates, the indifferent gonad becomes a testis in genetic males earlier than it becomes transformed into an ovary in the genetic female, one of the earliest signs of testis formation being the incorporation of the majority of germ cells into the developing medullary cords.

According to those authors who maintain that the medulla originates from the first proliferation of the germinal epithelium, the shift of the germ cells in the male entails little more than their inclusion within the proliferation during its advance into the underlying mesenchyme. Those who hold the view that the medulla originates from an extragonadal mesonephric "blastema" believe that, during differentiation of the testis, the germ cells migrate into the developing medulla across the preexisting albuginea. The latter process is said to be easily observed in amphibians, in which the albuginea is prominent (Witschi, 1914), and testicular differentiation is claimed to proceed similarly in man (Witschi, 1951a, 1956).

Regardless of its derivation, the medulla in the male consistently proliferates in the course of sex differentiation by the mitotic division of the germ cells and epithelial cells by which it is penetrated. The solid cords which arise from the latter cells become canalized and form a series of seminiferous tubules lined by supporting (future Sertoli) cells and germ cells. Connective tissue elements of mesenchymal origin further infiltrate between the tubules and also form a fibrous tunica albuginea which separates the tubules from the overlying epithelium, blood vessels following in their wake. Groups of interstitial cells commonly occur in the intertubular spaces.

The degree to which the cortex develops in the male is variable. In the majority of species it does not develop beyond the indifferent stage and undergoes regression as the medulla proliferates. In most reptiles, some birds, and a few mammals (also *Ambystoma*; Humphrey, 1929), the cortex may develop beyond the indifferent stage. This development is transitory, for sooner or later regression of the cortical region of the testis sets in, and any germ cells retained in it degenerate. As the testis enlarges, the cortex becomes attenuated and reacquires the appearance of a flat coelomic epithelium. The seminiferous tubules acquire open connections with the tubules of the urinogenital complex and, thus, with the Wolffian duct. These processes lead to the establishment of the full complement of structures characteristic of the immature testis.

It is generally recognized that the ovarian cortex is derived from a proliferation of the germinal epithelium (the "second proliferation" described by many authors), that the majority of germ cells remain situated in the peripheral region of the gonad (i.e., in the cortex), and that they do not migrate into the medullary region as in the male. The early phases of sex differentiation in the female are therefore essentially an extension of the indifferent gonad stage, and typical ovarian features (e.g., marked proliferation of the cortex, appearance of nests of oocytes, and formation of the definitive tunica albuginea) do not become fully apparent until somewhat later (except in urodeles; see Burns, 1925, 1928; Humphrey, 1929).

The rapid increase, through mitotic division, in the number of germ cells at early stages of ovarian differentiation has been recorded for elasmobranchs, amphibians, reptiles, birds, and mammals. Mitosis also occurs in the epithelial cells and follicle cells. These divisions result in an increase in the thickness of the cortex in amniotes, and localized areas of proliferation may give rise to cord-like ingrowths towards the medulla which are known as Pflüger's tubules (dog: Jonckheere, 1930; cat: de Winiwarter and Sainmont, 1909; rabbit: de Winiwarter, 1901; Allen, 1904; man: de Winiwarter, 1910). In other species, a solid epithelial "nucleus" later becomes split up into cords by the infiltration of connective tissue (Brambell, 1927; Hargitt, 1930). The cortical cords may be continuous with those of the medullary proliferation (see, e.g., Torrey, 1945), and while they retain their connection with the outer epithelial layers for a considerable time, even after their ingrowth has ceased, they eventually become separated from the surface by a tunica albuginea which, in amphibians, arises from the mesenchyme of the primary gonad cavity (Brambell, 1930, 1956; Willier, 1939; Witschi, 1951a). Whether or not the cortex is primarily arranged in cord-like structures or secondarily by the ingrowth of mesenchyme and connective tissue, the final result is the isolation of small nests of germ cells surrounded by follicular envelopes.

After their initial phase of mitotic proliferation, the cortical germ cells (oogonia) enter meiotic prophase and become, by definition, oocytes. (Corresponding stages in the male generally do not appear until after birth.) Not all oogonia enter meiotic prophase at the same time. Those situated relatively deeply develop earlier, so that in amphibians and birds, for example, oocytes in the more central areas of the ovary coexist with oogonia at the periphery which are still undergoing mitotic proliferation (D'Hollander, 1904; Witschi, 1914; Burns, 1925, among others). Oogonial divisions eventually cease, either permanently (as in birds and mammals) or temporarily (as in amphibians and most reptiles).

The nature of the mechanism underlying sex differentiation is still not fully understood. The indifference or bisexuality of gonads at early stages of development and the bipotential nature of the germ cells in lower vertebrates* show clearly that the differentiation of a testis or ovary depends on a complex sex-determining mechanism whose influence becomes manifest only gradually as development proceeds. Witschi's (1931) theory that differentiation is mediated by specific inductor substances, medullarin and cortexin, has received wide support, and shortly before his death, he returned to the theme in a paper (Witschi, 1970) in which he rechristened cortexin "corticin." Corticin, presumably elaborated by the gonadal cortex, is described as a weak inductor which has the capacity—but only in the absence of medullarin—to stimulate the cortical part of the undifferentiated gonad to become ovarian tissue. Medullarin is described as a powerful inductor of the gonadal medulla which it transforms into testicular tissue. Witschi contributed greatly to our knowledge of gonadal development and function, but it cannot be said that this swan-song of his, published at the end of an amazingly lengthy life in research, added to the far-reaching concept which he had advanced 40 years before.

The difficulties which beset enquiries into the mechanisms that underlie the sexual differentiation of the primordial gonads are also clearly revealed in other works on the subject, the results of which have been published since the appearance of the first edition of "The Ovary." Some have focused on the role of the sex chromosomes, which almost by definition play a significant part in the process. It is well known, for example, that anomalous gonadal development and infertility may be associated with a deficiency or excess of these chromosomes in comparison with the normal condition (XX or XY). The loss of one X chromosome in man results in Turner's syndrome (XO), and to the disappearance of all germ cells by the time of birth (Singh and Carr, 1966, 1967). XO mice, however, maintain a slightly reduced "stock" of oocytes and are fertile (Welshons and Russell, 1959; Cattenach, 1962; Morris, 1968). Moreover, owing to the presence of an autosomal gene for sex reversal, the XO condition can occur in male mice that undergo normal spermatogenesis and produce viable spermatozoa (Cattenach *et al.*, 1971).

* This has not as yet been confirmed for eutherian mammals.

The Y chromosome is clearly a very significant factor in testicular differentiation (see Jacobs, 1966), but many other factors (including autosomal genes for sex reversal and testicular feminization) can profoundly affect the fate of the presumptive gonad (see Crew, 1965; Short, 1972; Asayama and Furusawa, 1960; Boczkowski, 1971; Jost, 1970, 1973). When there are more than the normal number of sex chromosomes in the gonadal cells, gonadal differentiation seems to depend largely on the presence or absence of a Y chromosome. When one is present (XXY, XYY, XXXY, etc.) the gonad develops into a testis, even if this is imperfect and sterile. In the absence of the Y chromosome the primordial gonad differentiates into an ovary (XXX, XXXX, etc.) (see Jacobs, 1966).

The importance of the Y chromosome in gonadal sex differentiation is also revealed in experiments in which chimeric mice are produced by fusing two embryos at the 8-cell stage of cleavage. Chimeras of the genetic constitution XX/XY mainly develop testes, and germ cells within the seminiferous tubules behave like XY gonocytes. A proportion of the germ cells in the gonads of these chimeras progress through meiotic prophase before undergoing degeneration at the pachytene stage. In this respect, at least, they behave like XX germ cells in an ovary (Mystkowska and Tarkowski, 1970; McLaren *et al.*, 1972). Ohno and Gropp (1965) suggested that such chimerism could account for the freemartin condition in cattle, insofar as primordial germ cells from the male twin could pass across the placenta to the female cotwin, which then became masculinized (freemartin). Ohno (1969) subsequently withdrew this view (but see Herschler and Fechheimer, 1967; Jost *et al.*, 1972).

Experiments involving organ culture of the isolated embryonic ovary indicate that there is a critical moment in the process of differentiation before which the cells cannot survive *in vitro*; they either degenerate or show markedly retarded development (Wolff and Haffen, 1965; Borghese and Venini, 1956; Baker and Neal, 1973; Challoner, 1975a). In the mouse the critical time is day 13–14 postconception (p.c.), in the hamster day 15, and in man probably between the fifth and seventh weeks p.c. Gonads cultured after this critical point develop normally, and the germ cells progress through mitosis and meiosis until they reach the diplotene stage. Since gonadotropins and maternal serum when added to the cultures had no effect, these findings suggested that a substance ("sex differentiating factor" or "meiotic stimulating factor") is elaborated early in the differentiation of the gonad (Challoner, 1975a). A further suggestion is that the rete ovarii may be the source of this substance, and that the rete cells also contribute to the formation of the granulosa cells (Byskov, 1974, 1975; Byskov and Lintern Moore, 1973). It is a matter for speculation whether the meiotic stimulating factor bears any relation to Witschi's inductors (see above) or to the Müllerian duct inhibiting factor of the testis described by Josso (1971, 1972, 1975).

V. OOGENESIS

The central question about oogenesis discussed in the first edition of "The Ovary" in 1962, was whether or not oogenesis occurred in reproductively mature vertebrates. The primary oocytes formed by the final mitotic division of oogonia must of necessity undergo reduction division (meiosis*) to yield, first, a secondary (haploid) oocyte and a first polar body, and subsequently, by mitotic division of the secondary oocyte, the definitive gamete (the haploid ovum, or ovum for short) and a second polar body. This pattern occurs uniformly throughout the vertebrates. Oogenesis as such is therefore dependent in the first place upon the mitotic division of primordial germ cells and oogonia. The chromosomal changes characteristic of meiosis occur only in primary oocytes that have been formed from oogonia.

It should be noted that the term oogenesis has frequently been used incorrectly in the past, and occasionally still is, to apply to the growth of the oocyte and its follicular envelope, associated in some species with vitellogenesis—processes which are known to occur after part of the meiotic prophase has already taken place. Although there are some who would prefer to widen the term oogenesis to make it apply to the entire process of the formation and maturation of the female germ cells (e.g., Henderson and Henderson, 1960), it is used here as it was in the 1962 edition so as to avoid any possible ambiguity about what is meant. It becomes too vague a term if it is also used to comprehend such matters as the segregation of germinal elements in the blastula, their migration to the genital ridges, and stages in the cytological maturation of the primordial oogonia and oocytes. Terms such as "oocytogenesis," which have appeared in the literature since 1960 (Kennelly and Foote, 1966), are otiose since etymologically they add nothing to what is implied by the word oogenesis.

The reason why the process of oogenesis, *sensu stricto,* was selected as a major topic for review in the first edition of "The Ovary," was that over the years conventional wisdom had upturned "a basic doctrine," according to which, as Pearl and Schoppe put it in 1921, "during the life of the individual there neither is nor can be any increase in the number of primary oocytes beyond those originally laid down when the ovary is formed." This doctrine

* For a full discussion of the process of meiosis, see White (1973) and Baker (1972).

was originally associated with the name of Waldeyer. Ever since the early twenties, it had, however, been rejected (mainly as a result of publications by the late Edgar Allen and Herbert M. Evans, two highly distinguished American pioneers of modern reproductive physiology) in favor of the contrary belief that a cyclical proliferation of the germinal epithelium gives rise to a new generation of oocytes during each normal estrous or menstrual cycle. This view, which put oogenesis on a par with spermatogenesis as a process continuing throughout reproductive life, was based essentially on a subjective assessment of the microscopic appearance of the ovaries at different times of the cycle, and it remained unchallenged until the beginning of the 1950s, when the results of an extensive and critical review of the available evidence were published and shown to controvert the views which Allen and Evans had succeeded in disseminating (Zuckerman, 1951). A series of new enquiries was then embarked upon by the Birmingham School of Reproductive Physiology in order to subject the Waldeyer thesis, which had been in the shadows for some 30 years, to critical experimental test, and also to add to the purely histological evidence derived from studies of the avian ovary on the strength of which the hypothesis had originally been based. The 1962 edition of this chapter reviewed the new experimental evidence that had been gained, together with all other available data for cyclostomes, fishes, amphibians, reptiles, birds, and mammals, to determine the extent to which the vertebrates lend themselves to a single major generalization about the process of oogenesis. There were, for example, indications that a few mammalian species, such as the African and Asian lemuroids, were exceptions to the rule. The 1962 review ended with the following general statement:

> Although available information is anything but exhaustive, such observations as have been made on vertebrates and many invertebrates suggest that the process of oogenesis follows a uniform pattern, of which there are two main variants. In the one oogenesis appears to continue either uninterruptedly or cyclically throughout reproductive life (e.g. most teleosts, all amphibians, most reptiles and conceivably a few mammals). In the other meiotic prophase occurs very early in life, oogonia neither persisting nor dividing mitotically after the onset of sexual maturity (e.g. cyclostomes, elasmobranchs, a few teleosts, perhaps some reptiles, all birds, monotremes, and with a few possible exceptions, all eutherian mammals).
>
> It would be surprising indeed if the eutherian mammals and the marsupials, on the study of which much of the controversy about oogenesis has focused, were to deviate from such a uniform pattern. The observations listed in this chapter are illustrative of many others (see Zuckerman, 1951, 1956, 1960) which lead unequivocally to the conclusion that the female mammal begins its reproductive life furnished with a finite stock of oocytes. Conversely, the available facts are inconsistent with the alternative hypothesis that oogenesis in mammals continues uninterruptedly

> throughout reproductive life. It cannot be too strongly emphasized that (i) there is no possible way of ever proving the latter hypothesis experimentally, whereas it has been disproved, as already indicated, by a variety of experimental tests, and (ii) that views about oogenesis based purely on histological evidence are necessarily arbitrary. Thus, to take one example, the claim that either one or another constituent cell of the ovary can become directly transformed into an oocyte in effect completely dismisses the problem of how and when oogenesis occurs. A belief of this kind would demand attention only if it were supported by an unequivocal demonstration of a consecutive series of intermediate cellular stages. But no such demonstration has ever been offered. It is equally unscientific to base hypotheses about the formation of germ cells only on the histological observations of mitoses in the germinal epithelium and the topographical relationship between epithelial cells and oocytes in the dictyate phase (e.g., Allen, 1923; Bullough, 1942; Aron *et al.*, 1952, 1954a,b). In the last resort, the only acceptable histological evidence of neoformation of germ cells in the adult mammalian ovary would be the observation of mitotic divisions in oogonia*; and in the absence of such evidence for the overwhelming majority of eutherian mammals, the view that oogenesis persists after the onset of sexual maturity is clearly untenable.

These views have been substantiated by the vast majority of studies on the subject reported since the first edition of "The Ovary" was written in 1961.

Qualitative and quantitative studies of oogenesis have been carried out on a variety of mammals (e.g., mouse: Jones and Krohn, 1959; Borum, 1961; rat: Beaumont and Mandl, 1962; hamster: Challoner, 1974; guinea pig: Ioannou, 1964; rabbit: Teplitz and Ohno, 1963; Peters *et al.*, 1965; ferret: Deanesly, 1970; sheep: Mauléon, 1973; pig: Black and Erickson, 1968; cow: Erickson, 1966a,b; rhesus monkey: Baker, 1966; human: Baker, 1963). In these species the number of germ cells increases to a peak (usually before birth) as a result of the mitotic division of oogonia. The number then declines with increasing age due to cessation of mitotic divisions and to "waves" of degeneration affecting germ cells at all stages of their development (see Baker, 1972). In the hamster, rabbit, and ferret, oocytes at the leptotene stage of meiosis are first found on the day of birth, and meiosis is completed to the diplotene stage of prophase within about 9 days. With the exception of the prosimian primates, it has been confirmed that in most other mammals, meiosis starts early in fetal life, and that it is completed by the time of, or shortly after, birth. In the cat and pig some prediplotene stages can be detected in the hilus region of the ovary until shortly before puberty (e.g., Black and Erickson, 1968; Brambell, 1956).

In the prosimians (*Loris, Nycticebus, Daubentonia, Galago,* etc.) it would seem that "nests" of primoridal germ cells persist in the ovarian cortex

* [and of meiotic prophase changes in oocytes].

apparently throughout adult life (e.g., Butler, 1964; Duke, 1967; Ioannou, 1967; Petter-Rousseaux and Bourlière, 1965). Anand Kumar (1968) reports that in *Loris* the number of germ cells within these "nests" fluctuates with the phases of the reproductive cycle, being maximal at estrus and during pregnancy and minimal during anestrus and lactation. He suggests that this change in the population of germ cells (and the concomitant production of oocytes at the leptotene to diplotene stages of meiotic prophase) is under the control of gonadotropic hormones. It seems unlikely, however, that any of the oocytes which develop in the "nests" become part of the definitive (follicular) population of germ cells (David *et al.*, 1974).

The most significant technical advance which has affected the factual base from which views about oogenesis can be drawn derives from the use of tritiated thymidine as a "marker" of germ cell DNA. For example, Kennelly and Foote (1966) injected this radioisotope into rabbits on the day of birth, a time when only oogonia are found in the ovary, and then removed the ovaries for autoradiographic study at varying intervals. It was found that the ovarian somatic cells undergo repeated mitotic divisions during the ensuing 4–40 weeks. The germ cells on the other hand undergo at most one or two mitotic divisions before starting meiosis. If they had been derived from ovarian somatic cells, they would have rapidly lost their "label," but if they were derived from preexisting germ cells the "label" would have been retained. Kennelly and Foote, in fact, found that 90% of the oocytes in animals aged 12 to 20 weeks p.c. retained the label, and therefore concluded: ". . . (1) during this interval the desoxyribonucleic acid (DNA) is metabolically stable, and (2) significant *de novo* oocyte formation did not occur." Some of the rabbits which had been injected with [^{3}H]TDR on the day of birth were subsequently superovulated using LH (Kennelly and Foote, 1965): some 80% of the eggs which were recovered from the oviducts of these animals were labeled, providing strong evidence for the continuity of the germ cell line from oogonium to secondary oocyte and/or ovum. No paper which has so far been published in any way significantly controverts this view.

VI. MEIOSIS

The results of many studies of meiotic chromosomes have confirmed and extended the observations reported in the first edition of this book. DNA synthesis in preparation for meiosis takes place during the preleptotene stage, which in the mouse occurs between days 12 and 13 p.c. (Lima-de-Faria and Borum, 1962; Peters and Crone, 1967), and in the rabbit on the day of birth (Peters and Crone, 1967). The timing of the phases of meiosis in different species is discussed by Baker (1972), and the morphology of the chromosomes during human meiosis by many authors (e.g., Ohno *et al.*, 1961, 1962; Baker, 1963; Manotaya and Potter, 1963; Baker and Franchi, 1967a,b,c; Luciani and Stahl, 1971). It has been suggested that meiosis is initiated by contact of oogonia with either the granulosa cells (Ohno and Smith, 1964) or with the rete ovarii (Byskov, 1974, 1975; but, see also Deanesly, 1975).

The so-called "resting" phase (dictyate or dictyotene), which occurs shortly after the oocyte has entered the diplotene stage, is characterized by the presence within the oocytes of lampbrush chromosomes which actively produce ribonucleoprotein (Baker and Franchi, 1967a,b,c; Baker *et al.*, 1969; Zybina, 1969). The possible functions for the RNP produced by the lampbrush chromosomes in mammalian oocytes are discussed by Baker (1971, 1972), Davidson (1968), Burkholder *et al.* (1971), and Moore *et al.* (1974).

The resumption of meiosis by the oocyte within the mature Graafian follicle is under the control of gonadotropic hormones (see Lindner *et al.*, 1974). It is not known whether the action of follicle-stimulating hormone (FSH) and luteinizing hormone (LH) is direct or whether prostaglandins or cyclic AMP act as intermediates. Since steroid synthesis can be blocked with cyanoketone or aminoglutethamide (Lindner *et al.*, 1974), steroid hormones seemingly do not have to be *produced* in response to the gonadotropins for the oocytes to resume meiosis. But the possibility remains that these or other hormones, contained in large quantities in follicular fluid, can exert a direct effect (McNatty *et al.*, 1975).

VII. FOLLICULAR GROWTH

Peters and Levy (1966) have investigated the timing of follicular growth in the ovaries of mice. By labeling the granulosa cells with tritiated thymidine and observing in autoradiographic preparations the timing of cell divisions, it has been shown that it takes 19 days for a primordial follicle in the mouse ovary to become a Graafian follicle (Pedersen, 1970).

The results of many studies indicate that the early phases of follicular growth (up to four layers of granulosa cells) are not dependent on the action of gonadotropic hormones (see Rowlands and Williams, 1943; Brambell, 1956; Ross, 1974; Nakano *et al.*, 1975). It is generally believed that the final phases of follicular growth (multilayered stage) require the action of FSH, while antrum formation and preovulatory maturation of the oocyte are dependent on LH and possibly FSH (see Volume I, Chapter 6; Baird *et al.*, 1975).

It has recently been shown that there is a critical period in ovarian development when the process of follicular growth needs to be "switched on" by gonadotropic hormones. Treatment of mice on the day of birth with antisera to gonadotropins results in the failure of the follicles to form, but the subsequent treatment of these animals with FSH plus LH restores the normal process of follicular growth (Eshkol *et al.*, 1971). Similarly, mouse ovaries maintained in organ culture from day 14 p.c. to day 14 post-partum p.p. failed to develop follicles; the ooctyes remained clustered together with few intervening somatic cells. However, if these ovaries are treated with FSH and LH, follicular growth is restored (Baker and Neal, 1973). The critical time when goandotropins are required appears to be days 1 to 3 p.p. in the mouse (Baker and Neal, 1973), and days 5 to 15 p.p. in the hamster (Challoner, 1975b). Thereafter gonadotropic hormones have seemingly no effect on the early stages of follicular growth.

VIII. CONCLUSION

While a few gaps have been filled in our understanding of the processes whereby the ovary develops first, from one cell and then from its descendants in the segmenting egg, it cannot be claimed that the 15 years since the appearance of the first edition of this book have witnessed the emergence of any new major generalization on the subject. The same judgment applies to our knowledge of the dynamics of oogenesis. In addition to the new work cited above, mention may be made here of such general but essentially descriptive studies of the developing ovary as those of Van Wagenen and Simpson (1965, 1973) and Mossman and Duke (1973).

In seeking to understand why opinion is so uncertain on some aspects of ovarian development, in particular the origin of the primordial germ cells, it is useful to remember that the embryologist has essentially only three techniques at his disposal, of which two are inevitably subjective in their application. The method nearest to hand is the histological description of a series of presumed successive developmental stages, enabling him to recognize and trace individual cellular elements. Because of the complex morphogenetic movements which occur during ontogeny, this approach is necessarily arbitrary unless the cells that are being traced possess some characteristic feature or "marker" which is revealed by conventional histological techniques and which distinguishes them from other cellular elements in the embryo. For example, it must be certain that characteristic features like "pseudopodia" which are occasionally claimed to be visible in primordial germ cells are not the result of crenation of the cell membrane induced by inadequate fixation (but see Section II).

This kind of histological approach was the earliest to be used in the study of the origin of germ cells, and since the microscopic and chemical procedures required for its application had reached a high standard nearly a century ago, the earliest reports based on this method are in no way inferior to those published more recently. Indeed, the main elements in the story of oogenesis were firmly established by the time de Winiwarter and Sainmont (1909) published their classic observations on oogenesis in the cat. The more recent use of the electron microscope has largely confirmed and extended the earlier histological studies, especially in relation to chromosomes and other organelles in female germ cells and to the relationship between germ cells and somatic components of the ovary (e.g., Baker, 1971; Zamboni, 1972; Zamboni *et al.*, 1972; Gondos, 1974).

The second line of approach open to the embryologist who is enquiring into the origin of germ cells is to use histochemical reagents which specifically mark some particular structural, chemical or enzymic component of a lineage of cells which would not normally become distinguishable by standard staining methods. This method is of value only if the histochemical reaction is positive for one type of cell, and it is therefore also somewhat arbitrary in its application. Some limitations also apply to the technique of radioactive marking which was introduced to this field of study since the publication of the first edition of "The Ovary." Thus, the isotope and labeled compounds must be appropriate to the experimental investigation; the autoradiographic processing should be adequate to detect the label above background, and the "marker" must be stable and not move from the site of incorporation during processing (see Jacob, 1971; Rogers, 1973).

The third method is the experimental modification of the normal pattern of development by such procedures as extirpation, transplantation or irradiation of tissues, parabiosis, and administration of chemical substances (e.g., hormones). The techniques of extirpation and transplantation of gonadal primordia, which have been very successfully used in amphibian larvae and avian embryos, are hardly applicable to early embryos of placental animals.

In assessing the published data it is necessary to differentiate between conclusions that are based on possible arbitrary interpretations of simple histological observations and those founded on solid experimental proof. The brief survey given in this chapter suggests that there is substantial support for the view that, in all vertebrates, the primordial germ cells become segregated and are distinct from other embryonic cells at very early stages of development, and that frequently they are found in sites which are quite unrelated to those of the presumptive gonadal regions. There is also experimental evidence to indicate that the primordial germ cells migrate from their extragonadal sites of origin to the genital ridges, and that after multiplying, they differentiate at varying times during the life cycle of the individual into spermatogonia or oogonia. The method by which the germ cells migrate has not been fully established for the majority of vertebrates. Apart from a few decisive experiments such as those of Willier (1937), Simon (1957), and Mintz (1959), much of the available evidence for a particular mode of migration has been inferred from studies of histological material alone and is therefore open, as suggested above, to arbitrary interpretation.

A critical survey of the published reports on the embryological origin of nongerminal ovarian cells reveals an urgent need for experimental studies. It is plain that the conflicting statements on this subject which still appear in the literature are unlikely to be resolved until the results of properly controlled experiments become available.

REFERENCES

Allen, B. M. (1904). The embryonic development of the ovary and testis of the mammals. *Am. J. Anat.* **3,** 89–146.

Allen, B. M. (1906). The origin of the sex-cells in *Chrysemys. Anat. Anz.* **29,** 217–236.

Allen, E. (1923). Ovogenesis during sexual maturity. *Am. J. Anat.* **31,** 439–481.

Anand Kumar, T. C. (1968). Oogenesis in lorises: *Loris tardigradus lydekkerianus* and *Nycticebus coucang. Proc. R. Soc. London, Ser. B* **169,** 167–176.

Aron, C., Marescaux, J., and Petrovic, A. (1952). Formation d'ovogonies aux dépens de l'épithelium superficiel de l'ovaire chez le cobaye prémature ou mûr. *C. R. Ass. Anat.* **39,** 421–423.

Aron, C., Marescaux, J., and Petrovic, A. (1954a). Ovogenèse postnatale chez le cobaye prématurе ou mûr. *C. R. Soc. Biol. (Paris)* **148,** 388–390.

Aron, C., Marescaux, J., and Petrovic, A. (1954b). Etat actuel du problème del l'ovogenèse postnatale chez les mammifères. *Arch. Anat., Strasburg.* **37,** 1–46.

Asayama, S., and Furusawa, M. (1960). Culture *in vitro* of prospective gonads and gonad primordia of mouse embryos. *Zool. Mag.* **69,** 280–284.

Baird, D. T., Baker, T. G., McNatty, K. P., and Neal, P. (1975). Relationship between the secretion of the corpus luteum and the length of the follicular phase in the ovarian cycle. *J. Reprod. Fertil.* **45,** 611–619.

Baker, T. G. (1963). A quantitative and cytological study of germ cells in human ovaries. *Proc. R. Soc. London, Ser. B* **158,** 417–433.

Baker, T. G. (1966). A quantitative and cytological study of oogenesis in the rhesus monkey. *J. Anat.* **100,** 761–776.

Baker, T. G. (1971). Electron microscopy of the primary and secondary oocyte. *Adv. Biosci.* **5,** 7–23.

Baker, T. G. (1972). Oogenesis and ovarian development. *In* "Reproductive Biology" (H. Balin and S. R. Glasser, eds.), pp. 398–437. Excerpta Med. Found., Amsterdam.

Baker, T. G., and Franchi, L. L. (1967a). The fine structure of oogonia and oocytes in human ovaries. *J. Cell Sci.* **2,** 213–224.

Baker, T. G., and Franchi, L. L. (1967b). The structure of the chromosomes in human primordial oocytes. *Chromosoma* **22,** 358–377.
Baker, T. G., and Franchi, L. L. (1967c). The fine structure of chromosomes in bovine primordial oocytes. *J. Reprod. Fertil.* **14,** 511–513.
Baker, T. G., and Neal, P. (1973). Initiation and control of meiosis and follicular growth in ovaries of the mouse. *Ann. Biol. Anim., Biochim., Biophys.* **13,** 137–144.
Baker, T. G., Beaumont, H. M., and Franchi, L. L. (1969). The uptake of tritiated uridine and phenylalanine by the ovaries of rats and monkeys. *J. Cell Sci.* **4,** 655–675.
Balfour, F. M. (1878). On the structure and development of the vertebrate ovary. *Q. J. Microsc. Sci. [N.S.]* **18,** 383–438.
Bascom, K. F. (1923). The interstitial cells of the gonads of cattle with especial reference to their embryonic development and significance. *Am. J. Anat.* **31,** 223–259.
Beatty, R. A. (1960). Chromosomal determination of sex in mammals. *Mem. Soc. Endocrinol.* **7,** 45–48.
Beaumont, H. M., and Mandl, A. M. (1962). A quantitative and cytological study of oogonia and oocytes in the foetal and neonatal rat. *Proc. R. Soc. London, Ser. B* **155,** 557–579.
Black, J. L., and Erickson, B. H. (1968). Oogenesis and ovarian development in the prenatal pig. *Anat. Rec.* **161,** 45–55.
Blackler, A. W. (1958). Contributions to the study of germ cells in the Anura. *J. Embryol. Exp. Morphol.* **6,** 491–503.
Blackler, A. W. (1966). Embryonic sex cells of Amphibia. *Adv. Reprod. Physiol.* **1,** 9–28.
Blandau, R. J. (1969). Observations on living oogonia and oocytes from human embryonic and fetal ovaries. *Am. J. Obstet. Gynecol.* **104,** 310–319.
Blandau, R. J., White, B. J., and Rumery, R. E. (1963). Observations on the movements of the living primordial germ cells in the mouse. *Fertil. Steril.* **14,** 482–489.
Boczkowski, K. (1971). Sex determination and gonadal differentiation in man. A unifying concept of normal and abnormal sex development. *Clin. Genet.* **2,** 379–386.
Borghese, E., and Venini, M. E. (1956). Culture *in vitro* di gonadi embrionali di *Mus musculus*. *Symp. Genet.* **5,** 69–83.
Borum, K. (1961). Oogenesis in the mouse: A study of meiotic prophase. *Exp. Cell Res.* **24,** 495–507.
Bounoure, L. (1931a). Sur la nature golgienne d'un élément cytoplasmique caractéristique du germen dans les premiers stades du développement de la grenouille. *C. R. Acad. Sci.* (*Paris*) **193,** 297–300.
Bounoure, L. (1931b). Sur l'existence d'un déterminant germinal dans l'oeuf indivis de la grenouille rousse. *C. R. Acad Sci.* (*Paris*) **193,** 402–404.
Bounoure, L. (1935a). Sur la possibilité de réaliser une castration dans l'oeuf de la grenouille rousse; résultats anatomiques (avec projections). *C. R. Soc. Biol.* (*Paris*) **120,** 1316–1319.
Bounoure, L. (1935b). Une preuve expérimentale du rôle du déterminant germinal chez la grenouille rousse. *C. R. Acad. Sci.* (*Paris*) **201,** 1223–1225.
Bounoure, L. (1939). "L'origine des cellules reproductrices et le problème de la lignée germinale." Gauthiers-Villars, Paris.

Bounoure, L., Aubry, R. and Huck, M.-L. (1954). Nouvelles recherches expérimentales sur les origines de la lignée reproductrice chez la grenouille rousse. *J. Embryol. Exp. Morphol.* **2,** 245–263.

Brambell, F. W. R. (1927). The development and morphology of the gonads of the mouse. Part I. The morphogenesis of the indifferent gonad and of the ovary. *Proc. R. Soc. London, Ser. B* **101,** 391–409.

Brambell, F. W. R. (1930). "The Development of Sex in Vertebrates." Sidgwick & Jackson, London.

Brambell, F. W. R. (1956). Ovarian changes. *In* "Marshall's Physiology of Reproduction" (A. S. Parkes, ed.), Vol. 1, Part 1, pp. 397–544. Longmans, Green, London and New York.

Bullough, W. S. (1942). Oogenesis and its relation to the oestrous cycle in the adult mouse. *J. Endocrinol.* **3,** 141–149.

Burkholder, G. D., Comings, D. E., and Okada, T. A. (1971). A storage form of ribosomes in mouse oocytes. *Exp. Cell Res.* **69,** 361–371.

Burns, R. K. (1925). The sex of parabiotic twins in Amphibia. *J. Exp. Zool.* **42,** 31–89.

Burns, R. K. (1928). The transplantation of larval gonads in urodele amphibians. *Anat. Rec.* **39,** 177–191.

Burns, R. K. (1955). Urinogenital system. *In* "Analysis of Development" (B. H. Willier, P. A. Weiss, and V. Hamburger, eds.), pp. 462–491. Saunders, Philadelphia, Pennsylvania.

Butler, H. (1964). The reproductive biology of a Strepsirhine (*Galago senegalensis senegalensis*). *Int. Rev. Gen. Exp. Zool.* 241–296.

Byskov, A. G. (1974). Does the rete ovarii act as a trigger for the onset of meiosis? *Nature (London)* **252,** 396–397.

Byskov, A. G. (1975). The role of the rete ovarii in meiosis and follicle formation in different mammalian species. *J. Reprod. Fertil.* **45,** 201–209.

Byskov, A. G., and Lintern Moore, S. (1973). Follicle formation in the immature mouse ovary: the role of the rete ovarii. *J. Anat.* **116,** 207–217.

Cattenach, B. M. (1962). XO mice. *Genet. Res.* **3,** 487–490.

Cattenach, B. M., Pollard, C. E., and Hawkes, S. G. (1971). Sex-reversed mice: XX and XO males. *Cytogenetics* **10,** 318–337.

Challoner, S. (1974). Studies of oogenesis and follicular development in the golden hamster. 1. A quantitative study of meiotic prophase *in vivo*. *J. Anat.* **117,** 373–383.

Challoner, S. (1975a). Studies of oogenesis and follicular development in the golden hamster. 2. Initiation and control of meiosis *in vitro*. *J. Anat.* **119,** 149–156.

Challoner, S. (1975b). Studies of oogenesis and follicular development in the golden hamster. 3. The initiation of follicular growth *in vitro*. *J. Anat.* **119,** 157–162.

Chieffi, G. (1949). Ricerche sul differenziamento dei sessi negli embrioni di *Torpedo ocellata*. *Pubbl. Stn Zool. Napoli* **22,** 57–78.

Chieffi, G. (1959). Sex differentiation and experimental sex reversal in elasmobranchs. *Arch. Anat. Microsc. Morphol. Exp.* **48,** 21–36.

Chiquoine, A. D. (1954). The identification, origin and migration of the primordial germ cells in the mouse embryo. *Anat. Rec.* **118,** 135–146.

Christensen, K. (1930). Sex differentiation and development of oviducts in *Rana pipiens*. *Am. J. Anat.* **45,** 159–187.

Crew, F. A. E. (1965). "Sex Determination." Methuen, London.

Czolowska, R. (1969). Observations on the origin of 'germinal cytoplasm' in *Xenopus laevis*. *J. Embryol. Exp. Morphol.* **22,** 229–251.

Dantschakoff, V. (1931). Keimzelle und Gonade. I. A. Von der entodermalen Wanderzelle bis zur Urkeimzelle in der Gonade. *Z. Zellforsch. Mikrosk. Anat.* **13,** 448–510.

David, G. F. X., Anand Kumar, T. C., and Baker, T. G. (1974). Uptake of tritiated thymidine by primordial germinal cells in the ovaries of the adult slender loris. *J. Reprod. Fertil.* **41,** 447–451.

Davidson, E. H. (1968). "Gene Activity in Early Development." Academic Press, New York.

Deanesly, R. (1970). Oogenesis and the development of ovarian interstitial tissue in the ferret. *J. Anat.* **107,** 165–178.

Deanesly, R. (1975). Follicle formation in guinea pigs and rabbits: a comparative study with notes on the rete ovarii. *J. Reprod. Fertil.* **45,** 371–374.

de Winiwarter, H. (1901). Recherches sur l'ovogenèse et l'organogenèse de l'ovaire des mammifères (lapin et homme). *Arch. Biol.* **17,** 33–199.

de Winiwarter, H. (1910). Contribution à l'étude de l'ovaire humain. I. Appareil nerveux et phéochrome. II. Tissu musculaire. III. Cordons médullaires et corticaux. *Arch. Biol.* **25,** 683–756.

de Winiwarter, H., and Sainmont, G. (1909). Nouvelles recherches sur l'ovogenèse et organogenèse de l'ovaire des mammifères (chat). *Arch. Biol.* **24,** 1–142 and 165–276.

D'Hollander, F. (1904). Recherches sur l'oogénèse et sur la structure et la signification du noyau vitellin de Balbiani chez les oiseaux. *Arch. Anat. Microsc. Morphol. Exp.* **7,** 117–180.

Dodd, J. M. (1960). Genetic and environmental aspects of sex determination in cold-blooded vertebrates. *Mem. Soc. Endocrinol.* **7,** 17–44.

Dodds, G. S. (1910). Segregation of the germ-cells of the teleost, *Lophius*. *J. Morphol.* **21,** 563–610.

Duke, K. L. (1967). Ovogenetic activity of the fetal-type in the ovary of the adult slow loris, *Nycticebus coucang*. *Folia Primatol.* **7,** 150–154.

Erickson, B. H. (1966a). Development and radio-response of the pre-natal bovine ovary. *J. Reprod. Fertil.* **11,** 97–105.

Erickson, B. H. (1966b). Development and senescence of the postnatal bovine ovary. *J. Anim. Sci.* **25,** 800–805.

Eshkol, A., Lunenfeld, B., and Peters, H. (1971). Ovarian development in infant mice. *In* "Gonadotrophins and Ovarian Development" (W. R. Butt, A. C. Crooke, and M. Ryle, eds.), pp. 249–258. Livingstone, Edinburgh.

Everett, N. B. (1943). Observational and experimental evidences relating to the origin and differentiation of the definitive germ cells in mice. *J. Exp. Zool.* **92,** 49–91.

Firket, J. (1913). Recherches sur les gonocytes primaires (Urgeschlechts-zellen) pendant la période d'indifférence sexuelle et le développement de l'ovaire chez le poulet. *Anat. Anz.* **44,** 166–175.

Firket, J. (1920). Recherches sur l'organogénèse des glandes sexuelles chez les oiseaux. *Arch. Biol.* **30,** 393–516.

Gillman, J. (1948). The development of the gonads in man, with a consideration of the role of fetal endocrines and the histogenesis of ovarian tumors. *Contrib. Embryol. Carnegie Inst.* **32,** 81–131.

Goette, A. (1875). "Die Entwickelungsgeschichte der Unke (Bombinator igneus) als Grundlage einer vergleichenden Morphologie der Wirbeltiere." Leopold Voss, Leipzig.

Gondos, B. (1974). Differentiation and growth of cells in the gonads. *In* "Differentiation and Growth of Cells in Vertebrate Tissues" (G. Goldspink, ed.), pp. 169–208. Chapman & Hall, London.

Hardisty, M. W. (1967). The numbers of vertebrate primordial germ cells. *Biol. Rev. Cambridge Philos. Soc.* **42,** 265–287.

Hargitt, G. T. (1930). The formation of the sex glands and germ cells of mammals. III. The history of the female germ cells in the albino rat to the time of sexual maturity. IV. Continuous origin and degeneration of germ cells in the female albino rat. *J. Morphol.* **49,** 277–331 and 333–353.

Hegner, R. W. (1914). The Germ Cell Cycle in Animals. Macmillan, New York.

Henderson, I. F., and Henderson, W. D. (1960). "A Dictionary of Scientific Terms. Oliver & Boyd, Edinburgh.

Herschler, H. S., and Fechheimer, N. S. (1967). The role of sex chromosome chimerism in altering sexual development in mammals. *Cytogenetics* **6,** 205.

Heys, F. (1931). The problem of the origin of germ cells. *Q. Rev. Biol.* **6,** 1–45.

Humphrey, R. R. (1925). The primordial germ cells of *Hemidactylium* and other Amphibia. *J. Morphol.* **41,** 1–43.

Humphrey, R. R. (1928a). Sexual differentiation in gonads developed from transplants of the intermediate mesoderm of *Amblystoma. Biol. Bull.* (*Wood's Hole, Mass.*) **55,** 317–338.

Humphrey, R. R. (1928b). The developmental potencies of the intermediate mesoderm of *Amblystoma* when transplanted into ventrolateral sites in other embryos: the primordial germ cells of such grafts and their role in the development of a gonad. *Anat. Rec.* **40,** 67–101.

Humphrey, R. R. (1929). Studies on sex reversal in *Amblystoma.* 1. Bisexuality and sex reversal in larval males uninfluenced by ovarian hormones. *Anat. Rec.* **42,** 119–155.

Ioannou, J. M. (1964). Oogenesis in the guinea-pig. *J. Embryol. Exp. Morphol.* **12,** 673–691.

Ioannou, J. M. (1967). Oogenesis in adult prosimians. *J. Embryol. Exp. Morphol.* **17,** 139–145.

Jacob, J. (1971). The practice and application of electron microscope autoradiography. *Int. Rev. Cytol.* **30,** 91–181.

Jacobs, P. A. (1966). Abnormalities of the sex chromosomes in man. *Adv. Reprod. Physiol.* **1,** 61–91.

Jonckheere, F. (1930). Contribution à l'histogénèse de l'ovaire des mammifères. L'ovaire de *Canis familiaris. Arch. Biol.* **40,** 357–436.

Jones, E. C., and Krohn, P. L. (1959). Influence of the anterior pituitary on the ageing process in the ovary. *Nature* (*London*) **185,** 1155.

Josso, N. (1971). Interspecific character of the Müllerian-inhibiting substance: action of the human fetal testis, ovary and adrenal on the fetal rat Müllerian duct in organ culture. *J. Clin. Endocrinol. Metab.* **32,** 404–409.

Josso, N. (1972). Evolution of the Müllerian inhibiting activity of the human testis. *Biol. Neonate* **20,** 368–379.

Josso, N. (1975). L'Hormone antimullerienne: une foeto-proteine? *Arch. Fr. Pediatr.* **32,** 109–111.

Jost, A. (1970). Hormonal factors in the sex differentiation of the mammalian fetus. *Philos. Trans. R. Soc. London, Ser. B* **259,** 119–130.
Jost, A. (1973). Becoming a male. *Adv. Biosci.* **10,** 3–13.
Jost, A., Vigier, B., and Prépin, J. (1972). Freemartins in cattle: the first steps of sexual organogenesis. *J. Reprod. Fertil.* **29,** 349–379.
Jurand, A. (1957). Rozwój i róznicowanie gonad u *Xenopus laevis* Daud. *Folia Biol. (Krakow)* **5,** 123–149 (from *Biol. Abstr.*).
Kennelly, J. J., and Foote, R. H. (1965). Ovarian response and superovulation in pre- and post-pubertal rabbits treated with standard gonadotrophin preparations. *J. Reprod. Fertil.* **9,** 177–188.
Kennelly, J. J., and Foote, R. H. (1966). Oocytogenesis in rabbits. The role of neogenesis in the formation of the definitive ova and the stability of oocyte DNA measured with tritiated thymidine. *Am. J. Ant.* **118,** 573–590.
Lima-de-Faria, A., and Borum, K. (1962). The period of DNA synthesis prior to meiosis in the mouse. *J. Cell Biol.* **14,** 381–388.
Lindner, H. R., Tsafriri, A., Lieberman, M. E., Zor, U., Koch, Y., Bauminger, S., and Barnea, A. (1974). Gonadotrophin action on cultured Graafian follicles: induction of maturation division of the mammalian oocyte and differentiation of the luteal cell. *Recent Prog. Horm. Res.* **30,** 79–138.
Luciani, J. M., and Stahl, A. (1971). Etude des stades de début de la meiose chez l'ovocyte foetal humain. *Bull. Assoc. Anat.* **151,** 445–458.
McLaren, A., Chandley, A. C., and Kofman-Alfaro, S. (1972). A study of meiotic germ cells in the gonads of fetal mouse chimaeras. *J. Embryol. Exp. Morphol.* **27,** 515–524.
McNatty, K. P., Hunter, W. M., McNeilly, A. S., and Sawers, R. S. (1975). Changes in the concentration of pituitary and steroid hormones in the follicular fluid of human Graafian follicles throughout the menstrual cycle. *J. Endocrinol.* **64,** 555–571.
Manotaya, T., and Potter, E. L. (1963). Oocytes in phophase of meiosis from squash preparations of human fetal ovaries. *Fertil. Steril.* **14,** 378–392.
Maréchal, J. (1907). Sur l'ovogénèse des sélaciens et de quelques autres chordates. *Cellule* **24,** 7–239.
Matsumoto, T. (1932). On the early localization and history of the so-called primordial germ-cells in the chick embryo. (Preliminary report.) *Sci. Rep. Tohoku Imp. Univ., Ser. 4* **7,** 89–127.
Mauléon, P. (1973). Modification expérimentale de l'apparition et de l'evolution de la prophase meiotique dans l'ovaire d'embryon de brebis. *Ann. Biol. Anim., Biochim., Biophys.* **9,** 89–102.
Mintz, B. (1959). Continuity of the female germ cell line from embryo to adult. *Arch. Anat. Microsc. Morphol. Exp.* **48,** 155–172.
Moore, G. P. M., Lintern Moore, S., Peters, H., and Faber, M. (1974). RNA synthesis in the mouse oocyte. *J. Cell Biol.* **60,** 416–422.
Morris, T. (1968). The XO and OY chromosome constitutions in the mouse. *Genet. Res.* **12,** 125–137.
Mossman, H. W., and Duke, K. L. (1973). "Comparative Morphology of the Mammalian Ovary." Univ. of Wisconsin Press, Madison.
Muratori, G. (1937). Embryonal germ-cells of the chick in hanging-drop cultures. *Contrib. Embryol. Carnegie Inst.* **26,** 59–69.
Mystkowska, E. T., and Tarkowski, A. K. (1970). Behaviour of germ cells and

sexual differentiation in late embryonic and early postnatal mouse chimeras. *J. Embryol. Exp. Morphol.* **23,** 395–405.

Nakano, R., Muzuno, T., Katayama, K., and Tojo, S. (1975). Growth of ovarian follicles in the absence of gonadotrophins. *J. Reprod. Fertil.* **45,** 545–546.

Nieuwkoop, P. D. (1949). The present state of the problem of the "Keimbahn" in the vertebrates. *Experientia* **5,** 308–312.

Nussbaum, M. (1880). Zur Differenzierung des Geschlechts im Theirreich. *Arch. Mikrosk. Anat.* **18,** 1–120.

Ohno, S. (1969). The problem of the bovine freemartin. *J. Reprod. Fertil., Suppl.* **7,** 53–62.

Ohno, S., and Gropp, A. (1965). Embryological basis for germ cell chimerism in mammals. *Cytogenetics* **4,** 251–261.

Ohno, S., and Smith, J. B. (1964). Role of fetal follicular cells in meiosis of mammalian oocytes. *Cytogenetics* **3,** 324–333.

Ohno, S., Makino, S., Kaplan, W. D., and Kinosita, R. (1961). Female germ cells of man. *Exp. Cell Res.* **24,** 106–110.

Ohno, S., Klinger, H. P., and Atkin, N. B. (1962). Human oogenesis. *Cytogenetics* **1,** 42–51.

Okkelberg, P. (1921). The early history of the germ cells in the brook lamprey *Entosphenus wilderi* (Gage), up to and including the period of sexual differentiation. *J. Morphol.* **35,** 1–151.

Pearl, R., and Schoppe, W. F. (1921). Studies on the physiology of reproduction in the domestic fowl. XVIII. Further observations on the anatomical basis of fecundity. *J. Exp. Zool.* **34,** 101–118.

Pedersen, T. (1970). Follicle kinetics in the ovary of the cyclic mouse. *Acta Endocrinol.* (*Copenhagen*) **64,** 304–323.

Peters, H. (1970). Migration of gonocytes into the mammalian gonad and their differentiation. *Philos. Trans. R. Soc. London, Ser. B* **259,** 91–101.

Peters, H., and Crone, M. (1967). DNA synthesis in oocytes of mammals. *Arch. Anat. Microsc. Morphol. Exp.* **56,** Suppl. 3–4, 160–170.

Peters, H., and Levy, E. (1966). Cell dynamics of the ovarian cycle. *J. Reprod. Fertil.* **11,** 227–236.

Peters, H., Levy, E. and Crone, M. (1965). Oogenesis in rabbits. *J. Exp. Zool.* **158,** 169.

Petter-Rousseaux, A., and Bourlière, F. (1965). Persistance des phénomènes d'ovogénèse chez l'adulte de *Daubentonia madagascariensis* (Prosimii, Lemuriformes). *Folia Primatol.* **3,** 241–244.

Rogers, A. W. (1973). "Techniques of Autoradiography," 2nd ed. Elsevier, Amsterdam.

Ross, G. T. (1974). Gonadotropins and preantral follicular maturation in women. *Fertil. Steril.* **25,** 522–543.

Rowlands, I. W., and Williams, P. C. (1943). Production of ovulation in hypophysectomized rats. *J. Endocrinol.* **3,** 310–315.

Short, R. V. (1972). Germ cell sex. *In* "The Genetics of the Spermatozoon" (R. A. Beatty and S. Gluecksohn-Waelsch, eds.)," pp. 325–345. Genet. Dep., Edinburgh University.

Simon, D. (1957). La migration des cellules germinales de l'embryon de poulet vers les ébauches gonadiques; preuves expérimentales. *C. R. Soc. Biol.* (*Paris*) **151,** 1576–1580.

Singh, R. P., and Carr, D. H. (1966). The anatomy and histology of XO human embryos and fetuses. *Anat. Rec.* **155**, 369–383.

Singh, R. P., and Carr, D. H. (1967). Anatomic findings in human abortions of known chromosomal constitution. *Obstet. Gynecol.* **29**, 806–818.

Smith, L. D., and Williams, M. A. (1975). Germinal plasm and determination of the primordial germ cells. *Symp. Soc. Dev. Biol.* **33**, 3–24.

Stanley, A. J., and Witschi, E. (1940). Germ cell migration in relation to asymmetry in the sex glands of hawks. *Anat. Rec.* **76**, 329–342.

Stolk, A. (1958). Extra-regional oocytes in teleosts. *Nature* (*London*) **182**, 1241.

Swift, C. H. (1914). Origin and early history of the primordial germ-cells in the chick. *Am. J. Anat.* **15**, 483–516.

Swift, C. H. (1915). Origin of the definitive sex-cells in the female chick and their relation to the primordial germ-cells. *Am. J. Anat.* **18**, 441–470.

Swift, C. H. (1916). Origin of the sex-cords and definitive spermatogonia in the male chick. *Am. J. Anat.* **20**, 375–410.

Tarkowski, A. K. (1970). Germ cells in natural and experimental chimeras in mammals. *Philos. Trans. R. Soc. London, Ser. B* **259**, 107–112.

Teplitz, R., and Ohno, S. (1963). Postnatal induction of ovogenesis in the rabbit (*Oryctolagus cuniculus*). *Exp. Cell Res.* **31**, 183–189.

Torrey, T. W. (1945). The development of the urinogenital system of the albino rat. II. The gonads. *Am. J. Anat.* **76**, 375–397.

Van Wagenen, G., and Simpson, M. E. (1965). "Embryology of the Ovary and Testis *Homo sapiens* and *Macaca mulatta*." Yale Univ. Press, New Haven, Connecticut.

Van Wagenen, G., and Simpson, M. E. (1973). "Postnatal Development of the Ovary in *Homo sapiens* and *Macaca mulatta* and Induction of Ovulation in the Macaque." Yale Univ. Press, New Haven, Connecticut.

von Berenberg-Gossler, (1913). Die Urgeschlechtszellen des Hühner-embryos am 3. und 4. Bebrütungstage, mit besonderer Berücksichtigung der Kern- und Plasmastrukturen. *Arch. Mikrosk. Anat.* **81**, Abt. II, 24–72.

Waldeyer, W. (1870). "Eierstock und Ei." Engelmann, Leipzig.

Waldeyer, W. (1906). Die Geschlechtszellen. *In* "Handbuch der vergleichenden und experimentellen Entwicklungslehre der Wirbeltiere" (O. Hertwig, ed.), pp. 86–476. Fischer, Jena.

Weismann, A. (1885). "Die Continuität des Keimplasmas als Grundlage einer Theorie der Vererbung." Jena (cited from Hegner, 1914).

Welshons, W. J., and Russell, L. B. (1959). The Y-chromosome as the bearer of male determining factors in the mouse. *Proc. Natl. Acad. Sci. U.S.A.* **45**, 560–566.

White, M. J. D. (1973). "The Chromosomes," 6th ed. Chapman & Hall, London.

Willier, B. H. (1937). Experimentally produced sterile gonads and the problem of the origin of germ cells in the chick embryo. *Anat. Rec.* **70**, 89–112.

Willier, B. H. (1939). The embryonic development of sex. *In* "Sex and Internal Secretions" (E. Allen, ed.), 2nd ed., pp. 64–144. Williams & Wilkins, Baltimore, Maryland.

Witschi, E. (1914). Experimentelle Untersuchung über die Entwicklungsgeschichte der Keimdrüsen von *Rana temporaria*. *Arch. Mikrosk. Anat.* **85**, 9–113.

Witschi, E. (1929). Studies on sex differentiation and sex determination in amphibians. I. Development and sexual differentiation of the gonads of *Rana sylvatica*. *J. Exp. Zool.* **52**, 235–265.

Witschi, E. (1931). Studies on sex differentiation and sex determination in amphibians. V. Range of the cortex-medulla antagonism in parabiotic twins of Ranidae and Hylidae. *J. Exp. Zool.* **58,** 113–145.

Witschi, E. (1935). Origin of asymmetry in the reproductive system of birds. *Am. J. Anat.* **56,** 119–141.

Witschi, E. (1939). Modification of the development of sex in lower vertebrates and mammals. *In* "Sex and Internal Secretions" (E. Allen, ed.), 2nd ed., pp. 145–226. Williams & Wilkins, Baltimore, Maryland.

Witschi, E. (1948). Migration of the germ cells of human embryos from the yolk sac to the primitive gonadal folds. *Contrib. Embryol. Carnegie Inst.* **32,** 67–80.

Witschi, E. (1951a). Embryogenesis of the adrenal and the reproductive glands. *Recent Prog. Horm. Res.* **6,** 1–27.

Witschi, E. (1951b). Génétique et physiologie de la différenciation du sexe. *In* "Colloque sur la différenciation sexuelle chez les vertébrés," pp. 33–64. CNRS, Paris.

Witschi, E. (1956). "Development of Vertebrates." Saunders, Philadelphia, Pennsylvania.

Witschi, E. (1970). Embryology of the testis. *In* "The Human Testis" (E. Rosenburg and C. A. Paulsen, eds.), pp. 3–10. Plenum, New York.

Wolff, E., and Haffen, K. (1965). Germ cells and gonads. *In* "Cells and Tissues in Culture" (E. N. Willmer, eds.), Vol. 2, pp. 696–743. Academic Press, New York.

Woods, F. A. (1902). Origin and migration of the germ-cells in *Acanthias*. *Am. J. Anat.* **1,** 307–320.

Zamboni, L. (1972). Comparative studies on the ultrastructure of mammalian oocytes. *In* "Oogenesis" (J. D. Biggers and A. W. Schuetz, eds.), pp. 5–46. Univ. Park Press, Baltimore, Maryland.

Zamboni, L., and Merchant, H. (1973). The fine morphology of mouse primordial germ cells in extragonadal locations. *Am. J. Anat.* **137,** 299–336.

Zamboni, L., Thompson, R. S., and Smith, D. M. (1972). Fine morphology of human oocyte maturation *in vitro*. *Biol. Reprod.* **7,** 425–457.

Zuckerman, S. (1951). The number of oocytes in the mature ovary. *Recent Prog. Horm. Res.* **6,** 63–109.

Zuckerman, S. (1956). The regenerative capacity of ovarian tissue. *Ciba Found. Colloq. Ageing* **2,** 31–54.

Zuckerman, S. (1960). Origin and development of oocytes in foetal and mature mammals. *Mem. Soc. Endocrinol.* **7,** 63–70.

Zybina, E. G. (1969). Behaviour of the chromosomal nucleolar apparatus during the growth period of the rabbit oocytes. *Tsitologiya* **11,** 25–31.

3

Sexual Differentiation of the Ovary

Katy Haffen

I. INTRODUCTION

This chapter deals with the process whereby the indifferent gonad becomes an ovary (the associated problems of sex reversal are discussed in Volume I, Chapter 8) and focuses more on birds and mammals than on other vertebrates, particularly with respect to the differentiation of the interstitial and germ cells. Section II includes a summary of our basic knowledge of sexual differentiation of the ovary, together with some new experimental data. The experiments described in Section III show that the developing germ cells in birds and mammals can be affected by changes in their environment. Section IV deals with the enzyme biochemistry of, and steroid synthesis in, embryonic gonads. The extragonadal origin of the germ cells and their mode of migration is considered in Chapter 2 of this volume and in the detailed publication of Wolff (1962).

II. MORPHOLOGY

The sexual differentiation of the gonads and of the genital ducts implies a process of organogenesis occurring in successive stages (see also Volume II, Chapter 2). Sex determination is established at the moment of fertilization, and the genetic constitution depends upon the distribution of the X and Y or the W and Z chromosomes in the egg. In many vertebrate species (mammals and some fish) the genetic constitution is homogametic in the female (XX) and heterogametic in the male (XY). In birds and in certain fishes and amphibians the situation is reversed in that the male formula is ZZ and the female ZW (Gallien, 1965). The genetic mechanism of sex determination was established by the consideration of sex ratios, sex linkage, hybridizations occurring in inversed sexual phenotypes, and more recently by direct study of the sex chromosomes. These subjects have been reviewed (Beatty, 1964, 1970; Gallien, 1965) and will not be dealt with in this chapter.

The second stage of sex determination is that of "indifference," or sexual bipotentiality. Apart from certain fishes, every vertebrate embryo goes through a stage where it is furnished with a double set of rudiments which carry the elements necessary for the production of a male and a female genital system. The duration of this phase varies according to the species and is succeeded by the stage of sexual differentiation. During this third stage one of the genital systems develops and the other atrophies.

In the present chapter, only differentiation of the gonads will be discussed; differentiation of the genital ducts is considered in Volume II, Chapter 2.

A. Fishes

Fish have more forms of sexuality than any other vertebrate (Atz, 1964; Chan, 1970). It is only in the selachians (Chieffi, 1959; Thiebold, 1964) that gonadal development seems to occur in the same manner as in amphibians and amniotes. Cortical and medullary areas of the gonad are differentiated, and colonization of the medulla or cortex by gonocytes results either in a male or female gonad.

Gonadal development in bony fishes and cyclostomes differs from that of other vertebrates. In place of a cortex and a medulla, which together form the gonad, there is only a single primordium that corresponds to the cortex and is derived more from the peritoneal epithelium than from the mesonephritic blastema (d'Ancona, 1950, 1952; Dodd, 1965).

A period of "indetermination" follows the undifferentiated state in the

Muraenidae, the Symbrenchidae, the Salmonidae, the Cyprinodontidae, and the Cyprinidae (see d'Ancona, 1950; Atz, 1964). Several authors, including d'Ancona, regard the indeterminate state of the gonad in these particular teleosts, e.g., the eel, as constituting a female phase. This interpretation is based on the fact that the first germinal elements to be distinguished are oocytes; the differentiated characters of the male elements appear later. D'Ancona (1950) believes that the gonads should be considered as undifferentiated, even when oocytes are present, because male gonocytes may not have differentiated.

Differentiation follows another pattern in two hermaphroditic families, the Spariidae and the Serranidae; the gonad always contains distinct testicular and ovarian zones (d'Ancona, 1950). For example, in the sparid *Sparus auratus*, the germinal elements of the two sexes coexist in adjacent but separate zones, the male zone being in the ventral part of the gonad. In this fish, the dorade, hermaphroditism is always protandric. Testicular maturation is complete at the end of the second year of life, and ovarian maturity occurs at the end of the third year. During the period of testicular activity, the ovarian zone remains quiescent, but the oocytes develop as soon as the male zone becomes nonfunctional. Atrophy of the testicular zone proceeds progressively as the ovarian zone matures. Other members of the Sparidae are proterogynic hermaphrodites (Atz, 1964). In some serranids, the male and female zones mature almost simultaneously but gonochoric species also exist (d'Ancona, 1950; Atz, 1964).

Owing to the lack of experimental and embryological information, the factors responsible for sexual differentiation in fishes are not known (Chan, 1970). Working with *Torpedo ocellata* and *Scyliorhinus caniculus*, Chieffi (1959, 1967b) suggests that in selachians there may be an antagonistic corticomedullary mechanism analogous to that postulated by Witschi (1956) for amphibians. D'Ancona (1949) has suggested that, in the hermaphrodite teleosts, a sex differentiator, "gynogenine" or "androgenine," is produced in the two areas of the heterosexual gonad. In the gonochoric species *Oryzias latipes*, in which the sex can be reversed experimentally, Yamamoto (1962) suggests the existence of a "gynotermone" and an "androtermone," and considers that androgenic steroids may act as "androinductors" and estrogens as "gynoinductors."

B. Amphibians

The following description of the differentiation of amphibian gonads is based on a wealth of studies that are now classic (Ponse, 1949; Witschi, 1956).

The gonadal primordium develops on the ventral surface of the mesonephros and is formed from peritoneal epithelium, the peritoneal cavity containing the gonocytes which have arrived by the dorsal mesenteric route. In his studies of sexual differentiation in *Rana temporaria* (1914) and *R. sylvatica* (1929a), Witschi has shown that, at the moment when the genital ridge becomes suspended, buds of renal blastema penetrate the peduncle at intervals and constitute the medulla. A mesonephritic origin of the medulla has also been found in urodeles (Houillon, 1956). If development of the Wolffian duct is prevented, mesonephritic agenesia occurs and a gonad fails to develop on the side deprived of mesonephros. When agenesia of the mesonephros is incomplete, mesonephritic nodules differentiate and a gonad is formed only at the level of these vestigial nodules. This vestigial material forms a small medulla which appears to maintain the cortex. Houillon's (1956) experiments show that the medulla controls the development of the undifferentiated gonad and acts in synergism with the cortex.

Thus, the undifferentiated gonad formed in the genital ridge comprises three elements of different origins: the germ cells, the cortex which is developed from peritoneal epithelium, and the medulla which is separated from the cortex by a loose mesenchyme. In the frog, sexual differentiation into testis or ovary occurs during the second period of larval life.

1. Testicular Development

A genetically male gonad is converted into a testis by development of the medulla. The medullary cells undergo active proliferation and increase in size. They coalesce in fours or sixes to form a testicular cord in which somatic and germ cells derived from the epithelium become incorporated. The cortex is reduced to a fine peritoneal membrane. The layer of mesenchymatous cells of the undifferentiated gonad takes part in the formation of the tunica albuginea, which contains most of the testicular vasculature. At the time of metamorphosis, the male gonad comprises seminiferous tubules which surround the central blastema of the rete. This blastema does not develop into a rete testis in the frog until the end of the first or second year, when spermatogenesis begins. Little multiplication of spermatogonia occurs during early development of the gonad.

2. Ovarian Development

In larvae that are genetically female, the cortex plays the dominant role in the differentiation of the ovary. It becomes thickened by rapid multiplication of the germ cells. In young larvae of stage 26 (Witschi, 1956), some oogonia are transformed into oocytes, while others continue to divide. The somatic cells also multiply, and after metamorphosis they constitute the follicular cells around the growing occyte. The cords of the medulla become involuted and hollow, the lumina become confluent, and, finally, a single ovarian cavity is formed. Thus, the juvenile ovary is a sac whose walls correspond to the cortex. The ovarian cavity is transitory; it is obliterated by growth of the cortex necessitated by multiplication of oogonia and by developing oocytes. The connections with the mesonephros disappear after medullary involution and a vestigial rete ovarii persists in the region of the hilum. This pattern of sexual development occurs in races of frog that are described as "differentiated" (Witschi, 1930, 1942), since there is a normal (50:50) sex ratio at metamorphosis.

3. Special Cases

There are races of frogs in which all individuals possess ovaries at metamorphosis. The male sex appears up to as much as a year later, and the normal sex ratio is established only in the adult population. Such races of frogs are called "undifferentiated." The development of their gonads has been studied in detail in *Rana temporaria* (Witschi, 1914, 1921) and in *R. esculenta* (Galgano, 1935, 1941). At metamorphosis, the ovaries are smaller than those of differentiated races. During gonadal development, the ovarian cavity is obliterated by a late proliferation of medullary elements. Some oocytes degenerate, but undifferentiated gonocytes at the periphery reach the medullary zone and develop into spermatogonia. A testis is thus formed, although there is a stage of development at which the gonad contains both an ovarian cortex and a testicular medulla. This phase of juvenile hermaphroditism does not occur in all members of a species but only in certain races which have a localized geographical distribution. From his study of the factors which influence the distribution of these races, Witschi (1942) has shown that the differentiated races occur more frequently in the mountains, the undifferentiated ones more in the plains. Furthermore, Witschi (1929b), Piquet (1930), Uchida (1937), and Hsü *et al.* (1971) have shown that if larvae of *R. temporaria, Bufo bufo, Hynobius retardatus,* and *R. catesbeiana* are reared up to metamorphosis at high or low temperatures, differentiation of the gonads is modified and the sex ratio is altered. A high temperature (25°C) favors differentiation of the medulla and ovaries are transformed into testes. Lowering the temperature to 10°C retards medullary differentiation and favors development of the cortex.

There are also races of frogs and urodeles which are called "semi-differentiated" because females and hermaphrodites are produced at metamorphosis (Witschi, 1933).

Male and female toads possess small rudimentary structures (Bidder's organ) which cap their gonads. These organs resemble aberrant ovaries and are packed with oogonia. Ponse (1924, 1943) and Izadi (1943) have shown that the Bidder's organs correspond to regions of precocious differentiation which occur in the cranial portion of the genital ridge. These organs arise during the larval stage by proliferation of the cranial and middle segments of the genital ridge, and a progonad and mesogonad, which consist only of cortex, develop. The metagonad, or definitive gonad, arises after metamorphosis from the caudal end of the genital ridge and is the only gonadal part to differentiate in the male or female direction. Harms (1923, 1926) and Ponse (1924, 1927, 1949) demonstrated that Bidder's organ is ovarian in nature and that it is prevented from developing because of the presence of testes. If these are removed, the Bidder's organ continues to grow. After 3 to 7 years, and depending on the strain of toad, Bidder's organ develops into a functional ovary which, although compact, is capable of producing normal eggs. The rudimentary oviducts also resume their growth and are able to store and shed ripe eggs.

C. Reptiles

There are several distinct evolutionary pathways within this class of vertebrates. As Dufaure (1966) has observed, sexual differentiation varies throughout the Chelonia, the Crocodilia, and the Squamata: differentiation in the latter order is particularly variable.

The descriptive and experimental studies of Forbes (1938a,b, 1939, 1940a,b) on *Alligator mississippiensis*, of Risley (1933a,b, 1941) on the chelonian *Sternotherus odoratus*, and of Vivien (1959) and Stefan (1963) on another chelonian *Emys leprosa* (*Clemmys caspica leprosa*), show that, in crocodilians and chelonians, sexual bipotentiality continues for several months after birth. This bipotential state is more fleeting in the Squamata and does not extend beyond the embryonic period. Sexual differentiation of reptiles has been studied in *Anguis fragilis* by Raynaud (1960) and Raynaud and Pieau (1966), in *Lacerta vivipara* by Dufaure (1966), and in *Vipera aspis* by Dufaure and Gil-Alvaro (1967). In these species the gonads consist of a distinct cortex and medulla. Differentiation to the male type occurs earlier than does ovarian formation. The bipotential state is thus transitory and does not extend beyond the embryonic period. Like other snakes, the viper does not pass through a bipotential stage, and differentiation in these reptiles is precocious compared with that in other vertebrates.

The administration of sex steroids and hypophysial hormones to the embryo are discussed in detail in Volume I, Chapter 8. Conflicting results have been obtained by Raynaud (1965, 1967) working with *Lacerta viridis*, and by Dufaure (1966) with *L. vivipara*. According to Raynaud, estrogenic hormones provoke feminization of male gonads, but Dufaure states that this result is obtained only with hypophysial hormones. Dufaure further observed that, although estradiol does not cause cortical development of the embryonic testis, it does stimulate the cortex that is provoked after administration of gonadotropins. The mechanism of action and the significance of the effect of the gonadotropins are unknown.

D. Birds

In most birds, the female genital apparatus is asymmetrical (Witschi, 1935). The right ovary is usually rudimentary and only the left ovary and the oviduct of that side is well developed and functional. Exceptions do, however, occur, and right and left ovaries are frequently found in the Falconiformes. Wolff and Pinot (1961) studied 300 fowl embryos between 16 and 19 days of incubation and found 30% of the right gonads had rudiments of tissue and 1.3% had both ovaries complete. Thus, although there are exceptions, gonadal asymmetry occurs at an early stage of embryonic growth.

At the undifferentiated stage (4 days of incubation in the chick embryo) the gonadal primordia is composed of two elements: the germinal epithelium and the underlying mesenchyme. The germinal epithelium is essential for further development of the gonad, giving rise to the primary sex cords of the medulla (Firket, 1914; Swift, 1915, 1916). Wolff and Haffen (1959) showed that the isolated germinal epithelium of a male or female gonad of a duck embryo ($7\frac{1}{2}$ to 9 days of incubation) is capable of reconstituting the essential part of a male or female gonad in culture. It has been suggested that the medulla is derived from an adrenal blastema (Vannini, 1949) or from a mesonephric blastema (Witschi, 1956). Calame (1961) induced agenesia of the mesonephros of the chick by preventing descent of the Wolffian duct. A gonad developed on the operated side and its growth and differentiation was similar to that of a normal gonad. Calame therefore concluded that the mesonephros does not play an essential role in gonadal development. Nevertheless, a comparative study of the formation of the gonad in normal and experimental conditions has shown that there are slightly fewer germ cells in the ovary on the operated side and less stroma at the genital ridge stage. The stromal deficiency is manifested in the smaller size of male gonads and the right ovary. According to Stahl and Carlon (1973), the mesenchyme present in the undifferentiated gonad that gives rise to the urogenital connexions (Firket, 1914) plays a prominent part in the formation of the primary sex cords.

1. The Male

The germinal epithelium proliferates centripetally in the chick embryo after 5 days of incubation. It forms cellular tracts, or sex cords, that invade the subjacent stroma, carrying germ cells with them. About the end of the sixth day, a layer of mesenchymatous cells separates the sex cords from the germinal epithelium, and a testis can be distinguished histologically from an ovary at 7 days of incubation. The germinal epithelium gradually disappears, but parts of it persist until the eleventh day. When the germinal epithelium has completely disappeared, the medullary cords enlarge and contain multiplying spermatogonia. These sex cords do not become hollow and transformed into testicular tubules until the end of incubation, and spermatogenesis does not start until sexual maturity is reached.

The mesenchymatous stroma which surrounds the testicular cords remains undifferentiated until the ninth day, and then develops quickly, partly forming the tunica albuginea. The interstitial cells, which produce the hormonal secretion of the testis, are situated within the intertubular stroma. Swift (1916) and Firket (1920) have suggested a mesenchymal origin for the interstitial cells, but other authors (Nonidez, 1922; Benoit, 1923, 1929, 1950) believe that these are derived from the germinal epithelium. According to Benoit, the interstitial cells are vegetative cells of the sex cords, and their migration into the stroma can be observed from day 12 of incubation until hatching. Scheib (1970a) has confirmed an epithelial and cord origin of the interstitial cells of the fowl. In an ultrastructural study, she has shown that differentiation of primary interstitial cells, which are incorporated into the sex cords, begins on day 8, and further differentiation and migration into the intertubular spaces increases at later embryonic stages.

2. The Female

a. Left Gonad. Differentiation starts the same way as in the male. Beginning on the seventh day of incubation, the germinal epithelium thickens and secondary sex cords, which constitute the ovarian cortex, grow into the medulla from the germinal epithelium, where they obliterate the primary medullary elements.

The ovarian cortex contains germ cells which multiply and enter into the stage of premeiosis. Callebaut and Dubois (1965) have studied the development of the germ cells by noting the variations with age of the gonad in the rate of incorporation of tritiated thymidine into oogonia *in vitro*. The rates of incorporation were very high between the ninth and twelfth days of incubation, and this period corresponded to the phase of exponential multiplication of the oogonia. The percentage of radioactive oogonia reached a minimum at the fourteenth day and then increased sharply between days 17 and 18. Callebaut (1967a,b) investigated the significance of this late peak using injections of tritiated thymidine *in vivo*. The increased rate of incorporation corresponded to synthesis of premeiotic DNA, and Callebaut concluded that the majority of oogonia in the fowl enter into premeiosis at about day 17 of incubation. In the quail, meiosis starts at the tenth day of incubation (Callebaut, 1968). The remaining cells of the cortex become the follicular cells that later invest the oocytes: this further development is described in Chapter 2 of this Volume.

Two fates befall the primary medullary cords: (1) those cords nearest the hilum develop irregular lumina which are lined with flattened epithelial cells and (2) the cords in the subcortical zone break up after hatching and release cells which become interstitial cells. This process has been described in the quail by Scheib (1970c), who states that interstitial cells are present at day 7 of incubation. In the fowl, interstitial cells have been identified under the light microscope at day 10 and at day 8 by their ultrastructural characteristics of smooth endoplasmic reticulum, numerous free ribosomes, and mitochondria with tubular cristae and irregular vacuoles corresponding to extracted liposomes (Narbaitz and Adler, 1966; Simone-Santoro, 1968).

b. Right Gonad. There is a difference between the right and left gonad, even at the genital ridge stage. The right indifferent gonad is shorter and thinner than the left gonad. In the male, but not the female, there is progressive compensation for this initial asymmetry by growth.

Several authors have asserted that there is an unequal distribution of gonocytes in the fowl embryo. Dantschakoff and Guelin-Schedrina (1933) recognized this inequality from the 24 to 26 somite stage.

Van Limborgh (1958, 1960) reported that, in the duck, a numerical asymmetry appeared at the 38 somite stage, the left gonad containing about 60% of the oocytes. Fargeix (1964) also noted an asymmetrical distribution of gonocytes in the gonads of twin duck embryos obtained by fissuration of the blastoderm of the nonincubated egg. Fargeix (1970) and Fargeix and Didier (1973) suggested in other experimental studies that the number of gonocytes fixed by a right or left gonad is determined by the primordia themselves. Simon (1960) considered that there is no inequality and that the gonocytes are distributed equally in the two primordia.

As in the left ovary, towards the fifth day of incubation, cords derived from the germinal epithelium invade the subjacent stroma taking with them the gonocytes. The oogonia can be seen for about 3 weeks after hatching, and then disappear. After proliferation, the germinal epithelium becomes reduced to a single layer of cuboidal cells and further proliferation appears impossible. Wolff and Pinot (1961) have shown that the cortex of the right gonad can be stimulated by injection of female hormones, provided that the injections are made before the fifth day of incubation. The primary sex cords of the medulla of the right gonad develop in a similar way to that described above for the left ovary: they either become transformed into large lacunae, or into interstitial cells.

Thus, the right ovary in birds is reduced to a medullary vestige which does not disappear completely. Wolff (1949) unilaterally castrated chick embryos and showed that in the absence of the left ovary the right does not develop, indicating that regression of the right ovary is not due to an inhibiting action of the left. If the left ovary is removed *after* hatching, the rudimentary right ovary may differentiate into testicular tissue, and any gonocytes present may undergo spermatogenesis (Caridroit and Pezard, 1925; Domm, 1927; Benoit, 1932, 1950). Since the male potential of the medulla is only apparent after hatching, it seems probable that this differentiation is controlled by gonadotropic hormones. Wolff and his colleagues, using grafts of ovarian medulla, have shown that the medulla of the right and left ovaries is the site of a feminizing secretion of the embryo (Mintz and Wolff, 1951, 1954; Wolff and Wolff, 1951). The cortex can produce medullary tissue until the twelfth to fifteenth day of incubation, after which its feminizing action is lost (Haffen, 1963).

All these experiments indicate that there is no medullocortical antagonism in the bird embryo.

3. Experimental Studies

Modifications of gonadal development are described in Volume I, Chapter 8, and only experiments relative to the autodifferentiation of the gonads *in vitro* and the differentiation of the germ cell line are discussed here.

a. Autodifferentiation of Gonads *in Vitro*. Wolff and his colleagues have been particularly active in this field of research since 1952. They have shown, by means of the organ culture technique of Wolff and Haffen (1952c), that undifferentiated gonads of duck embryos (6 to 7 days of incubation) will differentiate, in accordance with their genetic sex, into ovaries or into testes, and that the female right gonad undergoes atrophy as in the normal embryo (Wolff and Haffen, 1952a). In the quail, embryonic gonads removed at the fifth day of incubation differentiate into ovaries or testes, and the oogonia enter into premeiosis after 6 days of culture (Haffen, 1964). The differentiation *in vitro* of gonads of the fowl embryo, removed between days 6 and 8 of incubation (Wolff and Haffen, 1952b) or days 4 and 5 (Weniger, 1961), is normal for the male sex and aberrant for the female. The female left gonad does not differentiate into a typical ovary but assumes the structure of an ovotestis. The right gonad resembles a testis that is similar to those taken from a male embryo except in its shape and smaller size. This apparent intersexuality of the female gonad of the fowl does not correspond to a functional intersexuality. The medulla still secretes a feminizing hormone that can modify the differentiation of the testis with which it is associated (Weniger, 1961) or transform germinal epithelium isolated from a male duck gonad into an ovarian cortex (Haffen, 1960).

b. Differentiation of the Germ Cell Line. During early sexual organogenesis, as well as during later stages of sexual differentiation, the germ cells undergo histochemical, ultrastructural, and physiological changes (Dubois and Croisille, 1970). Up to the sixth day of incubation, primordial germ cells and primary gonocytes have the appearance of undifferentiated cells. They contain stocks of glycogen and cytoplasmic lipid complexes. Differences start to appear in the germ cells on the eighth day of incubation. Lipids persist in the spermatogonia but disappear in most oogonia (Dubois and Cuminge, 1968). The physiological changes which occur during migration of the germ cells have been studied in culture (Dubois, 1968, 1969). Primordial germ cells can penetrate various tissues such as skin, liver, mesonephros, or lung, but primary gonocytes of the undifferentiated gonad cannot (Dubois, 1969). However, primary gonocytes can still manifest a chemotactic ameboid activity in the presence of young germinal epithelium. Spermatogonia possess the same migratory ability as the primordial germ cells and the primary gonocytes. They retain their migratory properties and ability to colonize germinal epithelium at least until the twelfth day of incubation, but a large majority of the oogonia no longer respond to the presumptive sexual territories after the eighth day (Dubois, 1968). In similar experiments, Tachinante (1974a) has given further evidence of the more prolonged migratory properties of spermatogonia in comparison with oogonia in quail gonads associated with the germinal epithelium of a chick embryo. Furthermore, the attractive stimulus exerted by the germinal epithelium is not species specific: gonocytes of duck (Simon, 1960), turkey (Reynaud, 1969), and mouse (Rogulska *et al.*, 1971; Tachinante, 1974b) are able to colonize a chick embryonic germinal epithelium.

E. Mammals

As in all vertebrates, the genital ridge, which the germ cells colonize, develops in mammals at the ventral surface of the mesonephros as a thickening of the peritoneal epithelium. The undifferentiated gonad develops differently in different species, in accordance with one of two ways which have been reviewed by Raynaud (1969). In the first, the germinal epithelium proliferates over the whole surface and its cells penetrate the subjacent epithelium to form an epithelial nucleus which is continuous with the overlaying coelomic epithelium. This form of development occurs in the mouse (Brambell, 1927; Raynaud, 1942), rat (Torrey, 1945), and in man (Witschi, 1956; van Wagenen and Simpson, 1965). In the second, the germinal epithelium forms invaginations, the primary sex cords, which penetrate the subjacent epithelium. This occurs in carnivores such as the cat (von Winiwarter and Sainmont, 1909).

1. Testicular Development

In genetic males, the gonad differentiates suddenly into a testis, whereas in genetic females the undifferentiated phase is more prolonged. A testis can be distinguished from an ovary at about the eighth week in a human fetus, at 12 days in the mouse, 14½ days in the rat, 30 days in the guinea pig and 40 days in the calf. Testicular differentiation is accompanied by rapid growth of the organ (Mittwoch, 1970).

In those species in which a general proliferation of the germinal epithelium occurs, the epithelial nucleus becomes organized into distinct epithelial cords which develop into testicular cords that incorporate the germ cells which were disseminated in the epithelial nucleus. In species in which primary sex cords develop, these cords enlarge and become transformed into testicular cords which contain the germ cells that were present in the superficial epithelium. When the testicular cords differentiate, mesenchymatous tissue spreads beneath the germinal epithelium to form the primordium of the tunica albuginea. In this way, the testicular cords become separated from the superficial epithelium. The tunica albuginea is invaded by blood vessels which form an extensive network.

Jost (1972a,b) has expressed disagreement with the classic concept of testicular differentiation. His description in the rat suggests that one of the first processes in testicular development is the differentiation of Sertoli cells, which swell, make contact with each other and encompass the germ cells. The delineation of the future seminiferous cords, resulting in the seclusion of the male germ cells, would be an immediate achievement.

The centrally situated germ cells do not multiply throughout embryonic life but stop at a particular stage, for example, at 14 days, in the mouse. This has been demonstrated by Peters (1970) who injected tritiated thymidine into mouse embryos between days 12 and 18 of gestation. Soon after birth, the germ cells migrate from the center toward the periphery of the seminiferous tubules and mitotic activity is resumed (Franchi and Mandl, 1964; Novi and Saba, 1968). Meiosis starts at various stages according to the species: at 8 to 12 days in the mouse (Nebel *et al.*, 1961), 11 to 12 days in the rat (Clermont and Perey, 1957), and 75 days in the ram (Courot, 1962).

Another characteristic of testicular differentiation is the precocious development of interstitial tissue. This has been studied ultrastructurally in the testis of the fetal guinea pig (Black and Christensen, 1969), of man (Niemi *et al.*, 1967; Pelliniemi and Niemi, 1969), of the mouse (Scheib, 1970b, 1972; Russo and de Rosas, 1971), of the hamster (Gondos *et al.*, 1974), of the rat (Lording and de Kretser, 1972) and of the rabbit (Bjerregard *et al.*, 1974). In mammals, the interstitial cells differentiate from undifferentiated cells of the intertubular mesenchyme, but the cause of their induction is unknown. In the human fetal testis, differentiation starts at the eighth week, and the maximum number of completely differentiated Leydig cells is attained at the twelfth week. The structure of these cells is similar to that of adult Leydig cells, but they lack crystals of Reinke (Niemi *et al.*, 1967). In the testis of the fetal mouse, ultrastructural characteristics of the interstitial cells are established at 12½ days of gestation (Scheib, 1970b, 1972) and completely resemble those of the adult at the end of gestation. Signs of involution are detected in interstitial cells of the human fetal testis of more than 18 weeks, but not in those of the fetal mouse (Scheib, 1970b, 1972), rat (Narbaitz and Adler, 1967), and guinea pig (Black and Christensen, 1969). The differentiation of the Leydig cells in the fetal rabbit testis is nonsynchronous, but at the nineteenth day most of the cells can be classified as Leydig cells or fibroblasts (Bjerregaard *et al.*, 1974). The functional properties of the interstitial cells are considered below (see Section IV).

2. Ovarian Development

In genetic females, the primordial gonad remains undifferentiated for a longer period than in the male. In Raynaud's (1969) first category of differentiation, the germinal epithelium continues to proliferate in a homogeneous manner over the whole surface. Mesenchymatous cells from the hilum then divide the epithelial zone, so formed, into radial projections like the fingers of a glove, transforming the epithelial zone into ovarian cortex. Some mesenchymatous cells migrate beneath the germinal epithelium and form a thin tunica albuginea. In Raynaud's (1969) second category of differentiation, the primary sex cords cease developing but persist in the interior of the gonad as the medullary cords, which are homologous with the testicular cords. A second proliferation of the germinal epithelium, of the kind described in the bird ovary (p. 76), then occurs and gives rise to epithelial cords that are structurally continuous with the germinal epithelium, and that constitute the ovarian cortex. The secondary cords contain both somatic and germ cells and are separated from the medullary cords in the center of the ovary by a thin layer of mesenchymatous tissue, the tunica albuginea. Jost (1972a), referring to work on normal and freemartin fetal calves (Jost *et al.*, 1972, 1973), has again expressed disagreement with the classical concept of gonadal differentiation in mammals. He emphasized that "proliferation of sex cords from the germinal epithelium, has never been demonstrated experimentally by studies of mitotic rates or so on."

The germ cells in the ovarian cortex continue dividing and become definitive oogonia. These then enter into the prophase of meiosis to become oocytes. The transformation of oogonia into oocytes constitutes the critical feature of the sexual differentiation of the ovary. The problem has been analyzed by Peters (1970) in various mammalian species (see Volume I, Chapter 2). It may be accomplished entirely during fetal life, as, for example, in the rat (Beaumont and Mandl, 1962), the mouse (Brambell, 1927; Borum, 1961), the guinea pig (Ioannou, 1964), the cow (Erickson, 1966), the goat (Mauleon, 1962), the pig (Black and Erickson, 1968), the monkey (Baker, 1966), and man (Witschi, 1948; Baker, 1963; Baker and Franchi, 1967; Gondos *et al.*, 1971). In a few species, such as the rabbit (von Winiwarter, 1901; Teplitz and Ohno, 1963; Peters *et al.*, 1965), the golden hamster (Weakley, 1967; Greenwald and Peppler, 1968), and the cat (von Winiwarter and Sainmont, 1909), oogenesis starts after birth. The synthesis of meiotic DNA has been studied in the mouse and rabbit by means of injections of tritiated thymidine (Peters and Levy, 1966; Borum, 1966). In the mouse, all oocytes become labeled if the injections are made between the thirteenth and sixteenth day of gestation, but no labeling occurs if the injections are made at later stages. In the rabbit, labeled oocytes are visible only after injections of tritiated thymidine between the first and tenth day of postnatal life. Some germ cells also degenerate during mammalian oogenesis (Baker, 1963; Baker and Franchi, 1967; Gondos *et al.*, 1971; Gondos, 1972). Degeneration affects oogonia undergoing mitosis and oocytes in meiotic prophase.

Interstitial cells with a secretory function, such as are found in the fetal testis of mammals and in the embryonic gonads of both sexes in birds, have only recently been identified in ovarian tissue of human fetuses (12 to 20 weeks of gestation) by Gondos and Hobel (1973). The cells are located in the medullary region just beneath the cortical cords. They contain abundant smooth endoplasmic reticulum, large spherical mitochondria, well developed Golgi complexes and scattered dense bodies. The cellular appearance is similar to that of Leydig cells in human fetal testis. The authors mention that the relative number of interstitial cells in the ovary is much lower than the number of Leydig cells in the testis during the corresponding period. The findings of this morphological study are compatible with the steroidogenic activity that has been reported to occur in the fetal ovary of several mammalian species (see Section IV).

3. The Mole and the Desman

The ovary develops in a peculiar way in these members of the order Insectivora. In the female mole embryo at the 18- to 20-mm stage, interstitial cells develop around the primary sex cords (Godet, 1950). At birth, the medullary interstitium is an ellipsoid body distinct from the functional part of the ovary. Peyre (1962) reported that there was no development of medullary interstitial tissue in the fetal ovary of the desman, but that medullary cords developed after birth and were prominent in immature animals. The interstitial cells in the immature desman ovary extend from the hilum to the cortex. The germinal epithelium involutes slowly in the male desman embryo, and its presence during fetal life results in the gonad resembling an ovotestis.

4. Experimental Studies

a. Autodifferentiation of Grafts. Mammalian fetal gonads have developed when grafted into various sites in adults or into the coelomic cavity of chick embryos. Gonadal primordia of the rat have been grafted under the kidney capsule (Buyse, 1935; Turner and Asakawa, 1962), under the tunica albuginea of the testis (Turner, 1969), on the omentum (Holyoke, 1949), subcutaneously (Moore and Price, 1942), and in the anterior chamber of the eye (Torrey, 1950). Testes usually differentiate normally, but ovaries develop in an aberrant way and remain rudimentary or become ovotestes. In the rat, differentiation of the grafted testis starts about the twelfth day (Torrey, 1950), but differentiation into a typical ovary starts about the seventeenth day (Moore and Price, 1942). Turner (1969) showed that when fetal rat ovaries are removed between 12½ and 15½ days of gestation, they continue to develop normally if grafted beneath the kidney capsule, but 80% of the grafts form sterile seminiferous tubules and the rest differentiate into ovotestis if they are grafted beneath the tunica albuginea of the testis. In experiments in which one of the ovaries was grafted into the testis, and the other placed as a control under the kidney capsule, testicular tubules formed only in the first situation, never in the second. Nine fetal ovaries transplanted into cryptorchid testes also differentiated normally. Turner (1969) suggested that proximity to male hormones secreted by the testis may have caused masculinization of the fetal ovaries. Salzgeber (1960, 1963) has grafted the ovaries of mouse fetuses at 12 to 19 days of gestation into the coelomic cavity of the chick. They developed normally. Weniger (1967), also using the chick coelomic cavity as a graft site, obtained differentiation into ovary and testis of mouse gonadal primordia of 9 to 11 days of gestation.

b. Autodifferentiation *in Vitro*. Testes and ovaries from an already differentiated fetuses, of 16 to 20 days gestation or from neonatal rats and mice, continue to develop normally *in vitro* (Martinovitch, 1938, 1939; Borghese and Venini, 1956; van de Kerckhove, 1959). Depending on the stage of removal of the ovary, the oogonia enter into meiosis, oocytes grow, and primary and even secondary follicles form. Wolff (1952) explanted fetal mouse gonads of 10 to 13 days of gestation and found that normal differentiation occurred at 12 to 13 days. When gonads at 10 to 11½ days of gestation were cultured with the mesonephros, sexual differentiation took place but the resulting ovary or testis was sterile. The male gonads developed tubules which converged on the hilum, and several mesonephric tubules became associated with the testicular tubules and formed a rete testis. The sterile female gonad consisted of a mass of tissue continuous with the germinal epithelium that had become lobulated. The medulla consisted of cords associated with tubules of the mesonephros. As in a normal ovary, these represented the rete ovarii. Peyre (1961) cultured gonads from fetal desmans (20 to 22 mm) and found that neither the testis nor the ovary exhibited any aberrant differentiation.

In 1950, Gaillard began a series of studies in which he cultured fragments of human ovaries removed at different stages between 14 weeks of gestation and birth. Survival in culture depended mainly on the presence of the germinal epithelium. In the youngest explants, the whole of the central parenchyma, including the young oocytes, disappeared. In explants from older embryos, some cells of primordial follicles survived. In all cultures, a reticular stroma persisted after other structures had degenerated. In explants taken from 14- to 21-week-old fetuses, new cords were formed by the germinal epithelium after 10 to 12 days of culture. New germ cells appeared in these cords and showed all the intermediate stages between undifferentiated gonocytes and completely differentiated oocytes. In explants taken from fetuses aged 24 to 36 weeks, a second type of cord was formed by follicular cells that had persisted. When all preexisting oocytes had degenerated, oogonia differentiated in the new cords and were surrounded by follicular cells. These experiments suggest that all the tissues of the ovarian cortex can be regenerated from germinal epithelium.

III. ROLE OF THE SOMA IN THE DIFFERENTIATION OF GERM CELLS

As explained in Chapter 8 of this volume, the gonads of fish and amphibia can be readily changed, by various treatments with sex hormones, from being typical of one sex to that of the other. Sexual inversions of the gonads brought about in this way are only temporary in birds. In mammals, sex steroids have little or no effect on the gonads. It is clear that gametogenesis does not necessarily proceed according to the genetic constitution of the primary gonocytes and that the hormonal environment of the germ cells may influence their differentiation.

1. Development of Germ Cells in Chick Embryo Chimeras

Using the technique of Wolff and Simon (1955) and Simon (1960), Haffen (1968, 1969) has produced AB chimera chick embryos (9 to 15 pairs of somites) in which the fertile, agonadic, anterior part of a blastoderm (A) was associated with the sterile, gonadic, posterior part of a blastoderm (B). Inverse association produced the BA chimera embryo. The sex of the two embryos was unknown at the time of the association and could only be established after development of a gonad from each of the two chimera embryos. The gonocytes of the anterior region migrate to the gonadal region which is in the posterior part. Thus, in about 50% of embryos, gonocytes of one sex from the germinal crescent colonized the gonad of the other sex. The chimera embryos (AB and BA) were grown for 48 hours on the culture medium and reached a stage corresponding to that of a 3- to 3½-day embryo. The genital primordia colonized by the gonocytes were then removed and grafted into the coelom of a 3-day embryo for 10 days. The graft was "passed" into a second, third, and fourth host and left to develop for 10 days in each. At each change, a fragment of the graft was examined histologically; the age of the grafts was estimated at 13 days at the end of the first graft, and 23, 33, and 43 days after the second, third, and fourth grafts, respectively.

a. Female Gonocytes into Female Gonad. At 23 days, the cortex of the graft was packed with oocytes in the leptotene stage of premeiosis. At 33 days, the primordial follicles contained oocytes in the diplotene stage, and further growth of the follicles had occurred by 43 days (Figs. 1a–c).

b. Male Gonocytes into Female Gonad. The ovarian cortex degenerated (Figs. 1d–1f) in a way similar to that described for male intersexual gonads (Volume I, Chapter 8). The degeneration was apparent after the second graft (23 days) because of the large number of "oocytes" in which the nuclear chromatin was condensed. At 33 days, the degeneration of the "male oocytes" was even more pronounced: chromatin finally disappeared and the nucleus appeared empty. At 43 days the cortical zone was represented only by some plaques of degenerating cells. The follicle cells were grouped at the medullary surface. In four of the ten chimeras, occasional primary follicles occurred in the cortex. Haffen attributed their appearance to female gonocytes that may be present in the posterior region of the chimera. Fargeix and Theillieux (1967) and Dubois (1967) have shown that some gonocytes migrate more slowly than others. Another explanation is that

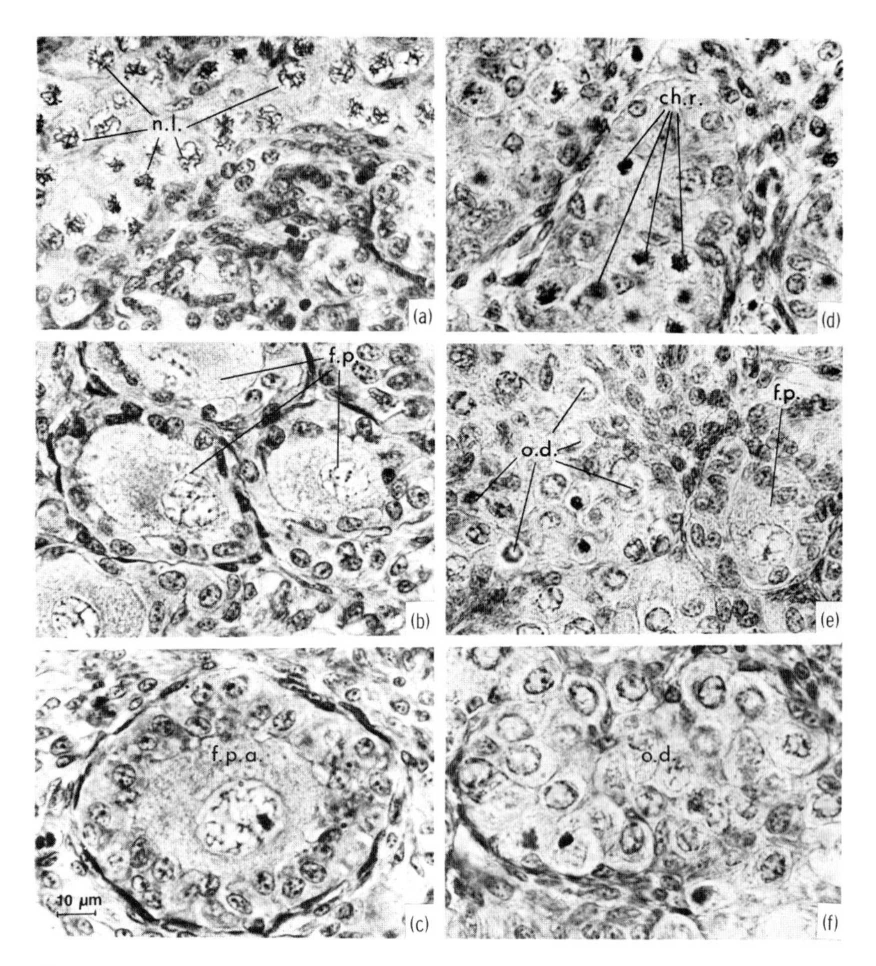

Fig. 1. Fragments of chick ovaries colonized by male and female germ cells after grafting into the coelom of three successive hosts. *Differentiation of female germ cells:* (a) Oogonia in meiotic prophase in a graft aged 23 days. The nuclear chromosomes (n.l.) are in leptotene. (b) Formation of primary follicles (f.p.) with diplotene nuclei in a 33-day-old graft. (c) An enlarged, primary follicle with diplotene nucleus at 43 days (f.p.a.). *Degeneration of male germ cells:* (d) Numerous nuclei with retracted chromatin (ch.r.) occur in the male gonocytes at the start of meiosis in a graft of 23 days. (e) In a graft at 33 days, the male gonocytes are degenerating (o.d.). Note the primary follicle (f.p.) developing with an oocyte derived from an oogonium which had colonized the ovary at the same time as the male germ cells. (f) The nuclear chromatin and the cytoplasm of the male gonocytes (o.d.) have almost disappeared at 43 days. (From Haffen, 1969.)

dispersion from the germinal crescent may have started in embryos with 12 to 15 pairs of somites. Thus, the presence in the same ovarian cortex of normal follicles and degenerating "male oocytes" is not necessarily an anomaly.

These experiments suggest that the somatic tissues of an ovary may initiate female sexual differentiation of genetically male germ cells, but that they do not maintain subsequent development according to the female pattern because the male "oogonia" degenerate after premeiotic prophase. The factors that lead the oogonia to start meiosis are unknown. Erickson (1974) presented evidence that 12-day-old germ cells do enter zygotene when cultured with pieces of 12-day-old cortex, suggesting that the differentiation of the female germ cells is regulated by the somatic cells of the cortex.

c. Female Gonocytes into Male Gonad. The gonocytes from the female developed in the same way as spermatogonia. Gametogenesis in the male is normally slower than in the female, but the experiment did not continue long enough for any differences in the rate of meiosis to be detected. Benoit (1932) has shown that if the left ovary of a hen is removed, any gonocytes present in the rudimentary right ovary will develop into spermatogonia.

2. Fate of Germ Cells in Mouse Chimeras

Tarkowski (1964) has perfected a technique of blastomere fusion for obtaining chromosomal chimeras in the mouse. These chimeras have been used (1969, 1970a,b) to determine whether a germ cell of the genetic constitution of one sex can undergo the pattern of gametogenesis typical of the opposite sex when located in a gonad of that sex. Chimeric individuals of male phenotype, from strains CBA-p and CBA-T6-T6 were crossed with CBA-p females to determine the genotype of the gametes produced by the chimeras (Mystkowska and Tarkowski, 1968, 1970). Two phenotypes appeared in the offspring when the male parents were of the same genotype, and a single phenotype appeared if the males were chimeras of different genotypes. In two chimeras that were phenotypic males and that produced only a single type of offspring, the constitution of bone marrow cells was 51% XX 41% XY and 19% XX 81% XY. Mystkowska and Tarkowski (1968, 1970) also analyzed the karyotypes of the spermatocytes present in the testes of these two chimeras and those in the testicular part of the gonad of an hermaphroditic chimera. The primary spermatocytes of all three animals were found to have the chromosomal constitution XY. The bone marrow cells of the hermaphrodite were 85% XX. These observations, confirmed by Mintz (1968), suggest that the XX germ cells do not develop into spermatogonia and do not develop as far as the first meiotic division.

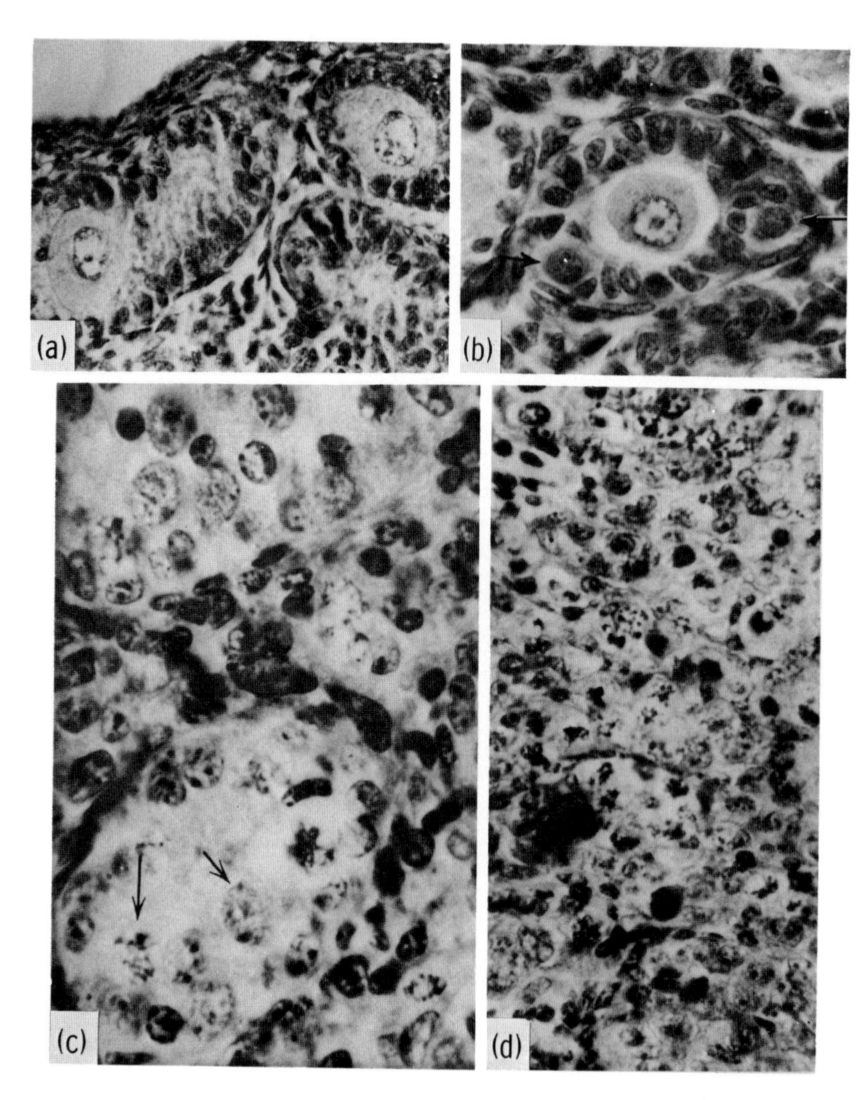

Fig. 2. Sections of testis from a 5-day-old mouse chimera derived from blastocyst fusion (a and b). High power magnification of typical sex cords containing developing oocytes. (From Mystkowska and Tarkowski, 1968.) Sections of fetal testes from 16- and 17-day-old mouse chimeras (c and d). Prespermatogonia and cells in meiotic prophase (arrows) (c). Testis without clearly defined sex cords containing germ cells in meiotic prophase (d). (From Mystkowska and Tarkowski, 1970.)

Growing oocytes (Figs. 2a and b) were observed in the testis of a male chimera at 5 days postpartum (Mystkowska and Tarkowski, 1968). Mystkowska and Tarkowski (1970) also examined the gonads of fetal (16 to 17 days of gestation) and postnatal (8 to 20 days) chimeric mice. Oocytes and spermatogonia were seen in the gonads of five of eleven male fetuses. Karyotypes of hepatic tissue in three of the embryos showed 7 XX and 13 XY, 18 XX and 31 XY, and 23 XX and 27 XY. No oocytes were found in postnatal male testes. In female chimeras with more XX than XY liver cells, oocyte development appeared normal. Mystkowska and Tarkowski inclined to the view that it is the genetic constitution of the somatic tissue of the testis rather than of the germ cell itself that determines entry of the germ cell into meiosis. McLaren *et al.* (1972) examined the gonads of 53 fetuses (16♀, 37♂) derived from embryo fusions as described by Bowman and McLaren (1970) and came to the same conclusion, i.e., that the meiotic cells seen in the testes of fetal chimeras 16½- to 18½-days-old were XX cells, but they considered that such cells only enter meiotic prophase under the influence of neighboring XX somatic tissue.

These studies suggest that oogonia in a male gonadal environment do not proceed beyond meiotic prophase, but degenerative stages, such as those described by Haffen (1968, 1969) in birds, have been described in a small number of cells in two out of ten fetal testes at 18½ days (McLaren *et al.*, 1972).

IV. BIOCHEMICAL ACTIVITY IN EMBRYONIC GONADS

The chemical nature of substances produced by the embryonic gonads during and after sexual differentiation has been the subject of much research, which has been reviewed by Chieffi (1965, 1972), Bloch (1967), Witschi (1967), Gallien (1967), Ozon (1969), and Haffen (1970). The main lines of enquiry have been (1) the ability of embryonic gonads to synthesize sex steroids from radioactive precursors; (2) the histoenzymological detection of enzymes involved in the biosynthetic pathways, particularly Δ^5-3β-hydroxysteroid dehydrogenase (Δ^5-3β-HSDH); and (3) the demonstration of endogenous secretions of embryonic gonads and their characterization.

A. Fishes, Amphibians, and Reptiles

1. Localization of Δ^5-3β-Hydroxysteroid Dehydrogenase

This enzyme is present in the gonads of adult fish (Della Corte *et al.*, 1961; Chieffi and Botte, 1963a; Chieffi *et al.*, 1963; Collenot and Ozon,

1965; Bara, 1965) but has not been found in embryonic gonads, either before or after their differentiation (Chieffi and Botte, 1963a).

In amphibians, Δ^5-3β-HSDH does not occur in the larval gonads of either sex in *Rana esculenta* (Chieffi and Botte, 1963b), but it does exist in the glandular tissue of the differentiated testis in *R. esculenta* and *Pleurodeles waltlii* (Chieffi and Botte, 1963b; Botte, 1964; Certain *et al.*, 1964) and in the ovary after metamorphosis in anurans and urodeles (Botte and Cottino, 1964; Gallien *et al.*, 1964; Joly, 1965). Collenot (1964, 1965) has demonstrated the presence of Δ^5-3β-HSDH in undifferentiated gonads of *P. waltlii* larvae up to developmental stage 53 (Gallien and Durocher, 1957). The Δ^5-3β-HSDH activity is localized in the medullary cords.

The localization of Δ^5-3β-HSDH activity is similar in reptiles and birds. In *Lacerta sicula* (Botte and Delrio, 1964), in *L. vivipara* (Dufaure and Mesure, 1967; Morat-Mesure, 1971), and in *L. viridis* (Raynaud and Pieau, 1971), the enzyme appears in the ovarian medulla, the medullary cords of the testis in the embryo, and in the interstitial cells of the testis in the adult (Chieffi and Botte, 1966; Dufaure and Mesure, 1967). This is also the case in *Vipera aspis* (Morat-Mesure, 1969, 1971) where the appearance of significant activity coincides with the histological sexual differentiation of the gonads.

In all these classes, Δ^5-3β-HSDH activity occurs earlier in the adrenal than in the gonad.

2. Hormonal Secretions

No information is available for reptiles or fish, but the embryonic gonads of amphibians have been investigated. Using fluorometric and chromatographic methods, Gallien and Chalumeau-Le Foulgoc (1960) analyzed the estrogen content of the ovaries of *Xenopus laevis*; at respectively 8 and 11 months after metamorphosis, estrogen levels of 0.005 μg and 0.1 μg/animal were detected.

3. Biosynthesis of Sex Steroids

Ozon (1969) studied the biosynthesis of progesterone, androstenedione, and testosterone from [6,7-^{3}H]pregnenolone in embryonic testes (stage 55) and juvenile testes (1 month after metamorphosis) in *Pleurodeles waltlii*. The ability of the gonads of various larval stages of *Xenopus laevis* to convert 17α-hydroxyprogesterone to sex steroids was studied by Rao *et al.* (1969). At stage 29, androstenedione was the only identifiable metabolite. The yield of androstenedione increased in the testes of older embryos but remained low in ovaries. Testosterone and estradiol were not found.

B. Birds

1. Localization of Δ^5-3β-Hydroxysteroid Dehydrogenase

The activity of this enzyme in the embryonic gonads of birds has been the subject of several studies, but there is some disagreement about the stage at which it appears and about its localization. Woods and Weeks (1969) described Δ^5-3β-HSDH activity at the genital ridge stage in the 2-day-old chick embryo. Scheib and Haffen (1967, 1969) detected enzymatic activity in the undifferentiated gonads at 6½ days in the chick and at 4½ days in the quail. According to Narbaitz and Kolodny (1964) and Chieffi *et al.* (1964b), Δ^5-3β-HSDH is not active until the eighth day; Boucek *et al.* (1966) suggest the ninth day.

In the differentiating testis Scheib and Haffen (1968, 1969) have described two types of cell clusters in which the Δ^5-3β-HSDH activity is very intense; they are located at different levels in some sex cords and also in the stroma. They appeared at 8½ days in the chick, and at 13½ days in the quail. In the stroma, the cell clusters that represent the interstitial cells enlarge and their number increases during development. These observations are not in agreement with those of Chieffi *et al.* (1964b) who state that Δ^5-3β-HSDH activity appears only in a diffuse fashion starting at the eighth day in the sex cords where it remains localized during the whole embryonic development. Narbaitz and Kolodny (1964) state that the testis has no Δ^5-3β-HSDH activity and that the reaction observed in the cords at 10 to 14 days is nonspecific.

In the ovary, the Δ^5-3β-HSDH activity is first confined to the medullary cords and then to the interstitial cells derived from the medullary cords. Furthermore, the reaction is similar at every stage in the rudimentary right gonad which has no cortex (Narbaitz and Kolodny, 1964; Chieffi *et al.*, 1964a; Scheib and Haffen, 1968, 1969).

2. Hormonal Secretions

Gallien and Le Foulgoc (1957, 1958) investigated the sex hormones of the embryonic gonads by fluorometric and chromatographic methods. Estrone–estradiol type hormones were found in chick embryos of 10, 13, and 21 days of age. The amounts per gonad were low at first but increased during incubation: 0.05 μg at 10 days, 0.1 μg at 13 days, and 0.7 μg at 23 days. These hormones have been identified by Ozon (1965, 1966) in the amniotic and allantoic fluids and in the peripheral blood of female embryos of the same age.

Wolff and Haffen (1952a,b) showed that female duck gonads, removed at 6 to 7 days of incubation, feminized male gonads of the same age when they were cultured together. The same transformation occurred in 5-day-old male chick embryos (Weniger, 1961). Close contact of the gonads was not necessary, and diffusion of ovarian hormone(s) in the medium was postulated (Weniger, 1962). The culture media of male gonads, female gonads, and other embryonic organs (liver, mesonephros, heart, and Müllerian duct) were extracted biochemically and tested by the method of Allen and Doisy (Weniger, 1964, 1965, 1966). Positive results were found for the estrone–estradiol fraction extracted from media in which female gonads had been cultured.

3. Biosynthesis of Sex Steroids

The ovaries of avian embryos are able to convert several radioactive precursors into estrogens. After addition of sodium [1-^{14}C]acetate, estrone, and estradiol-17β (Weniger *et al.*, 1967; Cedard *et al.*, 1968), estriol and epiestriol (Weniger, 1969) were isolated from media in which ovaries had been cultured for 24 or 48 hours. When the same precursor was supplied to 6-day-old undifferentiated gonads, radioactive estrone and estradiol-17β were found in the culture medium after 24 hours (Weniger and Zeis, 1969, 1971). The formation of estrone and estradiol-17β from [4-^{14}C]progesterone has been demonstrated in 5-day-old undifferentiated gonads (Weniger, 1968) and in embryonic chick or quail ovaries of 7 to 18 days (Noumura, 1966; Galli and Wassermann, 1972, 1973; Guichard *et al.*, 1973a,b). [4-^{14}C]Dehydroepiandrosterone is transformed into estrogens by chick embryonic ovaries of 6½ to 18 days (Cedard and Haffen, 1966; Haffen and Cedard, 1968) and by quail ovaries of 10 to 15 days (Scheib *et al.*, 1974). The testes of 7- to 18-day-old chick embryos and of 10- and 15-day-old quail embryos are able to transform [4-^{14}C]progesterone and [4-^{14}C]dehydroepiandrosterone into testosterone (Noumura, 1966; Guichard *et al.*, 1973a,b; Galli and Wassermann, 1972, 1973; Scheib *et al.*, 1974). Androgen synthesis has also been detected by immunohistochemistry in developing chick embryonic gonads of both sexes (Woods and Podczaski, 1974). These results contradict those of Weniger (1970) who did not observe testosterone production from any of the radioactive sex hormone precursors.

C. Mammals

1. Localization of Δ^5-3β-Hydroxysteroid Dehydrogenase

Activity of Δ^5-3β-HSDH has been demonstrated in the Leydig cells of the fetal testis (Fig. 3b) in many mammalian species (see Haffen, 1970, for references). Enzymatic activity has been found before sex differentiation in the genital ridge of a human fetus of 6 weeks (Baillie *et al.*, 1966a), in the gonads of mice at 12 to 14 days of gestation (Hart *et al.*, 1966), and in the testicular primordium on the twelfth day of gestation (Scheib and Lombard,

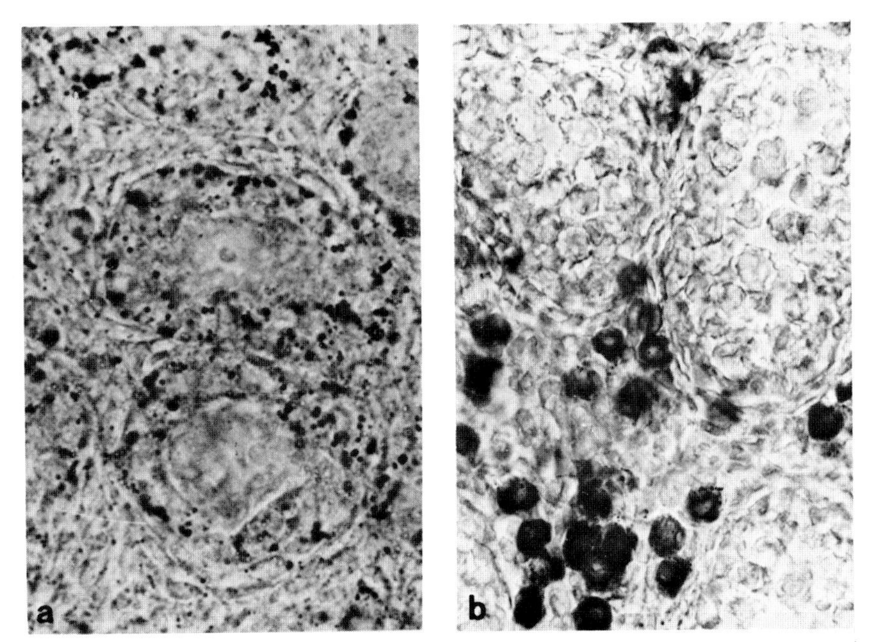

Fig. 3. Comparison of the activity of Δ^5-3β-hydroxysteroid-dehydrogenase in the ovary and testis of the fetal mouse. (a) The deposition of formazan is obvious in the follicular cells and in the interstitial tissue of a 3-day-old postnatal ovary. (b) The intense activity in the fetal testis at 14 days of gestation is localized only in the interstitial tissue (Leydig cells). (By courtesy of D. Scheib and M. N. Lombard.)

1971). Differences in histochemical features of the Leydig cells during fetal and postnatal development have been mentioned several times. As for the human testis (Mancini *et al.*, 1963), the existence of two generations of interstitial cells is proposed for the rat by Lording and de Kretser (1972). A discontinuity in the activity of the Leydig cells during fetal life and after birth in relation to the development of the gonadotropic cells has been suggested for the mouse.

The occurrence of enzyme activity in the fetal ovary is doubtful. Several authors have reported its absence during embryonic development. The first signs of enzymatic activity in the rat are in the interstitial cells of the stroma 8 to 9 days after birth (Chieffi *et al.*, 1964a; Presl *et al.*, 1965; Tramontana *et al.*, 1967; Schlegel *et al.*, 1967). Goldman and Kohn (1970) cultured fetal rat ovaries, removed at term (21 days), and found signs of Δ^5-3β-HSDH activity after 9 to 10 days of culture, and that activity was localized as in normal ovaries of the same age (Goldman *et al.*, 1966a). In the mouse fetus, Hart *et al.* (1966) detected enzymatic activity in the ovarian stroma beginning at 12 to 14 days of gestation, but they did not define its localization. No trace of Δ^5-3β-HSDH activity was found in the mouse ovary before 16 days of gestation (Scheib and Lombard, 1972). It occurred after the oogonia entered meiosis and was localized only in the medulla. Enzymatic activity occurred in the follicular cells (Fig. 3a) enveloping the oocytes to form primary follicles at 3 days postpartum. In the fetal rabbit ovary, Δ^5-3β-HSDH activity appears on day 29 and remains at a very low level for the remaining 2 days of gestation (Goldman *et al.*, 1972). Δ^5-3β-HSDH activity has been observed in the follicular cells in the fetal ovary of man at 3 months of gestation (Motta *et al.*, 1970), at 4 months (Goldman *et al.*, 1966b), and up to 5 months (Baillie *et al.*, 1966b).

2. Biosynthesis of Sex Steroids

Several authors have demonstrated that the fetal testis is able to synthesize androstenedione and testosterone from radioactive sodium acetate, pregnenolone, or progesterone at the time of differentiation of the male genital tract (see Bloch, 1967 for earlier references; Serra *et al.*, 1970; Warren *et al.*, 1972; Wilson and Siiteri, 1973) or even earlier (Weniger, 1972). The bovine fetal ovary (a month or more before term) has been shown to effect the conversion of [4-^{14}C]progesterone into androgens, and aromatization of [4-^{14}C]androstenedione into estradiol-17β. No sex steroids were formed from sodium acetate by bovine fetal ovaries (2½ to 7½ months of gestation) in organ culture (Weniger *et al.*, 1972). The human fetal ovary (9 and 11 weeks) is able to reduce [4-^{14}C]progesterone in the 20α position (Bloch, 1964; Bloch *et al.*, 1965). Jungmann and Schweppe (1968) have identified cholesterol synthesis and small but definite quantities of pregnenolone and progesterone formed from sodium [1-^{14}C]acetate by human fetal ovaries of 20 and 22 weeks. The capacity of human fetal ovaries (88, 121, and 124 days) to utilize [^{3}H]pregnenolone sulfate as precursors of free steroids has been studied by Payne and Jaffe (1974). Radioactive pregnenolone, 17α-hydroxypregnenolone, dehydroepiandrosterone, and androstenedione, but no progesterone, could be identified as products of pregnenolone sulfate metabolism. The formation of testosterone, estrone, and estradiol-17β could not be demonstrated. [4-^{14}C]Androstenedione can be converted into testosterone by the fetal rabbit ovary at day 16 of gestation (Lipsett and Tullner, 1965) and at day 23 and 27 (Wilson and Siiteri, 1973), but a quantitative improvement does not occur with development of the ovary.

3. Hormonal Secretions

a. Steroids. The fetal testis of the guinea pig (from the time of sex differentiation, 22 to 30 days of gestation) and man (9½ to 20 weeks) secrete androgens when cultured (Price *et al.*, 1963; Ortiz *et al.*, 1966; Zaaïjer and Price, 1971). Androgens are also secreted by the fetal ovary of the guinea pig—low levels at 41 to 46 days of gestation increasing until the sixty-second day just before parturition. In the sheep, testosterone and androstenedione have been detected in the testis at 30 days of gestation, 5 days before sex differentiation takes place (Attal, 1969). Estrone was found at 70 days. Huhtaniemi *et al.* (1970) analyzed 33 testes of human fetuses of 8.5 to 20 cm crown–rump length and found that most of the free steroid was pregnenolone. Reyes *et al.* (1973) measured similar concentrations of testosterone in the testes of human fetuses. Testosterone concentrations reached their highest values at 7–10 cm crown–rump length and then declined. This decline has also been observed by Resko (1970) in pooled fetal monkey testes obtained at 13–14 weeks and 20 weeks of gestation. Testosterone concentration was negligible in the ovaries of human fetuses and little or no estradiol was detected in testes or ovaries. Fetal human testes of 7.8 to 21.1 cm crown–rump length maintained in organ culture produced testosterone. Stimulation of cultured testes of fetuses (16.9 and 21.1 cm crown–rump length) with HCG caused a significantly greater secretion of testosterone, and Leydig cell hyperplasia was also maintained (Abramovich *et al.*, 1974).

b. Müllerian Duct Inhibiting Hormone. Josso (1970, 1973) has shown that development of the Müllerian duct is inhibited by a hormone secreted by the cultured human fetal testis. This hormone is interspecific, nondialyzable, and its synthesis is restricted to the fetal or perhaps early neonatal period. From experiments with whole fetal calf testicular tissue, isolated seminiferous tubules, and interstitial tissue, Josso (1973) suggested that the Müllerian duct inhibiting hormone may be produced by the fetal Sertoli cells.

V. CONCLUSIONS

The primordium of the undifferentiated gonad consists of a cortex formed by a thickened germinal epithelium and a central zone, the medulla. The primordium is colonized by extragonadal germ cells, but the mechanisms which determine the histogenesis of the gonad as an ovary or as a testis remain unresolved. The most generally accepted theory is that the sex genes initiate chemical syntheses which lead to the production of the substances that initiate sexual differentiation (Witschi, 1971).

In 1914, Witschi suggested that the sexual differentiation of the amphibian gonad is effected by special inductor substances which are produced by two types of cells. The cortex acts as an inductor of differentiation into an ovary because the follicular cells influence the development of germ cells into oogonia. The medulla is an inductor for male differentiation through the interstitial cells which affect differentiation of germ cells into spermatogonia. Witschi (1921) suggested that induction occurred in response to specific morphogenetic substances which were termed (Witschi, 1931) "cortexine" and "medullarine." In *Xenopus laevis,* these substances are produced between stages 21 and 27 (Witschi, 1967). An antagonism exists between the two inductor systems: in the male, the medullary system is dominant and a testis develops, and "cortexine" results in cortical development in the female. The interstitial and follicular cells secrete the antagonistic substances and they are these cells that later elaborate the sex steroids necessary for differentiation of the secondary sex characters after metamorphosis.

This theory of antagonism has been illustrated with experiments in which the dominant component of the gonad has been eliminated. The ablation of the testes in the male toad (Ponse, 1949) results in the development of Bidder's organ into an ovary. The right, rudimentary gonad of the chick becomes a testis if the left ovary is removed after hatching (Benoit, 1932). In urodeles, if the medulla does not develop because the renal blastema is destroyed (Gallien, 1954; Bruner and Witschi, 1954), the remaining cortical material develops, regardless of the genetic sex of the embryo. Witschi (1929a,b) considered that various external factors, such as temperature and exogenous sex hormones, influenced the inductor substances (and, therefore, histogenesis of the gonad) rather than stimulated directly one of the gonadal constituents.

Wolff and Ginglinger (1935) suggested that the somatic cells of the undifferentiated gonads produce substances that are similar to the sex hormones secreted by adult gonads. The stages of normal organogenesis of the gonads are similar to those of the inversion produced by exogenous sex hormones of the adult and by the secretions of embryonic gonads (grafts, parabiosis). Wolff considers that the same hormones are involved in the differentiation of the gonads, germ cells, and genital ducts, and that they act throughout development. The hormones are secreted by the medulla of the gonad, and male and female hormones originate from the same tissue (Wolff, 1965).

There are difficulties in using either theory as a generalization applicable to all vertebrates. Corticomedullary antagonism does not occur in the avian embryo or in the mammalian fetus, and only one secretory pattern has been demonstrated. In the avian embryo there is a feminizing secretion that stimulates development of the ovarian cortex and that is produced by the medulla of the female gonad. The fetal testis in mammals secretes a substance which inhibits development of the cortex in freemartin heifers. The adult sex hormones, which cause complete and permanent sex reversal in some amphibians and temporary reversal in birds, do not affect development of mammalian fetal gonads. Jost (1965, 1970) postulated that "during gonadal differentiation a genetically controlled masculinizing signal imposes testicular organogenesis on a gland which otherwise would become an ovary. The process of testicular organogenesis is probably controlled by one or more diffusible substance(s) produced by some cells so as to impose male organogenesis. The identity of the leading cells and the nature of the inducing substance(s) remains unknown." The mechanism that stimulates the oogonia to enter meiosis is also unknown.

In mammals, the results of biochemical studies suggest that the differentiation of the genital tract of the male is controlled by androgens produced by the fetal testis. The inability of the fetal ovary to synthesize estrogen corroborates results from castration experiments which indicate that differentiation of the female genital tract occurs in the absence of any secretion from the fetal ovary. In birds, histochemical and biochemical findings suggest that the embryonic gonads of both sexes may synthesize sex steroids even before gonadal differentiation is apparent (Wolff *et al.*, 1966; Weniger, 1968). The presence of enzymes involved in steroid biosynthesis has not been demonstrated in the embryonic gonads of most amphibians, reptiles, and fish, but Δ^5-3β-HSDH activity has been shown in *Pleurodeles waltlii* and *Vipera aspis* embryonic gonads.

It is not possible to make a generalization for all vertebrates on the mechanism of cytodifferentiation of the germ cells. In amphibians, gonocyte differentiation is controlled by the somatic tissue where they develop. Sex differentiation of the gonocytes can be completely reversed only in fish and amphibians. In birds and mammals, gonocytes of one sex do not complete gametogenesis in the somatic tissues of the other sex. For example, Haffen (1968) has shown in the chick that male gonocytes, when made to develop in an ovary, proceed as do oogonia and enter meiosis but cannot develop further and degenerate just before hatching. Mystkowska and Tarkowski (1968, 1970) and McLaren *et al.* (1972) have shown that meiotic germ cells occur only in fetal testes of chromosome chimeras (XX/XY). Their experiments strongly suggest that XX germ cells are unable to survive in a male environment.

REFERENCES

Abramovich, D. R., Baker, T. G., and Neal, P. (1974). Effect of human chorionic gonadotrophin on testosterone secretion by the fetal human testis in organ culture. *J. Endocrinol.* **60,** 179–185.

Attal, J. (1969). Levels of testosterone, androstenedione, estrone and estradiol-17β in the testis of fetal sheep. *Endocrinology* **85,** 280–289.

Atz, J. W. (1964). Intersexuality in fishes. *In* "Intersexuality in Vertebrates Including Man" (C. N. Armstrong and A. J. Marshall, eds.), pp. 145–232. Academic Press, New York.

Baker, T. G. (1963). A quantitative and cytological study of germ cells in human ovaries. *Proc. R. Soc. London, Ser. B* **158,** 417–433.
Baker, T. G. (1966). A quantitative and cytological study of oogenesis in the rhesus monkey. *J. Anat.* **100,** 761–776.
Baker, T. G., and Franchi, L. L. (1967). The fine structure of oogonia and oocytes in human ovaries. *J. Cell Sci.* **2,** 213–224.
Baillie, A. H., Ferguson, M. M., and Hart, D. McK. (1966a). Histochemical evidence of steroid metabolism in the human genital ridge. *J. Clin. Endocrinol. Metab.* **26,** 738–741.
Baillie, A. H., Ferguson, M. M., and Hart, D. McK. (1966b). "Developments in Steroid Histochemistry." Academic Press, New York.
Bara, G. (1965). Histochemical localization of Δ^5-3β-hydroxysteroid dehydrogenase in the ovaries of a teleost fish, *Scomber scomber* L. *Gen. Comp. Endocrinol.* **5,** 284–296.
Beatty, R. A. (1964). Chromosome deviations and sex in vertebrates. *In* "Intersexuality in Vertebrates Including Man" (C. N. Armstrong and A. J. Marshall, eds.), pp. 17–143. Academic Press, New York.
Beatty, R. A. (1970). Genetic basis for the determination of sex. *Philos. Trans. R. Soc. London, Ser. B* **259,** 3–14.
Beaumont, H. M., and Mandl, A. M. (1962). A quantitative and cytological study of oogonia and oocytes in the foetal and neonatal rat. *Proc. R. Soc. London Ser B,* **155,** 557–579.
Benoit, J. (1923). Sur l'origine des cellules interstitielles dans le testicule du coq domestique. *C. R. Acad. Sci.* (*Paris*) **177,** 412–414.
Benoit, J. (1929). Structure, origine et fonction des cellules interstitielles du testicule chez le coq domestique. *Arch. Anat. Microsc. Morphol. Exp.* **25,** 173–188.
Benoit, J. (1932). L'inversion sexuelle de la poule déterminée par l'ablation de l'ovaire gauche; *Arch. Zool. Exp. Gen.* **73,** 1–112.
Benoit, J. (1950). Différenciation sexuelle chez les oiseaux au cours du développement normal et de l'intersexualité expérimentale par ovariectomie. *Arch. Anat. Microsc. Morphol. Exp.* **39,** 395–410.
Bjerregard, P., Bro-Rasmussen, F., and Reumert, T. (1974). Ultrastructural development of fetal rabbit testis. *Z. Zellforsch. Mikrosk. Anat.* **147,** 401–413.
Black, J. L., and Erickson, B. M. (1968). Oogenesis and ovarian development in the prenatal pig. *Anat. Rec.* **161,** 45–56.
Black, V. H., and Christensen, A. K. (1969). Differentiation of interstitial cells and Sertoli cells in fetal guinea pig testes. *Am. J. Anat.* **124,** 211–238.
Bloch, E. (1964). Metabolism of 4-^{14}C-progesterone by human fetal testis and ovaries. *Endocrinology* **74,** 833–845.
Bloch, E. (1967). In vitro steroid synthesis by gonads and adrenals during mammalian foetal development. *Proc. Int. Congr. Horm. Steroids, 2nd, 1966* Excerpta Med. Int. Congr. Ser. No. 132, pp. 675–679.
Bloch, E., Romney, S. L., Klein, M., Lipiello, L., Cooper, P., and Goldring, I. P. (1965). Steroid synthesis by human fetal adrenals and ovaries maintained in organ culture. *Proc. Soc. Exp. Biol. Med.* **119,** 449–452.
Borghese, E., and Venini, M. A. (1956). Culture in vitro de gonades embryonnaires de *Mus musculus*. *Symp. Gen.* **5,** 69–83.
Borum, K. (1961). Oogenesis in the mouse. A study of the meiotic prophase. *Exp. Cell Res.* **24,** 495–507.

Borum, K. (1966). Oogenesis in the mouse. A study of the origin of the mature ova. *Exp. Cell Res.* **45,** 39–47.

Botte, V. (1964). I. Lipidi e la Δ^5-3β-HSDH nel tessuto interstiziale del testicolo di *Rana esculenta* durante il ciclo sessuala. Ricerche istochimiche. *Atti Soc. Peloritana Sci. Fis., Mat. Nat.* **10,** 521–528.

Botte, V., and Cotino, E. (1964). Ricerche istochimiche sulla distribuzione del colesterolo e, di alcuni enzimidella steroidogenesi nei follicoli ovarici e post ovulatori di *Rana esculenta* e *Triturus cristatus. Boll. Zool.* **31,** 491–499.

Botte, V., and Delrio, G. (1964). Ricerche istochimiche sulla distribuzione dei 3- e 17-chetosteroidi e di alcuni enzimi della steroidogenase nell'ovario di *Lacerta sicula. Boll. Zool.* **32,** 191–197.

Boucek, R. J., Gyori, E., and Alvarez, R. (1966). Steroid dehydrogenase reactions in developing chick adrenal and gonadal tissues. A histochemical study. *Gen. Comp. Endocrinol.* **7,** 292–303.

Bowman, P., and McLaren, A. (1970). Viability and growth of mouse embryos after *in vitro* culture and fusion. *J. Embryol. Exp. Morphol.* **23,** 693–704.

Brambell, F. W. R. (1927). The development and morphology of the gonads of the mouse. Part I. The morphogenesis of the indifferent gonad and of the ovary. *Proc. R. Soc. London, Ser. B* **101,** 391–409.

Bruner, J. A., and Witschi, E. (1954). Pluripotentiality of the mesonephric blastema and the mechanism of feminization of male salamanders by androgenic hormones. *Anat. Rec.* **120,** 99–117.

Buyse, A. (1935). The differentiation of transplanted mammalian gonad primordia. *J. Exp. Zool.* **70,** 1–30.

Calame, S. (1961). Contribution expérimentale à l'étude du développement du système urogénital de l'embryon d'oiseau. *Arch. Anat., Histol. Embryol.* **44,** Suppl., 45–65.

Callebaut, M. (1967a). Premeiosis and premeiotic DNA synthesis in the left ovary of the female chick embryo. *J. Embryol. Exp. Morphol.* **18,** 299–304.

Callebaut, M. (1967b). La synthèse de DNA dans les cellules germinales de l'ovaire de poulet pendant la dernière semaine de la vie embryonnaire. *Experientia* **23,** 419–420.

Callebaut, M. (1968). Extra corporal development of quail oocytes. *Experientia* **24,** 1242–1243.

Callebaut, M., and Dubois, R. (1965). Sur l'incorporation de thymidine tritiée par les cellules germinales de l'ovaire embryonnaire de poulet, en culture *in vitro. C. R. Acad. Sci. (Paris)* **261,** 5215–5218.

Caridroit, F., and Pezard, J. (1925). Poussée testiculaire autonome à l'intérieur des greffes ovariennes et autoplastiques chez la poule domestique. *C. R. Acad. Sci. (Paris)* **180,** 2067.

Cedard, L., and Haffen, K. (1966). Transformation de la dehydroépiandrostérone par les gonades embryonnaires de poulet cultivées *in vitro. C. R. Acad. Sci Ser. D* **263,** 430–433.

Cedard, L., Haffen, K., and Guichard, A. (1968). Influence de l'hormone gonadotrope chorionique sur la production d'oestrogènes à partir d'acetate de Na et de déhydroepiandrostérone radioactifs par les gonades embryonnaires de poulet, cultivées *in vitro. C. R. Acad. Sci. Ser. D* **267,** 118–120.

Certain, P., Collenot, G., Collenot, A., and Ozon, R. (1964). Mise en évidence biochimique et histochimique d'une Δ^5-3β HSDH dans le testicule du Triton *Pleurodeles waltlii. C. R. Soc. Biol.* **158,** 1040–1043.

Chan, S. T. H. (1970). Natural sex reversal in vertebrates. *Philos. Trans. R. Soc. London, Ser. B* **259,** 59–72.
Chieffi, G. (1959). Sex differentiation and experimental sex reversal in elasmobranch fishes. *Arch. Anat. Microsc. Morphol. Exp.* **48,** 21–36.
Chieffi, G. (1965). Onset of steroidogenesis in the vertebrate embryonic gonad. *Organogenesis* [*Int. Conf.*], *1964* pp. 653–671.
Chieffi, G. (1967a). Occurrence of steroids in gonads of non mammalian vertebrates and sites of their biosynthesis. *Proc. Int. Congr. Horm. Steroids, 2nd, 1966* Excerpta Med. Found. Int. Congr. Ser. No. 132, pp. 1047–1057.
Chieffi, G. (1967b). The reproductive system of elasmobranchs: developmental and endocrinological aspects. *Pharmacol. Endocrinol. Immunol.* **37,** 553–580.
Chieffi, G., and Botte, V. (1963a). Comportamento istochimico della steroide 3β-olo deidrogenasi nell'interrenale e nei corpusculi di stannius di *Anguilla anguilla. Atti Accad. Naz. Lincei, ci. Sci. Fig., Mat. Nat., Rend.* [8] **34,** 570–572.
Chieffi, G., and Botte, V. (1963b). Observazioni istochimiche della steroide 3β-olo deidrogenasi nell'interrenale e nelle gonadi di girini e adulti di *Rana esculenta. Riv. Istochim. Norm. Pathol.* **9,** 172–17.
Chieffi, G., and Botte, V. (1966). Il differenziamento istochimico dell'interrenale e dei tessuti somatici della gonade embrionale di *Lacerta sicula. Atti Accad. Naz. Lincei, Cl. Sci. Fis., Mat. Nat., Rend.* [8] **39,** 589–592.
Chieffi, G., Botte, V., and Visca, T. (1963). Attenttiva steroid 3β-olo deidrogenasica nell'interrenale di alcuni selacei. *Acta Med. Romana* **1,** 108–116.
Chieffi, G., Materazzi, G., and Botte, V. (1964a). Comportamento istochimico della 17β-idrossisteroide deidrogenasi e dello 17 idrossisteroide deidrogenasi (DPN et TPN-dipendenti) nel corticosurrene, nel testicolo e nell'ovario de embrioni e neonati di ratto. *Atti Soc. Peloritana Sci. Fis., Mat. Nat.* **10,** 515–520.
Chieffi, G., Manelli, H., Botte, V., and Mastrolia, L. (1964b). Il differenzia-mento istochimico dell'interrenale e dei tessuti somatici della gonade embrionale di pollo: comportamento della 3β-olo deidrogenasi. *Acta Embryol. Morphol. Exp.* **7,** 89–91.
Clermont, Y., and Perey, B. (1957). Quantitative study of the cell population of seminiferous tubules in immature rats. *Am. J. Anat.* **100,** 241–267.
Collenot, A. (1964). Mise en évidence histochimique d'une Δ^5-3β HSDH dans les gonades non différenciées et en cours de différenciation de mâles génétiques de l'urodèle *Pleurodeles waltlii. C. R. Acad. Sci.* (*Paris*) **259,** 2535–2537.
Collenot, A. (1965). Recherches comparatives sur l'inversion sexuelle par les hormones stéroïdes chez les amphibiens. *Mem. Soc. Zool. France* **33,** 1–141.
Collenot, A., and Ozon, R. (1965). Mise en évidence biochimique et histochimique d'une Δ^5-3β hydroxystéroide deshydrogénase dans le testicule de *Scylliorhinus canicula* L. *Bull. Soc. Zool. France* **89,** 577–585.
Courot, M. (1962). Développement du testicule chez l'agneau. Etablissment de la spermatogénèse. *Ann. Biol. Anim., Biochim., Biophys.* **2,** 25–41.
d'Ancona, V. (1949). Il differenziamento delle gonade e l'inversione sessualle degli sparidi. *Arch. Oceanogr. Limnol.* **6,** 97–163.
d'Ancona, V. (1950). Détermination et différenciation du sexe chez les poissons. *Arch. Anat. Microsc. Morphol. Exp.* **39,** 92–112.
d'Ancona, V. (1952). Territorial sexualization in the gonads of teleosteans. *Anat. Rec.* **114,** 666–667.
Dantschakoff, V., and Guelin-Schedrina, A. (1933). Assymetrie der gonaden beim

Huhn. A. Primäre quantitative assymetrie der gonadenanlagen. *Z. Zellforsch. Mikrosk. Anat.* **19,** 50–78.

Della Corte, F., Botte, V., and Chieffi, G. (1961). Ricerca istochimica dell' attivita della steroide 3β-olo deidrogenasi nel testicolo di *Torpedo marmorata Risso* e di *Scylliorhinus stellaris. Atti Soc. Peloritana Sci. Fis., Mat. Nat.* **7,** 393–397.

Dodd, J. M. (1965). Endocrine patterns in the reproduction of lower vertebrates. *Proc. Int. Congr. Endocrinol., 2nd, 1964* Excerpta Med. Found. Int. Congr. Ser. No. **83,** pp. 124–130.

Domm, L. V. (1927). New experiments on ovariectomy and the problem of sex inversion in the fowl. *J. Exp. Zool.* **48,** 31–119.

Dubois, R. (1967). Sur l'origine et l'amoeboïdisme des cellules germinales de l'embryon de poulet en culture in vitro et leur localisation dans le germe non incubé. *Arch. Anat. Microsc. Morphol. Exp.* **56,** 245–264.

Dubois, R. (1968). La colonisation des ébauches gonadiques par les cellules germinales de l'embryon de poulet, en culture *in vitro. J. Embryol. Exp. Morphol.* **20,** 189–213.

Dubois, R. (1969). Le mécanisme d'entrée des cellules germinales primordiales dans le réseau vasculaire, chez l'embryon de poulet. *J. Embryol. Exp. Morphol.* **21,** 255–270.

Dubois, R., and Croisille, Y. (1970). Germ-cell line and sexual differentiation in birds. *Philos. Trans R. Soc. London, Ser. B* **259,** 73–89.

Dubois, R., and Cuminge, D. (1968). Sur l'aspect ultrastructural et histochimique des cellules germinales de l'embryon de poulet. *Ann. Histochim.* **13,** 33–50.

Dufaure, J. P. (1966). Recherches descriptives et expérimentales sur les modalités et facteurs du développement de l'appareil génital chez le Lézard vivipare (*Lacerta vivipara Jacquin*). *Arch. Anat. Microsc. Morphol. Exp.* **55,** 437–537.

Dufaure, J. P., and Gil-Alvaro, M. C. (1967). Observations sur l'organogénèse de la gonade et ses particularités chez l'embryon de Vipère aspic (*Vipera aspis* L.) *C. R. Soc. Biol.* **161,** 2471–2473.

Dufaure, J. P., and Mesure, M. (1967). Données préliminaires sur l'activité steroido-3β-ol déshydrogénasique chez l'embryon de Lézard vivipare (*Lacerta vivipara Jacquin*). *C. R. Acad. Sci. Ser. D* **265,** 1215–1218.

Erickson, B. M. (1966). Development and radio-response of the prenatal bovine ovary. *J. Reprod. Fertil.* **11,** 97–105.

Erickson, G. F. (1974). The control of the differentiation of female enbryonic germ cells in the bird. *Dev. Biol.* **36,** 113–129.

Fargeix, N. (1964). Répartition des cellules germinales chez des embryons jumeaux de canard. *C. R. Soc. Biol.* **158,** 1507–1510.

Fargeix, N. (1970). La colonisation des gonades par les cellules germinales chez des embryons de régulation. *Ann. Embryol. Morphog.* **3,** 107–131.

Fargeix, N., and Didier, E. (1973). Colonisation des ébauches gonadiques de l'embryon de poulet par les cellules germinales après agénésie unilatérale du mésonephros. *C. R. Acad Sci. Ser. D* **276,** 2067-2069.

Fargeix, N., and Theilleux, D. (1967). Les cellules germinales chez de jeunes embryons de canard issus de moitiés antérieures et postérieures de blastodermes, isolées après 6 et 9h d'incubation. *C. R. Acad. Sci. Ser. D* **265,** 2019–2022.

Firket, J. (1914). Recherches sur l'organogenèse des glandes sexuelles chez les oiseaux. *Arch. Biol.* **29,** 200–351.

Firket, J. (1920). Recherches sur l'organogenèse des glandes sexuelles chez les oiseaux. *Arch. Biol.* **30,** 393–516.

Forbes, T. R. (1938a). Studies on the reproductive system of the alligator. II. The effects of prolonged injections of oestrone in the immature alligator. *J. Exp. Zool.* **78,** 335–367.

Forbes, T. R. (1938b). Studies on the reproductive system of the alligator. III. The action of testosterone on the accessory sex structures of recently hatched female alligators. *Anat. Rec.* **72,** 87–95.

Forbes, T. R. (1939). Studies on the reproductive system of the alligator. V. The effects of injections of testosterone propionate in immature alligators. *Anat. Rec.* **75,** 51–57.

Forbes, T. R. (1940a). Studies on the reproductive system of the alligator. IV. Observations on the development of the gonad, the adrenal cortex and the müllerian duct. *Contrib. Embryol. Carnegie Inst.* **14,** 129–155.

Forbes, T. R. (1940b). Studies on the reproductive system of the alligator. VI. Further observations on heterosexual structures in the female alligator. *Anat. Rec.* **77,** 343–365.

Franchi, L. L., and Mandl, A. M. (1964). The ultrastructure of germ cells in foetal and neonatal male rats. *J. Embryol. Exp. Morphol.* **12,** 289–308.

Gaillard, P. J. (1950). Sex cell formation in explants of the foetal human ovarian cortex. *Proc. K. Ned. Akad. Wet.* **53,** 1300–1347.

Galgano, M. (1935). Cellule a sessualita intermedia nelle gonadi di alcuni giovani essemplari *Rana esculenta* L. *Arch. Ital. Anat. Embriol.* **36,** 1–51.

Galgano, M. (1941). La variazone dei chiasmi nei mashi negli intersessuati di *"Rana esculenta" L. Arch. Ital. Anat. Embriol.* **46,** 127–164.

Galli, F., and Wassermann, G. F. (1972). Steroid biosynthesis by testes and ovaries of 15 day-old chick embryos. *Gen. Comp. Endocrinol.* **19,** 509–514.

Galli, F., and Wassermann, G. F. (1973). Steroid biosynthesis by gonads of 7 and 10 day-old chick embryos. *Gen. Comp. Endocrinol.* **21,** 77–83.

Gallien, L. (1954). Inversion expérimentale du sexe sous l'action des hormones sexuelles chez le triton *Pleurodeles waltlii Michah.* Analyse des conséquences génétiques. *Bull. Biol. Fr. Belg.* **88,** 1–51.

Gallien, L. (1965). Genetic control of sexual differentiation in vertebrates. *Organogenesis* [*Int. Conf.*], *1964,* Vol. 23, pp. 583–610.

Gallien, L. (1967). Developments in sexual organogenesis. *Adv. Morphog.* **6,** 259–317.

Gallien, L., and Chalumeau-Le Foulgoc, M. T. (1960). Mise en évidence de stéroïdes oestrogènes dans l'ovaire juvénile de *Xenopus laevis* Daudin et cycle des oestrogènes au cours de la ponte. *C. R. Acad. Sci.* (*Paris*) **251,** 460–462.

Gallien, L., and Durocher, M. (1957). Table chronologique du développement chez *Pleurodeles waltlii Michah. Bull. Biol. Fr. Belg.* **91,** 97–114.

Gallien, L., and Le Foulgoc, M. T. (1957). Détection par fluorimétrie et colorimétrie de stéroides sexuels dans les gonades embryonnaires de Poulet. *C. R. Soc. Biol.* **51,** 1088–1089.

Gallien, L., and Le Foulgoc, M. T. (1958). Activité oestrogène dans l'ovaire embryonnaire de poulet. Vérification par le test biologique de rate castrée (Allen et Doisy) de l'identification chimique d'hormones stéroides oestrogénes dans l'ovaire gauche du poulet à 21 jours. *C. R. Acad. Sci.* (*Paris*) **247,** 1776–1778.

Gallien, L., Certain, P., and Ozon, R. (1964). Mise en évidence d'une Δ^5-3β-

hydroxysteroide déshydrogénase dans le tissu interrenal de l'urodèle *Pleurodeles waltlii*, aux divers stades du développement. *C. R. Acad. Sci.* (*Paris*) **258**, 5729–5731.

Godet, R. (1950). La différenciation du sexe chez la Taupe. *Arch. Anat. Microsc. Morphol. exp.* **39**, 445–451.

Goldman, A. S., and Kohn, G. (1970). Rat ovarian 3β-hydroxysteroid dehydrogenase: normal developmental appearance in tissue culture. *Proc. Soc. Exp. Biol. Med.* **133**, 475–478.

Goldman, A. S., Bongiovanni, A. M., and Yakovac, W. C. (1966a). Production of congenital hyperplasia, hypospadias and clitoral hypertrophy (adrenogenital syndrome) in rats by inactivation of 3β-hydroxysteroid dehydrogenase. *Proc. Soc. Exp. Biol. Med.* **121**, 757-766.

Goldman, A. S., Yakovac, W. C., and Bongiovanni, A. M. (1966b). Development of activity of 3β-hydroxysteroid dehydrogenase in human fetal tissues and in two anencephalic newborns. *J. Clin. Endocrinol. Metab.* **26**, 14–22.

Goldman, A. S., Baker, M. K., and Stanek, A. E. (1972). Development of Δ^5-3β-hydroxysteroid and glucose-6-phosphate-dehydrogenase in the testes, adrenals and ovaries of the rabbit fetus. *Proc. Soc. Exp. Biol. Med.* **140**, 1486–1492.

Gondos, B. (1972). Germ cell degeneration in the developing rabbit ovary. *In* "Cell Differentiation" (R. Harris, P. Allin, and D. Viza, eds.), pp. 306–310. Munksgaard, Copenhagen.

Gondos, B., and Hobel, C. J. (1973). Interstitial cells in the human fetal ovary. *Endocrinology* **93**, 736–739.

Gondos, B., Bhiraleus, P., and Hobel, C. J. (1971). Ultrastructural observations on germ cells in human fetal ovaries. *Am. J. Obstet. Gynecol.* **110**, 644–000.

Gondos, B., Paup, D. C., Ross, J., and Gorski, R. A. (1974). Ultrastructural differentiation of Leydig cells in the fetal and postnatal hamster testis. *Anat. Rec.* **178**, 551–566.

Greenwald, G. S., and Peppler, R. D. (1968). Prepubertal and pubertal changes in the hamster ovary. *Anat. Rec.* **161**, 447–461.

Guichard, A., Cedard, L., and Haffen, K. (1973a). Aspect comparatif de la synthèse de stéroides sexuels par les gonades embryonnaires de poulet à differents stades du développement. (Etude en culture organotypique à partir de precurseurs radioactifs.) *Gen. Comp. Endocrinol.* **20**, 16–28.

Guichard, A., Cedard, L., Haffen, K., and Scheib, D. (1973b). Métabolisme de la pregnenolone et de la progesterone radioactives par les gonades embryonnaires de caille (*Coturnix coturnix japonica*) en culture organotypique. *Gen. Comp. Endocrinol.* **21**, 478–484.

Haffen, K. (1960). La culture *in vitro* de l'epithelium germinatif isolé des gonades mâles et femelles de l'embryon de canard. II. Influence de la medullaire sur la différenciation de l'épithélium germinatif. *J. Embryol. Exp. Morphol.* **8**, 414–424.

Haffen, K. (1963). Sur l'évolution en greffes coelomiques du constituant cortical isolé des gonades femelles et intersexuées de l'embryon de poulet. *C. R. Acad. Sci.* (*Paris*) **256**, 3755–3758.

Haffen, K. (1964). Sur la culture *in vitro* des glandes génitales des embryons de caille (*Coturnix coturnix*). Obtention de la différenciation sexuelle normale et de l'intersexualité expérimentale des gonades explantées. *C. R. Acad. Sci.* (*Paris*) **259**, 882–884.

Haffen, K. (1968). Sur la greffe prolongée d'ovaires d'embryons de poulet, colonisés expérimentalement par des cellules germinales du sexe mâle. *C. R. Acad. Sci. Ser. D* **267,** 511–513.

Haffen, K. (1969). Incompatibilité entre détermination chromosomique et stimulation hormonale dans la différentiation sexuelle. *Ann. Embryol. Morphog., Suppl.* **1,** 223–235.

Haffen, K. (1970). Biosynthesis of steroid hormones by the embryonic gonads of vertebrates. *Adv. Morphog.* **8,** 285–306.

Haffen, K., and Cedard, L. (1968). Etude, en culture organotypique *in vitro* du métabolisme de la dehydro-épiandrostérone et de la testostérone radioactives, par les gonades normales et intersexuées de l'embryon de poulet. *Gen. Comp. Endocrinol.* **11,** 220–234.

Harms, J. W. (1923). Die Physiologie des Bidderschen Organs und die experimentall-physiologische Umdiffrenzierung von Männchen in Weibchen. *Z. Anat. Entwicklungsgesch.* **69,** 598–629.

Harms, J. W. (1926). Beobachtungen über Geschlechtumwandlung reifer Tiere und deren F_2 Generation. *Zool. Anz.* **67,** 67–79.

Hart, D. McK., Baillie, A. H., Calman, K. C., and Ferguson, M. (1966). Hydroxysteroid dehydrogenase development in mouse adrenal and gonads. *J. Anat.* **100,** 801–812.

Holyoke, E. A. (1949). The differentiation of embryonic gonads transplanted to the adult omentum in the albino rat. *Anat. Rec.* **103,** 675–699.

Houillon, C. (1956). Recherches expérimentales sur la dissociation médullo-corticale dans l'organogénèse des gonades chez le triton *Pleurodeles waltlii M. Bull. Biol. Fr. Belg.* **90,** 359–444.

Hsü, C. Y., Yü, N. W., and Liang, H. M. (1971). Induction of sex reversal in female tadpoles of *Rana catesbeiana* by temperature treatment. *Endocrinol. Jpn.* **18,** 243–251.

Huhtaniemi, I., Ikonen, M., and Vihko, R. (1970). Presence of testosterone and other neutral steroids in human fetal testes. *Biochem. Biophys. Res. Commun.* **38,** 715–720.

Ioannou, J. M. (1964). Oogenesis in the guinea pig. *J. Embryol. Exp. Morphol.* **12,** 673–691.

Izadi, D. (1943). Développement de l'organe de Bidder du crapaud. *Rev. Suisse Zool.* **50,** 395–447.

Joly, G. (1965). Mise en évidence histochimique d'une Δ^5-3β-hydroxystéroid deshydrogénase dans l'ovaire de l'Urodèle *Salamandra salamandra* à différents stades du cycle sexuel. *C. R. Acad. Sci.* (*Paris*) **261,** 1569–1571.

Josso, N. (1970). Action du testicule humain sur le canal de Müller de foetus de rat en culture organotypique. *C. R. Acad. Sci. Ser. D* **271,** 2149–2152.

Josso, N. (1973). *In vitro* studies of mullerian inhibiting hormone by seminiferous tubules isolated from the calf fetal testis. *Endocrinology* **93,** 829–834.

Jost, A. (1965). Gonadal hormones in the sex differentiation of the mammalian fetus. *Organogenesis* [*Int. Conf.*], *1964* pp. 611–628.

Jost, A. (1970). Hormonal factors in the sex differentiation of the mammalian foetus. *Philos. Trans. R. Soc. London, Ser. B* **259,** 119–130.

Jost, A. (1972a). A new look at the mechanisms controlling sex differentiation in mammals. *Johns Hopkins Med. J.* **130,** 38–53.

Jost, A. (1972b). Données préliminaires sur les stades initiaux de la différenciation du testicule chez le rat. *Arch. Anat. Microsc. Morphol. Exp.* **61,** 415–438.

Jost, A., Vigier, B., and Prépin, J. (1972). Freemartins in cattle: the first steps of sexual organogenesis. *J. Reprod. Fertil.* **29,** 349–379.

Jost, A., Vigier, B., Prépin, J., and Perchellet, J. P. (1973). Studies on sex differentiation in mammals. *Recent Prog. Horm. Res.* **29,** 1–41.

Jungmann, R. A., and Schweppe, J. S. (1968). Biosynthesis of sterols and steroids from acetate-^{14}C by human fetal ovaries. *J. Clin. Endocrinol. Metab.* **28,** 1599–1604.

Lipsett, M. B., and Tullner, W. W. (1965). Testosterone synthesis by the fetal rabbit gonad. *Endocrinology* **77,** 273–277.

Lording, D. W., and de Kretser, D. M. (1972). Comparative ultrastructural and histochemical studies on the interstitial cells of the rat testis during fetal and postnatal development. *J. Reprod. Fertil.* **29,** 261–269.

McLaren, A., Chandley, A. C., and Kofman-Alfaro, S. (1972). A study of meiotic germ cells in the gonads of foetal mouse chimaeras. *J. Embryol. Exp. Morphol.* **27,** 513–524.

Mancini, R. E., Vilar, O., Lavieri, J. E., Andrada, J. A., and Heinrich, J. J. (1963). Development of Leydig cells in the normal human testis. A cytological and quantitative study. *Am. J. Anat.* **112,** 203–214.

Martinovitch, P. N. (1938). The development *in vitro* of the mammalian gonad. Ovary and ovogenesis. *Proc. R. Soc. London, Ser. B* **125,** 232–249.

Martinovitch, P. N. (1939). The effect of subnormal temperatures on the differentiation and survival of cultivated *in vitro* embryonic and infantile rat and mouse ovaries. *Proc. R. Soc. London, Ser. B* **128,** 138–143.

Mauléon, P. (1962). Deroulement de l'ovogénèse comparé chez différents mammifères domestiques. *Proc. Int. Congr. Anim. Reprod., 4th, 1961* Vol. 2, p. 348.

Mintz, B. (1968). Hermaphroditism, sex chromosomal mosaicism and germ cell selection in allophenic mice. *J. Anim. Sci.* **27,** Suppl., 51–60.

Mintz, B., and Wolff, Et. (1951). Sur les greffes coelomiques de la medullaire ovarienne d'embryons de poulet. L'évolution des greffons et leur action féminisante sur les hôtes mâles. *C. R. Soc. Biol.* **146,** 494–495.

Mintz, B., and Wolff, Et. (1954). The development of embryonic chick ovarian medulla and its feminizing action in intracoelomic grafts. *J. Exp. Zool.* **126,** 511–536.

Mittwoch, U. (1970). How does the Y chromosome affect gonadal differentiation? *Philos. Trans. R. Soc. London, Ser. B* **259,** 113–117.

Moore, K. L., and Price, D. (1942). Differentiation of embryonic reproductive tissues of the rat after transplantation into post-natal hosts. *J. Exp. Zool.* **90,** 229–265.

Morat-Mesure, M. (1969). Sur l'activité stéroide—3β-hydroxystéroide deshydrogenasique au cours de l'organogénèse de la gonade chez la vipère (*Vipera aspis* L.). *C. R. Acad. Sci. Ser. D* **268,** 546–549.

Morat-Mesure, M. (1971). Activité Δ^5-3β-hydroxysteroide deshydrogénase des glandes génitales et interrénales chez 2 reptiles, *Lacerta vivipara* et *Vipera aspis* L. *Ann. Embryol. Morphog.* **4,** 5–17.

Motta, P., Takeva, Z., and Bourneva, V. (1970). A histochemical study of Δ_5-3β-hydroxysteroid dehydrogenase activity in the interstitial cells of the mammalian ovary. *Experientia* **26,** 1128–1129.

Mystkowska, E. T., and Tarkowski, A. K. (1968). Observations on CBA-p/CBA-T6T6 mouse chimaeras. *J. Embryol. Exp. Morphol.* **20,** 33–52.

Mystkowska, E. T., and Tarkowski, A. K. (1970). Behaviour of germ cells and sexual differentiation in late embryonic and early post-natal mouse chimaeras. *J. Embryol. Exp. Morphol.* **23,** 395–405.

Narbaitz, R., and Adler, R. (1966). Submicroscopical observations on the differentiation of chick gonads. *J. Embryol. Exp. Morphol.* **16,** 41–47.

Narbaitz, R., and Adler, R. (1967). Submicroscopical aspects in the differentiation of rat fetal Leydig cells. *Acta Physiol. Lat. Am.* **17,** 286–291.

Narbaitz, R., and Kolodny, L. (1964). Δ^5-3β-hydroxysteroid dehydrogenase in differentiating chick gonads. *Z. Zellforsch. Mikrosk. Anat.* **63,** 612–617.

Nebel, B. R., Amarose, A. P., and Hackett, E. M. (1961). Calendar of gametogenic development in the prepuberal male mouse. *Science* **134,** 832–833.

Niemi, M., Ikonen, M., and Hervonen, A. (1967). Histochemistry and fine structure of the interstitial tissue in the human foetal testis. *Ciba Found. Colloq. Endocrinol.* [*Proc.*] **16,** 31–52.

Nonidez, J. F. (1922). The origin of the so-called luteal cells in the testis of hen-feathered cocks. *Am. J. Anat.* **31,** 109–124.

Noumura, T. (1966). Steroid biosynthesis by the gonads of the chick embryo. *Am. Zool.* **6,** 598.

Novi, A. M., and Saba, P. (1968). An electron microscopic study of the development of rat testis in the first 10 postnatal days. *Z. Zellforsch. Mikrosk. Anat.* **86,** 313–326.

Ortiz, E., Price, D., and Zaaijer, J. J. P. (1966). Organ culture studies of hormone secretion in endocrine glands of fetal guinea pigs. II. Secretion of androgenic hormone in adrenals and testis during early stages of development. *Proc. K. Ned. Akad. Wet.* **69,** 400–408.

Ozon, R. (1965). Mise en évidence d'hormones stéroïdes oestrogènes dans le sang de la poule adulte et chez l'embryon de poulet. *C. R. Acad. Sci.* (*Paris*) **261,** 5664–5666.

Ozon, R. (1966). Isolement et identification de steroides chez les vertébrés inférieurs et les oiseaux. *Ann. Biol. Anim., Biochim. Biophys.* **6,** 537–551.

Ozon, R. (1969). Steroid biosynthesis in larval and embryonic gonads of lower vertebrates. *Gen. Comp. Endocrinol.* **2,** 135–140.

Payne, A., and Jaffe, R. B. (1974). Androgen formation from pregnenolone sulfate by the human fetal ovary. *J. Clin. Endocrinol. Metab.* **39,** 300–304.

Pelliniemi, L. J., and Niemi, M. (1969). Fine structure of the human foetal testis. *Z. Zellforsch. Mikrosk. Anat.* **99,** 507–522.

Peters, H. (1970). Migration of gonocytes in the mammalian gonad and their differentiation. *Philos. Trans. R. Soc. London, Ser. B* **259,** 91–101.

Peters, H., and Levy, E. (1966). Cell dynamics of the ovarian cycle. *J. Reprod. Fertil.* **11,** 227–236.

Peters, H., Levy, E., and Crone, M. (1965). Oogenesis in rabbits. *J. Exp. Zool.* **158,** 169–179.

Peyre, A. (1961). Culture *in vitro* de gonades embryonnaires de Desman (*Galemys pyrenaicus G.*—Mammifère—Insectivore). *C. R. Acad. Sci.* (*Paris*) **252,** 605–607.

Peyre, A. (1962). Recherches sur l'intersexualité chez *Galemys pyrenaicus G. Arch. Biol.* **73,** 1–174.

Piquet, J. (1930). Détermination du sexe chez les batraciens en fonction de la température. *Rev. Suisse Zool.* **37,** 173–281.

Ponse, K. (1924). L'organe de Bidder et le déterminisme des caractères sexuels secondaires chez le crapaud (*Bufo vulgaris*). *Rev. Suisse Zool.* **31**, 177–336.
Ponse, K. (1927). L'évolution de l'organe de Bidder et la sexualité chez le crapaud. *Rev. Suisse Zool.* **34**, 217–220.
Ponse, K. (1943). L'organe de Bidder des crapauds est-il un territoire cortical depourvu d'ébauches médullaires? *C. R. Seances Soc. Phys. Hist. Nat. Geneve* **60**, 114–118.
Ponse, K. (1949). "La Différenciation du Sexe et l'Intersexualité chez les Vertébrés." F. Rouge, Lausanne.
Presl, J., Jirasek, J., Horsky, J., and Henzl, M. (1965). Observations on steroid 3β-ol-dehydrogenase activity in the ovary during early postnatal development in the rat. *J. Endocrinol.* **31**, 293–294.
Price, D., Ortiz, E., and Zaaïjer, J. J. P. (1963). Secretion of androgenic hormone by testes and adrenal glands of fetal guinea pigs. *Am. Zool.* **3**, 553–554.
Rao, G. S., Breuer, H., and Witschi, E. (1969). *In vitro* conversion of 17α-hydroxyprogesterone to androstenedione by mashed gonads from metamorphic stages of *Xenopus laevis*. *Gen. Comp. Endocrinol.* **12**, 119–123.
Raynaud, A. (1942). Recherches embryologiques et histologiques sur la différenciation sexuelle normale de la souris. *Bull. Biol. Fr. Belg., Suppl.* **29**, 1–114.
Raynaud, A. (1960). Sur la différenciation sexuelle des embryons d'orvet (*Anguis fragilis L.*). *Bull. Soc. Zool. Fr.* **85**, 210–230.
Raynaud, A. (1965). Effets d'une hormone oestrogène sur la différenciation sexuelle de l'embryon de lézard vert (*Lacerta viridis* Laur.). *C. R. Acad. Sci.* (*Paris*) **260**, 4611–4614.
Raynaud, A. (1967). Effets d'une hormone oestrogène sur le développement de l'appareil génital de l'embryon du lézard vert (*Lacerta viridis* Laur.). *Arch. Anat. Microsc. Morphol. Exp.* **56**, 63–122.
Raynaud, A. (1969). La différenciation sexuelle des voies génitales, des organes génitaux externes et d'autres ébauches de l'embryon. *In* "Traité de Zoologie" (P.-P. Grassé, ed.), Vol. 14, pp. 271–309. Masson, Paris.
Raynaud, A., and Pieau, C. (1966). Nouvelles observations sur la formation des glandes génitales chez l'embryon d'orvet (*Anguis fragilis* L.). *Bull. Soc. Zool. Fr.* **91**, 83–98.
Raynaud, A., and Pieau, C. (1971). Evolution des canaux de Müller et activité enzymatique de la Δ^5-3β-HSDH dans les glandes génitales chez les embryons de lézard vert (*Lacerta viridis* Laur.). *C. R. Acad. Sci. Ser. D* **273**, 2335–2338.
Resko, J. A. (1970). Androgen secretion by fetal and neonatal rhesus monkey. *Endocrinology* **87**, 680–687.
Reyes, F. I., Winter, J. S. D., and Faiman, C. (1973). Studies on human sexual development. I. Fetal gonadal and adrenal steroids. *J. Clin. Endocrinol. Metab.* **37**, 74–78.
Reynaud, G. (1969). Transfert de cellules germinales primordiales de dindon à l'embryon de poulet par injections intravasculaires. *J. Embryol. Exp. Morphol.* **21**, 485–507.
Risley, P. L. (1933a). Contributions on the development of the reproductive system in the musk turtle, *Sternotherus odoratus* (la treille). I. The embryonic origin and migration of the primordial germ cells. *Z. Zellforsch. Mikrosk. Anat.* **18**, 459–492.
Risley, P. L. (1933b). Contributions on the development of the reproductive system

in the musk turtle, *Sternotherus odoratus* (la treille). II. Gonadogenesis and sex differentiation. *Z. Zellforsch. Mikrosk. Anat.* **18,** 493–543.

Risley, P. L. (1941). A comparison of effects of gonadotropic and sex hormones on the urogenital systems of juvenile terrapins. *J. Exp. Zool.* **87,** 477–508.

Roberts, J. D., and Warren, J. C. (1964). Steroid biosynthesis in the fetal ovary. *Endocrinology* **74,** 846–853.

Rogulska, T., Ozdzenski, W., and Komar, A. (1971). Behaviour of mouse primordial germ cells in the chick embryo. *J. Embryol. Exp. Morphol.* **25,** 155–164.

Russo, J., and de Rosas, J. C. (1971). Differentiation of the Leydig cells of the mouse testis during fetal period. An ultrastructural study. *Am. J. Anat.* **130,** 461–480.

Salzgeber, B. (1960). Action inhibitrice du testicule de l'embryon de poulet sur le développement de l'ovaire embryonnaire de souris. *C. R. Acad. Sci.* (*Paris*) **251,** 1576–1577.

Salzgeber, B. (1963). Modification expérimentale du développement de l'ovaire de souris sous l'influence du testicule embryonnaire de poulet. *J. Embryol. Exp. Morphol.* **11,** 91–105.

Scheib, D. (1970a). Sur la présence de cellules "interstitielles primaires" dans les cordons du testicule de l'embryon de poulet. *C. R. Acad. Sci. Ser. D* **270,** 123–125.

Scheib, D. (1970b). Structure fine des cellules interstitielles du testicule foetal de la souris blanche. *C. R. Hebd. Seances Acad. Sci.* **271,** 423–425.

Scheib, D. (1970c). Origine et ultrastructure des cellules sécrétrices de l'ovaire embryonnaire de caille (*Coturnix coturnix japonica*). *C. R. Acad. Sci. Ser. D* **271,** 1700–1703.

Scheib, D. (1972). Ultrastructure et fonction des cellules interstitielles du testicule de la souris: évolution foetale. *Ann. Embryol. Morphog.* **5,** 121–133.

Scheib, D., and Haffen, K. (1967). Etude histochimique de la 3β-hydroxysteroide deshydrogenase des jeunes gonades embryonnaires de poulet. *C. R. Acad. Sci. Ser. D* **264,** 161–164.

Scheib, D., and Haffen, K. (1968). Sur la localisation histoenzymologique de la 3β-hydroxysteroide deshydrogénase dans les gonades de l'embryon de poulet; apparition et spécificité de l'activité enzymatique. *Ann. Embryol. Morphog.* **1,** 61–72.

Scheib, D., and Haffen, K. (1969). Apparition et localisation des hydroxy-stéroïde-deshydrogenases (Δ^5-3β et 17β) dans les gonades de l'embryon et du poussin de la caille (*Coturnix coturnix japonica*). Etude histoenzymologique et comparaison avec le poulet (*Gallus gallus domesticus*). *Gen. Comp. Endocrinol.* **12,** 586–697.

Scheib, D., and Lombard, M. N. (1971). Etude histoenzymologique de l'activité de la Δ^5-3β-hydroxystéroide deshydrogénase du testicule de souris blanche (avant et après la naissance); testicule normal et testicule explanté en présence d'hypophyse. *Arch. Anat. Microsc. Morphol. Exp.* **60,** 205–218.

Scheib, D., and Lombard, M. N. (1972). Apparition et évolution des cellules stéroïdogènes de l'ovaire de souris: étude cytoenzymologique. *Arch. Anat. Microsc. Morphol. Exp.* **61,** 439–452.

Scheib, D., Haffen, K., Guichard, A., and Cedard, L. (1974). Transformation de la dehydroépiandrostérone-4-^{14}C par les gonades embryonnaires de la caille japonaise (*Coturnix coturnix japonica*) explantées *in vitro*. *Gen. Comp. Endocrinol.* **23,** 453–459.

Schlegel, R. J., Farias, E., Russo, N. C., Moore, J. R., and Gardner, I. (1967). Structural changes in the fetal gonads and gonoducts during maturation of an enzyme, steroid 3β-ol dehydrogenase, in the gonads, adrenal cortex and placenta of fetal rats. *Endocrinology* **81,** 565–572.

Serra, G. B., Perez Palacios, G., and Jaffe, R. B. (1970). *De novo* testosterone synthesis in the human fetal testis. *J. Clin. Endocrinol. Metab.* **30,** 128–130.

Simon, D. (1960). Contribution à l'étude de la circulation et du transport des gonocytes primaires dans les blastodermes d'oiseau cultivés *in vitro. Arch. Anat. Microsc. Morphol. Exp.* **49,** 93–176.

Simone-Santoro I. de (1968). Morphologie ultrastructurale de l'ovaire de l'embryon de poulet au cours de sa différenciation et de son organogénèse. *J. Microsc. (Paris)* **8,** 739–752.

Stahl, A., and Carlon, N. (1973). Morphogenèse des cordons sexuels et signification de la zone medullaire de la gonade chez l'embryon de poulet. *Acta Anat.* **85,** 248–274.

Stefan, Y. (1963). Contribution à l'étude expérimentale de l'intersexualité chez un chélonien: *Emys leprosa* S. *Bull. Biol. Fr. Belg.* **97,** 363–467.

Swift, C. H. (1915). Origin of the definitive sex-cells in the female chick and their relation to the primordial germ cells. *Am. J. Anat.* **18,** 441–470.

Swift, C. H. (1916). Origin of sex-cords and definitive spermatogonia in the male chick. *Am. J. Anat.* **20,** 375–410.

Tachinante, F. (1974a). Sur les échanges interspécifiques de cellules germinales entre le poulet et la caille, en culture organotypique et en greffes coelomiques. *C. R. Acad. Sci. Ser. D* **278,** 1894–1898.

Tachinante, F. (1974b). Sur l'activité migratrice des cellules germinales de souris soumises à l'attraction de l'épithélium germinatif de poulet en culture *in vitro. C. R. Acad. Sci. Ser. D* **278,** 3135–3138.

Tarkowski, A. K. (1964). The hermaphroditism in chimaeric mice. *J. Embryol. Exp. Morphol.* **12,** 735–757.

Tarkowski, A. K. (1969). Consequences of sex chromosome chimerism for sexual differentiation in mammals. *Ann. Embryol. Morphog., Suppl.* **1,** 211–222.

Tarkowski, A. K. (1970a). Germ cells in natural and experimental chimaeras in mammals. *Philos. Trans. R. Soc. London, Ser. B* **259,** 107–111.

Tarkowski, A. K. (1970b). Are the genetic factors controlling sexual differentiation of somatic and germinal tissues of a mammalian gonad stable or labile? *Fogarty Int. Cent. Proc.* **2,** 49–60.

Teplitz, R. L., and Ohno, S. (1963). Postnatal induction of ovogenesis in the rabbit. *Exp. Cell Res.* **31,** 183–189.

Thiebold, J. J. (1964). Contribution à l'étude de l'organogénèse urogénitale et de son déterminisme chez un poisson elasmobranche: la petite roussette *Scyliorhinus caniculus. Bull. Biol. Fr. Belg.* **98,** 253–347.

Torrey, T. W. (1945). The development of the urogenital system of the albino rat. II. The gonads. *Am. J. Anat.* **76,** 375–397.

Torrey, T. W. (1950). Intraocular grafts of embryonic gonads of the rat. *J. Exp. Zool.* **115,** 37–58.

Tramontana, S., Botte, V., and Chieffi, G. (1967). Il differenziamento istochimico del corticosurrene e delle gonadi di coniglio. *Arch. Ostet. Ginecol.* **72,** 434–439.

Turner, C. D. (1969). Experimental reversal of germ cells. *Embryologia* **10,** 206–230.

Turner, C. D., and Asakawa, H. (1962). Differentiation of fetal rat ovaries following transplantation to kidneys and testes of adult male hosts. *Am. Zool.* **2,** Abstr. 270, p. 565.

Uchida, T. (1937). Studies on the sexuality of Amphibia. III. Sex transformation in *Hynobius retardatus* by the function of high temperature. *J. Fac. Sci., Hokkaido Imp. Univ., Ser. 0* **6,** 59–70.

van de Kerckhove, D. (1959). L'ovaire périnatal de la Souris blanche en culture organotypique. *C. R. Assoc. Anat.* **104,** 754–759.

Van Limborg, J. (1958). Number and distribution of the primary germ cells in duck embryos in the 28 to 36 somite stages. *Acta Morphol. Neerl. Scand.* **2,** 119–133.

Van Limborg, J. (1960). Number and distribution of the primary germ cells in duck embryos in the 37 to 44 somite stages. *Acta Morphol. Neerl. Scand.* **3,** 263–282.

Vannini, E. (1949). A proposito dell'origine interrenale del tessuto midollare della gonade negli Anfibi e negli Ucelli. *Atti Accad. Naz. Lincei, Cl. Sci. Fis., Mat. Nat., Rend.* [8] **6,** 511–518.

van Wagenen, G., and Simpson, M. E. (1965). Embryology of the ovary and testis in *Homo sapiens* and *Macaca mulatta*. *In* "Embryology of the Ovary and Testis" (G. van Wagenen and M. E. Simpson, eds.), p. 276. Yale Univ. Press, New Haven, Connecticut.

Vivien, J. H. (1959). Réactivité particulière du cortex gonadique et de l'épithélium du canal de Müller a l'action des hormones sexuelles chez le jeune mâle *d'Emys leprosa S.*, traité après l'éclosion. *Arch. Anat. Microsc. Morphol. Exp.* **48,** 297–312.

von Winiwarter, H. (1901). Recherches sur l'ovogénèse et l'organogénèse de l'ovaire des mammifères (Lapin et Homme). *Arch. Biol.* **17,** 33–199.

von Winiwarter, H., and Sainmont, Y. (1909). Nouvelles recherches sur l'ovogénèse et l'organogénèse de l'ovaire des mammifères (Chat). *Arch. Biol.* **24,** 165–276.

Warren, D. W., Haltmeyer, G. C., and Eik-Nes, K. B. (1972). Synthesis and metabolism of testosterone in the fetal rat testis. *Biol. Reprod.* **7,** 94–99.

Weakley, B. S. (1967). Light and electron microscopy of developing germ cells and follicle cells in the ovary of the golden hamster: twenty four hours before birth to eight days post partum. *J. Anat.* **101,** 435–459.

Weniger, J. P. (1961). Activité hormonale des gonades morphologiquement indifférenciées de l'embryon de poulet. *Arch. Anat. Microsc. Morphol. Exp.* **50,** 269–288.

Weniger, J. P. (1962). Diffusion des hormones gonadiques de l'embryon de poulet dans le milieu de culture. *Arch. Anat. Microsc. Morphol. Exp.* **51,** 325–336.

Weniger, J. P. (1964). L'ovaire d'embryon de poulet cultivé *in vitro* sécrète une hormone oestrogène. *C. R. Soc. Biol.* **158,** 175–178.

Weniger, J. P. (1965). L'hormone sexuelle de l'embryon de poulet femelle est-elle un stéroide phénolique? *C. R. Acad. Sci. (Paris)* **261,** 1427–1429.

Weniger, J. P. (1966). Activité oestrogène de la fraction "oestrone-oestradiol" de milieux incubés avec des gonades d'embryons de poulet. *C. R. Acad. Sci. Ser. D* **262,** 578–580.

Weniger, J. P. (1967). Transplantations d'ébauches gonadiques de Souris dans la cavité coelomique de l'embryon de poulet. *Experientia* **23,** 225–227.

Weniger, J. P. (1968). Sur la précocité de la sécrétion d'oestrogènes par les gonades embryonnaires de poulet cultivées *in vitro*. *C. R. Acad. Sci. Ser. D* **266,** 2277–2279.

Weniger, J. P. (1969). Recherches sur la nature chimique des hormones sexuelles embryonnaires de poulet. *Ann. Embryol. Morphog.* **2,** 433–444.

Weniger, J. P. (1970). Le testicule embryonnaire de poulet sécrète-t-il de la testostérone? *Arch. Anat. Histol. Embryol.* **53,** 97–105.

Weniger, J. P. (1972). Sur la sécrétion précoce de testostérone par le testicule embryonnaire de souris. *C. R. Acad. Sci. Ser. D* **275,** 1431–1433.

Weniger, J. P., and Zeis, A. (1969). Formation d'oestrone et d'oestradiol radioactifs à partir d'acétate de Na-1-^{14}C par les ébauches gonadiques d'embryon de poulet de 5 à 6 jours. *C. R. Acad. Sci. Ser. D* **268,** 1306–1309.

Weniger, J. P., and Zeis, A. (1971). Biosynthèse d'oestrogènes par les ébauches gonadiques de poulet. *Gen. Comp. Endocrinol.* **16,** 391–397.

Weniger, J. P., Ehrhardt, J. P., and Frittig, B. (1967). Sur la formation d'oestrone et d'oestradiol par les gonades de l'embryon de poulet femelle cultivées *in vitro. C. R. Acad. Sci. Ser. D* **264,** 838–841.

Weniger, J. P., Chouraqui, J., and Zeis, A. (1972). Sur l'activité sécrétoire de l'ovaire embryonnaire de mammifère. *Ann. Endocrinol.* **33,** 243–250.

Wilson, J. D., and Siiteri, P. K. (1973). Developmental pattern of testosterone synthesis in the fetal gonad of the rabbit. *Endocrinology* **92,** 1182–1191.

Witschi, E. (1914). Experimentelle Untersuchungen über die Entwicklungsgeschichte der Keimdrüsen von *Rana temporaria. Arch. Mikrosc. Anat.* **85,** 9–113.

Witschi, E. (1921). Der Hermaphroditismus der Frösche und seine Bedeutung für das Geschlechts problem und die Lehre von der inneren Sekretion der Keimdrüsen. *Arch. Entwicklungmech. Org.* **49,** 316–358.

Witschi, E. (1929a). Studies on sex differentiation and sex determination in amphibians. I. Development and sexual differentiation of the gonads of *Rana sylvatica. J. Exp. Zool.* **52,** 235–266.

Witschi, E. (1929b). Studies on sex differentiation and sex determination in amphibians. II. Sex reversal in female tadpoles of *Rana sylvatica* following the application of high temperature. *J. Exp. Zool.* **52,** 267–292.

Witschi, E. (1930). Studies on sex differentiation and sex determination in amphibians. IV. The geographical distribution of the sex races of the European grass frog (*Rana temporia* L.). A contribution to the evolution of sex. *J. Exp. Zool.* **56,** 149–165.

Witschi, E. (1931). Range of the cortex–medulla antagonism in parabiotic twins of *Ranidae* and *Hylidae. J. Exp. Zool.* **58,** 113–145.

Witschi, E. (1933). Studies on sex differentiation and sex determination in amphibians. VII. Sex in two local races of the spotted salamander, *Amblystoma maculatum shaw. J. Exp. Zool.* **65,** 215–241.

Witschi, E. (1935). Origin of asymmetry in the reproductive system of birds. *Am. J. Anat.* **56,** 119–143.

Witschi, E. (1942). Temperature factors in the development and the evolution of sex. *Biol. Symp.* **6,** 51–70.

Witschi, E. (1948). Migration of germ cells of human embryos from yolk sac to the primitive gonadal folds. *Contrib. Embryol. Carnegie Inst.* **32,** 67–80.

Witschi, E. (1956). Amphibia—sex glands. *In* "Development of Vertebrates" (E. Witschi, ed.), pp. 151–163. Saunders, Philadelphia, Pennsylvania.

Witschi, E. (1967). Biochemistry of sex differentiation in vertebrate embryos. *In* "The Biochemistry of Animal Development" (R. Weber, ed.), Vol. 2, pp. 193–225. Academic Press, New York.

Witschi, E. (1971). Mechanisms of sexual differentiation. *In* "Hormones in Development" (M. Hamburgh and E. J. W. Barrington, eds.), Chapter 46, pp. 601–618. Appleton, New York.

Wolff, Em., and Pinot, M. (1961). Stimulation du cortex de la gonade droite de l'embryon d'oiseau. *Arch. Anat. Microsc. Morphol. Exp.* **50,** 487–506.

Wolff, Et. (1949). Les premiers resultats de la castration chez l'embryon d'oiseau. *C. R. Assoc. Anat.* **59,** 703–712.

Wolff, Et. (1952). Sur la différenciation sexuelle des gonades de souris explantées *in vitro. C. R. Acad. Sci.* (*Paris*) **234,** 1712–1714.

Wolff, Et. (1962). "L'origine de la lignée germinale," Semin. Coll. Fr., Hermann, Paris.

Wolff, Et. (1965). Problèmes généraux et problème spécial de la différenciation sexuelle. *Ann. Fac. Sci. Univ. Clermont* **26,** 17–25.

Wolff, Et., and Ginglinger, A. (1935). Sur la production expérimentale d'intersexués par l'injection de folliculine à l'embryon de poulet. *C. R. Acad. Sci.* (*Paris*) **200,** 2118–2121.

Wolff, Et., and Haffen, K. (1952a). Sur l'intersexualité expérimentale des gonades embryonnaires de canard cultivées *in vitro. Arch. Anat. Microsc. Morphol. Exp.* **41,** 184–207.

Wolff, Et., and Haffen, K. (1952b). Sur la différenciation sexuelle des gonades embryonnaires de l'embryon de poulet en culture *in vitro. Ann. Endocrinol.* **13,** 724–731.

Wolff, Et., and Haffen, K. (1952c). Sur une méthode de culture d'organes embryonnaires *in vitro. Tex. Rep. Biol. Med.* **10,** 463–472.

Wolff, Et., and Haffen, K. (1959). La culture in vitro de l'épithelium germinatif isolè des gonades mâles et femelles de l'embryon de canard. *Arch. Anat. Microsc. Morphol. Exp.* **48,** 331–346.

Wolff, Et., and Simon, D. (1955). L'explanation et la parabiose *in vitro* de blastodermes incubés d'embryons de poulet. L'organisation de la circulation extra-embryonnaire. *C. R. Acad. Sci.* (*Paris*) **241,** 1994–1996.

Wolff, Et., and Wolff, Em. (1951). Mise en évidence d'une action féminisante de la gonade chez l'embryon femelle des oiseaux par les expériences d'hémi-castration. *C. R. Soc. Biol.* **145,** 1218–1219.

Wolff, Et., Haffen, K., and Scheib, B. (1966). Sur la détection et le rôle d'hormones sexuelles dans les jeunes gonades embryonnaires des oiseaux. *Ann. Histochim.* **11,** 353–368.

Woods, J. E., and Weeks, R. L. (1969). Ontogenesis of the pituitary gonadal axis in the chick embryo. *Gen. Comp. Endocrinol.* **13,** 242–254.

Woods, J. E. and Podczaski, S. (1974). Androgen synthesis in the gonads of the chick embryo. *Gen. Comp. Endocrinol.,* **24,** 413–423.

Yamamoto, T. (1962). Hormonic factors affecting gonadal sex differentiation in fish. *Gen. Comp. Endocrinol., Suppl.* **1,** 341–345.

Zaaïjer, J. J. P., and Price, D. (1971). Early secretion of androgenic hormones by human fetal gonads and adrenal glands in organ culture and possible implications for sex differentiation. *In* "Hormones in Development" (M. Hamburgh and E. J. W. Barrington, eds.), pp. 537–546. Appleton, New York.

4

Structure of the Mammalian Ovary

R. J. Harrison and Barbara J. Weir

I. INTRODUCTION

Much attention has been paid to the morphology and fine structure of the mammalian ovary since the first edition of this chapter (Harrison, 1962). The electron microscope has provided a refined means for investigating cell types, their activities, and their altered appearances under experimental conditions. In general, the investigations of ultrastructural detail have shown that many types of ovarian cell secrete actively, but have not revealed exactly what is secreted. Experimental effects of administration of hormones have indicated an increasing complexity in ovarian responses and the situation has been clarified only to a limited extent by studies of individual cell types *in vitro*.

This chapter first considers briefly the principal cellular components and general morphology of the mammalian ovary. Then follows a review of the salient characteristics and ultrastructural appearances of ovarian cell types in the mammalian orders. This taxonomic review in the 1962 edition was limited to considerations of the formation and fate of the corpus luteum; in this edition the coverage includes more structural and functional features in a greater number of forms. Much of the historical background and early references of the 1962 edition have been omitted unless they provide the only extant account of ovarian morphology for a particular Order. Further references will be found in reviews by Eckstein and Zuckerman (1956), Brambell (1956), Watzka (1957), Strauss (1966), and Mossman and Duke (1973).

The mammalian ovaries are paired organs, approximately equal in size (apart from functional changes), except in the duckbill platypus (*Ornithorhynchus*), in which the right ovary is atrophied; certain bats (e.g., Vespertilionidae), in which the left one shows atrophy; the mountain viscacha (*Lagidium*), in which only the right ovary is involved in ovulation; and some Odontoceti, especially Delphinidae, in which ovulation occurs more frequently, or consistently, in the left ovary. The ovaries are found within the abdominal or pelvic cavity as far cranial as the lower pole of the kidney (duckbill) and as caudal as the lesser pelvis (man); ectopic ovaries may be present in the inguinal canal. The organs are attached by a mesentery (mesovarium) to the dorsal aspect of the broad ligament, the medial and lateral edges of which may be condensed as suspensory or infundibulopelvic "ligaments."

The ovaries may be enclosed in a peritoneal capsule or ovarian bursa (see Beck, 1972) which varies considerably in development and in the extent to which it communicates with the coelomic cavity (Insectivora, Chiroptera, Carnivora, Rodentia). There is a periovarian space between the ovary and the peritoneal lining of the bursa which may become distended at estrus with fluid of unknown origin (e.g., in carnivores). The periovarian space is closed in species like the weasel, otter, and shrew, but communicates with the peritoneal cavity by a narrow passage, which allows interchange of fluid, in the rat and mouse (Alden, 1942; Wimsatt and Waldo, 1945) as well as in dogs, bears, and several other carnivores (see Kellogg, 1941). A wide capsule, communicating freely with the abdominal cavity, is present in the cat, hyaena, guinea pig, mole, pig, ewe, and rabbit. In catarrhine and lower primates the bursa is of variable depth and wholly or partially encloses the ovary; in catarrhine monkeys it is generally represented by a "peritoneal recess." The bursa is poorly developed in man (see Kellogg, 1941; Eckstein and Zuckerman, 1956) and is variable in its structure in cetaceans (Rankin, 1961). In the mare the bursa is a cleft-like structure which conceals the ovulation fossa; it is probably the result of a closer attachment of the fimbriated end of the Fallopian tube to ovarian tissue than occurs in other mammals. In the dugong (Sirenia) the ovaries are hidden in pouches in the dorsal abdominal wall (Hill, 1945).

The quiescent ovary of most mammals is a small flattened ovoid or spherical organ, compact in structure and with a relatively smooth surface. At the onset of the breeding season, the organ begins to swell owing to the growth of follicles. The mature follicles may protrude at the surface or may just be visible as translucent bodies within the substance of the ovary. After ovulation the follicles become converted into corpora lutea which are usually, but not always, pigmented in a range of colors from pink through yellow and orange to brown, according to age. In animals in which multiple ovulation is the rule (e.g., sow, rat, rabbit) the pedunculate follicles and corpora lutea give the ovary a grape-like appearance; in others, such as the horse, sheep, and primates, the regular outline of the ovary is only locally disturbed. In cetaceans the ovary may present a scarred appearance (Fig. 9) because of corpora albicantia (Harrison *et al.*, 1972). In the plains viscacha (*Lagostomus*) the ovary is leaf-like at all stages of the cycle (Weir, 1971c; Weir and Rowlands, 1974).

The mammalian ovary is covered by a "germinal" or covering epithelium beneath which lies a connective tissue tunica albuginea of varying thickness. The ovary contains the follicular apparatus, corpora lutea of various ages, other luteal bodies, stromal and connective tissue, interstitial tissue to a varying degree, vascular, nervous, and lymphatic tissues, and certain embryological remnants such as rete ovarii and tubules of the epoöphoron may or may not be encountered.

II. THE "GERMINAL" OR SURFACE EPITHELIUM

Mammalian ovaries are covered by a continuous sheet, usually single layered, of cuboidal or low columnar epithelium (Fig. 1) which can phagocytose dye-stuff and India ink particles (see Harrison, 1962). This surface covering, or "germinal" epithelium is supported on a distinct basement membrane and invests the contours of the ovary whether smooth or lobulated. It is only disrupted when follicles ovulate; repair occurs over the stigmata within 2 to 4 days. Mitotic figures are alleged to be seen in the epithelial cells at estrus and may be due to a direct effect of estrogen on the cells (Bullough, 1946) or to proliferation in response to the increased ovarian volume (Zuckerman, 1951).

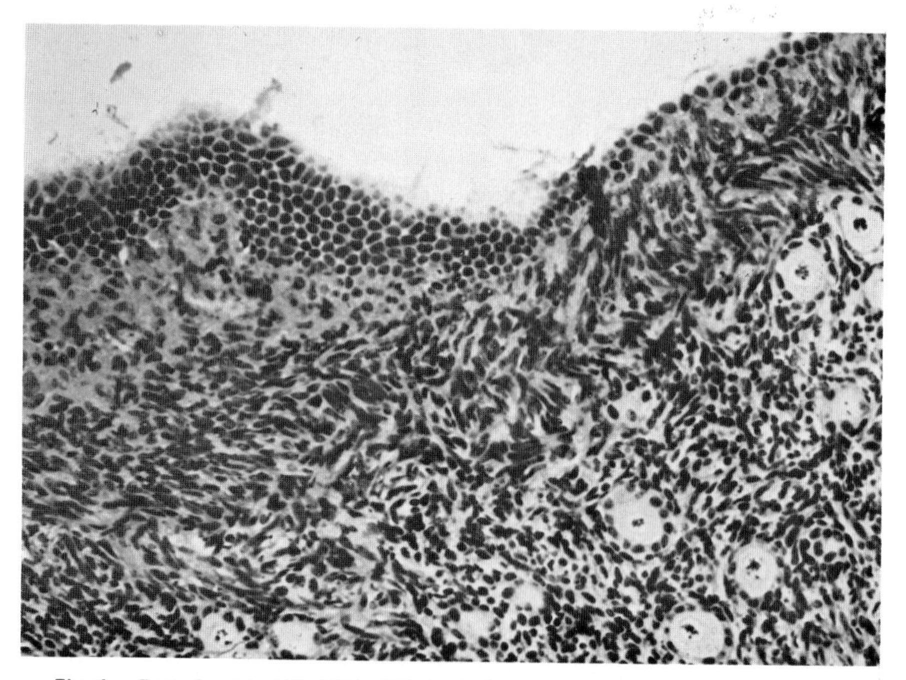

Fig. 1. Part of a chinchilla (*Chinchilla laniger*) ovary showing primary oocytes and the germinal epithelium which varies from one-cell-layer to several-cell-layers thick. (Bouin, hemalum, and eosin; ×200.)

The part played by the surface epithelium in ovarian development is discussed elsewhere (Volume I, Chapter 2). It is now generally agreed that in the adult mammalian ovary the surface epithelium is not a source of new oocytes and is not a store of oocytes disguised as epithelial cells. Ultrastructural (Wischnitzer, 1965; Weakley, 1967) and histochemical (O'Shea, 1966) studies also suggest that it is unlikely that the adult surface epithelium or associated subsurface structures (see later) contribute to the follicle envelope.

The surface epithelium is often invaginated into the subjacent tunica albuginea to form small folds, pits, or subsurface crypts (Harrison and Matthews, 1951). These formations are particularly developed in Pinnipedia (Harrison *et al.*, 1952; Laws, 1956; Craig, 1964) and Proboscidea (Perry, 1953). Variation in the number and extent of the subsurface crypts is not obviously associated with the estrous cycle. In some carnivores the crypts lead to islands of granulosa-like cells, resembling anovular follicles, lying deep to the tunica albuginea (O'Shea, 1966). Neal and Harrison (1958) describe an extensive intraovarian system of tubules in the badger (*Meles*) that appears to be continuous with well-developed surface invaginations lined by covering epithelium. The relationship of subsurface epithelial structures to other ovarian elements such as secondary medullary cords is still obscure (Guraya and Greenwald, 1964a). A mucin secretion has been demonstrated in these structures but its functional significance is unknown (O'Shea, 1966). Mossman (1938) has pointed out that follicular epithelium is to be regarded as modified coelomic epithelium and follicular antra as isolated portions of peritoneal cavity filled with somewhat modified peritoneal fluid. The intraovarian crypts may represent a vestige of a primitive pattern of ovarian structure (Harrison and Matthews, 1951), or else their development may reflect a natural propensity of the surface epithelium to contribute epithelial elements to the cortical region that, in some forms, is continued into adult life. The ovary of *Lagostomus* (Fig. 5) is not invaginated by crypts; the indentations are true clefts because the tunica albuginea is continuous under the extensive covering epithelium (Weir, 1971c).

In the horse, the originally complete covering of germinal epithelium becomes modified well before birth and in the mature mare is restricted to a narrow groove known as the ovulation fossa, where follicles rupture; the remainder of the surface is covered by serosa (Küpfer, 1928; Hammond and Wodzicki, 1941; O'Shea, 1968). An ovulation fossa is also formed in the caviid *Galea musteloides* (Weir and Rowlands, 1974).

III. THE FOLLICLE

Primordial follicles consist of a single layer of flattened epithelial cells surrounding each oocyte (Figs. 1, 5, and 12). They lack a connective tissue or thecal investment. They may be as small as 16.5 μm in diameter (in mice) but are usually somewhat larger. They may form a pronounced peripheral zone throughout the cortex, as in the rabbit and cat, or be more widely distributed.

A primordial follicle may contain more than one oocyte. Such polyovular follicles are more common in fetal and immature ovaries and have been described in a wide variety of mammals, including man [see Brambell (1956) for references, and Bocharov (1966) for a useful table]. They are frequently found in the opossum (*Didelphis*) in which 100 oocytes were found in a single follicle. They could result either from division of an original single oocyte or from coalescence of adjacent primordial follicles.

More probably, however, they result from failure of groups of germ cells to become separated by epithelial and connective elements. There are few examples of polyovular follicles reaching maturity and rupturing (Allen *et al.*, 1947, but see Section VIII,B); the majority undergo atresia or all but one of the oocytes die. Kent (1959, 1962) reported a reduction in the incidence of polyovular follicles and polynuclear ova to 17–18% of the control values after treatment of young golden hamster females with estradiol monobenzoate. He suggested that the results were due to alleviation of an estrogen deficiency in the young animal (Kent and Mandel, 1968). Graham and Bradley (1971) found 30% polyovular follicles in squirrel monkeys treated with diethylstilbestrol. Correct maturation of the oocyte is believed to be dependent upon the surrounding follicle cells (Ohno and Smith, 1964). Donahue and Stern (1968) find that the oocyte *in vitro* will only form pyruvate if follicular cells are also present.

As they increase in size, the follicles gradually sink deeper into the cortex of the ovary. The single layer of flattened cells enveloping the oocyte increases in thickness and its cells become cuboidal or columnar to form a distinct membrana granulosa. The layer rapidly becomes several cells thick. The enlargement of the granulosa layer is soon accompanied by the development of an outer encapsulating sheath derived from the stroma. This constitutes the theca, later to become divided into a glandular, well-vascularized, inner layer, the theca interna, and an outer one, or theca externa, composed of connective tissue (Fig. 2). The latter may contain smooth muscle fibres (O'Shea, 1970, 1971a; Burden, 1972a) although Claesson (1947) found none in a number of mammalian types, and elastic fibres are usually also absent. Numerous blood vessels and lymphatics penetrate the theca externa to communicate with a fine plexus of vessels in the theca interna. Fink and Schofield (1971), in ultrastructural studies in the cat, have found partly ensheathed axons in the perifollicular region of primary follicles close to the membrana propria (Figs. 2, 5, and 12) which circumscribes the granulosa layer. Light microscope preparations suggested that some nerve fibers penetrated the granulosa. The granulosa layer is avascular until after ovulation. A thick membrana propria separates it from the theca interna, which was previously considered to be the source of the membrane. Light microscope studies suggested that the membrana propria was two layered but in general this has not been confirmed by electron microscopy. Hope (1965) and Hertig and Adams (1967) have described ultrastructural characteristics of developing follicles.

The follicle enlarges as a result of proliferation of granulosa and thecal layers and, in the majority of mammals, one or more cavities soon form in the granulosa. The cavities enlarge, coalesce into an antrum, and the fluid-filled follicle becomes surrounded by a remarkably uniform wall, except at the point of attachment of the oocyte. Here the oocyte is surrounded by an irregular cluster of granulosa cells destined to become the corona radiata, the whole conglomerate being attached to cells forming the discus proligerus or cumulus oophorus. Experimental puncture of follicles (Harrison, 1946) has allowed investigation of early changes in luteinization. Using this technique and that of ovectomy, El-Fouly *et al.* (1970) consider that the oocyte directly prevents luteinization of the granulosa cells. Techniques have also been developed to harvest granulosa and theca interna cells from follicles at different stages of the cycle in order to culture them *in vitro* (Harrison, 1948a) and for assessment of the steroidogenic potential (Channing, 1969, 1970a,b,c).

The general account given by Harrison (1962) contains details of the early work on follicular characteristics in mammalian ovaries. The follicles of monotremes contain a fluid (called proalbumin) around the oocyte. The amount of fluid is small because the large oocyte of monotremes almost fills the follicle; in *Echidna* and *Ornithorhynchus* the fluid completely surrounds the oocyte. The formation of an antrum in the bat *Myotis* is foreshadowed by accumulation of a watery secretion within and between the granulosa cells. There is pronounced hypertrophy of the granulosa and a reduction in intrafollicular tension which results in disappearance of the antrum just before ovulation. A unique specialization has been described in the follicles of hibernating vespertilionid bats (Figs. 3 and 4); marked hypertrophy and vesiculation is present in the granulosa cells of the discus proligerus and enormous quantities of glycogen accumulate (Wimsatt and Kallen, 1957). This may provide a source of energy to facilitate the long survival of the oocyte during hibernation with reduced body metabolism. Ripe follicles of *Corynorhinus rafinesquei* differ from those of most other mammals in the relative smallness of the antrum because of the enormous "balls" of cumulus cells surrounding the oocyte (Pearson *et al.*, 1952). The authors suggest that "secretions" of the follicle cells are retained and not released into the antral fluid. There is no true antrum in the follicles of certain insectivores such as *Setifer* and *Hemicentetes*. The granulosa cells swell and prevent formation of an antrum; later they become loosened to form a sponge-like structure.

A. Membrana Granulosa

Early in follicular development the immature oocytes are in contact with the granulosa cells. Later a jellylike substance containing polysaccharides

(Braden, 1952) appears between the plasma membrane of the oocyte and the granulosa cells. This jelly eventually constitutes the zona pellucida. Sotelo and Porter (1959) and Franchi (1960) have described long processes passing from the granulosa cells of rat follicles through the zona pellucida to mingle with microvillous projections from the oocytes. The two sets of processes were observed to be in contact in places, but no evidence of fusion, as claimed by Moricard (1958), has been seen. Both types of processes are withdrawn from the zona during maturation and fertilization. Trujillo-Cenoz and Sotelo (1959) and others have made similar observations and presented evidence that the amorphous material of the zona pellucida is elaborated in the cytoplasm of follicle cells and later extruded to the zonal layer. Chiquoine (1960) has suggested that the zona material is elaborated in the form of a monomer by the follicle cells and that it is presumably the presence of the oocyte which is responsible for polymerization of the

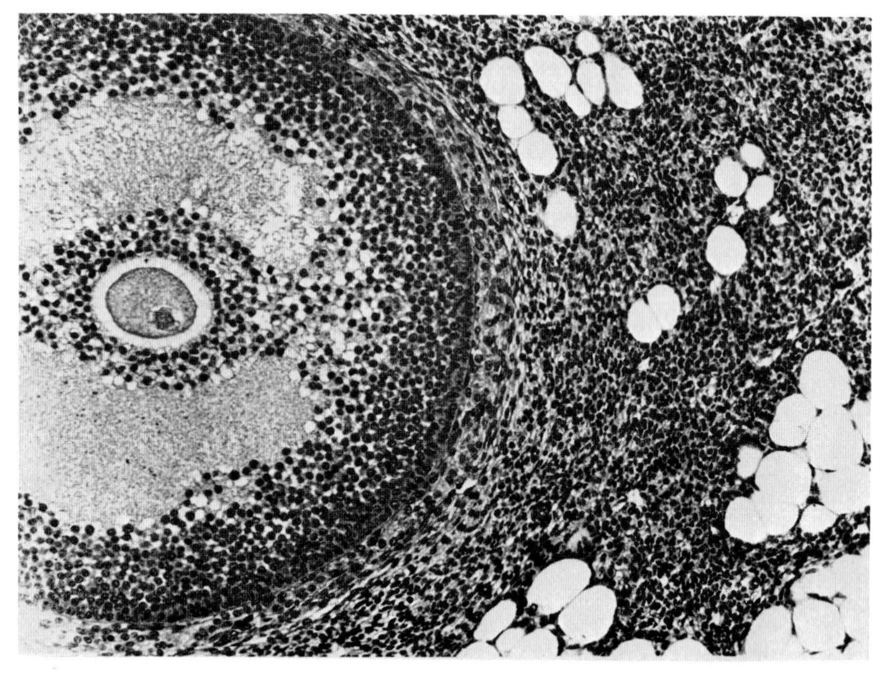

Fig. 2. Part of the ovary of a cuis (*Galea musteloides*). The maturing follicle on the left shows early cumulus formation, a distinct membrana propria and thecae interna and externa. The large vacuoles in the stroma represent an invasive condition known as lipophanerosis that is typical in the ovary of this rodent. (Bouin, hemalum, and eosin; ×150.)

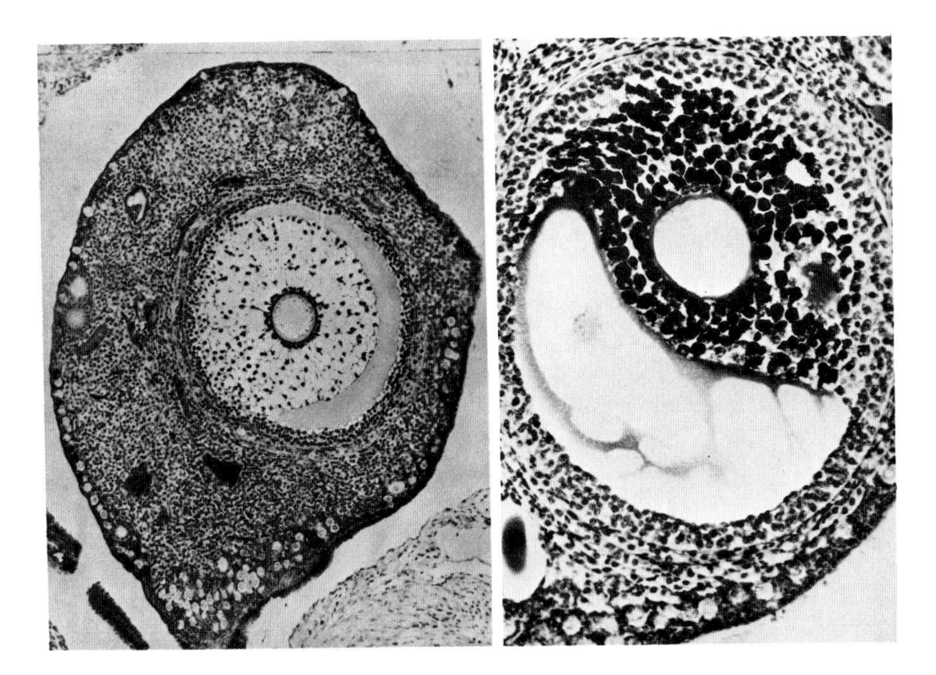

Fig. 3. (left) Section of the ovary containing the single ripe winter follicle of the bat, *Myotis lucifugus*. The glycogen in the hypertrophied granulosa cells has been removed during routine fixation. (Helly, eosin, and methylene blue; ×65.) (Reproduced by permission from Wimsatt and Kallen, 1957.)

Fig. 4. (right) The ripe follicle of the hibernating bat (*Myotis lucifugus*) showing the accumulation of glycogen in the inner granulosa cells. (PAS and hematoxylin; ×130.) (Reproduced by permission of W. A. Wimsatt.)

monomer. Kang (1974) gives a good survey of the literature on the origin of the zona and believes it to be produced by the oocyte in the rat. Björkman (1962), Odor (1965), Hope (1965), Weakley (1967), Stegner (1967), Hertig and Adams (1967), Gondos (1970), and other authors quoted in later sections have given descriptions of the ultrastructure of the follicular cell–germ cell relationship.

The granulosa cells of multilayered follicles are small, polygonal, or cuboidal, with granular cytoplasm and densely staining nuclei (Figs. 2, 5, 12, 13, and 14). Cilia have been observed in the granulosa cells of several species (Motta *et al.*, 1971). In the rabbit, Lipner and Cross (1968) state that all the granulosa cells project from the membrana propria; thus, the arrangement is like that of a pseudostratified epithelium. It is unlikely that this is true for all, or even any, other species.

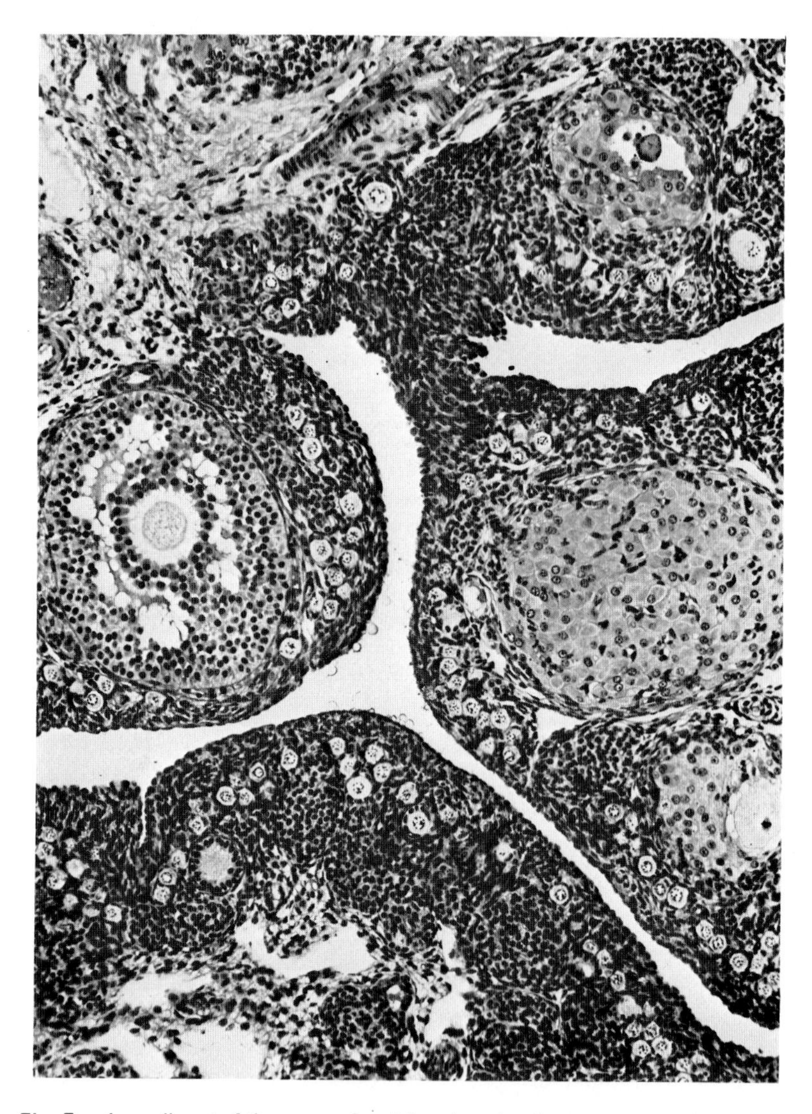

Fig. 5. A small part of the ovary of a plains viscacha, *Lagostomus maximus.* Note the extensive surface area, numerous oocytes, a mature follicle with a small antrum (left center), a mature true corpus luteum (right center) and two small accessory corpora lutea. (Bouin, hemalum, and eosin; ×135.)

Marked folding occurs in the granulosa layer of the mature follicles of most carnivores. Each complicated fold of mural granulosa projects into the antrum and contains delicate connective tissue and blood vessels from the thecal plexus. Accessory granulosa cells with long processes that make contact with the zona pellucida have been described in the follicular wall in kittens (Wotton and Village, 1951). Growth changes and cytological characteristics of the granulosa of rats have been described by Deane (1952) and in hamsters by Knigge and Leathem (1956), who state that modified granulosa cells of preantral follicles are the source of interstitital tissue in the hamster ovary.

The cluster of granulosa cells that surrounds the oocyte shows evidence of dispersion at the periphery as the follicle nears maturity and some cells may show signs of degeneration. The inner cells become columnar, radially arranged with their nuclei situated peripherally and form the corona radiata which is present after ovulation until fertilization. A corona is not formed in some marsupials and there is no corona at ovulation in *Setifer* and *Elephantulus* (van der Horst and Gillman, 1942). The granulosa cumulus cells at the site of attachment of the oocyte to the follicle wall may form a broad based mound or a slender stalk. The position of the cumulus on the follicle wall is probably fortuitous but some authors consider that just before ovulation it lies in a specific position relative to the impending site of rupture. We do not agree that the cumulus moves round the follicle wall as ovulation approaches as a result of some intrinsic movement in the granulosa layer. Numerous recent references to granulosa cells, especially their ultrastructure (e.g., Bjersing, 1967c; Dahl, 1971b) are quoted below (see Section VIII). The antrum appears after the first phase of follicular growth has ended; the follicle is then less than 400 μm in large mammals and about 150 μm to 200 μm in small mammals.

The antrum is sometimes lined at its periphery by an amorphous internal limiting membrane that has been considered to have arisen either as a condensation of the antral, or follicular, fluid containing cellular debris, or from the inner granulosa cells, or even from invading theca interna cells (Brambell, 1956). Mitotic figures occur frequently in the granulosa cells during follicular growth and particularly when the antrum develops (Bullough, 1942). Lane and Davis (1939) have measured the changes in volume of the granulosa and the occurrence of mitosis therein during the estrous cycle of the rat.

Small cavities, first described by Call and Exner in 1875 in rabbit follicles, are sometimes seen in the granulosa before or after antrum formation in the follicles of man, dogs, cats, whales, several hystricomorph rodents, and some other species (Motta, 1967; Motta and Nesci, 1969). Call–Exner bodies have a diameter of up to 50 μm and contain a reticulated coagulum; sometimes a fine membrane surrounds the contents and the granulosa cells are radially arranged outside it. These bodies may either be formed from degeneration and "liquefaction of a central cell or cells, or they form as the result of secretory activity of the surrounding cells" (Brambell, 1956).

The follicular fluid (or liquor folliculi) is not of consistent appearance throughout follicular growth. Light microscope investigations suggest that a primary liquor is formed by the granulosa cells to give rise to the antrum. It is secreted between the cells, exudes into the coalescing central cavities, and exhibits a marked reticulum. Secondary liquor is probably less viscous and is secreted rapidly just before ovulation, possibly in the region of the cumulus. Follicular growth at this time is very rapid and Burr and Davis (1951) consider that secondary liquor is a transudate from blood. Tertiary liquor is formed after ovulation and plugs the cavity of the ruptured follicle. Umbaugh (1949) states that immediately after ovulation in the cow there is first an oozing of serous fluid and then a discharge of viscous, mucoid material. Follicular fluid possesses slight thromboplastic activity (von Kaulla and Shettles, 1956), and the follicular proteins have been found to be similar to those of the blood plasma (Caravaglios and Cilotti, 1957). In the pig, there are ten different antigens in the follicle and seven of these are the same as those in the blood (Shivers *et al.*, 1964). Radioactive sulfur, as sulfate, is incorporated into the follicular mucopolysaccharide of the rabbit ovary and is transferred from granulosa cells to the follicular fluid (Zachariae, 1957).

Anovular follicles, apparently normal in structure other than lack of an oocyte, have been described in many mammals (Harrison and Matthews, 1951; Brambell, 1956) and have been produced experimentally by means of X-rays. These structures do not reach maturity and seldom develop an antrum. They may have epithelial connections with the covering epithelium, and Brambell (1956) considers that they may well be homologous with medullary cords such as those described by Wilcox and Mossman (1945) in ovaries of adult shrews.

B. Theca Interna

The theca differentiates around the outside of the granulosa from the surrounding stromal tissue; it is said to be induced by the growing granulosa (Dubreuil, 1957). The theca first forms a concentric sheath of compressed cells; later, at the time of antrum formation, it differentiates into theca interna and theca externa (Figs. 2, 12, and 13). The theca interna cells rapidly enlarge and assume a polygonal shape with vacuolated cytoplasm and vesicular nuclei. They are enmeshed in a reticular and fibrous network and the layer contains a plexus of capillaries and numerous lymphatics. The degree of development of the theca interna varies in the different placental mammals, but it is often maximal just before ovulation. This is surprising because as the follicle grows the thecal covering must extend over a larger area and yet mitotic figures are rarely seen. Mossman (1937) considered that the hypertrophied theca interna in *Geomys* near ovulation resembled an endocrine gland and called it thecal gland. Hypertrophy of the theca interna during estrus (Fig. 12) has been described in other mammals (see Harrison, 1962), and the layer is now considered a site of steroid formation. Nishizuka (1954) states that the cells produce estrogens, and McKay and Robinson (1947) describe the characteristics of theca interna cells in human ovaries. White *et al.* (1951) distinguished a second type of cell in mature follicles which has a small, dense, irregular, hyperchromatic nucleus and strikingly eosinophil cytoplasm; they called it a "K" cell. Electron microscopy clearly indicates steroidogenic activity in theca interna cells; recent findings are discussed in detail below (see Section VIII,C,8, 13, and 14).

The theca interna is not always of equal thickness all round the follicle. It may show a hypertrophied region in the form of a thecal cone projecting toward the ovarian surface and this cone may play a part as a "pathmaker" for the ascent of the growing follicle to the ovarian surface (Strassmann, 1941; Harrison, 1948a). The theca interna also thins or disappears at the site of the impending rupture in mature follicles and may also be hypertrophied in the region of the cumulus (Brambell, 1956). The permeability of thecal capillaries increases before ovulation in mated rabbits (Christiansen *et al.*, 1958).

The mature follicle of the marsupial *Dasyurus* and of the rodents *Lagostomus* and *Dasyprocta* appears to lack a theca interna, but a few cells may be present in small follicles. There is no differentiation of the theca into a cellular inner layer and a fibrous outer layer in the bats *Myotis* and *Eptesicus* (Wimsatt, 1944). The entire theca is composed of fusiform cells but includes some larger ones that are indistinguishable from interstitial cells. Culiner (1946) has considered the role of the thecal cells in irregularities of the baboon menstrual cycle, and Shippel (1950) has reviewed the functional relationships between thecal and other ovarian cell types.

C. Growth

The growth of the follicle is divided into two phases relative to that of the oocyte (Brambell, 1956). During the first phase the oocyte grows rapidly until almost adult size, while the follicle increases only slowly in size to a diameter of 110 μm (lesser shrew) to 300 μm (pig). The size of the oocyte in the adult bears no relationship to the body size and is remarkably uniform (60 to 120 μm) throughout the placental mammals. The size of the follicle at the end of the first phase displays only a slight tendency to increase with body size in different mammals. The follicle is a solid sphere or ovoid during the first phase; the theca starts to develop and the zona pellucida forms rapidly.

The follicle grows rapidly in size during the second phase while the oocyte increases only slowly in diameter. If y = the mean diameter of the oocyte and x = that of the follicle, then the two phases of growth can be expressed by the linear regression $y = a + bx$, where a and b are constants. During the second phase the follicle develops an antrum and the theca interna enlarges. The ultimate size of the follicle at the end of the second phase is directly related to the body size (Parkes, 1931) and the major part of this follicular growth is associated with enlargement of the follicular antrum. Harrison (1962) comments that the above concepts hold good for all mammals so far examined from the mouse to the whale, except for the Centetidae in which no antrum is formed. But the mature follicle of the 3 to 4 kg rodent *Lagostomus* is only 300 μm in diameter (Weir, 1971c; Fig. 5). This appears to be an adjustment for the 200 to 800 eggs that are ovulated by this species at each estrus. The number of follicles ovulating at maturity differs widely in different mammals. The number that rupture is not apparently related to body size. Bats, women, and elephants are usually monovular; rodents, insectivores, some large carnivores, and ungulates are polyovular. There is, however, a tendency for each species to ovulate approximately the same number of follicles at each estrus. This holds true even if one, or part of one, ovary is removed, and is known as the law of follicular constancy (Lipschütz, 1928; see also Volume I, Chapter 6). There is a tendency for the number of follicles that rupture at estrus within a particular species to be proportional to body weight (Brambell and Rowlands, 1936; Brambell, 1944).

IV. THE CORPUS LUTEUM

The corpus luteum is the endocrine gland which normally develops from the cellular components of the ovarian follicle after ovulation. In certain circumstances, follicles that do not rupture may become transformed into luteal-like masses which still may retain the oocyte. In this chapter the word luteal is used to refer to any structure or cell which pertains to the corpus luteum, such as granulosa luteal cells (also called granulosa lutein cells by many authors). The term luteinization is used to describe the several changes in appearance which occur during the transformation of a follicular granulosa cell into a luteal cell. Mossman *et al.* (1964) have taken to task those authors who use luteinization "to refer to any enlargement or hypertrophy of cells associated with the ovary, regardless of whether or not there is the slightest evidence of the relations of the hypertrophied cells to the luteal gland." There is also a danger that luteinization may become a synonym for steroidogenesis. This usage may be inappropriate but it is feared that it is now too late to dissuade authors from using the term descriptively for similar changes of appearance that may occur in thecal cells and sometimes in interstitial cells.

Several types of corpora lutea have been described and given names based on morphological or functional criteria. A corpus luteum of the cycle (Figs. 5, 6, 8, 10), i.e., in the nonpregnant female, has been termed a false corpus luteum or a corpus luteum of menstruation (in man). A corpus luteum of pregnancy has been called a true corpus luteum or a corpus luteum graviditatis. A secondary corpus luteum is one which develops from a follicle ovulating in pregnancy (e.g., in the mare). A corpus hemorrhagicum (Fig. 7) develops when hemorrhage occurs from the thecal vessels shortly after ovulation and results in a corpus luteum with a central blood clot. Corpora lutea of pseudopregnancy are found in many species and have been distinguished by their smaller size and lack of vascularization. Somewhat similar changes have been noted in corpora lutea associated with prolonged periods of delayed implantation. Accessory corpora lutea (accessoria) develop by luteinization of unruptured follicles and have been described in many hystricomorph rodents (Fig. 6), mares, elephants, cetaceans, and some primates. When a retained ovum cannot be found, it is suggested that the resulting corpus be termed a "luteal body" rather than an accessory or secondary corpus luteum. A corpus atreticum is formed from luteinized thecal cells in a follicle where the granulosa degenerates. During regression, glandular elements are lost and the corpus luteum becomes a corpus albicans (Fig. 8). Corpora aberrantia are derived from corpora lutea of the cycle that reach full maturity and, instead of gradually degenerating, endure for a period.

A mature corpus luteum may be contained within the confines of the ovary, or partly or completely everted onto the ovarian surface (macaque, fox, some bats), or extruded in a horseshoe shape (elephant shrew), or become pedunculated (tenrec, elephant shrew, dolphin) to various degrees (Harrison, 1948a; Nalbandov, 1970). The corpus luteum may be hollow or solid in its early development, may have a simply folded wall, or exhibit more complex folding: such arrangements may be reflected later in the presence or absence of a central cavity and in the degree of lobulation or septation which is manifest.

The different views held in the past about the histogenesis of the corpus luteum have been reviewed several times and are today only of historical interest. It is generally accepted that in eutherian mammals the granulosa cells become transformed into the luteal cells of the corpus luteum. The fate of the theca interna cells is less clear, and there appear to be species differences as to whether they are incorporated as secretory elements in the corpus luteum or whether they undergo some other fate. The belief that theca interna cells luteinize and become incorporated into the corpus

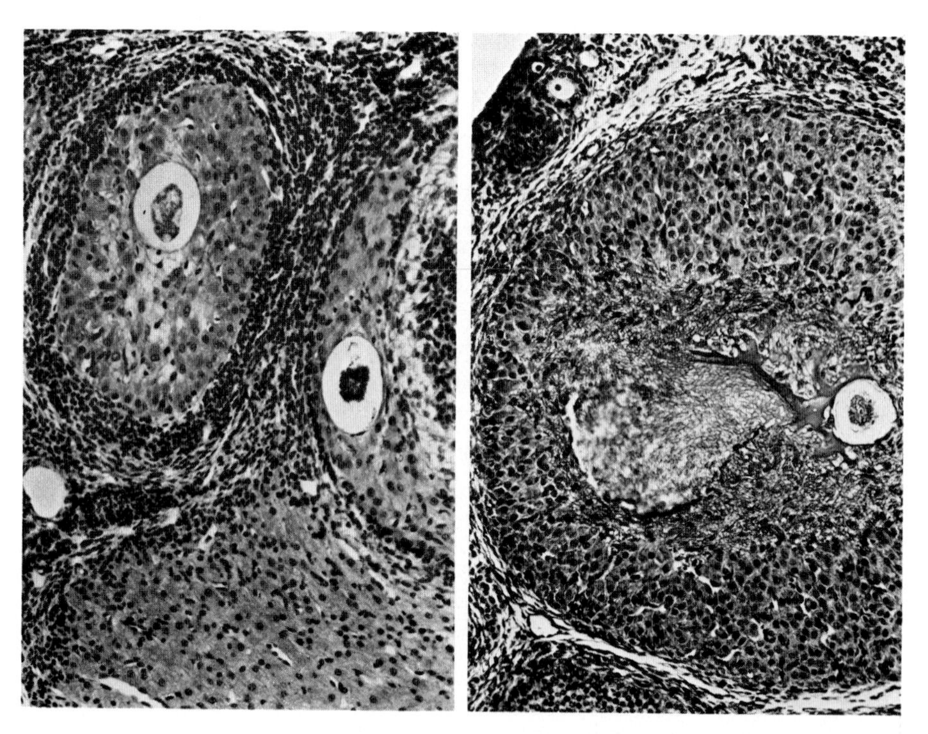

Fig. 6. (left) Corpora lutea in the ovary of an acouchi (*Myoprocta pratti*). Part of the true corpus luteum of pregnancy can be seen and two accessory corpora lutea are identified by their size and the retained ovum. (Bouin, hematoxylin, and eosin; ×150.)

Fig. 7. (right) A corpus hemorrhagicum accessorium of a 2-week-old degu (*Octodon degus*) identified by the persistent ovum and the extravasated blood. (Bouin, hemalum, and eosin; ×120.)

luteum derives from many studies which show that two cell populations, usually small and dark or large and light, can be recognized in the corpus luteum. Whether or not this is the case, there is little doubt that the theca interna cells have important functions before, during, and just after ovulation.

Ovulation in most mammals is associated with collapse of the follicle and disruption of the basement membrane, the membrana propria. Preovulatory changes occur in some species; many rodents and insectivores show early luteinization of the granulosa cells, and in *Elephantulus myurus* and the Centetidae the granulosa cells swell to fill the antrum (van der Horst and Gillman, 1940; Strauss, 1938, 1939).

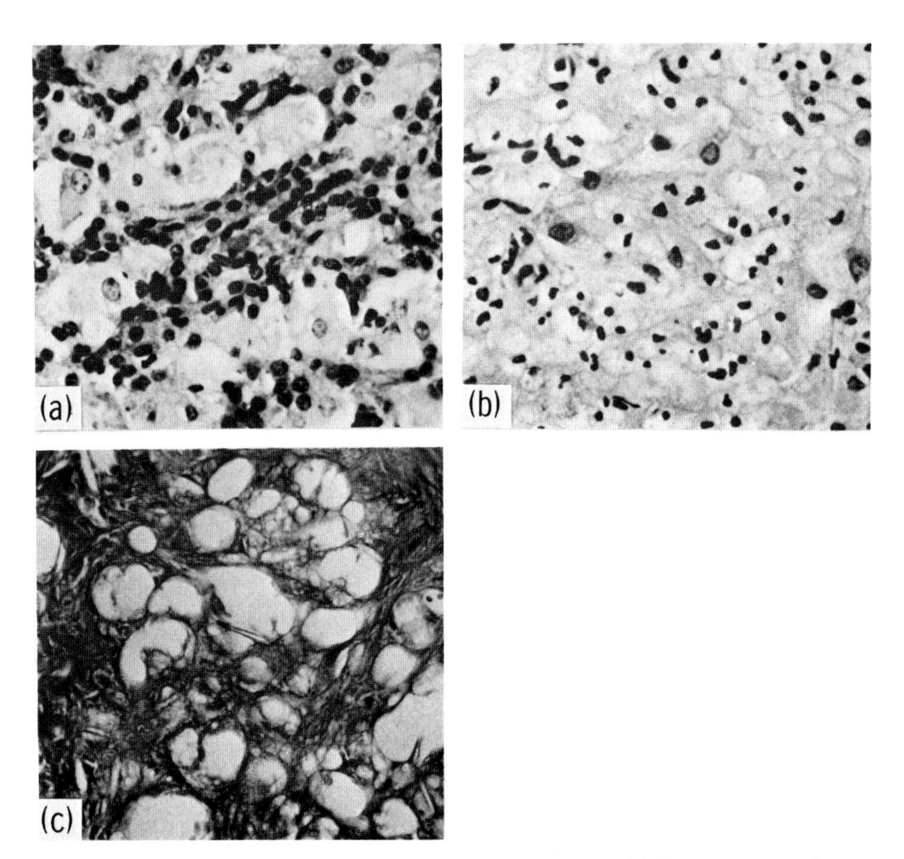

Fig. 8. Stages in the development of corpora albicantia: (a) A recent corpus luteum of a muntjac deer (*Muntiacus* sp.) (×300); (b) an older corpus luteum of the same deer (×300); (c) a corpus albicans in a casiragua (*Proechimys guairae*) (×200.)

After ovulation there may be a redistension of the ruptured follicle by continued secretion of fluid from the follicle wall (see Brambell, 1956, for references). Extravasation of blood from thecal capillaries may occur frequently (cow, guinea pig), occasionally (man), or rarely (most other mammals), and leucocytes may be present. The changes which occur in the granulosa cells converting them into luteal cells are collectively described as those of luteinization. The transformation is exhibited in the great majority of mammals but all the features are not always apparent (e.g., hedgehog: Deanesly, 1934). The characteristics of luteinization should include hypertrophy and/or hyperplasia of the granulosa cells, the appearance of lipid droplets, and increase in nuclear size, in the cytoplasmic–nuclear ratios, in eosinophilia, and in granularity of the cytoplasm. Luteinization has been studied in cultured granulosa cells of several species (Channing, 1969, 1970a,b). Channing (1969) argues that whenever the basement membrane of the ovarian follicle remains anatomically and functionally intact, the granulosa cells do not usually luteinize. When the "barrier" of the membrane is broken, as in ovulation, artificial rupture (Harrison, 1946), and when granulosa cells are removed and maintained *in vitro,* in the anterior chamber of the eye (Falck, 1959) or under the kidney capsule (Ellsworth and Armstrong, 1971), luteinization then follows.

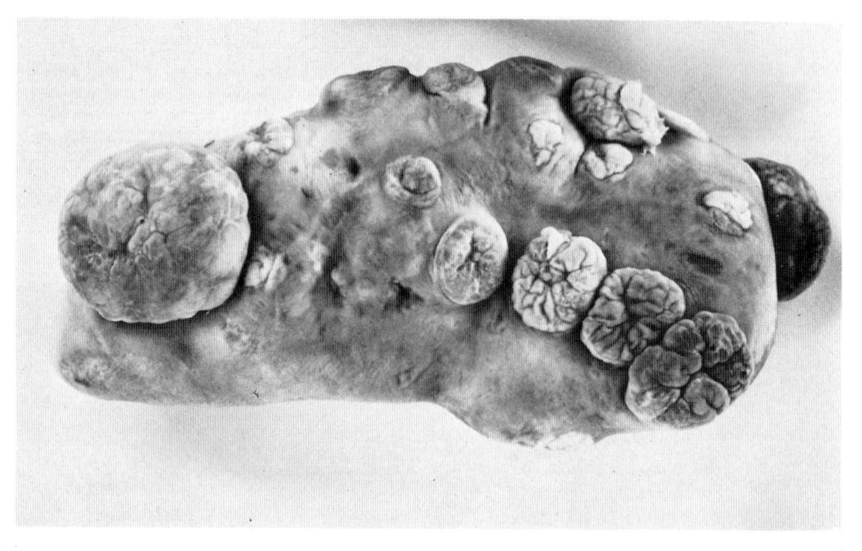

Fig. 9. The left ovary of a 236-cm Pacific white-sided dolphin (*Lagenorhynchus obliquidens*) which was lactating. Some of the twenty-four corpora albicantia in this ovary can be seen as excrescences on the surface; there was none in the right ovary. (×2.) (From Harrison *et al.,* 1972.)

The induction of luteinization by gonadotropin may be mediated by an increase of cyclic 3′,5′-AMP (Channing and Seymour, 1970). El-Fouly *et al.* (1970) have punctured and ovectomized mature follicles and shown that oocytes of intact mature rabbit follicles may exert an inhibiting effect on the formation of corpora lutea. A similar suggestion is made by Nekola and Nalbandov (1971) who cultured rat ova in association with granulosa cells and found that only the follicle cells farthest from the ovum luteinized. An inhibitory influence of the ovum may not be exercised in those species which frequently form accessory corpora lutea, and many authors have asserted that there is no detectable change in the oocyte before alteration in the granulosa cells is apparent.

During luteinization there is a significant increase in cell size. Cisternae of granular endoplasmic reticulum are gradually replaced by increasing quantities of tubular smooth-surfaced reticulum; the mitochondria become

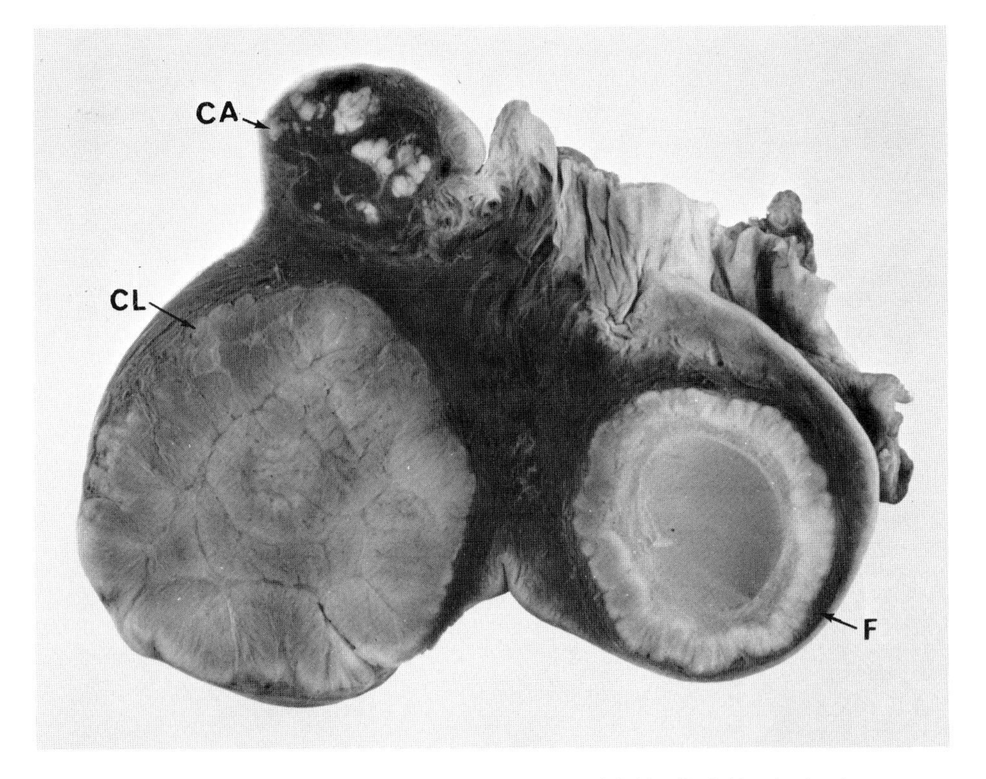

Fig. 10. Slice through the ovary of a common dolphin (*Delphinus*) showing the recent corpus luteum (CL), a corpus albicans (CA) and a luteinized follicle (F). (×2.) (From Harrison *et al.*, 1972.)

more numerous and develop lamellar cristae; the Golgi complex becomes enlarged and more widely dispersed; numerous lipid droplets also accumulate (Enders, 1962; Christensen and Gillim, 1969; and Section VIII for studies on particular species). The most extensive work on the fine structure of corpora lutea (Fig. 11) has been reported for man; and the corpus luteum of menstruation has been compared with that of pregnancy. Further descriptions of luteal-cell ultrastructure are given in Section VIII,14.

The final size of the corpus luteum is related to the diameter of the mature follicle from which it arose, but there is variation from near equality to a twofold increase over that of the follicle. Brambell (1956) gives a table showing the diameters of mature follicles and fully developed corpora lutea in thirteen mammalian species. Rossman (1942) had described pigment in the corpus luteum of rhesus monkeys. The yellowish color of corpora lutea in certain species is due to the presence in luteal cells of a carotenoid, lutein, similar to that of egg yolk. The bovine corpus luteum exhibits color changes from pale cream, yellow or orange to red or brown. This pigmentation should be distinguished from that resulting from hemorrhage and the breakdown products of hemoglobin. The lutein pigment appears to remain in the regressing corpus luteum in the cow and some other species and the term corpus rubrum has been used to describe such pigmented structures. Regressing corpora lutea and luteinized follicles in cetaceans and other species are often bright yellow or orange due to lutein granules. Corpora albicantia of most mammals eventually lose any evidence of pigment and appear as white or gray scars.

Details of the life span of corpora lutea of each type in different species, as well as some indication of what is known about regression, are given in Sections VIII,C,1–14.

Dubreuil and Rivière (1947) list four main methods by which a corpus luteum may degenerate, namely, a fibrohyalin method, lipoid degeneration, and rapid and slow necrobiosis. The authors are of the opinion that the only elements which take part in histiolytic degeneration are the epithelial cells, the remaining elements are said to dedifferentiate to the type of cell found in the ovarian stroma. Brambell (1956) concluded that the processes of regression are essentially similar in corpora lutea of ovulation, pseudopregnancy, pregnancy, or lactation. He also gives references to workers who studied changes during regression. Corner (1942) has studied the different types of regressive change in rhesus monkeys, one considered to be normal, the other giving rise to corpora aberrantia. At or near the menopause the stimulus for corpus albicans replacement by fibroblasts decreases and eventually ceases. The last few corpora albicantia formed during the

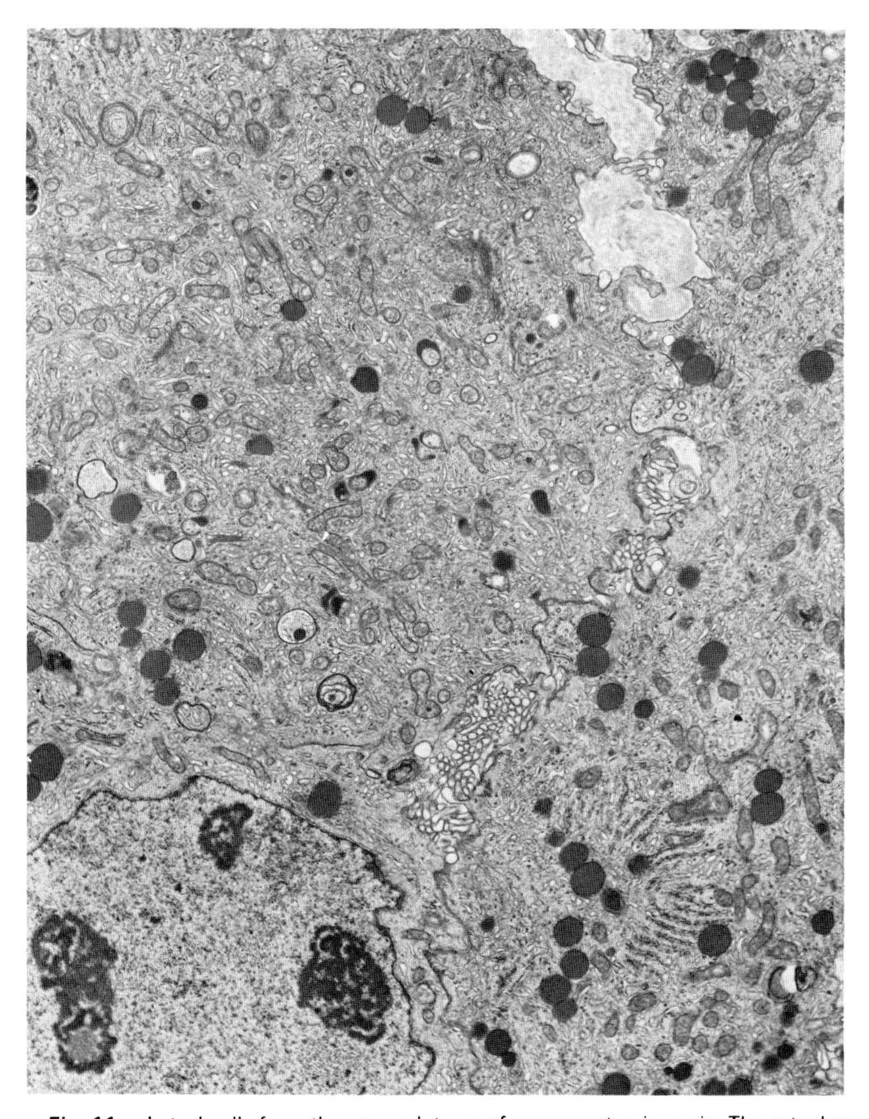

Fig. 11. Luteal cells from the corpus luteum of a pregnant guinea pig. The cytoplasm shows the abundant smooth endoplasmic reticulum that is characteristic of cells that make and secrete steroid hormones. This reticulum bears many of the enzymes involved in progesterone biosynthesis. (×11,400.) (Reproduced by permission from Christensen and Gillim, 1969.)

reproductive life of women do persist but show some hyalinization. In the majority of mammals in which they occur, corpora albicantia are assimilated relatively rapidly, either by the time of completion of the ensuing one or two cycles or during the postpartum period. In the cat at least, lactation appears to be associated with a rejuvenation of the corpus luteum. In all cetaceans studied to date, the corpora albicantia persist for life and several authors have endeavored to distinguish several types of corpus albicans in order to clarify the reproductive patterns of particular species (see Section VIII,C,5). Corpora albicantia also persist, for varying and unknown periods, in some fissipeds and pinnipeds, in certain cervids, and in proboscids. True hyalinized corpora albicantia are rarely found in rodents.

The cetacean corpus albicans usually appears as a wrinkled excrescence on the ovary (Fig. 9). It consists of a mass of connective tissue, collagenous fibres, and fibroblasts, embedded in an amorphous, sometimes hyaline, matrix which contains variable amounts of mucopolysaccaride. The quantity of cellular debris and the degree of cellularity (mainly fibroblasts) may give an indication of the relative ages. About the periphery and in the remnants of the original septa are numerous thick-walled, much-coiled blood vessels, usually arteries and often obliterated. The tunica intima of these vessels is either absent or much reduced to about 0.5 μm. It is suggested that the large cetacean corpus luteum is never effectively vascularized during its phase of functional activity and that regression is associated with collapse of the long luteal arterioles and capillaries. Excessive fibrosis and more or less marked hyalinization, together with a dramatic reduction in blood flow, are factors probably involved in the delayed assimilation of the cetacean corpus albicans (Harrison *et al.*, 1972).

V. INTERSTITIAL CELLS

Mammalian interstitial cells are large and polyhedral; their nuclei are polychromatic and vesicular and their cytoplasm contains lipid droplets (Figs. 12 and 13). It is probably unwise to consider that interstitial cells are never present in the ovaries of some species, but their prominence varies in different species and with the stages of the reproductive cycle (De Groodt-Lasseel, 1963; Mossman *et al.*, 1964). Interstitial cells are plentiful in certain Insectivora, Chiroptera, Rodentia, Carnivora, and Lagomorpha but are difficult to find in Cetacea, Artiodactyla, Perissodactyla, and Primates other than man. Collections of interstitial cells are often referred to as interstitial gland but it is preferable to use the term interstitial tissue as the word gland implies a more discrete entity than is usually found for groups of interstitial cells, even when they comprise practically the whole ovary as in the rabbit (Claesson and Hillarp, 1947) and the agouti (Weir, 1971b). In some species (see Harrison, 1962), the interstitial tissue becomes "luteinized" during pregnancy and may be difficult to distinguish from the corpora lutea.

In recent years the arguments about the origin of the interstitial cells (see Harrison, 1962) have been held in abeyance. It is now generally accepted that primary interstitial cells, which are present in the fetal ovary of many species and particularly in seals (Amoroso *et al.*, 1965), are formed from epithelial medullary cords or granulosa cell nests (Dawson and McCabe, 1951; Deanesly, 1970; Mori and Matsumoto, 1970), and secondary interstitial tissue is derived from the cells of the theca interna around a vesicular atretic follicle (see Mori and Matsumoto, 1973; and Volume I, Chapter 6). Interstitial cells are frequently found arranged in blocks surrounded by a band of connective tissue and with hyaline remains of the zona in the center, thus indicating their origin from follicles (Fig. 13). Mossman *et al.* (1964) distinguish clearly between the interstitial tissue so derived and the hypertrophy of the theca interna which occurs prior to ovulation in some species; they refer to this tissue as the thecal gland. However, there are some species which do not have a well developed theca interna (marsupials: O'Donoghue, 1916; agouti: Weir, 1971b, and Fig. 13; plains viscacha: Weir, 1971c, and Fig. 5), and the source of their interstitial tissue is problematical. It might be that a very few theca cells are sufficient to start the transformation or that the granulosa cells may be differentiated into cells of the interstitial type. There is some indication that the latter may happen in *Lagidium boxi* (Weir, 1971d). The fate of the interstitial cells has not yet been clearly established. They cannot persist indefinitely since the continuous accretion from atretic follicles would result in older animals having far more interstitial tissue than younger animals. This is not the case, although the cyclic fluctuations of interstitial tissue in some species has been attributed to the coincidence of increased rates of atresia at certain times of the cycle, mainly in proestrus and pregnancy (Mossman *et al.*, 1964; Koering, 1969). Therefore, the interstitial cells must either degenerate or dedifferentiate into stromal connective tissue (see Stafford and Mossman, 1945; Brambell, 1956). A yellowish-brown pigment has been described in the ovarian interstitial cells of old mice (Deane and Fawcett, 1952).

No histochemical studies have been reported which are comparable to those of Claesson and his co-workers (see Claesson, 1954, for references) on the rabbit, although Guraya has used histochemical techniques on the ovaries of a number of species (see Guraya, 1973). The interstitial cells are filled with droplets containing phospholipids, triglycerides, and cholesterol and its esters (Guraya and Greenwald, 1964a). The proportions of these lipids and sterols varies from species to species, from cell to cell, and from droplet to droplet. The activity of Δ^5-3β-hydroxysteroid dehydrogenase also varies from cell to cell (Motta *et al.*, 1970). In connection with a possible secretory role, many recent studies have been concerned with the ultrastruc-

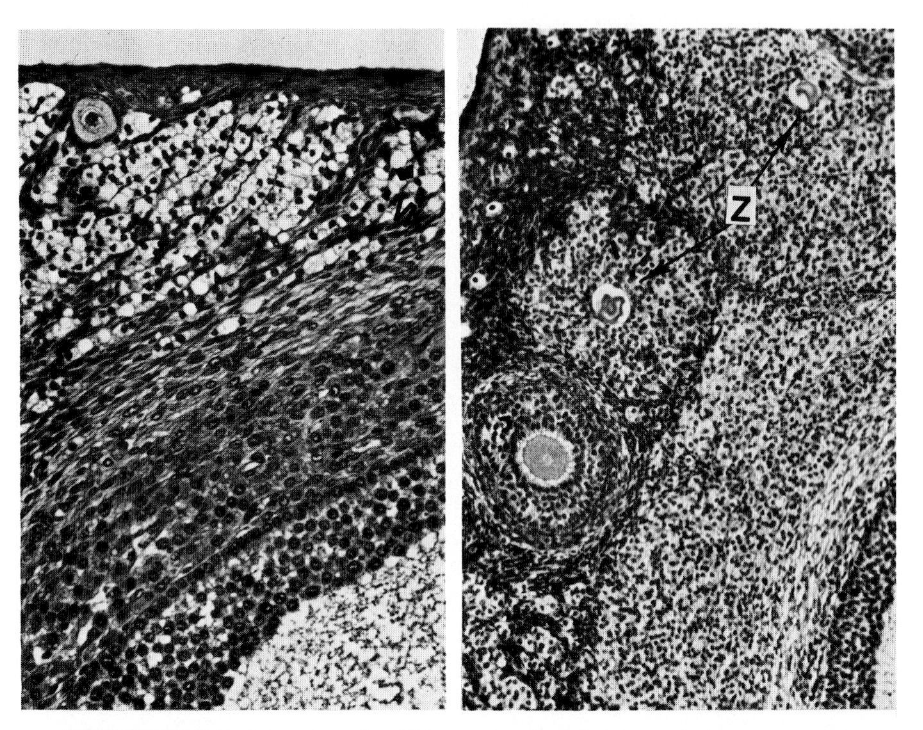

Fig. 12. (left) Part of the ovary of an African porcupine (*Hystrix cristata*) showing from top to bottom: a flat germinal epithelium and a narrow tunica albuginea, vacuolated interstitial tissue, theca externa, theca interna, membrana propria bounded by a close-set layer of granulosa cells, more loosely arranged granulosa cells, and the antrum of the follicle. (Bouin, hemalum, and eosin; ×160.)

Fig. 13. (right) Part of the ovary of an agouti (*Dasyprocta aguti*) showing the extensive formation of interstitial tissue. This is clearly derived from atretic follicles as zona pellucida remains are still visible (Z). The thick tunica albuginea can be seen and the large follicle (bottom right) appears to have only a theca externa. (Bouin, hemalum, and eosin; ×100.)

ture of the interstitial cells (mouse: Muta, 1958; rat: Merker and Diaz-Encinas, 1969; Dahl, 1971a; rabbit: Motta, 1966; Davies and Broadus, 1968; Mori, 1970; man: Laffargue *et al.*, 1968; Motta and Takeva, 1969). All these workers have reported that the cell organelles are typical in many respects of those of other steroid secreting cells; they show an abundance of smooth endoplasmic reticulum, spherical mitochondria with tubular cristae, lipid droplets, and conspicuous Golgi apparatus. The perivascular space is irregular (De Groodt *et al.*, 1957).

VI. OTHER CELL TYPES

The majority of these other cell types have been studied in man where they may be present regularly or sporadically; most of them have been seen in other mammals but identification has often been difficult because of lack of adequate references for comparison.

Epoöphoron (Fig. 14) and paroöphoron tubules may occur in the mesovarium and they are believed to be homologues of the mesonephric ducts (Franchi, 1970). Strauss and Bracher (1954) claimed that the epithelial cells of these tubules in the hamster showed a change of appearance with progression of the estrous cycle, but Beltermann (1965a) did not find any evidence of epoöphoron secretory cells in women.

The most clearly recognizable structure is the rete ovarii (Fig. 14) which is found in the hilus of the ovary. It is probably homologous with the rete testis of the male (Gropp and Ohno, 1966); Sauramo (1954a) describes the appearance, development, and the changes which can occur in the rete in man; and a brief report has appeared on the cow (Archbald *et al.*, 1971). It has been suggested that the rete ovarii tubules ramify throughout the ovary, and are necessary, at least in the mouse, for the initiation of follicle formation and meiosis (Byskov and Lintern-Moore, 1973; Byskov, 1974). Some species, such as the guinea pig, seem prone to development of cysts of these tubules, but their position and lack of an egg prevents confusion with cystic follicles.

Hilus cells are found in close relationship to nonmyelinated nerves and vascular spaces; they may also be called sympathicotropic cells, chromaffin cells, or Berger cells after Berger who first observed them (1922). They closely resemble Leydig cells in their nuclear and cytoplasmic detail, lipids, lipochrome pigment, and crystalloids of Reinke (Green and Maqueo, 1966). In man, they may be a normal component of the ovary and are spe-

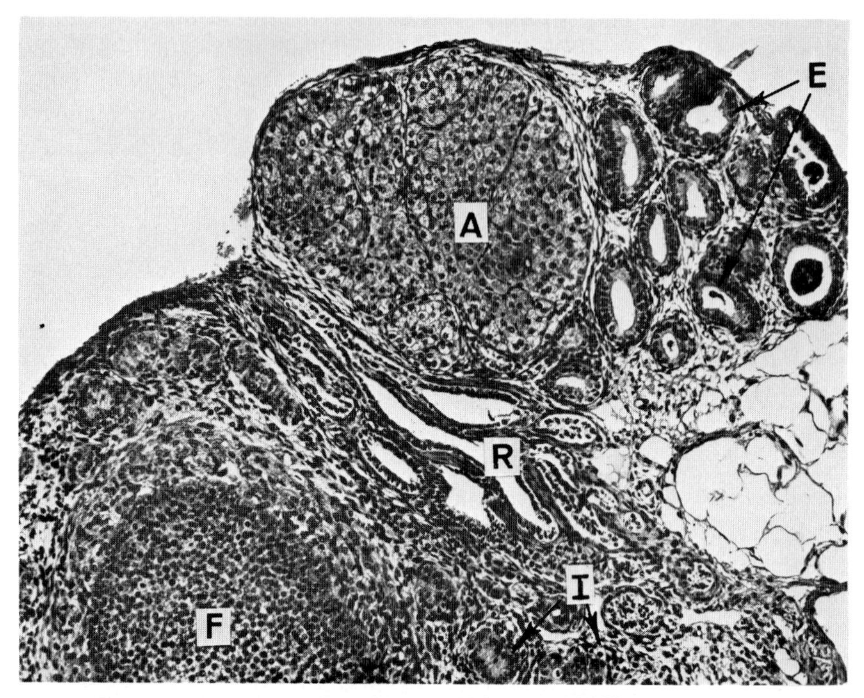

Fig. 14. Section of the ovary of a cuis (*Galea musteloides*) showing a follicle (F), "immature testis tubules" (I), rete ovarii (R), epoophoron tubules (E), and adrenal-like tissue (A). (Bouin, hemalum, and eosin; ×80.)

cially prominent at the end of the fetal life (Sauramo, 1954c), at puberty (Loubet and Loubet, 1961; Valdes-Dapena, 1967), during pregnancy, and at the menopause (Sternberg, 1949). The cells appear to be secretory and since tumors and hyperplasia of ovarian hilus cells are associated with masculinization, it is probable that they secrete androgens (MacRae *et al.*, 1971). As in other secretory cell types, they contain alkaline phosphatase, acid phosphatase, and nonspecific esterase (McKay *et al.*, 1961; Pinkerton *et al.*, 1961). Ultrastructural studies on the hilus cells of pig have confirmed their similarity to Leydig cells and that they contain organelles typical of secretory cells (Unsicker, 1970). Although single axons were seen passing to hilus cells, no synapses were observed.

Cell nests of adrenocortical-like tissue are rarely found within the ovary itself, but islands are often found in the vicinity of the ovary (Fig. 14). They are usually present in the hilar region (Sauramo, 1954b) and can be distinguished from hilus cells because they are encapsulated (Sternberg, 1949).

The cells are histologically identical with adrenal cells of the zonae glomerulosa and fasciculata and, in some species, the effects of bilateral adrenalectomy are remedied because the ovarian rests hypertrophy and become functional (Chester Jones and Henderson, 1963; Seth and Prasad, 1967). Squirrels seem particularly sensitive to such surgery, but adrenal rests have been found in moles, pikas, armadillos, cuis, the Patas monkey (Conaway, 1969), and the macaque (Koering, 1969).

Various cord-like structures have have found in the ovaries of many species; they may be medullary or cortical. Tubules which resemble those of the immature testis (Fig. 14) have been described in the shrew (Wilcox and Mossman, 1945), hippopotamus (Clough, 1970), acouchi (Weir, 1971a), agouti (Weir, 1971b), and pig (Unsicker, 1971). Unsicker believes that these cords are homologue remnants of the testis tubules, but Gropp and Ohno (1966), like League and Hartman (1925), suggest that in cattle they are follicular cell cord remnants.

VII. VASCULAR AND NERVOUS SUPPLY

The main arterial supply to the ovary is from the ovarian artery which normally arises from the aorta below the level of the renal vessels; in the elephant it branches from the renal artery (Perry, 1964). The ovarian artery passes by the infundibulopelvic ligament to enter the ovary at its mesovarian border where anastomoses occur with the uterine artery. The main vessel may branch before it reaches the ovary or only when it enters at the hilus. Before entering, the ovarian artery is often thrown into one or more loose coils. The degree of spiraling, which is said to be always anticlockwise (Greep, 1963), is less in the human ovary than in that of the rabbit (Delson *et al.*, 1949). The secondary and tertiary branches may also display spiraling in many species (Reynolds, 1950; Burr and Davies, 1951; Watzka, 1957; Rosati and Pelagalli, 1968). Extensive branching occurs within the ovary; the vessels are extremely tortuous and anastomosing ovarian veins wind about among their coils.

The ovary is bounded by a narrow nonvascular germinal epithelium and underlying connective tissue, beneath which there is a complete capillary network interrupted only where follicles or corpora lutea reach the surface. Primordial follicles have no special arterial supply but when an antrum develops, an envelope of capillaries is formed in the theca interna close to the membrana propria and may communicate with a similar one in the theca externa. No vessels enter the granulosa layer of the follicle although

Rosati and Pelagalli (1968) say that there are some which do so in the cow. A nonvascular area, the macula pellucida, develops in the follicle wall at the site of rupture. The vascular invasion of the granulosa cells and the formation of a corpus luteum has been described by many authors in many species (see Section VIII). The newly formed corpus luteum may be characterized by variable amounts of extravasated blood in the central lumen. The corpus albicans, when present, is virtually nonvascular. Arteries and their branches ramify throughout the rest of the stroma or interstitial tissue.

Changes in the ovarian arteries have been described for different reproductive states in man (see Harrison, 1962), sow (Bal and Getty, 1970), and rabbit (Gillet, 1971). Recent workers have emphasized the importance of the changes in blood flow through the ovary that occur during estrus in the sheep (Mattner and Thorburn, 1969), during ovulation in the rabbit (Smith and Waites, 1969), and at the time of regression of the corpora lutea in the pig (Rathmacher and Anderson, 1968). Further involvement in the secretory function of the ovary is suggested by Zawistowska (1956) who found higher alkaline phosphatase activity in the endothelial cells of vessels in glandular tissues than in those in the stromal connective tissue.

The lymphatic drainage of the follicle and of the corpus luteum has been implicated in the removal of hormones from the ovary (Burr and Davies, 1951; Jdanov, 1960; Wenzel, 1966; Morris and Sass, 1966; LeBrie, 1970), and of debris from atretic follicles (Hill and Gatenby, 1926). The rabbit ovary has been studied by several workers using the technique of a protein and water-color injection to cause congestion (Jdanov, 1960; Wenzel and Staudt, 1966). Täher (1964) describes large lymph channels around rabbit follicles with two or more granulosa cell layers, and Jdanov (1960) indicates that the lymph system changes with age and reproductive state. Morris and Sass (1966) describe the arrangement of the lymphatics, their ultrastructure, and the rate of lymph flow in ewes of different reproductive states; and suggest that the high rate of flow and the protein content of the lymph is due to discontinuity of the endothelium of the channels. Such discontinuity is evident in the corpus luteum and a change of capillary permeability may be related to the synthesis and secretion of steroid hormones.

The ovarian innervation has been studied in man, baboon, bushbaby, rat, cat, dog, rabbit, and monkey. The ovarian nerves are derived from the ovarian ganglion and the lumbar sympathetic chain via the uterovaginal ganglion. In the dog (Sato, 1955) and rat (O'Shea, 1971b), no evidence of contribution from the coeliac ganglion or of direct vagal innervation was found. The majority of the nerve structures of the ovary are adrenergic

(Rosengren and Sjöberg, 1967; Burden, 1972b) arising in or traversing the aorticorenal ganglia (Fink and Schofield, 1971), and norepinephrine is the main transmitter (Owman *et al.*, 1967). The density of the innervation varies according to the species; cat ovary contains many adrenergic nerves, there are moderate numbers in man and few in monkey (Jacobowitz and Wallach, 1967). The nerves are mainly found accompanying the blood vessels and smooth muscle (Kanagasuntheram and Vorzin, 1964; Le Pere *et al.*, 1966; Osvaldo-Decima, 1970) and are prominent near follicles (Owman and Sjöberg, 1966). In the neonatal ovary, there is a fine plexus of nerves immediately beneath the germinal epithelium and in the adult ovary the stroma may be randomly innervated. Neilson *et al.* (1970) and Fink and Schofield (1971) suggest that these stromal nerves can inhibit the secretory activity of the interstitial tissue and in this way can influence the rate of atresia, the selection of follicles for maturation, and the output of steroids.

VIII. TAXONOMIC SURVEY

A. Monotremata

No studies of the ovarian structure in *Tachyglossus* and *Ornithorhynchus* have been reported since the original descriptions of Solomons and Gatenby (1924), Hill and Gatenby (1926), and Flynn and Hill (1939). Both ovaries are functional in *Tachyglossus* but only the left is in *Ornithorhynchus*. The ovary is similar to that of reptiles in having a narrow cortical zone containing the oocytes and a central medulla which is filled with large blood vessels and lymph channels. The ripe follicle measures about 4.5 mm in diameter but is not fluid filled because of the large telolecithal egg. The three follicular layers typical of eutherians (membrana granulosa, theca interna, and theca externa) are recognizable, especially as the egg matures, but the granulosa layer (up to 30 μm across) is relatively thicker than in sauropsid mature follicles. Enlargement of the granulosa cells occurs just before ovulation and the corpus luteum is formed from these cells only. The theca interna cells, which are one to five layers thick, exhibit extensive mitosis after ovulation and form syncytial-like masses. These may remain at the periphery of the corpus luteum between the granulosa luteal cells and the theca externa, or be carried into the corpus luteum along the connective tissue ingrowths of the theca externa. They do not become luteinized. The diameter of the new corpus luteum is only about 2.5 mm, but proliferation, as well as hypertrophy, of the granulosa cells is presumed to occur in a limited period although mitosis was not found. Capillary invasion occurs from the thecal layers, but extravasation into the central cavity was not observed (Garde, 1930). Corpus luteum development is maximal at the time when the blastocyst is formed, but thereafter it regresses, and, when the egg is laid, regression is almost complete. The statement of Hill and Gatenby that no lipid or true fat is found in the luteal cells has not been confirmed.

B. Marsupialia

Changes in reproductive organs in marsupials have been reviewed by Sharman (1959) and by Sharman *et al.* (1966). Functional aspects of the corpus luteum are discussed by Hughes (1962) and Tyndale-Biscoe (1963). Reproductive patterns in diprotodont marsupials have been divided into four groups by Sharman *et al.* (1966) who suggest that *Macropus giganteus* is alone in having a hormonal secretion of pregnancy which prevents the recurrence of ovulation while a young kangaroo is contained in the pouch.

In the ovary of the common opossum, *Didelphis marsupialis virginiana,* Hartman (1926) found numerous polynuclear ova in 93% of ovaries, and polyovular follicles in 60% of ovaries, as well as accessory stalked outgrowths of ovarian tissue (Hartman, 1923). The ovum reaches the uterus more rapidly than in any other mammal and by this time (24 hours) the corpus luteum is hollow with a solid border of cells (Martinez-Esteve, 1942). The gland is solid by the third and fourth cleavages and mitotic figures are present in the luteal cells. Theca interna cells are said not to contribute glandular elements to the corpus luteum. Abundant diffuse lipoproteins develop throughout the cytoplasm in the luteinizing granulosa cells after ovulation. Regression of the corpus luteum is accompanied by increasing numbers of sudanophilic droplets in the luteal cells, and large vacuoles and early degenerative changes are seen in the corpus luteum at parturition (12½ days). Hartman (1923) states that the corpus luteum rapidly loses its vascularity and has almost disappeared by 20 days. Degeneration occurs several days earlier in pseudopregnancy. No trace can be found after 3 months when the pouch young leave the mother. The corpora lutea do not determine the length of the cycle (28 days).

Nothing has been added to the work of Sandes (1903), O'Donoghue (1912), and Hill and O'Donoghue (1913) on the almost extinct marsupial cat (*Dasyurus viverrinus*). Large numbers of ripe follicles (about 15 per ovary) are borne on projecting bosses; there is little intrafollicular hemorrhage after ovulation. Luteal cells become fully developed when the blastocysts in the uterus are 6.5 to 7 mm in diameter 3 days after ovulation. Sandes (1903) found it difficult to distinguish between the thecae interna and externa. The theca interna is poorly developed and forms only the vascular connective tissue of the corpus luteum. Luteal cells show some vacuolation in the peripheral cytoplasm. The corpus luteum remains fully developed until at least 7 weeks after the animal has started lactating. No trace of the gland remains 4 months after parturition.

The ovaries of the brush-tail possum, *Trichosurus vulpecula,* are described by Tyndale-Biscoe (1955) and Pilton and Sharman (1962). In the ring-tailed possum, *Pseudocheirus peregrinus,* follicles reach a diameter of nearly 4.0 mm before ovulation (Hughes *et al.,* 1965). In early pregnancy, the corpus luteum is not fully formed and has a central cavity that is replaced by a connective tissue nodule when the maximum size of 3.5 mm is reached late in pregnancy. The corpus luteum decreases in size to about 2.0 mm after parturition and contains much connective tissue; it shrinks to about 1.0 mm and is retained as a corpus albicans for a variable period after the young leave the pouch.

Ovarian characteristics of the Macropodidae are outlined in the red kangaroo, *Megaleia rufa,* by Sharman (1964), Sharman and Pilton (1964), and Sharman and Clark (1967). Tyndale-Biscoe (1965) give details of reproduction in *Lagostrophus fasciatus* and the same author (1968) briefly describes the ovaries of *Bettongia lesueur* and adds a discussion with useful references of the length of obligatory diapause in various marsupials. Poole and Pilton (1964) and Clark and Poole (1967) describe the ovaries of the grey kangaroo, *Macropus giganteus,* especially the histology of the corpus luteum of lactation. The luteal phase is defined, as by Sharman (1955a), as that stage of the estrous cycle or pregnancy at which the epithelium of the peripheral portions of the uterine glands consists of high columnar cells with basal nuclei. Corpora lutea of pregnant and nonpregnant animals are similar and reach a maximum size of up to 8 mm during the luteal phase. The gland decreases in size until it has a diameter of 6 mm at the time of birth and can seldom be observed macroscopically in the ovaries of females with pouch young older than 150 days.

In the quokka, *Setonix brachyurus,* the mature follicle reaches 2.8 mm and corpus luteum over 4.0 mm in diameter (Waring *et al.,* 1955). Ovulation is followed by a rapid invasion of blood vessels into the collapsed follicle. Theca interna cells can be distinguished 3 days after ovulation, but theca luteal cells do not turn into fibroblasts. The corpus luteum reaches its maximum size 5 days, and degenerative changes appear 18 days, after ovulation. The corpus luteum starts to degenerate towards the end of the 27-

day gestation period. The corpus luteum formed following a postpartum ovulation is small (2.5 mm in diameter); it resembles a 3- to 4-day-old corpus luteum of the normal cycle and remains in this arrested state for 4 to 6 months (Sharman, 1955a,b). Matthews (1947) has commented on the large number of atretic and luteinized follicles in the ovaries of tree kangaroos (*Dendrolagus*).

C. Eutheria

1. Insectivora

Very little new work has appeared on members of this order.

The ultrastructure of the lamellar bodies in the oocyte of *Erinaceus europaeus* has been described by Sydow (1968), but the shrunken condition of the granulosa cells of the postovulatory follicle in this species (Deanesly, 1934) has not been confirmed.

A study by Tripp (1971) on several species of elephant-shrews has confirmed much of the original descriptions given for *Elephantulus* by van der Horst and Gillman (1940, 1942, 1946). The family cannot be divided into polyovulators and oligo-ovulators; a continuum exists with *E. myurus* being at one extreme with a mean of forty-nine corpora lutea per ovary. Up to five luteinized follicles were found in some *E. myurus* ovaries but corpora lutea and egg counts generally tallied, indicating that corpora lutea are formed at ovulation and their numbers are not augmented in pregnancy. In the polyovulating macroscelidids the follicles tend to be smaller (300 to 350 μm) than usual and to have little or no antrum. Tripp's description of the ovary of *E. myurus* differs from that of van der Horst in three points. Vacuolation of the zona pellucida in van der Horst's material was considered to be caused by Bouin fixation. Tripp could not detect any hypertrophy of the proximal thecal wall in his only preovulatory specimen, and corpora albicantia were visible only for 2 months. Tripp (1971) suggests that eversion of the corpus luteum may be characteristic for the Macroscelididae, as well as being common in other insectivores such as *Hemicentetes, Setifer,* and *Suncus*. Small globules of yolk-lipid were found in the oocyte of *E. myurus* (Tripp, 1970). Interstitial tissue is uncommon in this family, but testis cords have been found.

Similar cords are present in abundance in the Talpidae. In moles (*Talpa europaea* and *Parascaptor leucurus*) the functional part of the ovary, which is similar in structure to that of other mammals except that the granulosa cells luteinize before ovulation, is confined to a cap of active tissue. The rest of the ovary is a mixture of epithelial cords derived from Pflüger's tubules (Deanesly, 1966) and interstitial tissue of the primary type (Duke, 1966, and Section V). The state of this tissue fluctuates with the breeding condition of the animal and enlargement occurs during anestrus. The fine structure of the ovary has been studied by Aumüller (1974). In *Galemys pyrenaicus* the medullary tissue is most prominent before puberty (Peyre, 1962) and, like Matthews (1935) and Godet (1949), Peyre considers that the alternating peaks of activity of the two parts of the ovary represent a form of intersexuality.

Preovulatory luteinization of the granulosa cells occurs in moles (Matthews, 1935), in *Blarina brevicauda* (Pearson, 1944) and in *Sorex palustris* (Conaway, 1952). In the mature follicle of *Suncus murinus* (Dryden, 1969) two distinct layers of granulosa cell have been recognized; the one next to the oocyte being more lightly staining. The follicle ovulates when it is 350 to 400 μm in diameter and the antrum is just a small slit between the two types of granulosa cells. The germinal epithelium regenerates over the everted corpus luteum in about 54 hours and intraovarian fusion of corpora lutea (as in *Neomys*, Price, 1953) is common. The pregnancy corpora lutea persist to term and can be recognized during the succeeding pregnancy.

In the centetids *Hemicentetes* and *Setifer* the granulosa cells swell before ovulation and occlude the central cavity (Strauss, 1938, 1939). Fertilization is said to be intrafollicular and the ovum is extruded by further swelling of the granulosa cells. The everted corpus luteum is similar to that of *Elephantulus*.

2. Chiroptera

The structure of a bat ovary at any one time is largely determined by the reproductive pattern of each species (see Volume II, Chapter 6). The primary variable is whether or not the species hibernates.

In the hibernating species, mainly temperate-zone vespertilionid bats, copulation usually occurs in the autumn and ovulation in the spring. During the winter only one ripe follicle is present in the ovary (Guthrie and Jeffers, 1938a,b; Guthrie *et al.*, 1951; Sluiter and Bels, 1951; Pearson *et al.*, 1952; Anand Kumar, 1965; Saint Girons *et al.*, 1969). This follicle is characterized by hypertrophied cumulus (Figs. 3 and 4) cells which nearly fill the antrum and which contain glycogen (Wimsatt and Kallen, 1957). The ultrastructural characteristics of the glycogen-laden cells and their relationship to the oocyte in *Myotis lucifugus* have been studied by Wimsatt and Parks (1966) who suggest that glycogen is transferred to the oocyte by follicle cell processes that pass through the zona and penetrate the oocyte surface. There are a few short oocyte microvilli and, when the follicle is activated, retraction of these occurs before the first polar body is extruded. The histological appearance of the granulosa cells and progestational reactions of the endometrium suggest that progesterone may be secreted before ovulation (Guthrie and Jeffers, 1938a,b; Wimsatt, 1944; Guthrie *et al.*, 1951). The corpus luteum of *Corynorhinus rafinesquei* reaches maximum diameter shortly after attachment of the blastocyst (Pearson *et al.*, 1952), and vacuoles are present very early in pregnancy. No mitotic figures have been seen in the rapidly developing corpus luteum of *Desmodus rotundus* (Wimsatt and Trapido, 1952). In *Rhinopoma kinneari* the corpus luteum is extroverted and does not persist until term (Anand Kumar, 1965). An unusual pattern of corpus luteum longevity has been reported in *Rousettus leschenaulti* (Gopalakrishna, 1969), because the corpus luteum of the first pregnancy lasts for 6 months and that of the second for 10 months. Regression occurs half-way through the next pregnancy and is rapid.

In tropical bats, follicle growth is similar to that of other mammals (Gopalakrishna and Moghe, 1960) although early formation of the antrum has been reported in the bilaminar follicle of *Megaderma lyra*. Glycogen is not stored by the granulosa cells in *Cynopterus sphinx* and *Rhinolophus rouxi* (Modak and Kamat, 1968). Large numbers of accessory corpora lutea have been found in late pregnancy in some bats, especially in *Scotophilus wroughtoni.* The presence of corpora lutea in the ovaries of bats with uterine blastocysts has led to the suggestion that some bats (*Megaderma lyra:* Ramaswamy, 1961; *Miniopterus schrebersii:* Peyre and Herlant, 1963; *Eidolon helvsum:* Mutere, 1965) experience a delay of implantation rather than of fertilization.

Interstitial tissue has been found in most bat ovaries and Pearson *et al.* (1952) suggest that cyclic variation occurs in *Corynorhinus,* especially during lactation.

3. Carnivora

The ponderal growth of the ovaries of the fetal bitch and their development from birth to 6 months are described by Latimer (1956) and Raps (1948). Barton (1945) suggests that anovular follicles in the tunica albuginea of the bitch ovary are produced from epithelial cords derived from the covering epithelium and not by degeneration of the oocytes in primary follicles. A single ovary may undergo several waves of epithelial proliferation:

small cycles of activity apparently occur throughout the estrous cycle. Mucin secretion has been demonstrated from various subsurface epithelial structures categorized by O'Shea (1966). Follicles develop marked folds in the granulosa layer with vascular cores of theca interna cells. The folds reach remarkable complexity at estrus. The early corpus luteum has an open lacework appearance but becomes closer packed by the eighth day and compact by the eighteenth day (Evans and Cole, 1931; Mulligan, 1942). Theca interna cells can be recognized by the presence of alkaline phosphatase until shortly after ovulation (Corner, 1948). The preovulatory folding of the granulosa layer in the mature follicle of *Vulpes fulva* allows two main types of corpus luteum to be distinguished (O. P. Pearson and Enders, 1943). The appearance of the young corpus luteum is either of an open lacework character, the folding of the mural granulosa having proceeded so rapidly that the cells are loosely arranged, or the changes of luteinization have progressed enough to give a compact appearance. The corpus luteum of *Vulpes vulpes* persists throughout pregnancy and is still recognizable, though inactive, 20 weeks after parturition (Bonnin-Laffargue and Canivenc, 1970). In the dog, pigmented islands of regressed luteal cells are said to be visible for about 1½ years after pregnancy (Anderson and Simpson, 1971).

Descriptions of the corpora lutea associated with delayed implantation in bears, especially *Ursus americanus*, are given by Wimsatt (1963) who noted a radial arrangement of connective tissue septa in the corpora lutea. The septation is lost when the corpus luteum enlarges and undergoes dramatic vascular changes after implantation. Other details of the black bear ovary are given by Erickson *et al.* (1964).

Three types of luteal cells are recognized in the pregnant racoon (Sinha *et al.*, 1971a). In type I, endoplasmic reticulum is agranular and tubular; its morphological features suggest that this type secretes most of the steroids. Many lipid droplets are present in type II cells. Fine filaments and microtubules are seen and plasma membranes of opposing luteal cells are much folded except in relation to perivascular spaces, these cells secrete steroids and store lipids. Type III cells contain numerous lipid droplets but few cytoplasmic organelles; they store lipids but do not secrete steroids.

The corpus luteum in *Mustela putorius furo* reaches its greatest diameter 3 to 5 weeks after ovulation, and the ovaries consist almost entirely of luteal tissue (Robinson, 1918). The corpus luteum regresses 3 to 4 days before parturition (Hammond and Marshall, 1930). Deanesly (1967) has shown that the corpora lutea in the ovaries of unilaterally pregnant ferrets, in which pregnancy was terminated 8 days before removal of the ovaries, resembled those of normal pregnancy. The development of the ferret ovary from the thirty-second day of pregnancy to the thirty-fifth day *postpartum* is described by Deanesly (1970) who records a gradual degeneration and absorption of oocytes from day 14 onward. Degenerating oocytes are also seen in the lumen of rete tubules which originate in the ovarian medulla. The corpora lutea of infertile cycles in *Mustela erminea* are smaller than those of pregnancy (Deanesly, 1935). The corpora lutea of ovulation are well vascularized but the luteal cells remain small. Wright (1942) and Watzka (1940) suggest that the small corpora lutea associated with so-called sterile matings indicated delayed implantation. Four to five stages of the development of the corpus luteum of *Mustela vison* have been recognized (Hansson, 1947). The corpus luteum enters an inactive phase with small luteal cells during the period of delayed implantation. The size and state of development of the corpus luteum during the breeding season are more closely related to the length of day than to their age (Enders, 1952). In mink (Enders, 1952; Enders and Enders, 1963), changes of luteinization do occur, contrary to earlier reports, in developing corpora lutea but the luteal cells do not hypertrophy until implantation. Ribosomes are present in variable numbers in luteal cells during delayed implantation, but only in small areas are they associated with membranes of the endoplasmic reticulum. No changes with reproductive condition were found in the fine structure of the interstitial cells (Møller, 1973). The corpora lutea of *Mustela frenata* and *M. cicognani, Martes americana,* and *M. caurina* are small and inconspicuous when associated with unimplanted blastocysts (Wright, 1942). In the martens, highly vacuolated luteal cells, comparable to those of the cat in late pregnancy and to those found in the armadillo in similar circumstances, are present during the long period when blastocysts remain unimplanted in the uterus. Some details of ovarian structure in North American mustelids are reported by Wright (1963) who gives a useful bibliography.

The corpora lutea of pregnancy in *Meles meles* regress rapidly after parturition, and a postpartum ovulation occurs within a few weeks (Canivenc, 1957, 1966; Neal and Harrison, 1958). The corpora lutea of the long period of delay (2 to 10 months) are poorly vascularized and appear to exhibit reduced secretory activity, as indicated by electron microscope studies (Canivenc *et al.,* 1967). Additional ovulations occur during the period of delay but it is not possible, once the additional corpora lutea have become solid, to distinguish them from those of earlier ovulations. Corpora lutea may exhibit degenerative changes during the later part of the delay, and in early pregnancy may contain an unusual number of leucocytes. Changes have also been described in luteal cells and in nuclear inclusions which may be related to secretory activity of the corpus luteum (Canivenc *et al.,* 1966a,b). The American badger, *Taxidea taxus,* ovulates from two to four follicles; the corpora lutea of pregnancy are twice the diameter of those before implantation and luteal cells are larger (Wright, 1966). No evidence of ovulation during delay is found. Tubules are found in the medulla of skunks; they are surrounded by a basement membrane and contain numerous oocytes. These tubules probably give rise to the polyovular follicles which are common in *Spilogale* and *Mephitis* (Leach and Conaway, 1963). In *Spilogale*, Mead (1968a,b) noted corpora lutea atretica, which are formed during estrus by invasion of stromal and thecal cells, but corpora lutea accessoria were formed only after HCG treatment.

The covering epithelium is multilayered in thickened areas and in fissures of the ovary of the sea otter, *Enhydra lutris*, and epithelial tubes extend from the fissures into the cortex (Sinha and Conaway, 1968). Interstitial gland cells of theca interna origin are abundant and appear to be secretory during estrus (Sinha *et al.*, 1966). In the preimplantation corpus luteum, an antrum is present but it is obliterated by the time of implantation. Luteal cells progressively hypertrophy during delay. Secondary cavities develop in the corpora lutea after implantation and coalesce to form larger ones by midpregnancy. The corpus luteum degenerates rapidly after parturition but persists as a corpus albicans for at least 2 years.

In the palm civet, *Paradoxurus*, the granulosa cells disappear in atretic follicles and are replaced by connective tissue (Duke, 1969). Stromal cells convert directly into interstitial cells of at least three types; stromal cells also change to thecal cells and perhaps *vice versa*. Pearson and Baldwin (1953) describe the ovary of the mongoose, *Herpestes*, as being almost completely encapsulated. Ripe follicles (1 to 4 mm) protrude from the ovarian surface and the mature corpus luteum is nearly twice as large (3 mm) late in pregnancy but regresses rapidly at parturition.

Vascular patterns in cat ovaries and related structures are probably not involved in the spontaneous contractions noted in isolated ovaries (Rocereto *et al.*, 1969); adrenergic receptors are believed to be responsible for the increase of contractions which occur near to ovulation. Fink and Schofield (1971) also discuss the innervation of the cat ovary in relation to functional activities. Pinocytosis in follicular cells has been described by Liss (1964) who suggests this as a mechanism for nutrition of the oocyte. Shehata (1974) discusses polyovuly in kittens and adults. No new studies on the cat corpus luteum have been published since those reviewed by Harrison (1962). As in most carnivores, the wall of the ruptured follicle is deeply plicated and the folds involve the theca interna cells (Dawson, 1941). These cells hypertrophy at estrus and soon after ovulation move into the granulosa layer where by day 3 they appear fibroblastic and lay down collagenous tissue between the luteinizing granulosa cells. The theca interna cells remaining at the periphery of the corpus luteum do not develop into theca-luteal cells (Dawson and Friedgood, 1940). The corpus luteum reaches maximum size 10 to 16 days after mating and regression starts about day 20 although corpora lutea of pseudopregnancy appear viable up to day 28. At day 27 of pregnancy there is a marked decrease in size of the corpus luteum. Vacuoles in the 50-day-old corpus luteum are considered to indicate degeneration, but the corpus luteum is presumably functional because ovariectomy results in abortion (Courrier and Gros, 1935, 1936). The life span of the corpus luteum may be as long as 6 to 8 months from the time of mating because, during lactation, pregnancy corpora lutea display increased vascularity (Dawson, 1946). The smaller vacuoles in the luteal cells disappear rapidly, but some "giant" vacuoles persist. The functional capacity of this rejuvenated corpus luteum is not known, but it is suggested that the phenomenon is either a secondary effect from released pituitary luteotrophin or due to a general depletion of fat reserves in response to the needs of lactation. Histochemical features of the ovary in the cat and the dog are compared by Guraya (1969). The corpora lutea of *Lynx baileyi* and *L. vinta* are of two types according to the size of the luteal cells (Duke, 1949). They appear to persist for months and probably years after parturition. Rowlands and Sadleir (1968) illustrate the folded wall of the lion, *Panthera leo,* follicle after rupture with PMSG and HCG.

4. Pinnipedia

The pinniped ovary has been particularly investigated because of the abundance of primary interstitial tissue in near-term fetal ovaries and the delay of implantation which has been demonstrated in most species.

Few otariids have been studied. In *Otaria byronia* the follicles ovulate at 2.5 cm diameter and the corpus luteum, which regresses rapidly after parturition, measures 3.2 cm (Hamilton, 1934, 1939). The mature follicle of *Arctocephalus pusillus* is 1.7 cm across (Rand, 1955); the corpus luteum becomes well vascularized at the time of implantation and the rapid regression after parturition is followed by signs of vascular engorgement. The corpus luteum is an irregular scar at the next implantation and disappears after 2 years. Ovarian changes in *Callorhinus ursinus* have been described in detail by Enders *et al.* (1946), A. K. Pearson and Enders (1951), Harrison *et al.* (1952), and most recently by Craig (1964) and Okamoto (1970). Ovulation occurs from follicles 1.0 to 1.3 cm in diameter and is probably spontaneous. Initial luteal activity is associated with progestational uterine changes, and is followed by a regressive phase associated with delayed implantation. Luteal cells become markedly vacuolated but do not die (Harrison, 1948a). Follicles develop in the ovary containing the corpus luteum and the reproductive tract shows indications of estrogenic stimulation. There is a revival of luteal activity just before implantation. The corpus luteum remains active during this period of pregnancy. Craig (1964) suggests that placental secretions supersede those of the ovary until parturition in July. The corpus albicans persists for about 10 months after parturition. The corpus luteum of *Odobenus rosmarus* involutes more rapidly and disappears within weeks of parturition (Fay, 1955). Associated with the long lactation period of this species, the ovaries remain quiescent for at least a year after parturition.

The fetal ovaries of the phocids *Halichoerus, Phoca,* and *Mirounga* and other antarctic forms exhibit marked enlargement toward term (Harrison *et al.*, 1952; Bonner, 1955; Harrison, 1960; Amoroso *et al.*, 1965). The changes involve an increase in vascularity and a hypertrophy of the primary interstitial tissue, and in *Halichoerus* the near-term fetal ovary may be heavier than the maternal. The follicular apparatus does not develop in phocid fetuses but does in near-term fetuses of the otariids *Zalophus* and *Callorhinus*. The fetal ovarian hypertrophy is accompanied by stimulation of the fetal uterine mucosa but not of mammary tissue. The ovarian interstitial tissue regresses during the first 7 to 10 days after parturition. Possible causes of the fetal ovarian hypertrophy, whether stimulated by placental hormones and/or by those of maternal and fetal pituitaries, are discussed by Harrison (1969).

Information is now available about the adult ovary in several phocid species. Details of appearances in *Phoca, Halichoerus,* and antarctic forms given by Harrison *et al.* (1952) have been added to by Laws (1956) for *Mirounga leonina,* by Mansfield (1958) for *Leptonychotes,* and by Harrison (1960), Amoroso *et al.* (1965), and Bigg (1969) for *Phoca vitulina.* In the ovary which does not contain the pregnancy corpus luteum, follicles mature at the end of pregnancy or a few days after parturition; this may indicate the existence of a postpartum estrus (Harrison, 1962). The well marked theca interna could be a source of estrogens in late pregnancy (Amoroso *et al.*, 1965). The diameter of the oocyte in *Mirounga* increases rapidly from 25 μm to 90 μm while the follicle grows to a diameter of 150 μm. Growth of the oocyte then occurs more slowly to a diameter of 140 μm; the follicle develops an antrum and reaches a preovulatory size of 15.0 mm. In *Mirounga,* and probably in other phocids, there is an initial development of follicles during the period of delay and also when the embryo is 15 to 20 mm long. The follicular activity during delay may be associated with an estrogen surge which primes the uterine mucosa prior to implantation.

Ovulation is usually stated to be spontaneous in phocids and has been observed in *Phoca vitulina* isolated in captivity (Harrison, 1960). Nonpregnant females with recent corpora lutea could be pseudopregnant or undergoing cyclical activity of limited type (Laws, 1956). There are several distinct phases in the life history of the phocid corpus luteum of pregnancy: a short phase after ovulation, a phase which can last for up to 3 months during delay, a short phase related to implantation, one after implantation and during early pregnancy, another including the rest of pregnancy, and a postparturient phase as a corpus albicans. The corpus luteum develops in the normal manner but is small and poorly vascularized and does not exhibit the changes in the luteal cells characteristic of full activity. The corpus luteum of delay is even smaller than that of the first week, is still poorly vascularized, and the luteal cells show marked cytoplasmic vacuolation (Harrison *et al.*, 1952; Laws, 1956; Harrison, 1960; Amoroso *et al.*, 1965). Implantation is pressaged by a disappearance of the vacuoles and the gland enlarges with a return of functional activity and increased vascularization. After implantation, minor cellular adjustments can be detected which suggest transitions in activity; these involve accumulation of fluid, more pronounced intercellular spaces, appearance of small vacuoles, increase of connective tissue elements, and thickening of the walls of blood vessels. Depending on the species, these changes become more apparent, even to the extent of being regressive, but in all species the corpus luteum appears to persist until term without marked shrinkage in size. In *Phoca,* at least, some enlargement of the gland occurs during the early lactational period (Harrison, 1960); it becomes a small pale tan-colored corpus albicans by the next ovulation and is resorbed in 1 to 4 years (Bigg, 1969). If pregnancy is missed in a particular year, the corpus luteum and corpus albicans can regress at rates differing from those had pregnancy occurred. The corpus luteum of pseudopregnancy can be detected after at least 4 to 5 months.

Subsurface crypts, lined by the covering epithelium of the ovary, have been found, in varying complexity, in many pinnipeds (Harrison and Matthews, 1951; Harrison *et al.*, 1952; Laws, 1956). Their possible significance is discussed by Harrison (1969).

5. Cetacea

Many cetacean ovaries have been examined macroscopically, but there have been few histological or histochemical studies. The cetacean ovary has been of particular interest because the persistence of corpora albicantia (Fig. 9) provides a permanent record of ovulation in mysticetes (Mackintosh and Wheeler, 1929; Peters, 1939; Mackintosh, 1946), and can be used as a relative measure of age (for references, see Laws, 1958; Mackintosh, 1965; Harrison, 1969).

Laws (1961) has analyzed ovarian events in *Balaenoptera physalus*. The diameter of the mature follicle is about 7 cm and ovulation most often occurs near the anterior pole of the ovary. Follicular activity occurs during pregnancy and at the start and end of lactation, but these follicles do not mature. The corpus luteum of ovulation is 8.28 ± 0.82 cm across and weighs between 1.5 and 3.4 kg; the pregnancy corpus luteum measures 11.44 ± 0.15 cm and weighs 2.2 to 2.4 kg. The terms everted, cavitated, meandrine, vesicular, and nonvesicular have been used to describe luteal morphology. Vesicular forms account for 17.1% of all corpora lutea.

The problem of deciding whether there are distinguishing differences between corpora albicantia derived from a corpus luteum of ovulation and one of pregnancy has still to be solved both in mysticetes and odontocetes. Peters (1939) claimed that there was a histological difference in his material from *B. musculus* and *B. physalus,* but Laws (1961) suggests that he was confused by the stages of regression in the corpus luteum. Dempsey and Wislocki (1941) suggested that the large size of the corpus luteum in *Megaptera* (5 cm) imposed limitations on the vascular supply so that luteal tissue never developed throughout the entire gland. The central region would remain avascular and excessive fibrosis and hyalinization would follow. Van Lennep (1950) postulated differences in size and arrangement of connective tissue trabeculae in the corpora lutea of ovulation and pregnancy. Robins (1954) suggested that in *Megaptera* a corpus luteum with a central cavity was one of pregnancy, as was a corpus albicans with a central core of connective tissue. In fact, solid corpora lutea are apparently more common in *Megaptera,* as indicated by Chittleborough (1954) in his work on the growth curve of the corpus luteum in this form. The corpora lutea of *B. physalus,* having become corpora albicantia after parturition, regress to a constant proportion of their size when fully developed and remain visible in the ovary for the rest of the whale's life. The number of corpora albicantia accumulate as the whale ages. The various types of corpora albicantia reflect the structure of the morphological forms of corpora lutea, but no significant difference can be detected in the final mean diameters of corpora albicantia from pregnant, lactating, and resting females. Corpora albicantia have a mean diameter of 2.5 cm and a mean weight of 10 gm (Laws, 1961), and young, medium, and old forms can be detected. The mean annual increment of corpora albicantia is 1.4 to 1.5 (but see Laws' paper on possible errors). Laws also describes certain corpora aberrantia, with an incidence of 1.7%, which are colored yellow buff and exhibit a form of hyaline degeneration. These could be another type of corpus albicans. Accessory corpora lutea are not uncommon in cetaceans and in odontocete ovaries about 5% of pigmented corpora atretica are found. These bodies persist for varying periods depending on their size at formation.

In *Physeter catodon,* Chuzhakina (1961) has found evidence of age changes in the ovary involving loss of the follicular elements, thinning of the cortex, and increased fibrosis. Best (1967) has estimated follicles to be 5.3 cm in diameter just prior to ovulation. In small atretic follicles, hemorrhage occurs from thecal vessels into the central cavity and there is subsequent fibrosis. The corpora lutea of pregnancy (7.49 ± 0.86 cm) may be distinguished from those of ovulation (5.55 ± 0.95 cm) by their larger size and more compact structure, but they do not have a central, jelly-filled cavity. The fully developed corpus luteum remains the same size throughout pregnancy but regresses rapidly after parturition. Groups of corpora albicantia with mean diameters of 2.86 ± 0.75 cm, 1.86 ± 0.33 cm, and 1.35 ± 0.26 cm represent successive stages in regression of the corpus luteum. After parturition the corpus luteum shrinks by more than one third of its diameter in less than 8 weeks. All indications are that corpora albicantia are retained throughout life and the modal diameter eventually stabilizes. Most corpora albicantia are found in the anterior half of the ovary.

Ovarian characteristics in some of the smaller odontocetes have become better known partly as a result of the popularity of dolphins as exhibits in "marinelands." Most is known about *Globicephala,* but accurately timed histological detail is still lacking for all species. In *Globicephala* the follicle reaches a diameter of 30 mm before ovulation and shrinks to 19 mm after rupture. The corpus luteum attains a diameter of 22 to 40 mm; early corpora lutea are loosely arranged (Fig. 10) becoming compact later, and some have a jelly-filled center (Fig. 10). Luteal cells are vacuolated early in pregnancy and many may be multinucleated. Theca interna cells persist at the periphery of the gland and in relation to trabeculae (Harrison, 1949; Sergeant, 1962). Three types of corpora albicantia have been recognized and in *Globicephala* the majority persist, except perhaps some resulting from ovulations not followed by pregnancy. Up to sixteen corpora albicantia have been counted in the ovaries of the oldest females. The presence of three corpora albicantia in a similar early state of regression in one animal could be accounted for by regression being at different rates in different types of corpora lutea.

In odontocetes the ovulation site may be anywhere on the ovarian surface and the corpus luteum is often pedunculated (Slijper, 1966; Harrison, 1969). Sleptsov (1940) considered that a distinction could be made between a corpus albicans derived from ovulation and one associated with pregnancy in *Delphinus.* Differences in types of corpora albicantia have also been noted in *Delphinapterus* (Kleinenberg *et al.,* 1964), in *Lagenorhynchus* (Harrison *et al.,* 1969), in *Phocoena* (Fisher and Harrison, 1970), but not, as yet, in *Tursiops* (Harrison and Ridgway, 1971). The essential distinguishing feature in the two types of corpus albicans is whether or not the characteristic, amorphous, relatively acellular, hyaline material accumulates in the corpus, but other features such as degrees of vascularization, cellularity, and also size and shape have been considered. A feature characteristic of almost all corpora albicantia (Figs. 8 and 10) is the resistance to regressive changes of the connective tissue elements in the walls of the large coiled and obliterated arterial remnants of the functional arteries of the original corpus luteum.

Corpora albicantia also persist in some platanistids (Harrison and Brownell, 1971); in *Inia* the ovaries are bulkier structures than in delphinids with the corpora lutea incorporated in the ovarian structure and not pedunculated as in *Pontoporia*. In general, ovulation in the smaller odontocetes appears to be induced but there is some evidence in several species of the likelihood of one or more ovulations occurring before the supervention of pregnancy. The reproductive behavior of some odontocetes in captivity, especially *Tursiops*, indicates that ovarian activity can be suppressed even in the presence of a fertile male (Harrison and Ridgway, 1971).

Ohsumi (1964) has classified cetaceans into three main groups according to the pattern of accumulation of corpora lutea in the ovaries. In type I Ohsumi maintains that ovulation occurs equally in right and left ovaries (e.g., *Balaenoptera, Megaptera, Physeter, Kogia*). Type II displays a slightly greater accumulation in the left than in the right (e.g., *Delphinapterus, Globicephala, Orcinus*). In type III only the left ovary is active for a period and then the right (e.g., *Phocoenoides, Lagenorhynchus, Tursiops, Stenella, Delphinus*). Subsequent accounts have not altogether confirmed such distributions of corpora lutea, but there is no doubt that several odontocetes (*Pontoporia, Sotalia, Lagenorhynchus, Delphinus, Stenella* and *Phocoena*) ovulate consistently from the left ovary, at least when young (Harrison, 1969). No satisfactory explanation has been advanced to account for this functional asymmetry.

A rete ovarii, of varying tubular development, is common in odontocetes and in several species pseudohermaphrodite and intersexual forms have been reported (Harrison, 1969).

6. Edentata

The distribution of oocytes and the structure of the cortex of the ovary in the armadillo (*Dasypus*) are described by Enders (1960). Follicles form mainly at the periphery of a mass of oocytes. This pattern of growth results in the incorporation of older oocytes into follicles. Neoformation of oocytes ceases at the end of the juvenile stage except in rare instances. Meiotic activity persists in medullary cords or "intersex tubules" in many ovaries but probably does not result in neoformation of oocytes in the adult.

The corpus luteum develops rapidly until the volume may equal that of the rest of the ovarian tissue. The ovary appears as a small cap of cells resting on the corpus luteum (Hamlett, 1932). The luteal cells are large and fully luteinized by the time the blastocyst reaches the uterus (Enders, 1962, 1966). The average weight of the ovary and corpus luteum during implantation delay is 276 ± 55 mg, and after implantation it is 356 ± 82 mg. Some follicular enlargement accounts for part of the weight difference but no mitotic figures are seen in the corpora lutea at implantation and there is no apparent further hypertrophy of the luteal cells. The corpora lutea of delay and of the postimplantation period appear active by ultrastructural criteria. Their luteal cells contain endoplasmic reticulum, mitochondria with tubular cristae, and extensive marginal folding in relation to perivascular spaces. Hamlett (1935) considered the secretory activity of the corpus luteum of delay to be suppressed and Labhsetwar and Enders (1968) have shown that activity may be lower during the delay period than in the early postimplantation period. However, Talmage and Buchanan (1954) report that serum progesterone levels increase just before ovulation, remain high, and increase again after implantation. Activity of the corpus luteum decreases during the last third of pregnancy; the corpus luteum continues to degenerate during the postparturient period and in a few months is entirely replaced by connective tissue.

7. Lagomorpha

Studies on the rabbit ovary in recent years have followed three main lines. Since it is an induced ovulator, the rabbit has been used extensively for investigations into the mechanism of ovulation (see Volume I, Chapter 6). As in most other laboratory species, the fine structure of the oocyte has been described in great detail, and the prominent interstitial gland has been investigated with combined biochemical and experimental techniques.

The early studies of Burr and Davies (1951) on the vascular system have been repeated (Gillet *et al.*, 1968), confirming that there is a fine capillary network in the theca interna of the follicle. In the last quarter of pregnancy, the capillaries of the corpus luteum become grouped and large central sinuses are formed. The lymphatic system has been described as a network around the tertiary follicles, corpora lutea, atretic follicles, and interstitial tissue masses (Wenzel and Staudt, 1966) and is thought to be the system for removal of the ovarian secretions (Wenzel, 1966). Small nerves, which are independent of smooth muscle, run to the follicles, but most of the, mainly adrenergic, nerves are associated with the smooth muscle cells that are found in the vascular system and theca externa (Owman and Sjöberg, 1966; Owman *et al.*, 1967).

The oocyte of the rabbit appears to be much like that of other mammals (Blanchette, 1961; Zamboni and Mastroianni, 1966) and no neogenesis occurs (Kennelly and Foote, 1966). The relationship of the oocyte surface to the follicular cells has been found to be one of interdigitating processes which pass through the zona pellucida (Trujillo-Cenoz and Sotelo, 1959; Gondos and Zamboni, 1967; Gondos, 1970). Intercellular bridges are absent between the germ cell and the granulosa cell (Gondos, 1970), but bridges between germ cells within the same nest are common (Zamboni and Gondos, 1968) and possibly represent the mechanism that organizes synchronous development or degeneration of apposite oocytes (Gondos and Zamboni, 1969). Later stages of development of the oocyte are described by Trujillo-Cenoz and Sotelo (1959), Zamboni and Mastroianni (1966), and Moricard and Sallusto (1968). An obvious increase in the number of mitochondria occurs during growth of the oocyte and Blanchette (1961) suggests that dumbell-shaped mitochondria may be involved in this mechanism, presumably by budding. The structure of the oocyte mitochondria changes under the influence of PMSG but that of the other ovarian components is not altered (Montemagno and Caramia, 1965).

The electron microscope has been used for detailed study of the granulosa cells; at birth, when they are distinguishable from germ cells by their irregular nuclei and lipid inclusions (Gondos, 1969, 1970), and later (Hashimoto *et al.*, 1960; Merker, 1961; Franceschini *et al.*, 1965; Lipner and Cross, 1968). Motta *et al.* (1971) demonstrated cilia and centrioles in the granulosa cells. The ultrastructural changes in the cumulus cells between coitus and ovulation are described by Moricard *et al.* (1969); the Golgi apparatus, the extent of endoplasmic reticulum, and the type of mitochondria change 1 to 6 hours after copulation, and the authors consider these alterations to be essential for ovulation to take place. Histochemical identification of lipids (Guraya, 1968a,b) and proteins (Brandau *et al.*, 1967) show that there are changes in their distribution before and after ovulation. An increased blood supply at ovulation (Smith and Waites, 1969) is probably important for bringing the gonadotropins to the ovary rather than for increasing the substrates for protein synthesis. Espey and his colleagues have investigated many ripe follicles to determine the ovulatory mechanism; the ultrastructural appearance of the apex of the follicle as it approaches rupture has been described (Espey *et al.*, 1965; Espey, 1967), and decomposition of connective tissue by release of proteins from multivesicular bodies of thecal fibroblasts is thought to occur (Espey, 1971a,b).

The time of appearance of Δ^5-3β-hydroxysteroid dehydrogenase in the granulosa suggests that no progesterone can be secreted until about 8 hours before ovulation (Brandau *et al.*, 1967). The conversion of granulosa cells to luteal cells has been studied with the electron microscope (Blanchette, 1966a). The change is typical for that of a nonsecreting cell becoming a secretory one. The theca interna cells (Santoro, 1964) are probably only involved as they enter the corpus luteum along the vascular network, although there is one study which suggests that if HCG is administered to a rabbit the corpus luteum is formed entirely from thecal cells (Yasuda and Doi, 1953).

The fine structure of the mature luteal cells is described by Blanchette (1966b) and it is similar to that of other mammalian luteal cells (Christensen and Gillim, 1969). However, Deane and Rubin (1965) reported that Δ^5-3β-hydroxysteroid dehydrogenase could not be demonstrated in the rabbit corpus luteum. Cell numbers in the developing corpus luteum have been counted by Rennie *et al.* (1965), and they find that between 24 and 192 hours *postcoitum* the total cell numbers increase from 10×10^4 to 67×10^4 cells, mostly due to invasion of endothelium and blood cells as the proportion of luteal cells drops from 65% to 21%. A useful reference for a start to the voluminous literature on factors affecting the life span of the corpus luteum is Scott and Rennie (1970). The rabbit corpus luteum appears to have an inherent life span which is usually overridden by a luteolytic effect (see Volume III, Chapter 3). Keyes and Armstrong (1968) suggest that the presence of follicles is important for regulating luteal function, and the role of the interstitial gland tissue has been extensively studied by Hilliard and her colleagues.

The ultrastructure of the interstitium (Davies and Broadus, 1968) shows that the cells are typical secretory ones. Experimental work indicates that coitus or exogenous gonadotropin causes an increased synthesis of 20α-hydroxypregnen-4-en-3-one, and more of this progestin than progesterone is present in the blood (Hilliard *et al.*, 1963). The progestin is believed to act on the hypothalamus and is the effective agent for prolonging LH release after coitus in the rabbit (Hilliard *et al.*, 1967). Three distinct actions of LH on the interstitium can be mimicked by cyclic AMP (see Dorrington and Baggett, 1969; and Volume III, Chapter 6). Histochemical results have been published (Guraya and Greenwald, 1964b) and the formation of the primary interstitial gland from epithelial medullary cords or granulosa cell nests is described by Mori and Matsumoto (1970). Development of the secondary interstitial tissue is described by Deanesly (1972) and Mori and Matsumoto (1973).

Studies on other lagomorphs have been few. No new studies on the pika are known since that of Duke (1952). A brief comment on *Lepus americanus* suggests that the ovary in this species is fairly smooth on the surface; the mature follicle is 1 to 2 mm diameter and the corpora lutea reach a maximum of 5 mm at midpregnancy. Final regression occurs about 4 weeks after parturition but new and old sets of corpora lutea can be distinguished (Newson, 1964). Superfetation in *L. europaeus* has been confirmed (Bloch *et al.*, 1967) and ripe follicles must be present in the ovary while the corpora lutea are active. Some details of ovarian histology can be found in Raczynski (1964) but more information is contained in Bloch and Strauss (1958).

8. Rodentia

Since Harrison's 1962 account of the background to the morphological studies on which much laboratory rodent reproductive physiology is based, only one review of the rodent ovary has appeared. This is an excellent coverage by Mossman (1966) who indicates that most of the differences between the various rodents are of degree rather than kind, but his plea for morphological investigation of further species of this very large and diverse group has been generally ignored. Most recent studies on rodents have been ultrastructural, histochemical, and endocrinological ones and have been exhaustive on the few common laboratory rodents such as rat, mouse, hamster, and guinea pig.

A brief account of the beaver ovary (Provost, 1962) indicates that the membrana granulosa is folded until the follicles mature and ovulate at a diameter of 5 to 6.5 mm. No large follicles are present in the ovary when the new corpora lutea are present (about 6 to 8 mm diameter), but corpora lutea soon degenerate if pregnancy does not ensue. After pregnancy, non-pigmented corpora albicantia persist until the following breeding season and the rupture points are reported to be visible for at least a year. No mention was made of the prominent interstitial gland tissue that Mossman (1966) described for this species. The ovaries of *Citellus beldingi* and *C. lateralis* (McKeever, 1966) and *Funambulus pennanti* (Seth and Prasad, 1967) have been described. Ovulation is induced in all three species and functional corpora lutea persist until the end of pregnancy. Although ripe follicles appear in the ovary of *Funambulus* before parturition, a postpartum estrus and ovulation do not occur. Recruitment to the luteal-like interstitial tissue occurs in pregnancy from hypertrophied theca interna cells of small antral follicles that undergo atresia. The prairie dog ovary (Foreman, 1962) may contain apparently normal secondary and Graafian follicles with two to five oocytes, and a tubular medullary tissue is described as being sensitive to gonadotropins and similar to the interstitial tissue of *Talpa* and the anovular cords of *Sorex*. Like other squirrels (Chester Jones and Henderson, 1963), the palm squirrel, *Funambulus*, develops adrenocortical-like cells mainly in the hilar and medullary regions of the ovary after adrenalectomy (Seth and Prasad, 1967).

Although the Myomorpha comprises the greatest number of rodent species, very few have been investigated. New studies on several species of *Microtus* show that corpora lutea are only present in the ovaries of females kept with males (*M. pinetorum*, Valentine and Kirkpatrick, 1970; *M. pennsylvanicus*, Lee *et al.*, 1970; *M. ochrogaster*, Richmond and Conaway, 1969; and *M. agrestis*, Breed, 1967; Breed and Clarke, 1970a), but the only detailed work has been on *M. agrestis* (Breed, 1969; Breed and Clarke, 1970b; Breed and Charlton, 1971; Peters and Clarke, 1974). Luteinized follicles, which may reach nearly the same size as the corpora lutea, are formed during pregnancy and pseudopregnancy. Ripe follicles are present in the ovary about 3 days after ovulation and throughout pregnancy but no postpartum ovulation occurs, unlike *M. pennsylvanicus* (Lee *et al.*, 1970) and *M. ochrogaster* (Richmond and Conaway, 1969). Changes in the oocyte between the dictyate stage and diakinesis have been related to follicular growth rather than the ovulatory stimulus in *M. montanus* (Cross, 1971). Study of the ovary of the Skomer vole (Coutts and Rowlands, 1969) showed it to be similar to that of the mainland *Clethrionomys*, and some evidence of formation of accessory luteal tissue in a few animals was found. No such discrepancies between corpora lutea and embryos have been recorded in *Tatera indica* (Prasad, 1961; Bland, 1969) or in *Lemmus trimucronatus* (Mullen, 1968). A brief description of follicular and stromal elements in *Dicrostonyx* is given by Quay (1960). In *Peromyscus boylii* and *P. leucopus*, the corpora lutea of pregnancy (800 to 1300 μm in diameter) are reported to persist for about 2 months after the end of the breeding season with only slight decrease in size (1170 to 800 μm); the luteal cells appear healthy for some time and then develop into unilocular fat cells. The corpora lutea regress rapidly when reproduction starts again in the spring (Brown and Conaway, 1964). Two brief studies on Australian murids have been published. The one to four corpora lutea (mean maximum volume at day 8 about 13 mm^3) of *Mesembriomys* are probably formed from small thecal cells as well as from the larger granulosa luteal cells. Follicular atresia was common (Crichton, 1969) and all ovaries contained abundant interstitial tissue. Follicular atresia was not seen in *Leggadina* (Taylor and Horner, 1970). The numbers of corpora lutea were equivalent to the numbers of fetuses and were equally distributed between the two ovaries although 73% of the embryos were in the left uterine horn.

The ovary of *Pedetes* (Coe, 1969) is capsulated and only one corpus luteum of pregnancy was found. This is of interest because the pedetids have often been placed with the Hystricomorpha but no hystricomorph has been found to possess a bursa, and accessory corpora lutea are common in the group (Mossman, 1966; Weir and Rowlands, 1974). Since Mossman's (1966) statement that "the most extreme divergences from the typical rodent ovary occur among the Hystricomorpha" investigations on several species have confirmed the anticipated variety.

The mountain viscacha, *Lagidium peruanum*, has always been considered unique because although both ovaries appear similar until before puberty, ovulation nearly always occurs only from the right ovary (Pearson, 1949). Study of a pair of ovaries from a pregnant *L. boxi* (Weir, 1971d) indicated that the right ovary contained one large corpus luteum, presumably of the pregnancy, and twenty-two smaller corpora lutea which were probably luteinized follicles. Follicular atresia in the right ovary was less than that in the left which contained no normal follicles greater than 300 μm in diameter and was filled with interstitial tissue. Interstitial tissue (Figs. 12 and 13) is found in abundance in *Chinchilla, Dasyprocta* (although not in the closely related *Myoprocta*), *Hystrix, Lagostomus, Proechimys,* and *Octodon.* Accessory corpora lutea, formed from luteinized follicles (Figs. 5, 6, and 7), usually during pregnancy but sometimes during the cycle (*Chinchilla* and *Hystrix*) are common and have been shown to be functionally similar to true corpora lutea (Tam, 1970). The domestic guinea pig, *Cavia,* rarely develops them (Rowlands, 1956) but some individuals of *Galea* and *Myoprocta* (Weir, 1971a) have more corpora lutea than eggs or embryos. Accessory corpora lutea are infrequent in *Ctenomys, Octodon,* and *Proechimys,* all of which are suspected to be induced ovulators. Mossman (1966) suggested that the luteal-like masses described in *Myocastor* by Stanley and Hilleman (1960) were probably interstitial cell masses. Rowlands and Heap (1966) referred to these structures as accessory corpora lutea although they are not so numerous as in *Dasyprocta* (Weir, 1971b) or *Chinchilla* (Weir, 1966) in which they are indistinguishable in size and histological appearance from the true corpora lutea. Unilateral pregnancies do occur in these two species, but no differences in numbers or size of corpora lutea in the two ovaries have been found, so no light can be shed on the problem in *Erethizon* in which the accessory corpora lutea regress only from the ovary on the nonpregnant side (Mossman and Judas, 1949; Mossman, 1966). A high incidence of "immature testis tubules" is found in ovaries of *Myoprocta* and *Dasyprocta* (Weir, 1971a,b). However, the most unusual ovary yet described in the hystricomorphs is that of *Lagostomus,* a chinchillid (Fig. 5). The ovary appears to be modified to allow release of about 100 to 500 eggs at a time (Weir, 1971c). The neonatal ovary contains vast numbers of primordial oocytes and many were still present in a viscacha calculated to have lost at least 6000 eggs during estrous periods over 4 years. The surface of the ovary is very convoluted, increasing the surface area, and the mature, just antral follicle is no larger than 300 μm in diameter. The corpus luteum is about 400 μm, but it is estimated that up to 50% of the ripe follicles do not ovulate. Accessory corpora lutea are formed from these unovulated follicles and so the ovary of the pregnant plains viscacha is filled with very many small luteal bodies which are probably all functional as in *Chinchilla* (Tam, 1971) and *Myoprocta* (Rowlands *et al.*, 1970).

Large, presumably ripe follicles with more than one oocyte are frequently found in some adult hystricomorphs, but polyovuly in rodents is usually apparent in the early stages of postnatal development. This has been described in mice (Kent, 1960), guinea pigs (Collins and Kent, 1964), and hamsters (Bodemer and Warnick, 1961a,b; Kent, 1962), and an attempt has been made to correlate the incidence of polyovular follicles and polynuclear oocytes in the hamster with circulating estrogens and progestins (Kent and Mandel, 1968). It is suggested that such ovarian abnormalities are maximal just after the time when circulating estrogens are low. Estrogen has also been implicated in the occurrence of cystic follicles in guinea pigs which have been fed moldy lucerne hay (Schoenbaum and Klopper, 1969).

Light microscope studies on oogenesis and folliculogenesis have been reported in the rat (Beaumont and Mandl, 1962; Tuohimaa and Niemi, 1969; Vongpayabal and Callahan, 1970), guinea pig (Labhsetwar and Diamond, 1970; Greenwald, 1961; Greenwald and Peppler, 1968), and mouse (Yamada *et al.*, 1957; Peters, 1969). The rate of growth of medium-sized follicles in the immature mouse ovary is less than that of larger follicles (Pedersen, 1970a) and autoradiographic techniques have been used to follow cells through at least five cycles (Peters and Levy, 1966). Oocytes are not labeled but labeled theca interna cells move into the corpus luteum. The development time for a mouse follicle from the primordial stage to one ready for ovulation is estimated to be 19 days (Pedersen, 1970b), but large follicles may grow rapidly and regress without rupturing (Peters and Levy, 1966). A detailed proposal for classifying normal follicular stages in the mouse is suggested by Pedersen and Peters (1968). In the pregnant mouse the follicles develop at the same rate as in the normal cycle (Greenwald and Choudary, 1969), and Pedersen and Peters (1971) suggest that follicles which ovulate at parturition started growth at the beginning of pregnancy, although there may be many other follicles which grow at the same rate as those of the cycle but degenerate before term. Follicles are competent to ovulate for a shorter period in the rat (Brown-Grant, 1969), and evidence from induced ovulations for waves of follicular growth in pregnancy was found by Brown-Grant (1969) but not by Greenwald (1966). Follicles which can be ovulated by exogenous gonadotropins are also present in the ovaries of the pregnant guinea pig (Rowlands, 1956), hamster (Greenwald, 1964, 1967), and mouse (Greenwald and Choudary, 1969). Changes in the ovary of the lactating hamster vary with the presence or absence of suckling (Greenwald, 1965). At parturition all large antral follicles become atretic and the ovary is almost a solid mass of interstitial tissue, but if suckling ceases, follicular growth and ovulation occur in about 4 days (Greenwald *et al.*, 1967). Sato (1965) has given a comprehensive account of the ultrastructure of the rat ovary, but other authors have dealt with component tissues.

The fine structure of oocytes has been described for the rat (Sotelo, 1959; Sotelo and Porter, 1959; Odor, 1960; Franchi, 1960), mouse (Yamada *et al.*, 1957), guinea pig (Anderson and Beams, 1960; Adams and Hertig, 1964), and hamster (Weakley, 1966, 1967; Hadek, 1966). A description of the cytoplasmic lamellae in these four species (and the cat) is given by Weakley (1968), but differences may be attributable to different sensitivities to the procedure used. The hamster is reported to have unusual mitochondria which are aggregated into small groups and are separated by a dense vesicular substance (Odor, 1965); protein and RNA studies indicate that this substance may be involved in the synthesis of mitochondrial proteins (Weakley, 1971). As in other mammals, microvilli project from the surface of the oocyte and pass through the zona pellucida, and follicular cytoplasmic processes pass towards the oocyte. Franchi (1960) and Sato (1965) suggest that desmosome connections are present. Oocyte microvilli are present before the appearance of the zona pellucida and the oocytes may secrete part of the zona which appears while the follicles are still at the unilaminar stage (Odor, 1965; Weakley, 1966). The further development of the rat oocyte as it matures has been described by Sotelo and Porter (1959), Odor and Renninger (1960), and Mandl (1963). If ovulation is blocked centrally in the rat by pentobarbitone, meiosis is delayed in the oocyte but the follicle continues to grow (Toyoda and Chang, 1969; Freeman *et al.*, 1970). This seems strange since in the guinea pig (Perry and Rowlands, 1963) follicles can grow only up to the antral stage without pituitary support. The secretion of the liquor folliculi in the mouse primary follicle is associated with enlargement of the Golgi apparatus and the development of smooth-walled cisternae in the cortical areas of the cell (Hadek, 1963). The ultrastructure of the granulosa cells in larger follicles has been described for the rat (Björkman, 1962) and mouse (Byskov, 1969; Kimura *et al.*, 1970). Agranular endoplasmic reticulum develops in the granulosa cells before ovulation, but rat theca interna cells have an enlarged Golgi system (McDonald and Goldfien, 1965) and in the early stages of formation of corpora lutea invading thecal cells can be distinguished from the granulosa cells. These do not show Golgi enlargement until late diestrus (McDonald *et al.*, 1969).

The ultrastructure of the corpus luteum has been described for the mouse (Yamada and Ishikawa, 1960; Macaluso, 1965; Crisp and Browning, 1968), rat (Koizumi, 1965; Rennels, 1966; Cohéré *et al.*, 1967; Breinl *et al.*, 1967; Long, 1970), and guinea pig (Crombie *et al.*, 1971) and is similar to that described for other mammals (see pp. 182–184). In mice, two cell types are recognized and these may represent stages of development in one type or, more likely since light and dark forms of each are recognized, the theca and granulosa luteal cells. The eightfold increase of volume of the rat luteal cell is caused by proliferation of the endoplasmic reticulum; lysosomes accumulate near the cell periphery late in pregnancy (Long, 1973). This supports light microscopic and histochemical studies which correlate an increase of lysosomes with regression of the corpus luteum (Lobel *et al.*, 1961; Banon *et al.*, 1964). A detailed endocrinological account of luteolysis in the rat is given by Malven (1969) and blood supply variations are implicated by Clemens *et al.* (1968). Changes in the blood supply to the rat ovary in response to LH occur within 2 minutes and are maximal in 20 minutes (Wurtman, 1964). There are numerous papers on factors affecting the life of the corpus luteum, particularly in the guinea pig, which is susceptible to experimental studies (Sammelwitz *et al.*, 1961; Rowlands, 1961; Heap *et al.*, 1967; Greenwald and Rothchild, 1968; Deanesly and Perry, 1969) and the hamster (Duby *et al.*, 1969). Studies on corpus luteum formation in the mouse show that only 3 or 4 seconds of the ejaculatory reflex and the presence of the copulatory plug are sufficient to convert a nonfunctional cyclic corpus luteum into the functional one of pseudopregnancy or pregnancy (Land and McGill, 1967; McGill and Coughlin, 1970). A detailed light microscope study on luteinized follicles in the ovaries of two hamsters was described by Horowitz (1967) who compared these structures with normal corpora lutea and atretic follicles (see Volume I, Chapter 6). Atresia has also been considered at the ultrastructural level in the mouse (Beltermann, 1965b) and in histochemical studies in the guinea pig (Guraya, 1968a) and hamster (Guraya and Greenwald, 1965). Rat interstitial cells can be produced by causing atresia in follicles with gonadotropin treatment and their fine structure is described by Carithers (1969) and Merker and Diaz-Encinas (1969); changes typical of those in luteal cells are produced.

The germinal epithelium has been studied in the mouse (Wischnitzer, 1964, 1965) and the hamster (Weakley, 1969); several cell types from squamous to columnar are present and microvilli project on the free surfaces. Some peculiar tubular structures and epithelial crypts are reported in rats (Mozanska, 1968). Smooth muscle has been found in the rat ovary and it is connected mainly with blood vessels which pass to the theca externa of the developing follicles (O'Shea, 1970) and to the blocks of interstitial tissue (Osvaldo-Decima, 1970). The innervation of the rat ovary is briefly described by O'Shea (1971b). Mast cells have been found in the guinea pig ovary and in the early postnatal mouse ovary where they are probably involved with connective tissue formation (Skalko *et al.*, 1968). The postnatal and fetal ovarian structure has been studied electron microscopically (Odor, 1967; Odor and Blandau, 1969), and intercellular bridges, which may be implicated in synchrony of meiosis in oogenesis, have been described in the fetal mouse ovary (Ruby *et al.*, 1969).

9. Proboscidea

Perry (1953) found many, apparently active, corpora lutea in both ovaries of the pregnant African elephant (*Loxodonta africana*). He concluded that the elephant was polyestrous and polyovular and that at midpregnancy the first set of corpora lutea was replaced by a second set derived from luteinization of ovulated and unovulated follicles. On the basis of one female which was followed throughout estrus and had only one newly ruptured follicle at death, Short (1966) suggested that the elephant was polyestrous but monovular. To account for the corpora lutea present in the ovaries of early pregnant and nonpregnant, nulliparous females, a number of sterile estrous cycles before conception were postulated (see also Hanks and Short, 1972). Laws (1969) studied five different elephant populations and from the mass of data obtained he avers that there is no replacement or augmentation of corpora lutea in midpregnancy and that the elephant is seasonally polyestrous and polyovular. The small corpora lutea without stigmata found in early pregnancy are probably luteinized follicles.

The corpus luteum possesses a prominent system of channels with a well defined lining which are considered by Perry (1953) to be lymphatics. The luteal cell population is heterogenous; typical luteal cells are less abundant in the late stages of pregnancy (Smith, 1969). The presence of Golgi apparatus and cytoplasmic vacuolation was considered to indicate that the corpora lutea were active from the second to thirteenth month of gestation. The corpora lutea regress rapidly at term and form corpora albicantia which probably persist for life. Laws (1969) found an increase in corpus albicans number with increase of age; a 55-year-old female had sixty-one corpora albicantia.

Although Perry (1953) and Smith (1969) suggest that the corpus luteum appears active, Short (1966) disagrees, and progesterone secretion analyses have not been conclusive (Short and Buss, 1965; Smith *et al.*, 1969; Smith, 1969; Ogle *et al.*, 1973). Quantities recorded vary from 0 to 12.97 μg/gm of luteal tissue. Smith *et al.* (1969) suggest that either the elephant is extremely sensitive to progesterone or that this hormone may not be necessary for normal reproduction in this species.

The follicular system of the elephant has received little attention. In anestrous females the follicles are 4 to 6 mm in diameter and may be as large as 10 mm. In early pregnancy the follicles are 5 to 6 mm in diameter and decline about midpregnancy (Laws, 1969); very few follicles are visible in the second half of gestation (Perry, 1953; Laws, 1969; Smith, 1969).

10. Hyracoidea

The theca interna appears to be a hypertrophied, actively secreting tissue from the first sign of antrum formation in growing follicles of *Dendrohyrax* (O'Donoghue, 1963). Immediately prior to ovulation a follicle is 1.7 mm in diameter with the theca interna and membrana granulosa layers each 52 μm thick. Cyclic changes, related in some way to ovarian activity are discernible in the interstitial tissue which is centrally situated in the ovarian medulla and invariably separated from the cortex by medullary connective tissue. Interstitial cells are vacuolated during the follicular phase, the tissue is well vascularized and considered to have a secretory function. The fully formed corpus luteum measures 1.6 mm and increases to 2.1 mm in diameter during pregnancy. O'Donoghue maintains that the luteal tissue is formed solely from cells of the theca interna and that the granulosa cells degenerate. Thus, ". . . either the theca interna cells continue to secrete oestrogen, and 'luteinization' in this animal is merely hypertrophy . . . ; or else the thecal cells start to produce another hormone, progesterone, in addition to or in place of oestrogen, when they grow to form the corpus luteum." The corpus luteum maintains its size and activity throughout pregnancy; corpora albicantia are easily recognized and probably persist throughout life. No accessory corpora lutea are formed.

From a study of ovaries of *Procavia* in our possession, we do not consider that the theca interna of this hyrax resembles that of *Dendrohyrax*. No information on this point is given by Kayanja and Sale (1973).

11. Artiodactyla

A vast amount of work has been reported on the three main artiodactyl species of economic importance, the cow, the sheep, and the pig. A useful comparative review is that by Robertson (1969) and references to the early work will be found in Harrison (1962). The ewe and the sow are more alike in their physiology (see Volume II, Chapters 6 and 8) than either is to the cow, but for clarity the three are considered separately in this chapter.

a. Sow. The ovarian blood vessels are described by Andersen (1926) and Yamashita (1961, 1962). Bal and Getty (1970) review the changes from birth onward and variations in thickness of the tunica intima are specially noticeable.

Oogenesis and ovarian development in the prenatal pig are described by Black and Erickson (1968). Land (1970) has found variation in the number of oocytes at birth and in the number of antral follicles in pure and crossbred pigs, and variations in follicular development at different ages are found in four genetic groups of pigs (Bhalla *et al.*, 1969). The sequence of the maturation of oocytes after HCG injection has been followed by Hunter and Polge (1966).

The histochemical and ultrastructural appearances of follicles and isolated granulosa cells have been correlated with steroidogenic activity (Bjersing, 1967a). Some granulosa cells of large tertiary follicles contain agranular endoplasmic reticulum and whorls, which indicates that steroid synthesis may occur in follicular granulosa cells as well as in theca interna cells. After ovulation, further development of the agranular endoplasmic reticular membranes and whorls from the granular membranes occurs in luteal granulosa cells. This development, especially the formation of whorls, is most intense during the period of high progesterone secretion (Rathmacher and Anderson, 1968) and is maintained for a short time after cessation of progesterone secretion (Bjersing, 1967c).

Invasion of the folded granulosa layers by theca interna cells occurs between the first and third day after ovulation, but their subsequent fate is doubtful. Solomons and Gatenby (1924) suggested that they laid down the reticulum around the luteal cells, but Corner (1919, 1920) showed that the endothelial cells effected this. Theca interna cells can be recognized up to day 18 of pregnancy by their high phosphatase content, but thereafter the granulosa cells also acquire phosphatase granules (Corner, 1948). The granulosa cells are fully luteinized by the fifth to sixth day and the corpus luteum is generally solid by the eighth day, although corpora hemorrhagica are common. Regression of the corpus luteum in the nonpregnant animal commences at the thirteenth to sixteenth day. The degenerative process involves marked peripheral vacuolation, cytoplasmic shrinkage, fragmentation of nuclei, and subsequent disappearance of luteal cells. Corner (1921) finds numerous angular and elongated cells, with foamy cytoplasm and much osmiophilic material, that survive the degenerative process and that may be remnants of the theca interna. In nonpregnant pigs Mirecka (1969) claims to identify granulosa luteal cells and two populations of theca luteal cells.

In early pregnancy, the corpus luteum remains about the same size as it is at the twelfth to fourteenth day of the cycle and Corner (1915) distinguished a number of stages. The early developmental stages are divided into a preparatory stage of 25 days and then into two stages covering the next 15 days, in which there is a progressive appearance of vacuoles in the luteal cells. Perry and Rowlands (1962), in their study of early embryonic loss in the pig, found many distended, cystic corpora lutea in animals killed before the tenth day. Such forms appear to be transient as the corpus luteum reverts to a compact structure before the eighteenth day. From the fortieth to the seventy-fifth day there is a transitional period in which the vacuoles disappear. After day 75 diverse forms of endoplasmic vacuolation occur and regressive changes appear around the one hundred and tenth day. The number of corpora lutea at early and late stages of pregnancy have been counted by Longenecker *et al.* (1968); at day 40, 22% of the corpora lutea were new but it was not determined whether these were accessory or secondary corpora lutea (Reitmeyer and Sorensen, 1965).

Fine structural details of porcine luteal cells are given by Bjersing (1967b), Goodman *et al.* (1968), Cavazos *et al.* (1969), and Belt *et al.* (1970). At the periphery of the corpus luteum on day 1, luteal cells contain long cisternae of granular endoplasmic reticulum and many free ribosomes; in central cells the cisternae are short. Luteinization is essentially complete by day 4 and the hypertrophied luteal cells contain masses of agranular endoplasmic reticulum. Abundant small coated vesicles are associated with the Golgi complex and large coated vesicles are plentiful in relation to the plasmalemma. Regression, on days 8 to 12 of the cycle, is associated with an increase in cellular lipid in droplets, cytoplasmic disorganization, vacuolation of agranular endoplasmic reticulum, and an increase in the number of lysosomes. Rathmacher and Anderson (1968) consider that decreased blood flow does not precede luteal regression. Belt *et al.* (1970) correlate appearances with progesterone levels and describe changes after hysterectomy. During pregnancy and after hysterectomy, the granulosa luteal cells contained a larger Golgi apparatus, more agranular endoplasmic reticulum arranged as arrays of fenestrated cisternae, more granular endoplasmic reticulum, and a very much larger number of membrane-limited dense bodies than there are in the corpus luteum of the cycle. The increased granular endoplasmic reticulum has been associated with a more marked capacity for protein synthesis; the dense bodies may have a protein content related to relaxin.

The fine structure of four types of cell identified in the medullary epithelial cords of mature sows (Unsicker, 1971) suggests that some of these cells are steroidogenic.

b. Ewe. The maturation sequences of the oocyte are described by Dziuk (1965) and three phases of growth and atresia of follicles during the cycle are discussed by Smeaton and Robertson (1971). O'Shea (1971a) reports that cells in the theca externa of the sheep follicle contain cytoplasmic filaments and dense bodies characteristic of smooth muscle cells; he suggests that a contractile function may be involved in the collapse of the follicle after ovulation.

The growth stages of the corpus luteum (Thwaites and Edey, 1970) are similar to those of the sow except that bleeding is usually only slight. Most studies have been concentrated on the rapid structural and functional involution of the corpus luteum of the cycle (Edgar and Ronaldson, 1958; Restall, 1964; Hutchinson and Robertson, 1966). The appearances of the normally regressing corpora lutea and those after anti-bovine LH are described by Fuller and Hansel (1970) who state that LH is not only luteotropic in the ewe but that it is also necessary for follicular development and estrogen secretion. Deane *et al.* (1966) report that the first sign of cyclic regression is an alteration in size and density of mitochondria on days 12 to 13; this is followed, on days 13 to 14, by the appearance of large cytoplasmic lipid bodies in the large luteal cells and, on day 15, by a rapid decline in secretory activity as well as reduction in Δ^5-3β-hydroxysteroid dehydrogenase and diaphorase activities. The timing of the cyclic regression on day 12 may be related to the fact that at this time the corpus luteum must be stimulated by an embryo in the uterus to survive (Kaltenbach *et al.*, 1968). Dingle *et al.* (1968) considered lysosomal function in the corpus luteum; they demonstrated an increase in size and fragility of lysosomes late in the estrous cycle but not in the corpora lutea of pregnancy. Modification of lysosomal function, or of luteal cells, directly by a cell-specific lytic factor from the nonpregnant uterus is discussed in the light of observations that involution of the corpus luteum at the end of the cycle is dependent on the presence of a nonpregnant uterus (Wiltbank and Casida, 1956; Denamur and Mauléon, 1963; Moor and Rowson, 1966).

A detailed correlation of endocrine activity, histochemistry, and ultrastructure of corpora lutea is attempted by Bjersing *et al.* (1970a). Ultrastructural characteristics indicate that regression after hysterectomy (Bjersing *et al.*, 1970b) differs from cyclic regression in that the initial changes involve impaired hormone release rather than impaired synthesis.

c. Cow. A detailed study of prenatal and postnatal ovaries is reported by Erickson (1966a,b). Vesicular follicles are present at day 250 of gestation and normal vesicular follicles were found in ovaries of cows up to 10 years of age. Follicular growth and atresia are also described by Marion *et al.* (1968) who consider that there is a higher rate of follicular turnover in the cycle than in pregnancy (Choudary *et al.*, 1968). The absence of normal follicles of sizes above 2 mm during late pregnancy perhaps indicates a lack of follicle-stimulating hormone at this time (Labhsetwar *et al.*, 1964). The two main types of atresia (see Volume I, Chapter 6) of bovine follicles are discussed by Rajakoski (1960) and by Priedkalns *et al.* (1968). Rajakoski and Hafez (1963) have described interstitial cells in adult freemartin ovaries of bovine quintuplets, but this was associated with other signs of masculinization, such as the presence of a centrally located rete, epididymis, and the absence of a germinal epithelium. Normally the bovine ovary does not contain interstitial cells capable of steroidogenesis; the thecal cells of atretic follicles degenerate.

The period for development of the follicle from a primordial state is about 10 days (Priedkalns *et al.*, 1968). Asdell (1960) suggested that an elevated mitotic rate in follicle cells at days 8 to 10 of the cycle may be due to estrogenic stimulation as indicated by vaginal cornification. Rajakoski (1960) described two consecutive waves of follicular growth within one estrous cycle, but Priedkalns *et al.* (1968) find that more than two growth waves, or even a continuum of new follicles, may arise at any time in the cycle. The first wave generally reaches maturity at midcycle. There is an even distribution of the numbers of large and developing and atretic follicles after midcycle. The follicle destined to ovulate is therefore assumed to begin its development at midcycle.

Folding of the follicle wall occurs before ovulation and Lobel and Levy (1968) report "luteinization" in preovulatory follicles. They estimated luteinization by morphological characteristics and by cytochemically demonstrable enzyme activities which correlated with secretion. Periovulatory changes in granulosa and thecal cells have been described in ultrastructural detail by Priedkalns and Weber (1968a). During the 3 to 4 days before ovulation the granulosa cells cease growing in size, while the theca interna cells form two populations: large epithelioid cells with round nuclei and smaller fibroblast-type cells with spindle-shaped nuclei. Donaldson and Hansel (1965) and Hafs and Armstrong (1968) describe morphological changes in the corpus luteum of the cycle. Mitosis continues for about 2 days after ovulation in the granulosa luteal cells and for about twice that time in the theca cells. Then growth is mainly due to hypertrophy of the large luteal cells which are formed from granulosa cells about 3 to 4 days after ovulation. The spindle-shaped theca interna cells become rounded and together with some paraluteal and trabecular luteal cells (both probably of theca externa origin) contribute to the "small luteal cell" population. The epithelioid theca interna cells disperse into the ovarian stroma, and eosinophils and mast cells are commonly observed among the theca cells at this time. Lipid secretion signs and ultrastructural organelles (notably abundant agranular, tubular, branching endoplasmic reticulum, large round mitochondria with tubular cristae, and lipid bodies) are indicators of steroid synthesis, estrogen in the epithelioid theca interna cells during proestrus and estrus, and progesterone in the large luteal cells during metestrus and diestrus (Priedkalns and Weber, 1968a,b,c).

The young corpus luteum has a fluid-filled central cavity which often persists throughout its life span. One day after ovulation, luteal cells, whether of granulosa or thecal origin, contain demonstrable levels of Δ^5-3β-hydroxysteroid dehydrogenase and NADP and $NADPH_2$ diaphorases (Lobel and Levy, 1968). The enzyme levels increase during the period of luteal cell proliferation and growth, and are maximal at day 9 (Höflinger, 1948), especially in the large luteal cells. Hydrolytic enzymes increase with the development of the vascular system. The corpus luteum of pregnancy, which grows steadily during the first 3 months, is heavier than that of the cycle (Edwards, 1962) but cannot be distinguished by light microscopy. The only difference is that phosphorylase activity is observed in luteal cells of gestational corpora lutea but is confined to vascular walls in cyclic corpora lutea. The fine structure of luteal cells in cyclic and pregnant heifers is compared by Leland *et al.* (1969).

Involution of cyclic corpora lutea on day 14 begins with degenerative changes in blood vessels: pycnosis of endothelial cell nuclei and a sudden decline of hydrolytic enzyme activities in the vascular wall occur. The luteal cells may show large lipid bodies in association with lysosomes and granular bodies, and, subsequently, luteal cell dehydrogenase activities decline. Large epithelioid luteal cells which contain PAS-positive material are found in groups during regression of corpora lutea (Moss *et al.*, 1954). Luteotropic and luteolytic mechanisms affecting bovine corpora lutea are discussed by Hansel (1966).

d. Other Species. Some references to ovarian characteristics in less well known artiodactyls are given by Asdell (1964).

The ovary of the goat (*Capra*) is similar to that of the ewe and is described by Harrison (1948b) and Lyngset (1968a,b,c). Seasonal variation of ovarian activity is investigated in the buffalo (*Bubalus*) by Shalash and Salama (1962) and Roy and Mullick (1964) discuss the corpora lutea of this species.

In the peccary (*Tayassu angulatus*), old corpora lutea seem to persist longer than usual: larger (10 mm) and smaller ones (7 mm) were present in a pregnant female (Wislocki, 1931). Laws and Clough (1966) note equal incidence of corpora lutea in right and left ovaries in *Hippopotamus*. An almost linear increase in numbers of accessory corpora lutea occurs during pregnancy to reach a mean of 7.5 at term. In the warthog (*Phacochoerus*), Clough (1969) reports that ovulation takes place from follicles 7 to 8 mm in diameter. The corpora lutea grow rapidly to 10 mm in diameter and maintain this size throughout pregnancy. Regression of the corpus luteum is rapid and corpora albicantia persist for several months.

Several studies on Camelidae have been reviewed recently (Novoa, 1970). The left ovary of the dromedary (*Camelus dromedarius*) is more active than the right. The corpus luteum persists throughout pregnancy (Bezrukov and Schmidt, 1970), reaching a size of 2 cm, but regresses quickly after parturition (Tayes, 1948). Ovulation and corpus luteum formation are described in llamas (*Lama glama*) by England *et al.* (1969). In alpacas (*Lama pacos*), corpora lutea formed after mating or after HCG-induced ovulation attain a maximum size at 8 to 9 days and regressive changes are observed as early as day 12. There is a sharp decline in progesterone levels at day 13, indicating that the corpus luteum has a short life span and that regression in nonpregnant alpacas is earlier than in other domestic ungulates, except perhaps the pig. Corpora lutea in right ovaries regress more rapidly than those in left (Fernandez-Baca *et al.*, 1970).

Ovarian changes in Cervidae have been investigated in relation to twinning and delayed implantation in roe deer (*Capreolus capreolus*) by Stieve (1950), Hamilton *et al.* (1960), and Short and Hay (1966). There is an increase in the size and weight of the corpus luteum during or after the period of delay, and the corpus luteum does not regress as pregnancy advances. In *Muntiacus reevesi* both ovaries are equally active and the single corpus luteum persists until term but has regressed 10 days after birth. Follicular growth may occur in late pregnancy (Chaplin and Harrison, 1971). Elk, *Cervus canadensis,* display secondary corpus luteum formation at about the time the embryo is macroscopically visible. Both the original and the secondary corpora lutea persist until term; hemorrhage may occur from the follicle which ruptures during pregnancy (Halayon and Buechner, 1956; Morrison, 1960). No unusual features were described in reindeer ovaries (Borozdin, 1969).

Thomas (1970) describes changes in the ovaries of the deer *Odocoileus hemionus columbianus.* The cycle in this deer is peculiar in that pregnancy apparently never ensues from the first ovulation of the season, which is unaccompanied by estrus in the majority of females. Follicles are poorly synchronized at first ovulation and in some females one or more of the large follicles fails to rupture. These develop into accessory corpora lutea with large, fluid-filled cavities. These and smaller luteinized follicles occur in 47% of females after first ovulation. At second ovulation the follicles are more uniform in size and maturity resulting in a lower incidence (11%) of luteinized follicles, most of which are small. Secondary corpora lutea also develop from small, ruptured follicles and, as in primary corpora lutea, their cells are derived from the membrana granulosa and the theca interna. Hemorrhage from thecal vessels may occur several hours after ovulation, but a large proportion of follicles that rupture at first ovulation do not develop into corpora hemorrhagica. Considerable mitotic activity occurs during the first 4 days in corpora lutea of pregnancy, but cell division ceases after about 2 days in nonconceiving females. The fate of the cells from the three follicular layers could not be followed with certainty after 2 or 3 days. From 5 to 10 days after conception, luteal cells containing a central, acidophilic granular zone and a highly vacuolated outer zone are prominent. The corpora lutea of pregnancy attain their maximum size just prior to parturition, although their cells are greatly vacuolated. Regression of the corpora lutea is rapid after parturition and may begin before, but the resulting pigmented scars are composed almost entirely of thick-walled blood vessels and may persist for life. Trauger and Hauger (1965) have studied color and size variation of corpora lutea in *Odocoileus virginianus* and Hesselton (1967) has recorded three animals in which the number of corpora lutea was one less than the number of fetuses. Because the fetuses were of different sex, he concluded that the supernumerary fetuses arose from polyovular follicles. A similar situation has been observed in roe deer (R. E. Chaplin, personal communication). Sinha *et al.* (1971b) consider that theca interna and granulosa lutein cells contribute to the corpus luteum of pregnancy in *O. virginianus*.

In contrast to the white-tailed deer, in the pronghorn, *Antilocapra americana* (O'Gara, 1969), there are three to seven ovulations and all but two of the blastocysts are eliminated at implantation or at the thread stage.

In the Uganda kob (*Adenota kob thomasi*) the presence of small corpora lutea in adult estrous females suggests that silent ovulation occurs without fertilization before an overt estrus (Buechner *et al.*, 1966; Morrison and Buechner, 1971). The adolescent kob apparently ovulates two or three times before the first conception. The developing follicle reaches 8 to 9 mm in diameter and the corpus luteum of pregnancy measures 10 to 14 mm and degenerates after parturition (Morrison, 1971). Kellas (1955) gives some details of gross ovarian morphology in the dik-dik (*Rhynchotragus kirkii thomasi*), finding no increase of ovary weight with age in mature specimens. The largest follicle was 4 mm and the corpus luteum had a diameter of 5 to 7 mm. Spinage (1969) found the largest follicle in *Kobus defassa ugandae* measured 11 mm. The corpus luteum reached maximum size by the thirtieth day of pregnancy and remained constant in size until full term. No secondary corpora lutea developed and thecal luteal cells could not be discerned. Regression of the corpus luteum occurs rapidly after parturition. As many as ten corpora albicantia were found in a 10-year-old female but there was no correlation between the number of corpora lutea and the expected number of pregnancies. The impala (*Aepyceros melampus*) ovary is described by Kayanja (1969). The follicles rupture at 8 mm in diameter and the theca interna is still very prominent at this stage. The corpus luteum grows rapidly and regresses at parturition. Precocious development of antral follicles and formation of corpora lutea have been described in the fetal giraffe ovary (Kellas *et al.*, 1958; Kayanja and Blankenship, 1973).

12. Perissodactyla

The remarkable hypertrophy of the fetal gonads in the horse, with the associated increase in the size and numbers of primary interstitial cells, has been known for many centuries (see Davies *et al.*, 1957, and Volume I, Chapter 1). Hypertrophy begins at about the fourth month and reaches a maximum at the eighth month of fetal life, which correlates with the peak output of urinary estrogens by the mare. Most investigators ascribe the cause of the fetal gonadal enlargement in the mare to placental gonadotropins. The constituents of horse embryonic gonads have been analyzed by Pace (1966) especially with reference to serotonin and tryptophan.

A considerable extent of the ovulation fossa of the ovaries of adult mares is directly covered by columnar epithelium with many ciliated cells. This epithelium is frequently folded, with short clefts projecting into the ovarian surface; the remainder of the fossa is covered by a simple, unfolded cuboidal or squamous epithelium. Fossa cysts are common around the ovulation fossa, exhibiting varied structure; they and the epithelium may be of Müllerian duct origin (O'Shea, 1968).

The epoöphoron is frequently recognizable in equine ovaries and may also contain cysts (Arthur, 1958). Bergin and Shipley (1968) find that in over 900 ovaries from mares 1.5 to 26 years in age, all mature follicles and corpora lutea impinge on the ovulation fossa. Witherspoon and Talbot (1970) have used fiber optics, cinematography, surgical translocation, and injection of follicles with India ink to show that ovulation occurs only through the ovulation fossa. A sudden rupture in the tissue is followed by a rush of follicular fluid without any prior indications. The extensive connections and fine ramifications of the ovarian artery and its branches are described by Simić and Gadev (1968) using radiological techniques.

Küpfer (1928), Hammond and Wodzicki (1941), and Harrison (1946) find that the active stage of the corpus luteum of the mare is short compared with that of the cow and sow, and that the maximum diameter of the fully developed gland is less than that of the mature ovarian follicle. Benirschke and Sullivan (1966) have found oocytes and primordial follicles to be extremely rare in the ovaries of proven mules but cystic follicles and corpora lutea were found in about 10% of ovaries.

Four stages have been described for the corpora lutea of pregnancy by Cole *et al.* (1931). During the first 40 days there is only one corpus luteum in the ovary. Numerous follicles develop in one or both ovaries from the twenty-fifth day (Bain, 1967). About the fortieth day several follicles become fully luteinized; some are not true corpora lutea but luteinized theca interna cells that remain after the granulosa cells have degenerated (Kimura and Lyons, 1937). Regression of these corpora lutea commences about the one hundred and fiftieth day, and only minute vestiges of the glands are present during the later stages of pregnancy. Amoroso *et al.* (1948) have confirmed the observations of Kimura and Lyons (1937) that regression of the primary corpus luteum commences toward the end of the first month, and suggest that during the second month of pregnancy a succession of follicles ovulate to form new corpora lutea, a suggestion consonant with the reports of van Rensberg and van Niekerk (1968) and Allen (1971). Fallon (1967) describes follicle formation during pregnancy.

The possible steroidogenic activities of these interna and granulosa and luteal cells in the mare's ovary have been discussed by Short (1962), Ryan and Short (1965, 1966), and YoungLai and Short (1970). Although the ability of both granulosa and thecal cells to aromatize steroids has been demonstrated, other somewhat equivocal results still leave open the precise capabilities of the cell types either as single entities or in combination.

13. Primates Other than Man

Details of the reproductive organs of primates with accounts of the macroscopic appearances of the ovaries, and, where available, some histological considerations, are reviewed by Eckstein (1958).

The original reports by Gérard (1920, 1932; Gérard and Herlant, 1953) of the unique oogenetic activity in the ovary of the adult *Galago* has been followed up by more extensive investigation of a number of prosimians: *G. crassicaudatus* and *G. senegalensis, Galagoides demidovii, Loris tardigradus, Nycticebus coucang, Perodicticus potto,* and *Daubentonia madagascarensis.* Numerous cords of cells resembling germinal cords of embryonic cortex are seen in the adult ovarian cortex. Large nests of cells contain several dozens of germ cells enclosed in a thin basement membrane. Oogonia are at interphase and in all phases of mitosis, and oocytes at each stage of meiotic prophase are present. Large oocytes at diplotene are more usually outside the germinal nests and within a primary follicle. Some cells in the nests incorporated [^{3}H]thymidine, suggesting DNA synthesis and active oogenesis. Many of the oogonia undergo atresia and disappear. The new germ cells are probably formed by division of preexisting oogonia but it is still not known whether they ever constitute part of the definitive oocyte population (Herlant, 1961; Petter-Rousseaux, 1962; Petter-Rouseaux and Bourlière, 1965; Duke, 1967; Ioannou, 1967; Anand Kumar, 1968; Butler, 1969). Duke and Luckett (1965) have described the ovaries of several tree shrews; no suggestion of neogenesis was made.

The fine structure of the oocyte and the germ cell–follicle cell relationship in *Macaca mulatta* has been described by Hope (1965). The appearance of the zona as a two-layered structure, homogenous next to the oocyte and flocculent and electron-dense next to the follicle cells, leads Hope to suggest a dual origin for the zona in this species. Desmosome endings of the follicle cell processes were found and, unlike other mammals (see Section III,A), some processes penetrated into the cortical region of the oocyte as in amphibians.

The appearances of ovaries of *Macaca mulatta* from embryonic, fetal, and perinatal periods and also some from the postnatal period up to age 26 years are described and fully illustrated by van Wagenen and Simpson

(1965). The gonad is recognizable as an ovary at a crown-rump length of 25.5 mm and a conception age of 46 days: a definite cortex is established by 61 days and becomes lobulated with connective tissue outgrowths by 75 days when the nuclei of increasing numbers of oocytes are in the prophase of meiosis. By 110 days primordial follicles have appeared and by 153 days they constitute over half of the cortex, often widely separated by connective tissue and degeneration products. A few follicles, with cuboidal granulosa and enlarged, often degenerate, oocytes lie at the corticomedullary junction; an occasional vesicular follicle and a few polyovular follicles are also present. A table gives the shifts in the site of formation of primordial follicles from a conception age of 110 days until the ninth postnatal month and there is a comparison of ovarian development with that in human ovaries. The development of the monkey ovary between the fifth month after conception and the time of birth is much faster than in man. Oogenesis in this species has been studied by Green and Zuckerman (1951, 1954) and Baker (1966).

Folliculogenesis starts at 1 mm in *Macaca mulatta* when the corpus luteum of the cycle shows signs of degeneration; at day 7 the follicle is 4.5 mm and at days 12 to 14 it is 6.8 mm in diameter. The follicular fluid is thought to become viscous in preovulatory follicles (Koering, 1969) and the final swelling of the follicle is attributed to a dissociation of collagen about the periphery of the follicle (Espey, 1967).

Theca interna tissue of preovulatory follicles is sparse but well vascularized. Koering (1969) has discussed whether the interstitial tissue derived from the theca interna cells of atretic follicles (see Mossman *et al.*, 1964) could be a source of progestins (Johansson *et al.*, 1968). Another type of interstitial tissue, with glandular cells in cord-like arrangements and which tends to be in functional phase with atretic follicular thecal tissue, has been described by Koering (1969). Specialized cells in the hilar region in a few macaques were thought to be adrenal rests (Koering, 1969), like those found in *Erythrocebus patas* (Conaway, 1969).

In vitro culture studies (see Volume III, Chapter 6) have shown that granulosa cells from 6- to 8-mm-diameter follicles secrete 100- to 1000-fold more progestins than cells from small follicles harvested earlier in the cycle (Channing, 1970a,b,c).

Morphological changes on the surface of follicles of rhesus monkeys could not be detected within 1 or 2 days of ovulation by Betteridge *et al.* (1970), although Johansson *et al.* (1968) reported that the stigma can be found within 2 days before ovulation. Betteridge *et al.* suggest, however, that Johansson *et al.* were mistaking fresh ovulation points as follicles showing preovulatory stigma and that their monkeys were ovulating earlier than supposed. Even should this have been so, then the precise relationship between plasma progestin levels and ovulation still needs to be settled. Betteridge *et al.* also point out that the age of a corpus luteum in a rhesus monkey cannot be judged from the morphology of the ovulation point. Some corpora lutea persist from previous menstrual cycles in a form that resembles a recent ovulation point.

The corpus luteum of *Macaca mulatta* has been comprehensively described (see Corner *et al.*, 1945; Koering, 1969). Before rupture of the follicle the granulosa cells become loosened and radially arranged. Blood vessels begin to invade the granulosa cells on the second day and reach the inner part of the corpus luteum about the fourth day, when fibroblastic invasion is apparent. Up to day 4, theca cells can be recognized by their position, large size, sharp outline, and lipid-filled cytoplasm, but after the tenth day the granulosa cells exceed the thecal cells in size and irregularity of vacuoles.

In females that do not become pregnant degeneration of the corpus luteum starts about the thirteenth day after ovulation. Numerous cells can be seen with many closely packed vacuoles of uniform size evenly dispersed throughout the cytoplasm ("mulberry" cells). During menstruation almost every cell shows either the "mulberry" type of vacuolation, or possesses single "giant" vacuoles. The volume of the corpus luteum in nonpregnant animals increases until a few days before the beginning of menstruation and then decreases suddenly. There is shrinkage of degenerating luteal cells and a collapse of blood vessels. Production of progesterone continues to rise until about postovulatory day 10 to 12 and then decreases.

The corpus luteum of pregnancy cannot be distinguished from a cyclic corpus luteum of the same age until visible degeneration of the latter sets in on the thirteenth day. By the twenty-fourth day there is a reduction in size of the luteal cells, together with a disappearance of lipid vacuoles. The theca interna cells retain their lipid and can be seen as clumps of cells packed into the folds of the corpus luteum, especially at the bases of the folds. The period from the nineteenth to the twenty-fourth day is described as one of "transition," associated with the fact that gonadotropic substances in the urine of early pregnancy are disappearing at this time (Hamlett, 1937). There may also be a connection between the change occurring in the corpus luteum at this time and the "placental sign" between the twelfth and twentieth day (Hartman, 1929). The corpus luteum of pregnancy after the third week can be distinguished by its distinctly folded pattern, the small size of the luteal cells and absence of lipid vacuoles, the prominence of the peripheral capillary network, and the distinctness of the theca interna cells. This typical pregnancy state persists until the one hundred and forty-sixth day. Specimens obtained on day 154 and day 169 show progressively advancing signs of degeneration. Gulyas (1974) has reported an ultrastructural study of the corpus luteum in late pregnancy.

Corpora aberrantia (see Section IV) are commonly found in macaques.

There may be none to ten in an ovary, and certain females show a predisposition for this type of regression of the mature corpus luteum. The corpora aberrantia may survive for 23 weeks and do not produce progesterone (Koering, 1969). Some corpora aberrantia, or parts of them, may regress more rapidly than others; persistence of this atypical form therefore requires a stimulus from some external source, presumably the pituitary. That this transformation of the normal corpus luteum probably occurs during the earliest phase of regression is supported by the distribution of luteolipin (Rossman, 1942) and by the study of specimens removed at ovariectomy (Koering, 1969). Accessory corpora lutea (see Section IV) are formed in about 17% of ovulatory cycles. This form probably produces limited progesterone; thecal cells are poorly developed or absent. If a corpus luteum becomes a corpus aberrans, all accessory corpora lutea will become corpora aberrantia accessoria.

Zuckerman and Parkes (1932) described marked folding of the follicular wall in *Papio porcarius* and *P. hamadryas* after ovulation which gives the corpus luteum a lobulated appearance. The luteal cells reach their maximum development after 7 to 8 days. Theca cells form the vascular reticular tissue. By the thirteenth day, when menstrual flow commences, signs of regression are marked. When the next luteal phase is well established the corpus luteum of the previous phase is hardly discernible. The corpus luteum of pregnancy and the individual luteal cells are larger than those of the cycle. Maximum development occurs during the third week of pregnancy, but the volume of the corpus luteum and the size of the luteal cells falls after 4 to 5 weeks and remains fairly constant for at least 26 weeks. The corpus luteum disappears rapidly after parturition. Culiner (1945, 1946) finds that theca interna cells undergo changes similar to those of luteinization both in atretic and cystic follicles. They may give rise to theca luteal cysts and "yellow bodies" which resemble luteal cysts and corpora lutea respectively. Culiner suggests that this precocious activity of the theca interna may be related to irregularities of the menstrual cycle. Dempsey (1940) has described the ovary of *Hylobates lar* and has observed thecal luteinization during the luteal phase like that found in *Papio* and by Săglik (1938) in the orangutan.

14. Man

The characteristic structure and cell types of the human ovary from early embryonic stages until the menopause have been the subject of reviews too numerous to list (see Simkins, 1932; Sauramo, 1954a–d; Watzka, 1957; Young, 1961; van Wagenen and Simpson, 1965; Crisp *et al.*, 1970, for references). Recent investigations have made use of the electron microscope to describe the fine structure of human ovaries.

Cytological studies of germ cells of fetal, near term, and postnatal human ovaries include those of Ohno *et al.* (1961, 1962), Manotaya and Potter (1963), and the quantitative studies of Baker (1963). Fine structure of the oogonia and oocytes and their relation to follicle cells and the zona pellucida have been investigated by Wartenberg and Stegner (1960), Tardini *et al.* (1960, 1961), Stegner and Wartenberg (1963), Lanzavecchia and Mangioni (1964), Baker and Franchi (1967), Hertig and Adams (1967), and Gondos *et al.* (1971). In the nucleus of the oocyte, electron-dense chromosomal threads or "cores" are similar to those in the rat and similar, but not identical, variations in structure occur through the stages of differentiation of the oocyte. Human oocytes in primordial follicles contain lampbrush chromosomes (Baker and Franchi, 1967). The cytoplasm does not undergo profound changes as prophase advances; organelles become more numerous and the internal structure more complex. Oogonia which degenerate during mitosis exhibit chromosomal fusion into an irregular mass and abnormal membrane-bound areas are present in the cytoplasm. Oocytes in the primordial follicles of adult ovaries possess a complex paranuclear structure composed of a mass of mitochondria with associated endoplasmic reticulum, multiple compound aggregates forming a ring around the cytocentrum, and with a stack or coil of annulate lamellae either free or attached to the nuclear membrane. Hertig and Adams (1967) identify this structure as Balbiani's (1864) vitelline body and discuss the function of the vacuole-containing compound aggregates as reservoirs or waste deposits and the annulate lamellae which appear to be more prevalent when estrogen production is dominant in the normal menstrual cycle. Compound aggregates are also found in some cortical cells and may be evidence that such cells are related to those within the wall of the follicle. Hertig and Adams (1967) comment on the absence of mitotic activity in primitive follicle cells and suggest that these cells may be renewed by a process other than mitosis. Cortical cells might move into the crescentic cap of the follicle wall, slide round the oocyte and return to the cortex. Some follicle cells do indeed appear to overlap the periphery of their neighbors and are separated by a material similar to that of the basal lamina: furthermore, there is a resemblance in density between certain dark follicle cells and those of the cortex. The presence of oocyte microvilli and follicle cell processes ending in desmosomes on the oocyte plasma membrane has been clearly described and figured by Wartenberg and Stegner (1960) who also discuss the mucopolysaccharide nature of the layers in the zona pellucida.

In vitro incubation of theca and granulosa cells from preovulatory follicles (Ryan and Petro, 1966) indicated that the thecal cells could synthesize mostly estrogens and the granulosa cells mostly progesterone. However, it is still debated whether or not the granulosa cells contribute to estrogen secretion (Guraya, 1968c).

Follicle counts in nonpregnant women are given by Winter (1962). Govan (1968, 1970) has described the appearances of the human ovary in pregnancy and finds little follicular activity in mid-pregnancy: almost all follicles are atretic. From 33 weeks until term healthy follicles up to 4 mm in diameter appear in steadily increasing numbers; this was also noted by Maqueo and Goldzieher (1966). The failure of these follicles to grow beyond a certain size, in fact the size usually found in micropolycystic disease, suggests the intervention of an inhibitor or that the follicles have reached a critical size when a certain level of a particular gonadotropin(s) is needed for further development. Theca interna cells of such follicles show hypertrophy and when the follicle becomes atretic with disappearance of the antrum, a solid thecal mass is formed. The origin of such thecal masses may be from follicular elements, as Govan (1970) finds, or from stromal cells (Mossman *et al.*, 1964; Scully and Cohen, 1964; Novak *et al.*, 1965). Theca interna cells remain as distinct groups of cells about the periphery of the corpus luteum and in the bases of the folds formed in the wall of the collapsed follicle. These cells have been designated the "paraluteal cells," or the whole group has been collectively known as the "paraluteal" gland. Mossman *et al.* (1964) have endeavored to clarify the confusion that has developed in relation to thecal interstitial and paraluteal cells in human ovaries. They state that there are indeed luteal cells formed from the granulosa cells of ovulated follicles, and that there are thecal gland cells formed from the theca interna cells of ripening follicles, present only in individuals that are sexually mature or nearly so and only at or near the time of ovulation. They maintain that there are also interstitial gland cells formed from the theca interna of atretic follicles from infancy to old age. The paraluteal cells may be persistent thecal gland cells or intermediate stages in the differentiation of stromal cells into luteal cells which may occur after the disappearance of the thecal gland. White *et al.* (1951) and Hertig (1964) postulate a migration of theca interna cells into the granulosa layer of recently ruptured follicles. They call these stellate cells "K" cells because of their presumed high content of ketosteroids and believe that they produce progestins. Gillman and Stein (1941) considered that such cells represented different phases in secretory activity by thecal or granulosa cells, and Nelson and Greene (1953, 1958) thought them to be degenerating forms. Electron microscopy suggests that the sudanophilic stellate cells are probably macrophages (Adams and Hertig, 1969a; Gillim *et al.*, 1969) which lack histochemically demonstrable steroidogenic enzymes and contain lipid droplets which do not contain cholesterol and its esters (Fienberg and Cohen, 1965). The dense eosinophilic stellate cells described in the reports of White *et al.* (1951) and Hertig (1964) may correspond to the dense stellate cells with eccentric and irregular nuclei and other characteristics which are described by Adams and Hertig (1969a,b) in the 3- and 5-day postovulatory corpus luteum and which appear to undergo changes as pregnancy advances. Evidence of secretory exhaustion, cell shrinkage, and vacuolation occurs from days 9 to 12; so-called "mulberry" cells appear on day 13 and degenerating nuclei from day 14. Corpora lutea later in menstruation show a variable amount of cellular degeneration but this bears no relation to the duration of bleeding. Eventually the corpus luteum collapses entirely and the nuclei of the extremely concentrated cells display great variation in size and structure. Finally, the whole corpus luteum is replaced by a structureless hyaline body, the corpus albicans. It is suggested that complete regression of the corpus luteum takes between 7 and 10 months.

Ardent attempts have been made to differentiate under the light microscope a corpus luteum of the cycle from that of pregnancy (Marcotty, 1914; Dubreuil, 1944; Gillman and Stein, 1941; Nelson and Greene, 1958). A corpus luteum of pregnancy may be distinguished by its greater size and central cavity, by an increased amount of vascular and connective tissue, by the presence of colloid and/or calcium and certain lipoids. Presence of calcium or colloid material, not unlike that of Call–Exner bodies, in the luteal cells is said to be the best diagnostic characteristic. Accurate ageing of the corpus luteum of pregnancy, other than as early (1 to 3 months), middle (3 to 6 months), and late (6 to 9 months), is still virtually impossible using a light microscope. Polyploidy in human granulosa luteal cells is found between the third and ninth postovulatory day and during pregnancy; it may be related to chorionic gonadotropin and/or LH secretion (Stangel *et al.*, 1970).

Nelson and Greene (1958) consider that the corpus luteum of pregnancy functions only during the first 6 weeks and that on histological grounds it deteriorates rapidly during the eighth to sixteenth weeks (see also Gillman and Stein, 1941). During late pregnancy the ovary has been assumed to act only as an end organ responding to placental influences, and that the placenta is the source of the increasing quantity of circulating progesterone. However, Green *et al.* (1967) and Adams and Hertig (1969b) have demonstrated in electron microscope studies that massive regression of the corpus luteum in the later months of pregnancy does not take place and that luteal cells at term are capable of steroidogenic secretion. Nevertheless, there seems little doubt that Savard *et al.* (1965) are right in concluding, from their comparison of progesterone-forming abilities of corpora lutea of various ages, that the human corpus luteum undergoes profound metabolic changes during its life history. Electron microscope studies (see pp. 182–184) suggest that the presence of concentric whorls in human granulosa luteal cells of early pregnancy may distinguish them from those of the menstrual cycle (Vacek, 1967; Crisp *et al.*, 1970). Human corpora lutea of the menstrual cycle and early pregnancy are capable, *in vitro*, of synthesis of estrone and estradiol-17β as well as several progestagens (Huang and Pearlman, 1963; Hammerstein *et al.*, 1964; Savard *et al.*, 1965; Ryan and Smith, 1965; Arceo and Ryan, 1967).

Meyer (1911) made the fundamental statement on the life history of the human corpus luteum. Reviews by Pratt (1935), Gillman and Stein (1941), Brewer (1942), Harrison (1948a), and Nelson and Greene (1953) should be consulted for the historical background of observations based on the use of the light microscope. Meyer divided the life history into four phases, now well known as the phases of proliferation and hyperemia, vascularization, maturity or bloom, and regression. Corner (1956) finds it impossible to date the age of the corpus luteum of the cycle to within 1 day. Proliferation of the granulosa cells occurs during the first day after ovulation. Capillaries invade the granulosa on day 2 and reach the cavity on day 4. Few mitotic figures are seen in the granulosa and theca interna before day 4, none afterwards, Hemorrhage into the cavity may occur on any day, commonly on days 2, 3, and 9 and regularly at ovulation and menstruation. Fibroblasts appear in the central cavity on day 5 and from days 3 to 6 the granulosa and theca interna are indistinguishable. Capillary dilatation reaches a peak on day 7 and thin-walled venules appear along the border of the central cavity. By day 8 the corpus luteum has reached its secretory peak and connective tissue lines the central cavity.

At the end of the so-called stage of vascularization certain regressive changes appear in the blood vessels and both fatty degeneration and degeneration by simple atrophy take place in the luteal cells. There is an increase in the amount of visible lipids and cholesterol esters, accompanied by a diminution of the phospholipid content.

Electron microscope observations on active human corpora lutea (Carsten, 1965; Green and Maqueo, 1965; van Lennep and Madden, 1965; Tokida, 1965, Motta, 1969; Green *et al.* 1967, 1968; Pedersen and Larsen, 1968; Adams and Hertig, 1969a,b; Gillim *et al.*, 1969; Crisp *et al.*, 1970) have emphasized the abundant quantities of tubular smooth endoplasmic reticulum (strongly believed to be essential to steroid biosynthesis, Ryan and Petro, 1966), mitochondria with tubular cristae, free ribsomes, and lipid droplets in the luteal cells (Fig. 12). As well as existing in the form of anastomosing tubules, the smooth endoplasmic reticulum is also present as a folded membrane complex in granulosa luteal cells of early pregnancy and during the active progestational phase of the cycle. This complex may be unique to primates: it is not present in the theca luteal cells. Concentric membranous whorls are found in granulosa luteal cells of early pregnancy but not in those of the cycle (Vacek, 1967): they have been related to the increased steroidogenesis following LH stimulation (see Bjersing, 1967a,b,c, for pig) and also in man to a luteotropic effect of human chorionic gonadotropin (Crisp *et al.*, 1970). The large pleomorphic, and sometimes unusually contoured, mitochondria of luteal cells in early pregnancy are not found in those of the secretory phase of the cycle. The mitochondria exhibit osmiophilic inclusions within the matrix which might represent accumulation of substrates for subsequent conversion to steroid hormones. Rough endoplasmic reticulum occurs as isolated cisternae and as scattered stacks of cisternae. Golgi elements are present not only at one pole of the nucleus but scattered thoughout the cytoplasm. Microtubules of usual dimensions are prominent about the Golgi elements as well as membrane-bound, dense granules considered to be lysosomes, and multivesicular bodies often with a granular matrix. The Golgi elements of human luteal cells may well be a source of lysosomes. Fienberg and Cohen (1965) indicate that cholesterol and cholesterol esters are stored in cytoplasmic lipid droplets in luteal cells. Crisp *et al.* (1970) have discussed the significance of the types of granules found within human luteal cells. Membrane-bound granules, 150 to 200 nm in diameter and moderately electron dense, are closely associated with the rough endoplasmic reticulum. Other membrane-bound granules, twice the size and denser than the above, are frequently associated with the Golgi complex; it is argued (Rennels, 1966) that these granules represent a proteinaceous secretory product. Both granulosa and theca luteal cells of human corpora lutea contain lipofuscin pigment granules which may possess lysosomal activity. Crisp *et al.* (1970) postulate whether there is a mechanism involving these granules for the elaboration of relaxin during luteal cell regression towards term. Absence of obvious secretory structures and doubt about the occurrence of reversed pinocytosis and pinching off from microvilli have resulted in the view that steroids pass out of human luteal cells by active transport or by diffusion into the extracellular spaces (Gillim *et al.*, 1969). Luteal cells are separated by a space of 200 to 300 Å with occasional long tight junctions bringing portions of adjacent cell membranes into close proximity. Hertig (1964), and others, have noted that there is a broad, pericapillary space about the capillaries permeating the human corpus luteum. The free surface of the luteal cells bears numerous irregular microvilli beneath which is a peripheral, fibrillar terminal web. Bundles of fine filaments have been reported in human luteal cells even at term (Green *et al.*, 1967). They have been noted during the regressive phase of cyclic corpora lutea and seem to become prominent as pregnancy advances (Adams and Hertig, 1969a,b).

Extensions of the pericapillary space between neighboring luteal cells have been likened by Green *et al.* (1968) to an intercellular canalicular system for transfer of steroids. Intracellular canaliculi, lined by microvilli and containing accumulations of osmiophilic material, have been seen only in granulosa luteal cells (Crisp *et al.*, 1970). These authors point out that intracellular canaliculi could not only increase the facility of secretion but also increase the surface area for absorbtion of hormone precursors. An interrupted basal lamina (500 nm) of dense, amorphous material approximates to the cell surfaces of the pericapillary space and a similar lamina surrounds the endothelial cells of the capillaries. Macrophages within the pericapillary spaces probably represent the sudanophilic stellate "K" cells described above.

Electron microscopy confirms the distribution of theca luteal cells as seen under the light microscope. They are smaller, with less cytoplasm, and exhibit a wide range of cytoplasmic density. Luteal cells of the corpora lutea of early pregnancy are twice the size of the theca luteal cells at that stage. Luteinization of theca interna cells appears to commence at least 1 day before it occurs in luteal cells, and is indicated by much tubular smooth endoplasmic reticulum and mitochondria with tubular cristae. Rough endoplasmic reticulum is present as occasional cisternae and also in stacks, continuous with the smooth reticulum. Golgi elements are at one pole of each cell. The lipid cytoplasmic droplets often suffer extraction of the central region during processing, a change which does not seem to occur in luteal cells. Nuclei resemble those of luteal cells but contain large nucleoli and scanty heterochromatin. Variations in cytoplasmic density allow the differentiation of light, intermediate, and dark theca luteal cell types, probably determined by the degree of dilatation of the tubular smooth endoplasmic reticulum though this could well be artefactual in origin. Theca luteal cells lack surface microvilli and have no fibrillar terminal web, nor do they exhibit long processes projecting into neighboring cells. Desmosomes have not been observed between luteal cells and theca luteal cells.

REFERENCES

Adams, E. C., and Hertig, A. T. (1964). Studies on guinea-pig oocytes. I. Electron microscopic observations on the development of cytoplasmic organelles in oocytes of primordial and primary follicles. *J. Cell Biol.* **21,** 397–427.

Adams, E. C., and Hertig, A. T. (1969a). Studies on the human corpus luteum. I. Observations on the ultrastructure of development and regression of the luteal cells during the menstrual cycle. *J. Cell Biol.* **41,** 697–715.

Adams, E. C., and Hertig, A. T. (1969b). Studies on the human corpus luteum. II. Observations on the ultrastructure of luteal cells during pregnancy. *J. Cell Biol.* **41,** 716–735.

Alden, R. H. (1942). The periovarial sac in the albino rat. *Anat. Rec.* **83,** 421–433.

Allen, P., Brambell, F. W. R., and Mills, I. H. (1947). Studies on the sterility and prenatal mortality in wild rabbits. I. The reliability of estimates of prenatal mortality based on counts of corpora lutea, implantation sites and embryos. *J. Exp. Biol.* **23,** 312–331.

Allen, W. E. (1971). The occurrence of ovulation during pregnancy in the mare. *Vet. Rec.* **88,** 508–509.

Amoroso, E. C., Hancock, J. L., and Rowlands, I. W. (1948). Ovarian activity in the pregnant mare. *Nature* (*London*) **161,** 355–356.

Amoroso, E. C., Bourne, G. H., Harrison, R. J., Matthews, L. H., Rowlands, I. W., and Sloper, J. C. (1965). Reproductive and endocrine organs of foetal, newborn and adult seals. *J. Zool.* **147,** 430–486.

Anand Kumar, T. C. (1965). Reproduction in the rat-tailed bat *Rhinopoma kinneari. J. Zool.* **147,** 147–155.

Anand Kumar, T. C. (1968). Oogenesis in lorises; *Loris tardigradus* and *Nycticebus coucang. Proc. R. Soc. London, Ser. B* **169,** 167–176.

Andersen, D. H. (1926). Lymphatics and blood-vessels of the ovary of the sow. *Contrib. Embryol. Carnegie Inst.* **17,** 107–123.

Anderson, A. C., and Simpson, M. E. (1971). Life span of the corpus luteum in the dog (beagle). *Biol. Reprod.* **5,** 88.

Anderson, E., and Beams, H. W. (1960). Cytological observations on the fine structure of the guinea-pig ovary with special reference to the oogonium, primary oocyte and associated follicle cells. *J. Ultrastruct. Res.* **3,** 432–446.

Arceo, R. B., and Ryan, K. J. (1967). Conversion of androst-4-ene-3, 17-dione-4-^{14}C to oestrogens by subcellular fractions of the human corpus luteum of the normal cycle. *Acta Endocrinol.* (*Copenhagen*) **56,** 225–230.

Archbald, L. F., Schultz, R. H., Fahning, M. L., Kurtz, H. J., and Zemjanis, R. (1971). Rete ovarii in heifers: a preliminary study. *J. Reprod. Fertil.* **26,** 413–414.

Arthur, G. H. (1958). An analysis of the reproductive function of mares based on post-mortem examination. *Vet. Rec.* **70,** 682–686.

Asdell, S. A. (1960). Growth in the bovine Graafian follicle. *Cornell Vet.* **50,** 3–8.

Asdell, S. A. (1964). "Patterns of Mammalian Reproduction," 2nd ed. Cornell Univ. Press, Ithaca, New York.

Aumüller, G. (1974). The fine structure of the ovary in the mole (*Talpa europaea,* L.). *Z. Anat. Entwicklungsgesch.* **144,** 1–18.

Bain, A. M. (1967). Ovaries of the mare during early pregnancy. *Vet. Rec.* **80,** 229–231.

Baker, T. G. (1963). Quantitative and cytological study of germ cells in human ovaries. *Proc. R. Soc. London, Ser. B* **158,** 417–433.

Baker, T. G. (1966). A quantitative and cytological study of oogenesis in the rhesus monkey. *J. Anat.* **100,** 761–766.

Baker, T. G., and Franchi, L. L. (1967). Fine structure of oogonia and oocytes in human ovaries. *J. Cell Sci.* **2,** 213–224.

Bal, H. S., and Getty, R. (1970). Morphological changes in the vasculature of the ovaries in the sow as influenced by age from birth to 8 years. *J. Reprod. Fertil.* **22,** 311–320.

Balbiani, E. G. (1864). Sur la constitution du germe dans l'oeuf animal avant la fécondation. *C. R. Acad. Sci.* (*Paris*) **58,** 584–588.

Banon, P., Brandes, D., and Frost, J. L. (1964). Lysosomal enzymes in rat ovary and endometrium during estrous cycle. *Acta Cytol.* **8,** 416–425.

Barton, E. P. (1945). The cyclic changes of the epithelial cords of the dog ovary. *J. Morphol.* **77,** 317–349.

Beaumont, H. M., and Mandl, A. M. (1962). A quantitative and cytological study of oogonia and oocytes in the foetal and neonatal rat. *Proc. R. Soc. London, Ser. B* **155,** 557–579.

Beck, L. R. (1972). Comparative observations on the morphology of the mammalian periovarial sac. *J. Morphol.* **136,** 247–254.

Belt, W. D., Cavazos, L. F., Anderson, L. L., and Kraeling, R. R. (1970). Fine structure and progesterone levels in the corpus luteum of the pig during pregnancy and after hysterectomy. *Biol. Reprod.* **2,** 98–113.

Beltermann, R. (1965a). Elektronmikroskopische Untersuchungen an Epoophoron des Menschen. *Arch. Gynaekol.* **200,** 275–284.

Beltermann, R. (1965b). Elektronmikroskopische Befunde bei beginnender Follikelatresie im Ovar der Maus. *Arch. Gynaekol.* **200,** 601–609.

Benirschke, K., and Sullivan, M. M. (1966). Corpora lutea in proven mules. *Fertil. Steril.* **17,** 24–33.

Berger, L. (1922). Sur l'existence de glands sympathicotropes dans l'ovaire et le testicule humains; leur rapports avec la gland interstielle du testicule. *C. R. Acad. Sci.* (*Paris*) **175,** 907–909.

Bergin, W. C., and Shipley, W. D. (1968). Genital health in the mare. 1. Observations concerning the ovulation fossa. *Vet. Med. & Small Anim. Clin.* **63,** 362–365.

Best, P. B. (1967). The sperm whale (*Physeter catodon*) off the west coast of South Africa. 1. Ovarian changes and their significance. *Investl. Rep. Div. Fish., Union S. Afr.* **61,** 1–27.

Betteridge, K. J., Kelly, W. A., and Marston, J. H. (1970). Morphology of the rhesus monkey ovary near the time of ovulation. *J. Reprod. Fertil.* **22,** 453–460.

Bezrukov, N. I., and Schmidt, G. A. (1970). Certain age-dependent and functional changes in ovaries and uterus of female bactrian camels. *Arkh. Anat. Gistol. Embriol.* **58,** 64–73.

Bhalla, R. C., First, N. L., Chapman, A. B., and Casida, L. E. (1969). Quantitative variation in ovarian and follicular development in four genetic groups of pigs at different ages. *J. Anim. Sci.* **28,** 780–784.

Bigg, M. A. (1969). The harbour seal. *Bull. Fish. Res. Board Can.* **172,** 1–33.

Bjersing, L. (1967a). On the ultrastructure of follicles and isolated follicular granulosa cells of porcine ovary. *Z. Zellforsch. Mikrosk. Anat.* **82,** 173–186.

Bjersing, L. (1967b). On the ultrastructure of granulosa luteum cells in porcine corpus luteum with special reference to endoplasmic reticulum and steroid hormone synthesis. *Z. Zellforsch. Mikrosk. Anat.* **82,** 187–211.

Bjersing, L. (1967c). On the morphology and endocrine function of granulosa cells in ovarian follicles and corpora lutea. *Acta Endocrinol.* (*Copenhagen*), *Suppl.* **125,** 1–23.

Bjersing, L., Hay, M. F., Moor, R. M., Short R. V., and Deane, H. W. (1970a). Endocrine activity, histochemistry and ultrastructure of ovine corpora lutea. I. Further observations on regression at the end of the oestrous cycle. *Z. Zellforsch. Mikrosk. Anat.* **111,** 437–457.

Bjersing, L., Hay, M. F., Moor, R. M., and Short R. V. (1970b). Endocrine activity, histochemistry and ultrastructure of ovine corpora lutea. II. Observations on regression following hysterectomy. *Z. Zellforsch. Mikrosk. Anat.* **111,** 458–470.

Björkman, N. H. (1962). A study of the ultrastructure of the granulosa cells of the rat ovary. *Acta Anat.* **51,** 125–147.

Black, J. L., and Erickson, B. H. (1968). Oogenesis and ovarian development in the prenatal pig. *Anat. Rec.* **161,** 45–52.

Blanchette, E. J. (1961). A study of the fine structure of the rabbit primary oocyte. *J. Ultrastruct. Res.* **5,** 349–363.

Blanchette, E. J. (1966a). Ovarian steroid cells. 1. Differentiation of the lutein cell from the granulosa follicle cell during the preovulatory state and under the influence of exogenous gonadotrophins. *J. Cell Biol.* **31,** 501–516.

Blanchette, E. J. (1966b). Ovarian steroid cells. II. The lutein cell. *J. Cell Biol.* **31,** 517–542.

Bland, K. P. (1969). Reproduction in the female Indian gerbil (*Tatera indica*). *J. Zool.* **157,** 47–61.

Bloch, S., and Strauss, F. (1958). Die weiblichen Genitalorgane von *Lepus europaeus* Pallas. *Z. Saeugetierkd.* **23,** 66–80.

Bloch, S., Hediger, H., Lloyd, H. G., Müller, C., and Strauss, F. (1967). Beobachtungen zur Superfetation beim Feldhasen (*Lepus europaeus*). *Z. Jagdwiss.* **2,** 49–52.

Bocharov, W. S. (1966). Multiovular follicles in the ovaries of mammals. *Adv. Sov. Biol.* **62,** 139–147.

Bodemer, C. W., and Warnick, S. (1961a). Polyovular follicles in the immature hamster ovary. I. Polyovular follicles in the normal intact animal. *Fertil. Steril.* **12,** 159–169.

Bodemer, C. W., and Warnick, S. (1961b). Polyovular follicles in the immature hamster ovary. II. The effects of gonadotrophic hormones in polyovular follicles. *Fertil. Steril.* **12,** 353–364.

Bonner, W. N. (1955). Reproductive organs of foetal and juvenile elephant seals. *Nature (London)* **176,** 982–983.

Bonnin-Laffargue, M., and Canivenc, R. (1970). Biologie lutéale chez le Renard (*Vulpes vulpes* L.). Persistance du corps jaune après la mise bas. *C. R. Acad. Sci. Ser. D* **271,** 1402–1405.

Borozdin, E. K. (1969). [Histological structure of the ovary and development of oocytes in the reindeer.] *Tr. Nauchno-Issled. Inst. Sel'sk. Khoz. Krainego Sev.* **17,** 85–93.

Braden, A. W. H. (1952). Properties of the membranes of rat and rabbit eggs. *Aust. J. Sci. Res. B* **5,** 460–471.

Brambell, F. W. R. (1944). The reproduction of the wild rabbit, *Oryctolagus cuniculus* (L). *Proc. Zool. Soc. London* **114,** 1–45.

Brambell, F. W. R. (1956). Ovarian changes. *In* "Marshall's Physiology of Reproduction" (A. S. Parkes, ed.), 3rd ed., Vol. 1, pp. 397–542. Longmans Green, London and New York.

Brambell, F. W. R., and Rowlands, I. W. (1936). Reproduction of the bank vole (*Evotomys glareolus* Schreber). 1. The oestrous cycle of the female. *Phil. Trans. R. Soc. London, Ser. B* **226,** 71–120.

Brandau, H., Remmlinger, K., and Luh, W. (1967). Enzymmuster und progesteronsynthese. Eine histochemische Analyse der Enzymverteilung in der Granulosazelle des Kaninchenovars vor und nach dem Follikelsprung. *Acta Endocrinol. (Copenhagen)* **56,** 433–444.

Breed, W. G. (1967). Ovulation in the genus *Microtus*. *Nature (London)* **214,** 826.

Breed, W. G. (1969). Oestrus and ovarian histology in the laboratory vole (*Microtus agrestis*). *J. Reprod. Fertil.* **18,** 33–42.

Breed, W. G., and Charlton, H. M. (1971). Hypothalamo-hypophysial control of ovulation in the vole (*Microtus agrestis*). *J. Reprod. Fertil.* **25,** 225–229.

Breed, W. G., and Clarke, J. R. (1970a). Ovulation and associated histological changes in the ovary following coitus in the vole (*Microtus agrestis*). *J. Reprod. Fertil.* **22,** 173–175.

Breed, W. G., and Clarke, J. R. (1970b). Ovarian changes during pregnancy and pseudopregnancy in the vole, *Microtus agrestis. J. Reprod. Fertil.* **23,** 447–456.

Breinl, H., Andrzejewski, C., and Jonutti, E. (1967). Zur Feinstructur der Luteinzellen des Laktationsgerbkörpers der Ratte. *Z. Mikrosk. Anat. Forsch.* **77,** 442–452.

Brewer, J. I. (1942). Studies of the human corpus luteum. Evidence for the early onset of regression of the corpus luteum of menstruation. *Am. J. Obstet. Gynecol.* **44,** 1048–1059.

Brown, L. N., and Conaway, C. H. (1964). Persistence of corpora lutea at the end of the breeding season in *Peromyscus. J. Mammal.* **45,** 260–265.

Brown-Grant, K. (1969). The induction of ovulation during pregnancy in the rat. *J. Endocrinol.* **43,** 529–538.

Buechner, H. K., Morrison, J. A., and Leuthold, W. (1966). Reproduction in Uganda Kob with special reference to behaviour. *Symp. Zool. Soc. London* **15,** 69–88.

Bullough, W. S. (1942). The method of growth of the follicle and corpus luteum in the mouse ovary. *J. Endocrinol.* **3,** 150–156.

Bullough, W. S. (1946). Mitotic activity in the adult mouse, *Mus musculus L.* A study of its relation to the oestrous cycle in normal and abnormal conditions. *Phil. Trans. R. Soc. London, Ser. B* **231,** 453–516.

Burden, H. W. (1972a). Ultrastructural observations on ovarian perifollicular smooth muscle in the cat, guinea pig and rabbit. *Am. J. Anat.* **133,** 125–142.

Burden, H. W. (1972b). Adrenergic innervation in ovaries of rat and guinea pig. *Am. J. Anat.* **133,** 455–462.

Burr, J. H., and Davies, J. L. (1951). The vascular system of the rabbit ovary and its relationship to ovulation. *Anat. Rec.* **111,** 273–297.

Butler, H. (1969). Post pubertal oogenesis in Prosiminae. *Proc. Int. Congr. Primatol., 2nd, 1968* pp. 15–21.

Byskov, A. G. S. (1969). Ultrastructure studies on the preovulatory follicle in the mouse ovary. *Z. Zellforsch. Mikrosk. Anat.* **100,** 285–299.

Byskov, A.-G. S. (1974). Does the rete ovarii act as a trigger for the onset of meiosis? *Nature* (*London*) **252,** 396–397.

Byskov, A.-G., and Lintern-Moore, S. (1973). Follicle formation in the immature mouse ovary: the role of the rete ovarii. *J. Anat.* **116,** 207–217.

Call, E. L., and Exner, S. (1875). Zur Kenntniss des Graafschen Follikels und des Corpus luteum beim Kaninchen. *Sitzungsber. Akad. Wiss. Wien* **71,** 321–328.

Canivenc, R. (1957). Etude de la nidation differée du blaireau européen (*Meles meles*). *Ann. Endocrinol.* **18,** 716–736.

Canivenc, R. (1966). A study of progestation in the European badger (*Meles meles L.*). *Symp. Zool. Soc. London* **15,** 15–26.

Canivenc, R., Lajus-Boue, M., and Bonnin-Laffargue, M. (1966a). Les inclusions nucléaires des cellules du corps jaune chez le Blaireau européen. Nature, origine, évolution. *C. R. Soc. Biol.* **160,** 968–972.

Canivenc, R., Short, R. V., and Bonnin-Laffargue, M. (1966b). Etude histologique et biochimique du corps jaune du blaireau européen (*Meles meles L.*). *Ann. Endocrinol.* **27,** 401–414.

Canivenc, R., Cohéré, G., and Brechenmacher, C. (1967). Quelques aspects ultrastructuraux de la cellule lutéale chez le blaireau (*Meles meles L.*). *C. R. Acad. Sci. Ser. D* **264,** 1187–1189.

Caravaglios, R., and Cilotti, R. (1957). A study of the proteins in the follicular fluid of the cow. *J. Endocrinol.* **15,** 273–278.

Carithers, K. J. R. (1969). Ultrastructural responses of rat ovarian interstitial cells. *Diss. Abstr. B* **29,** 271–276.

Carsten, P. M. (1965). Elektronmikroskopische Probleme bei Strukturdentungen von Einschlusskörpern im menschlichen Corpus luteum. *Arch. Gynaekol.* **200,** 552–568.

Cavazos, L. F., Anderson, L. L., Belt, W. D., Henricks, D. M., Kraeling, R. R., and Melampy, R. M. (1969). Fine structure and progesterone levels in the corpus luteum of the pig during the estrous cycle. *Biol. Reprod.* **1,** 83–106.

Channing, C. P. (1969). The use of tissue culture of granulosa cells as a method of studying the mechanism of luteinization. *In* "The Gonads" (K. W. McKerns, ed.), pp. 245–275. Appleton, New York.

Channing, C. P. (1970a). Effects of the stage of the menstrual cycle and gonadotrophins on luteinization of rhesus monkey granulosa cells in culture. *Endocrinology* **87,** 49–60.

Channing, C. P. (1970b). Effect of stage of the estrous cycle and gonadotrophins upon luteinization of porcine granulosa cells in culture. *Endocrinology* **87,** 156–164.

Channing, C. P. (1970c). Influences of the *in vivo* and *in vitro* hormonal environment upon luteinization of granulosa cells in tissue culture. *Recent Prog. Horm. Res.* **26,** 589–622.

Channing, C. P., and Seymour, J. F. (1970). Effects of dibutyryl cyclic-3′,5′-AMP and other agents upon luteinization of porcine granulosa cells in culture. *Endocrinology* **87,** 165–169.

Chaplin, R. E., and Harrison, R. J. (1971). The uterus ovaries and placenta of the Chinese muntjac deer (*Muntiacus reevesi*). *J. Anat.* **110,** 147.

Chester Jones, I., and Henderson, I. W. (1963). The ovary of the 13-lined ground squirrel (*Citellus tridecemlineatus* Mitchell) after adrenalectomy. *J. Endocrinol.* **26,** 265–272.

Chiquoine, A. D. (1960). The development of the zona pellucida of the mammalian ovum. *Am. J. Anat.* **106,** 149–169.

Chittleborough, R. G. (1954). Studies on the ovaries of the humpback whale *Megaptera nodosa* (Bonnaterre) on the western Australian coast. *Aust. J. Mar. Freshwater Res.* **5,** 35–63.

Choudary, J. B., Gier, H. T., and Marion, G. B. (1968). Cyclic changes in bovine vesicular follicles. *J. Anim. Sci.* **27,** 468–471.

Christensen, K., and Gillim, S. W. (1969). The correlation of fine structure and function of steroid-secreting cells, with emphasis on those of the gonads. *In* "The Gonads" (K. W. McKerns, ed.), pp. 415–488. Appleton, New York.

Christiansen, J. A., Jensen, C. E., and Zachariae, F. (1958). Studies on the mechanism of ovulation. Some remarks on the effects of depolymerization of high polymers on the pre-ovulatory growth of follicles. *Acta Endocrinol. (Copenhagen)* **29,** 115–117.

Chuzhakina, E. S. (1961). [Morphological characterisation of the ovaries of the female sperm whale (*Physeter catodon* L., 1758) in connection with age determination.] *Tr. Inst. Morfol. Zhivotn., Akad. Nauk SSSR* **34,** 33–53.

Claesson, L. (1947). Is there any smooth musculature in the wall of the Graafian follicle? *Acta Anat.* **3,** 295–311.

Claesson, L. (1954). The intracellular localization of the esterified cholesterol in the living interstitial gland of the rabbit ovary. *Acta Physiol. Scand., Suppl.* **113,** 53–78.

Claesson, L., and Hillarp, V. A. (1947). The formation mechanism of oestrogenic hormones. I. The presence of an oestrogen-precursor in the rabbit ovary. *Acta Physiol. Scand.* **13,** 115–129.

Clark, M. J., and Poole, W. E. (1967). The reproductive system and embryonic diapause in the female grey kangaroo *Macropus giganteus. Aust. J. Zool.* **15,** 441–459.

Clemens, J. A., Minaguchi, H., and Meites, J. (1968). Relation of local circulation between ovaries and uterus to lifespan of corpora lutea in rats. *Proc. Soc. Exp. Biol. Med.* **127,** 1248–1251.

Clough, G. (1969). Some preliminary observations on reproduction in the warthog, *Phacochoerus aethiopicus* Pallas. *J. Reprod. Fertil., Suppl.* **6,** 323–337.

Clough, G. (1970). A record of 'testis cords' in the ovary of a mature hippopotamus (*Hippopotamus amphibius* Linn). *Anat. Rec.* **166,** 47–50.

Coe, M. J. (1969). The anatomy of the reproductive tract and breeding in the springhaas, *Pedetes surdaster kuvatis* Hollister. *J. Reprod. Fertil., Suppl.* **6,** 159–174.

Cohéré, G., Brechenmacher, C., and Mayer, G. (1967). Variations des ultrastructures de la cellule lutéale chez la ratte au cours de la grossesse. *J. Microsc. (Paris)* **6,** 657–670.

Cole, H. H., Howell, C. E., and Hart, G. H. (1931). The changes occurring in the ovary of the mare during pregnancy. *Anat. Rec.* **49,** 199–210.

Collins, D. C., and Kent, H. A. (1964). Polynuclear ova and polyovular follicles in the ovaries of young guinea-pigs. *Anat. Rec.* **148,** 115–119.

Conaway, C. H. (1952). Life history of the water shrew (*Sorex palustris navigator*). *Amer. Midl. Nat.* **48,** 219–248.

Conaway, C. H. (1969). Adrenal cortical rests of the ovarian hilus of the Patas monkey. *Folia Primatol.* **11,** 175–180.

Corner, G. W. (1915). The corpus luteum of pregnancy as it is in swine. *Contrib. Embryol. Carnegie Inst.* **5,** 69–94.

Corner, G. W. (1919). On the origin of the corpus luteum of the sow from both granulosa and theca interna. *Am. J. Anat.* **26,** 117–183.

Corner, G. W. (1920). On the widespread occurrence of reticular fibrils produced by capillary endothelium. *Contrib. Embryol. Carnegie Inst.* **9,** 85–93.

Corner, G. W. (1921). Cyclic changes in the ovaries and uterus of the sow and their relations to the mechanism of implantation. *Contrib. Embryol. Carnegie Inst.* **13,** 117–146.

Corner, G. W. (1942). The fate of the corpora lutea and the nature of the corpora aberrantia in the Rhesus monkey. *Contrib. Embryol. Carnegie Inst.* **30,** 85–96.

Corner, G. W. (1948). Alkaline phosphatase in the ovarian follicle and in the corpus luteum. *Contrib. Embryol. Carnegie Inst.* **32,** 1–8.

Corner, G. W., Hartman, C. G., and Bartelmez, G. W. (1945). Development, organization and breakdown of the corpus luteum in the rhesus monkey. *Contrib. Embryol. Carnegie Inst.* **31,** 117–146.

Corner, G. W., Jr. (1956). The histological dating of the human corpus luteum of menstruation. *Am. J. Anat.* **98,** 377–401.

Courrier, R., and Gros, G. (1935). Contribution à l'endocrinologie de la grossesse chez la chatte. *C. R. Soc. Biol.* **120,** 5–7.
Courrier, R., and Gros, G. (1936). Dissociation foeto-placentaire realisée par la castration chez la chatte. Action endocrinienne du placenta. *C. R. Soc. Biol. (Paris)* **121,** 1517–1520.
Coutts, R. R., and Rowlands, I. W. (1969). The reproductive cycle of the Skomer vole (*Clethrionomys glareolus skomerensis*). *J. Zool.* **158,** 1–25.
Craig, A. M. (1964). Histology of reproduction and the estrous cycle in the female fur seal *Callorhinus ursinus*. *J. Fish. Res. Board Can.* **21,** 773–812.
Crichton, E. (1969). Reproduction in the pseudomyine rodent *Mesembriomys gouldii* (Gray) (Muridae). *Aust. J. Zool.* **17,** 785–797.
Crisp, T. M., and Browning, H. C. (1968). The fine structure of corpora lutea in ovarian transplants of mice following luteotrophin stimulation. *Am. J. Anat.* **122,** 169–192.
Crisp, T. M., Dessouky, D. A., and Denys, F. R. (1970). The fine structure of the human corpus luteum of early pregnancy and during the progestational phase of the menstrual cycle. *Am. J. Anat.* **127,** 37–70.
Crombie, P. R., Burton, R. D., and Ackland, N. (1971). The ultrastructure of the corpus luteum of the guinea-pig. *Z. Zellforsch. Mikrosk. Anat.* **115,** 473–493.
Cross, P. C. (1971). Dictyate oocyte of *Microtus montanus*. *J. Reprod. Fertil.* **25,** 291–293.
Culiner, A. (1945). The relation of the theca cells to disturbances of the menstrual cycle. *J. Obstet. Gynecol.* **52,** 545–558.
Culiner, A. (1946). Role of the theca cell in irregularities of the baboon menstrual cycle. *S. Afr. J. Med. Sci.* **11,** Biol. Suppl., 55–70.
Dahl, E. (1971a). Studies of the fine structure of ovarian interstitial tissue. A comparative study of the fine structure of the ovarian interstitial tissue in the rat and the domestic fowl. *J. Anat.* **108,** 275–290.
Dahl, E. (1971b). The fine structure of the granulosa cells in the domestic fowl and the rat. *Z. Zellforsch. Mikrosk. Anat.* **119,** 58–67.
Davies, J., and Broadus, C. D. (1968). Studies on the fine structure of ovarian steroid-secreting cells in the rabbit. I. The normal interstitial cells. *Am. J. Anat.* **123,** 441–474.
Davies, J., Dempsey, E. W., and Wislocki, G. B. (1957). Histochemical observations on the fetal ovary and testis of the horse. *J. Histochem. Cytochem.* **5,** 584–590.
Dawson, A. B. (1941). The development and morphology of the corpus luteum of the cat. *Anat. Rec.* **79,** 155–169.
Dawson, A. B. (1946). The post-partum history of the corpus luteum of the cat. *Anat. Rec.* **95,** 29–51.
Dawson, A. B., and Friedgood, H. B. (1940). The time and sequence of preovulatory changes in the cat ovary after mating or mechanical stimulation of the cervix uteri. *Anat. Rec.* **76,** 411–429.
Dawson, A. B., and McCabe, M. (1951). The interstitial tissue of the ovary in infantile and juvenile rats. *J. Morphol.* **88,** 543–571.
Deane, H. W. (1952). Histochemical observations on the ovary and oviduct of the albino rat during the estrous cycle. *Am. J. Anat.* **91,** 363–413.
Deane, H. W., and Fawcett, D. W. (1952). Pigmented interstitial cells showing brown degeneration in the ovaries of old mice. *Anat. Rec.* **113,** 239–245.
Deane, H. W., and Rubin, B. L. (1965). Identification and control of cells that

synthesize steroid hormones in the adrenal glands, gonads and placentae of various mammalian species. *Arch. Anat. Microsc. Morphol. Exp.* **54,** 49–66.

Deane, H. W., Hay, M. F., Moor, R. M., Rowson, L. E. A., and Short, R. V. (1966). Corpus luteum of the sheep: relationships between morphology and function during the oestrous cycle. *Acta Endocrinol.* (*Copenhagen*) **51,** 245–263.

Deanesly, R. (1934). The reproductive processes of certain mammals. Part VI. The reproductive cycle of the female hedgehog. *Phil. Trans. R. Soc. London, Ser. B* **223,** 239–276.

Deanesly, R. (1935). The reproductive processes of certain mammals. Part IX. Growth and reproduction in the stoat (*Mustela erminea*). *Phil. Trans. R. Soc. London, Ser. B* **225,** 459–492.

Deanesly, R. (1966). Observations on reproduction in the mole, *Talpa europaea. Symp. Zool. Soc. London* **15,** 387–400.

Deanesly, R. (1967). Experimental observations on the ferret corpus luteum of pregnancy. *J. Reprod. Fertil.* **13,** 183–185.

Deanesly, R. (1970). Oögenesis and the development of the ovarian interstitial tissue in the ferret. *J. Anat.* **107,** 165–178.

Deanesly, R. (1972). Origin and development of interstitial tissue in ovaries of rabbit and guinea-pig. *J. Anat.* **113,** 251–260.

Deanesly, R., and Perry, J. S. (1969). Independent regression of normal and induced corpora lutea in hysterectomized guinea-pigs. *J. Reprod. Fertil.* **20,** 503–508.

De Groodt, M., Lagasse, A., and Seybruyns, M. (1957). L'espace perivasculaire dans le tissu interstitiel de l'ovaire vu au microscope électronique. *Bull. Microsc. Appl.* **7,** 101–103.

De Groodt-Lasseel, M. (1963). Functie en vorm van de interstitiële cellen van het ovarium. *In* "Arscia Uitgaven N.V. Brussel," pp. 1–215. Presses Acad. Eur., S.C., Brussels.

Delson, D., Lubin, S., and Reynolds, S. R. M. (1949). Vascular patterns in human ovaries. *Am. J. Obstet. Gynecol.* **57,** 842–852.

Dempsey, E. W. (1940). The structure of the reproductive tract in the female gibbon. *Am. J. Anat.* **67,** 229–253.

Dempsey, E. W., and Wislocki, G. B. (1941). The structure of the ovary of the humpback whale (*Megaptera nodosa*). *Anat. Rec.* **80,** 243–257.

Denamur, R., and Mauléon, P. (1963). Effets de l'hypophysectomie sur la morphologie et l'histologie du corps jaune des ovins. *C. R. Hebd. Seances Acad. Sci.* **257,** 264–267.

Dingle, J. T., Hay, M. F., and Moor, R. M. (1968). Lysosomal function in the corpus luteum of the sheep. *J. Endocrinol.* **40,** 325–336.

Donahue, R. P., and Stern, S. (1968). Follicular cell support of oocyte maturation: production of pyruvate *in vitro. J. Reprod. Fertil.* **17,** 395–398.

Donaldson, L., and Hansel, W. (1965). Histological study of bovine corpora lutea. *J. Dairy Sci.* **48,** 905–909.

Dorrington, J. H., and Baggett, B. (1969). Adenyl cyclase activity in the rabbit ovary. *Endocrinology* **84,** 989–996.

Dryden, G. L. (1969). Reproduction in *Suncus murinus. J. Reprod. Fertil., Suppl.* **6,** 377–396.

Dubreuil, G. (1944). De quelques caractères propres des corps gestatifs de la femme. *C. R. Soc. Biol.* (*Paris*) **138,** 699–700.

Dubreuil, G. (1957). La déterminisme de la glande thécale de l'ovaire. Induction morphogène à partier de la granulosa folliculaire. *Acta Anat.* **30,** 269–274.

Dubreuil, G., and Rivière, M. (1947). Morphologie et histologie des corps progestatifs et gestatifs (corps jaunes) de l'ovaire feminin. *Gynecologie* **43,** 1–101.

Duby, R. T., McDaniel, J. W., Spilman, C. H., and Black, D. L. (1969). Utero-ovarian relationships in the golden hamster. III. Influence of uterine transplants and extracts on ovarian function following hysterectomy. *Acta Endocrinol.* (*Copenhagen*) **60,** 611–620.

Duke, K. L. (1949). Some notes on the histology of the ovary of the bobcat (*Lynx*) with special reference to the corpora lutea. *Anat. Rec.* **103,** 111–132.

Duke, K. L. (1952). Ovarian histology of *Ochotona princeps,* the Rocky Mountain pika. *Anat. Rec.* **112,** 737–760.

Duke, K. L. (1966). Histological observations on the ovary of the white-tailed mole *Parascaptor leucurus. Anat. Rec.* **154,** 527–532.

Duke, K. L. (1967). Ovogenetic activity of the fetal type in the ovary of the adult slow loris, *Nycticebus coucang. Folia Primatol.* **7,** 150–154.

Duke, K. L. (1969). Ovary of *Paradoxurus,* the palm civet. *Anat. Rec.* **163,** 180.

Duke, K. L., and Luckett, W. P. (1965). Histological observations on the ovary of several species of tree shrews (Tupaiidae). *Anat. Rec.* **151,** 450.

Dziuk, P. J. (1965). Timing of maturation and fertilization of the sheep egg. *Anat. Rec.* **153,** 211–223.

Eckstein, P. (1958). A. Internal reproduction organs. II. Reproductive organs. *Primatologia III* **1,** 542–629.

Eckstein, P., and Zuckerman, S. (1956). Morphology of the reproductive tract. *In* "Marshall's Physiology of Reproduction" (A. S. Parkes, ed.), 3rd ed., Vol. 1, Part 1, Chapter 2, pp. 43–155. Longmans Green, London and New York.

Edgar, D. G., and Ronaldson, J. W. (1958). Blood levels of progesterone in the ewe. *J. Endocrinol.* **16,** 378–384.

Edwards, M. J. (1962). Weights of cyclic and pregnancy corpora lutea of dairy cows. *J. Reprod. Fertil.* **4,** 93–98.

El-Fouly, M. A., Cook, B., Nekola, M., and Nalbandov, A. V. (1970). Role of the ovum in follicular luteinization. *Endocrinology* **87,** 288–293.

Ellsworth, L. R., and Armstrong, D. T. (1971). Effect of LH on luteinization of ovarian follicles transplanted under the kidney capsule in rats. *Endocrinology* **88,** 755–762.

Enders, A. C. (1960). A histological study of the cortex of the ovary of the adult armadillo with special reference to the question of neoformation of oocytes. *Anat. Rec.* **136,** 491–499.

Enders, A. C. (1962). Observations on the fine structure of lutein cells. *J. Cell Biol.* **12,** 101–113.

Enders, A. C. (1966). The reproductive cycle of the nine-banded armadillo (*Dasypus novemcinctus*). *Symp. Zool. Soc. London* **15,** 295–310.

Enders, R. K. (1952). Reproduction in the mink (*Mustela vison*). *Proc. Am. Philos. Soc.* **96,** 691–755.

Enders, R. K., and Enders, A. C. (1963). Morphology of the female reproductive tract during delayed implantation in the mink. *In* "Delayed Implantation" (A. C. Enders, ed.), pp. 129–139. Univ. of Chicago Press, Chicago, Illinois.

Enders, R. K., Pearson, O. P., and Pearson, A. K. (1946). Certain aspects of reproduction in the fur seal. *Anat. Rec.* **94,** 213–223.

England, B. G., Foote, W. C., Matthews, D. H., Cardozo, A. G., and Riera, S. (1969). Ovulation and corpus luteum formation in the llama (*Lama glama*). *J. Endocrinol.* **45**, 505–513.

Erickson, A. W., Nellor, J., and Petrides, G. (1964). "The Black Bear in Michigan." Michigan State Univ. Exp. Stn., East Lansing.

Erickson, B. H. (1966a). Development and radio-response of the prenatal bovine ovary. *J. Reprod. Fertil.* **11**, 97–105.

Erickson, B. H. (1966b). Development of senescence of the postnatal bovine ovary. *J. Anim. Sci.* **25**, 800–805.

Espey, L. L. (1967). Ultrastructure of the apex of the rabbit Graafian follicle during the ovulatory process. *Endocrinology* **81**, 267–276.

Espey, L. L. (1971a). Decomposition of connective tissue in rabbit ovarian follicles by multivesicular structures of thecal fibroblasts. *Endocrinology* **88**, 437–444.

Espey, L. L. (1971b). Multivesicular structures in proliferating fibroblasts of rabbit ovarian follicles during ovulation. *J. Cell Biol.* **48**, 437–442.

Espey, L. L., Slagter, C., Weymouth, R., and Rondell, P. (1965). Ultrastructure of the rabbit Graafian follicle as it approaches rupture. *Physiologist* **8**, 161.

Evans, H. M., and Cole, H. H. (1931). An introduction to the study of the oestrous cycle in the dog. *Mem. Univ. Calif.* **9**, 65–118.

Falck, B. (1959). Site of production of oestrogen in rat ovary as studied in micro-transplants. *Acta Physiol. Scand.* **47**, Suppl. 163, 1–101.

Fallon, P. (1967). Ovarian follicle formation in relation to pregnancy in mares. *Aust. Vet. J.* **43**, 536–540.

Fay, F. H. (1955). "The Pacific Walrus." Ph.D. Thesis, University of British Columbia.

Fernandez-Baca, S., Hansel, W., and Novoa, C. (1970). Corpus luteum function in the alpaca. *Biol. Reprod.* **3**, 252–261.

Fienberg, R., and Cohen, R. (1965). A comparative histochemical study of the ovarian stromal lipid band, stromal theca cell and normal ovarian follicular apparatus. *Am. J. Ostet. Gynecol.* **92**, 958–969.

Fink, G., and Schofield, G. C. (1971). Experimental studies on the innervation of the ovary in cats. *J. Anat.* **109**, 115–126.

Fisher, H. D., and Harrison, R. J. (1970). Reproduction in the common porpoise (*Phocoena phocoena*) of the North Atlantic. *J. Zool.* **161**, 471–486.

Flynn, T. T., and Hill, J. P. (1939). The development of the Monotremata. IV. Growth of the ovarian ovum, maturation, fertilization, and early cleavage. *Trans. Zool. Soc. London* **24**, 445–622.

Foreman, D. (1962). The normal reproductive cycle of the female prairie dog and the effects of light. *Anat. Rec.* **142**, 391–405.

Franceschini, M. P., Santoro, A., and Motta, P. (1965). L'ultrastruttura delle cellule della granulosa nelle variefasi di maturazione del follicolo ooforo. Ricerche in *Lepus cuniculus* L. 1758. *Z. Anat. Entwicklungsgesch* **124**, 522–532.

Franchi, L. L. (1960). Electron microscopy of oocyte-follicle cell relationships in the rat ovary. *J. Biophys. Biochem. Cytol.* **7**, 397–398.

Franchi, L. L. (1970). The ovary. *In* "Scientific Foundations of Obstetrics and Gynaecology" (E. E. Philipp, J. Barnes, and M. Newton, eds.), pp. 107–131. Heinemann, London.

Freeman, M. E., Butcher, R. L., and Fugo, W. W. (1970). Alteration of oocytes and follicles by delayed ovulation. *Biol. Reprod.* **2**, 209–215.

Fuller, G. B., and Hansel, W. (1970). Regression of sheep corpora lutea after treatment with antibovine luteinizing hormone. *J. Anim. Sci.* **31,** 99–103.

Garde, M. L. (1930). The ovary of *Ornithorhynchus*, with special reference to follicular atresia. *J. Anat.* **64,** 422–453.

Gérard, P. (1920). Contribution à l'étude de l'ovaire des mammifères. L'ovaire de *Galago mossambicus* (Young). *Arch. Biol.* **30,** 357–391.

Gérard, P. (1932). Etudes sur l'ovogenèse et l'ontogenèse chez les lémuriens du genre *Galago*. *Arch. Biol.* **43,** 93–151.

Gérard, P., and Herlant, M. (1953). Sur la persistance de phenomènes d'oogenèse chez les lémuriens adultes. *Arch. Biol.* **64,** 97–111.

Gillet, J. Y. (1971). La microvascularisation de l'ovaire. *Gynecol. Obstet.* **70,** 251–272.

Gillet, J. Y., Koritke, J. G., Muller, P., and Jeliens, C. (1968). Sur la microvasularisation de l'ovaire chez la lapine. *C. R. Soc. Biol.* **162,** 762–766.

Gillim, S. W., Christensen, K., and McLennan, C. E. (1969). Fine structure of the human menstrual corpus luteum at its stage of maximum secretory activity. *Am. J. Anat.* **126,** 409–428.

Gillman, J., and Stein, H. B. (1941). The human corpus luteum of pregnancy. *Surg., Gynecol. Obstet.* **72,** 129–149.

Godet, R. (1949). Recherches d'anatomie, d'embryologie normale et expérimentale sur l'appareil génital de la taupe (*Talpa europaea* L.). *Bull. Biol. Fr. Belg.* **83,** 25–111.

Gondos, B. (1969). The ultrastructure of granulosa cells in the newborn rabbit ovary. *Anat. Rec.* **165,** 67–78.

Gondos, B. (1970). Granulosa cell-germ cell relationship in the developing rabbit ovary. *J. Embryol. Exp. Morphol.* **23,** 419–426.

Gondos, B., and Zamboni, L. (1967). Electron microscopic studies of the embryogenesis and post natal development in the rabbit ovary. *J. Ultrastruct. Res.* **21,** 162.

Gondos, B., and Zamboni, L. (1969). Ovarian development: the functional importance of germ cell interconnections. *Fertil. Steril.* **20,** 176–189.

Gondos, B., Bhiraleus, P., and Hobel, C. J. (1971). Ultrastructural observations on germ cells in human fetal ovaries. *Am. J. Obstet. Gynecol.* **110,** 644–652.

Goodman, P., Latta, J. S., Wilson, R. B., and Kadis, B. (1968). The fine structure of sow lutein cells. *Anat. Rec.* **161,** 77–90.

Gopalakrishna, A. (1969). Unusual persistence of the corpus luteum in the Indian fruit-bat, *Rousettus leschenaulti*. *Curr. Sci.* **38,** 388–389.

Gopalakrishna, A., and Moghe, M. A. (1960). Observations on the ovaries of some Indian bats. *Proc. Natl. Inst. Sci. India, Part B* **26,** 11–19.

Govan, A. D. T. (1968). The human ovary in early pregnancy. *J. Endocrinol.* **40,** 421–428.

Govan, A. D. T. (1970). Ovarian follicular activity in late pregnancy. *J. Endocrinol.* **48,** 235–241.

Graham, C. E., and Bradley, C. F. (1971). Polyovular follicles in squirrel monkeys after prolonged diethyl-stilboestrol treatment. *J. Reprod. Fertil.* **27,** 181–185.

Green, J. A., and Maqueo, M. (1965). Ultrastructure of the human ovary. I. The luteal cell during the menstrual cycle. *Am. J. Obstet. Gynecol.* **92,** 946–957.

Green, J. A., and Maqueo, M. (1966). Histopathology and ultrastructure of an ovarian hilar cell tumor. *Am. J. Obstet. Gynecol.* **96,** 478–485.

Green, J. A., Garcilazo, J. A., and Maqueo, M. (1967). Ultrastructure of the human ovary. II. The luteal cell at term. *Am. J. Obstet. Gynecol.* **99,** 855–863.
Green, J. A., Garcilazo, J. A., and Maqueo, M. (1968). Ultrastructure of the human ovary. III. Canaliculi of the corpus luteum. *Am. J. Obstet. Gynecol.* **102,** 57–64.
Green, S. H., and Zuckerman, S. (1951). The number of oocytes in the mature rhesus monkey (*Macaca mulatta*). *J. Endocrinol.* **7,** 194–202.
Green, S. H., and Zuckerman, S. (1954). Further observations on oocyte numbers in mature rhesus monkeys (*Macaca mulatta*). *J. Endocrinol.* **10,** 284–290.
Greenwald, G. S. (1961). Quantitative study of follicular development in the ovary of the intact or unilaterally ovariectomized hamster. *J. Reprod. Fertil.* **2,** 351–361.
Greenwald, G. S. (1964). Ovarian follicular development in the pregnant hamster. *Anat. Rec.* **148,** 605–609.
Greenwald, G. S. (1965). Histologic transformation of the ovary of the lactating hamster. *Endocrinology* **77,** 641–650.
Greenwald, G. S. (1966). Ovarian follicular development and pituitary FSH and LH content in the pregnant rat. *Endocrinology* **79,** 572–578.
Greenwald, G. S. (1967). Induction of ovulation in the pregnant hamster. *Am. J. Anat.* **121,** 249–258.
Greenwald, G. S., and Choudary, J. B. (1969). Follicular development and induction of ovulation in the pregnant mouse. *Endocrinology* **84,** 1512–1520.
Greenwald, G. S., and Peppler, R. D. (1968). Prepubertal and pubertal changes in the hamster ovary. *Anat. Rec.* **161,** 447–458.
Greenwald, G. S., and Rothchild, I. (1968). Formation and maintenance of corpora lutea in laboratory animals. *J. Anim. Sci.* **27,** Suppl., 139–162.
Greenwald, G. S., Keever, J. E., and Grady, K. L. (1967). Ovarian morphology and pituitary FSH and LH concentration in pregnant and lactating hamster. *Endocrinology* **80,** 851–856.
Greep, R. O. (1963). Histology, histochemistry and ultrastructure of adult ovary. *Monogr. Pathol.* **3,** 48–68.
Gropp, A., and Ohno, S. (1966). Presence of a common embryonic blastema for ovarian and testicular parenchymal (follicular, interstitial and tubular) cells in cattle, *Bos taurus*. *Z. Zellforsch. Mikrosk. Anat.* **74,** 505–528.
Gulyas, B. J. (1974). The corpus luteum of the rhesus monkey (*Macaca mulatta*) during late pregnancy. An electron microscope study. *Am. J. Anat.* **139,** 95–120.
Guraya, S. S. (1968a). Histochemical study of the guinea-pig ovary. *Acta Biol. Acad. Sci. Hung.* **19,** 278–288.
Guraya, S. S. (1968b). Histochemical study of pre-ovulatory and post-ovulatory follicles in the rabbit ovary. *J. Reprod. Fertil.* **15,** 381–387.
Guraya, S. S. (1968c). Histochemical study of granulosa and theca interna during follicular development, ovulation and corpus luteum formation and regression in the human ovary. *Am J. Obstet. Gynecol.* **101,** 448–457.
Guraya, S. S. (1969). Some observations on the histochemical features of developing follicle and corpus luteum in the cat and dog ovary. *Acta Vet. Acad. Sci. Hung.* **19,** 351–362.
Guraya, S. S. (1973). Interstitial gland tissue of mammalian ovary. *Acta Endocrinol. (Copenhagen), Suppl.* **171,** 5–27.
Guraya, S. S., and Greenwald, G. S. (1964a). A comparative histochemical study of

the interstitial tissue and follicular atresia in the mammalian ovary. *Anat. Rec.* **149,** 411–437.

Guraya, S. S., and Greenwald, G. S. (1964b). Histochemical studies on the interstitial gland in the rabbit ovary. *Am. J. Anat.* **114,** 495–520.

Guraya, S. S., and Greenwald, G. S. (1965). A histochemical study of the hamster ovary. *Am. J. Anat.* **116,** 257–268.

Guthrie, M. J., and Jeffers, K. R. (1938a). Growth of follicles in the ovaries of the bat *Myotis lucifugus lucifugus. Anat. Rec.* **71,** 477–496.

Guthrie, M. J., and Jeffers, K. R. (1938b). A cytological study of the ovaries of the bats *Myotis lucifugus lucifugus* and *Myotis grisescens. J. Morphol.* **62,** 523–558.

Guthrie, M. J., Jeffers, K. R., and Smith, E. W. (1951). Growth of follicles in the ovaries of the bat *Myotis grisescens. J. Morphol.* **88,** 127–144.

Hadek, R. (1963). Electron microscope study on the primary liquor folliculi secretion in the mouse. *J. Ultrastruct. Res.* **9,** 445–458.

Hadek, R. (1966). Cytoplasmic whorls in the golden hamster oocyte. *J. Cell Sci.* **1,** 281–282.

Hafs, H. D., and Armstrong, D. T. (1968). Corpus luteum growth and progesterone synthesis during the bovine estrous cycle. *J. Anim. Sci.* **27,** 134.

Halayon, G. C., and Buechner, H. K. (1956). Postconception ovulation in elk. *Trans. North Am. Wildl. Conf.* **21,** 545–554.

Hamilton, J. E. (1934). The southern sea lion, *Otaria byronia* (de Blainville). *'Discovery' Rep.* **8,** 269–318.

Hamilton, J. E. (1939). A second report on the southern sea lion (*Otaria byronia*). *'Discovery' Rep.* **19,** 121–164.

Hamilton, W. J., Harrison, R. J., and Young, B. A. (1960). Aspects of placentation in certain Cervidae. *J. Anat.* **94,** 1–33.

Hamlett, G. W. D. (1932). The reproduction cycle in the armadillo. *Z. Wiss. Zool.* **141,** 143–154.

Hamlett, G. W. D. (1935). Delayed implantation and discontinuous development in the mammals. *Q. Rev. Biol.* **10,** 432–447.

Hamlett, G. W. D. (1937). Positive Friedman tests in the pregnant rhesus monkey, *Macaca mulatta. Am. J. Physiol.* **118,** 664–666.

Hammerstein, J., Rice, B. F., and Savard, K. (1964). Steroid hormone formation in the human ovary. I. Identification of steroids formed *in vitro* from acetate-1-^{14}C in the corpus luteum. *J. Clin. Endocrinol. Metab.* **24,** 597–605.

Hammond, J., and Marshall, F. H. A. (1930). Oestrus and pseudo-pregnancy in the ferret. *Proc. R. Soc. London, Ser. B* **105,** 607–629.

Hammond, J., and Wodzicki, K. (1941). Anatomical and histological changes during the oestrous cycle in the mare. *Proc. R. Soc. London, Ser. B* **130,** 1–23.

Hanks, J., and Short, R. V. (1972). The formation and function of the corpus luteum in the African elephant, *Loxodonta africana. J. Reprod. Fertil.* **29,** 79–89.

Hansel, W. (1966). Luteotropic and luteolytic mechanisms in bovine corpora lutea. *J. Reprod. Fertil., Suppl.* **1,** 33–48.

Hansson, A. (1947). The physiology of reproduction in mink (*Mustela vison,* Schreb) with special reference to delayed implantation. *Acta Zool.* (*Stockholm*) **28,** 1–136.

Harrison, R. J. (1946). The early development of the corpus luteum in the mare. *J. Anat.* **80,** 160–165.

Harrison, R. J. (1948a). The development and fate of the corpus luteum in the vertebrate series. *Biol. Rev. Cambridge Philos. Soc.* **23,** 296–331.
Harrison, R. J. (1948b). The changes occuring in the ovary of the goat during the oestrous cycle and in early pregnancy. *J. Anat.* **82,** 21–48.
Harrison, R. J. (1949). Observations on the female reproductive organs of the Ca'aing whale, *Globiocephala melaena* Traill. *J. Anat.* **83,** 238–253.
Harrison, R. J. (1960). Reproduction and reproductive organs in common seals (*Phoca vitulina*) in the Wash, East Anglia. *Mammalia* **24,** 372–385.
Harrison, R. J. (1962). The structure of the ovary in mammals. *In* "The Ovary" (S. Zuckerman, ed.), Vol. 1, Chapter 2c, pp. 143–187. Academic Press, New York.
Harrison, R. J. (1969). Reproduction and reproductive organs. *In* "The Biology of Marine Mammals" (H. T. Andersen, ed.), Chapter 8, pp. 253–348. Academic Press, New York.
Harrison, R. J., and Brownell, R. L., Jr. (1971). The gonads of the South American dolphins *Inia geoffrensis, Pontoporia blainvillei* and *Sotalia fluviatilis. J. Mammal.* **52,** 413–419.
Harrison, R. J., and Matthews, L. H. (1951). Sub-surface crypts in the cortex of the mammalian ovary. *Proc. Zool. Soc. London* **120,** 699–712.
Harrison, R. J., and Ridgway, S. H. (1971). Gonadal activity in some bottlenose dolphins (*Tursiops truncatus*). *J. Zool.* **164,** 355–366.
Harrison, R. J., Matthews, L. H., and Roberts, J. M. (1952). Reproduction in some Pinnipedia. *Trans. Zool. Soc. London* **27,** 437–540.
Harrison, R. J., Boice, R. C., and Brownell, R. L., Jr. (1969). Reproduction in wild and captive dolphins. *Nature* (*London*) **222,** 1143–1147.
Harrison, R. J., Brownell, R. L., Jr., and Boice, R. C. (1972). Reproduction and gonadal appearances in some odontocetes. *In* "Functional Anatomy of Marine Mammals" (R. J. Harrison, ed.), Vol. 1, pp. 361–429. Academic Press, New York.
Hartman, C. (1923). The oestrous cycle in the opossum. *Am. J. Anat.* **32,** 353–421.
Hartman, C. G. (1926). Polynuclear ova and polyovular follicles in the opossum and other mammals, with special reference to the problem of fecundity. *Am. J. Anat.* **37,** 1–51.
Hartman, C. G. (1929). Uterine bleeding as an early sign of pregnancy in the monkey (*Macacus rhesus*), together with observations on the fertile period of the menstrual cycle. *Bull. Johns Hopkins Hosp.* **44,** 155–164.
Hashimoto, M., Kawasaki, T., Kosaka, M., Mori, Y., Komori, A., Shimoyama, T., and Akashi, K. (1960). Electron microscopic studies on the fine structure of the rabbit ovarian follicles. *J. Jpn. Obstet. Gynecol.* **7,** 228–235 and 267–275.
Heap, R. B., Perry, J. S., and Rowlands, I. W. (1967). Corpus luteum function in the guinea-pig; arterial and luteal progesterone levels, and the effects of hysterectomy and hypophysectomy. *J. Reprod. Fertil.* **13,** 537–553.
Herlant, M. (1961). L'activité genitale chez la femelle de *Galago senegalensis mohioli* (Geoffr.) et ses rapports avec la persistance de phénomènes d'ovogénèse chez l'adulte. *Ann. Soc. R. Zool. Belg.* **91,** 1–15.
Hertig, A. T. (1964). Gestational hyperplasia of endometrium. A morphologic correlation of ova, endometrium and corpora lutea during early pregnancy. *Lab. Invest.* **13,** 1153–1191.
Hertig, A. T., and Adams, E. C. (1967). Studies on the human oocyte and its follicle. 1. Ultrastructural and histochemical observations on the primordial follicle stage. *J. Cell Biol.* **34,** 647–676.

Hesselton, W. T. (1967). Two further incidents of polyovulation in white-tailed deer in New York. *J. Mammal.* **48,** 321.
Hill, J. P., and Gatenby, J. B. (1926). The corpus luteum of the Monotremata. *Proc. Zool. Soc. London* **47,** 715–763.
Hill, J. P., and O'Donoghue, C. H. (1913). The reproductive cycle in the marsupial *Dasyurus viverrinus. Q. J. Microsc. Sci.* [N.S.] **59,** 133–174.
Hill, W. C. O. (1945). Notes on the dissection of two dugongs. *J. Mammal.* **26,** 153–175.
Hilliard, J., Archibald, D., and Sawyer, C. H. (1963). Gonadotropic activation of preovulatory synthesis and release of progestin in the rabbit. *Endocrinology* **72,** 59–66.
Hilliard, J., Penardi, R., and Sawyer, C. H. (1967). A functional role for 20α-hydroxypregn-4-en-3-one in the rabbit. *Endocrinology* **80,** 901–909.
Höflinger, H. (1948). Das Ovar des Rindes in den verschiedenen Lebensperioden unter besonderer Berücksichtigung seiner funktionellen Feinstruktur. *Acta Anat.* **3,** Suppl. 5, 1–196.
Hope, J. (1965). The fine structure of the developing follicle of the rhesus ovary. *J. Ultrastruct. Res.* **12,** 592–610.
Horowitz, M. (1967). The anovular corpus luteum in the golden hamster (*Mesocricetus auratus,* Waterhouse) and comparisons with the normal corpus luteum and follicle of atresia. *Acta Anat.* **66,** 199–225.
Huang, W. Y., and Pearlman, W. H. (1963). The corpus luteum and steroid hormone formation. II. Studies on the human corpus luteum *in vitro. J. Biol. Chem.* **238,** 1308–1315.
Hughes, R. L. (1962). Role of the corpus luteum in marsupial reproduction. *Nature* (*London*) **194,** 890–891.
Hughes, R. L., Thomson, J. A., and Owen, W. H. (1965). Reproduction in natural populations of Australian ringtail possum, *Pseudocheirus peregrinus* (Marsupialia: Phalangeridae), in Victoria. *Aust. J. Zool.* **13,** 383–406.
Hunter, R. H. F., and Polge, C. (1966). Maturation of follicular oocytes in the pig after injection of human chorionic gonadotrophin. *J. Reprod. Fertil.* **12,** 525–531.
Hutchinson, J. S. M., and Robertson, H. A. (1966). The growth of the follicle and corpus luteum in the ovary of sheep. *Res. Vet. Sci.* **7,** 17–24.
Ioannou, J. M. (1967). Oogenesis in adult prosimians. *J. Embryol. Exp. Morphol.* **17,** 139–145.
Jacobowitz, D., and Wallach, E. E. (1967). Histochemical and chemical studies of the autonomic innervation of the ovary. *Endocrinology* **81,** 1132–1139.
Jdanov, D. A. (1960). Nouvelles données sur la morphologie fonctionelle du système lymphatique des glands endocrines. *Acta Anat.* **41,** 240–259.
Johansson, E. D. B., Neill, J. D., and Knobil, E. (1968). Periovulatory progesterone concentration in the peripheral plasma of the rhesus monkey with a methodologic note on detection of ovulation. *Endocrinology* **82,** 143–148.
Kaltenbach, C. C., Graber, J. W., Niswender, G. D., and Nalbandov, A. V. (1968). Effect of hypophysectomy on the formation and maintenance of corpora lutea in the ewe. *Endocrinology* **82,** 753–759.
Kanagasuntheram, R., and Vorzin, J. A. (1964). The intrinsic nerve supply of the female reproductive organs in the lesser bush baby (*Galago senegalensis senegalensis*). *Acta Anat.* **58,** 306–316.

Kang, Y.-H. (1974). Development of the zona pellucida in the rat oocyte. *Am. J. Anat.* **139,** 535–566.
Kayanja, F. I. B. (1969). The ovary of the impala, *Aepyceros melampus* (Lichtenstein, 1812). *J. Reprod. Fertil., Suppl.* **6,** 311–317.
Kayanja, F. I. B., and Blankenship, L. H. (1973). The ovary of the giraffe, *Giraffa camelopardalis. J. Reprod. Fertil.* **34,** 305–313.
Kayanja, F. I. B., and Sale, J. B. (1973). The ovary of rock hyrax of the genus *Procavia. J. Reprod. Fertil.* **33,** 223–230.
Kellas, L. M. (1955). Observations on the reproductive activities, measurements, and growth rate of the dikdik (*Rhynchotragus kirkii thomasi* Neumann). *Proc. Zool. Soc. London* **124,** 751–784.
Kellas, L. M., van Lennep, E. W., and Amoroso, E. C. (1958). Ovaries of some foetal and prepubertal giraffes (*Giraffa camelopardalis* Linnaeus). *Nature (London)* **181,** 487–488.
Kellogg, M. P. (1941). The development of the periovarial sac in the white rat. *Anat. Rec.* **79,** 465–477.
Kennelly, J. J., and Foote, R. H. (1966). Oocytogenesis in rabbits. The role of neogenesis in the formation of the definitive ova and the stability of oocyte DNA measured with tritiated thymidine. *Am. J. Anat.* **118,** 573–590.
Kent, H. A. (1959). Reduction of polyovular follicles and polynuclear ova by oestradiol monobenzoate. *Anat. Rec.* **134,** 455–462.
Kent, H. A. (1960). Polyovular follicles and multinucleate ova in the ovaries of young mice. *Anat. Rec.* **137,** 521–524.
Kent, H. A. (1962). Polyovular follicles and multinucleate ova in the ovaries of young hamsters. *Anat. Rec.* **143,** 345–349.
Kent, H. A., and Mandel, J. A. (1968). Correlation of ovarian abnormalities in young golden hamsters with blood estrogen content. *Anat. Rec.* **161,** 53–56.
Keyes, P. L., and Armstrong, D. T. (1968). Endocrine role of follicles in regulation of corpus luteum function in the rabbit. *Endocrinology* **83,** 509–515.
Kimura, J., and Lyons, W. R. (1937). Progestin in the pregnant mare. *Proc. Soc. Exp. Biol. Med.* **37,** 423–427.
Kimura, T., Yaoi, Y., and Kamasaki, T. (1970). Ovarian morphology following HMG treatment with special reference to histological, histochemical and electron microscopic studies. *Bull. Tokyo Med. Dent. Univ.* **17,** 39–51.
Kleinenberg, S. E., Yablokov, A. V., Bel'kovich, V. M., and Tarasevich, M. N. (1964). [Urogenital system.] *In* "Beluga—Results of Monographic Investigation of the Species," Chapter 7, pp. 79–117., Acad. Sci., USSR, Moscow.
Knigge, K. M., and Leathem, J. H. (1956). Growth and atresia of follicles in the ovary of the hamster. *Anat. Rec.* **124,** 679–707.
Koering, M. J. (1969). Cyclic changes in ovarian morphology during the menstrual cycle in *Macaca mulatta. Am. J. Anat.* **126,** 73–101.
Koizumi, T. (1965). [Electron microscopic study of the corpus luteum in rabbits and rats.] *Folia Endocrinol. Jpn.* **41,** 994–1009.
Küpfer, M. (1928). The sexual cycle of female domesticated mammals. *13th and 14th Rep. Vet. Res. S. Afr.* Part II, pp. 1209–1256, and Suppl.
Labhsetwar, A. P., and Diamond, M. (1970). Ovarian changes in the guinea-pig during various reproductive stages and steroid treatments. *Biol. Reprod.* **2,** 53–57.
Labhsetwar, A. P., and Enders, A. C. (1968). Progesterone in the corpus luteum and placenta of the armadillo, *Dasypus novemcinctus. J. Reprod. Fertil.* **16,** 381–387.

Labhsetwar, A. P., Collins, W. E., Tyler, W. J., and Casida, L. E. (1964). Some pituitary-ovarian relationships in the periparturient cow. *J. Reprod. Fertil.* **8**, 85–90.

Laffargue, P., Adechy-Benkoel, L., and Valette, C. (1968). Ultrastructure du stroma ovarien. (Signification fonctionelle). *Ann. Anat. Pathol.* **13**, 381–402.

Land, R. B. (1970). Number of oocytes present at birth in the ovaries of pure and Finnish Landrace cross Blackface and Welsh sheep. *J. Reprod. Fertil.* **21**, 517–521.

Land, R. B., and McGill, T. E. (1967). The effects of the mating pattern of the mouse on the formation of corpora lutea. *J. Reprod. Fertil.* **13**, 121–125.

Lane, C. E., and Davis, F. R. (1939). The ovary of the adult rat. 1. Changes in growth of the follicle and in volume and mitotic activity of the granulosa and theca during the estrous cycle. *Anat. Rec.* **73**, 429–442.

Lanzavecchia, G., and Mangioni, C. (1964). Etude de la structure et des constituants du follicle humain dans l'ovaire foetal. 1. Le follicle primordial. *J. Microsc. (Paris)* **3**, 447–464.

Latimer, H. B. (1956). The ponderal growth of the ovaries and uterus in the fetal dog. *Anat. Rec.* **125**, 731–744.

Laws, R. M. (1956). The elephant seal (*Mirounga leonina* Linn.). III. The physiology of reproduction. *Falkland Isl. Depend. Surv., Sci. Rep.* **15**, 1–66.

Laws, R. M. (1958). Recent investigations on fin whale ovaries. *Nor. Hvalfangst-tid.* **5**, 225–254.

Laws, R. M. (1961). Reproduction, growth and age of southern fin whales. *'Discovery' Rep.* **31**, 327–486.

Laws, R. M. (1969). Aspects of reproduction in the African elephant, *Loxodonta africana. J. Reprod. Fertil., Suppl.* **6**, 193–217.

Laws, R. M., and Clough, G. (1966). Observations on reproduction in the hippopotamus, *Hippopotamus amphibius* Linn. *Symp. Zool. Soc. London* **15**, 117–140.

Leach, B. J., and Conaway, C. H. (1963). Origin and fate of polyovular follicles in the striped skunk. *J. Mammal.* **44**, 67–74.

League, B., and Hartman, C. G. (1925). Anovular Graafian follicles in mammalian ovaries. *Anat. Rec.* **30**, 1–13.

LeBrie, S. J. (1970). Structure and function of the lymphatic system with emphasis on its role in hormone transport. *J. Reprod. Fertil., Suppl.* **10**, 123–138.

Lee, C., Horvath, D. J., Metcalfe, R. W., and Inskeep, E. K. (1970). Ovulation in *Microtus pennsylvanicus* in a laboratory environment. *Lab. Anim. Care* **20**, 1098–1102.

Leland, T. M., Dickey, J. F., and Hill, J. R. (1969). Fine structure of bovine luteal cells. *J. Anim. Sci.* **28**, 126.

Le Pere, R. H., Benoit, P. E., Hardy, R. C., and Goldzieher, J. W. (1966). Origin and function of the ovarian nerve supply in the baboon. *Fertil. Steril.* **17**, 68–75.

Lipner, H., and Cross, N. L. (1968). Morphology of the membrana granulosa of the ovarian follicle. *Endocrinology* **82**, 638–641.

Lipschütz, A. (1928). New developments in ovarian dynamics and the law of follicular constancy. *J. Exp. Biol.* **5**, 283–291.

Liss, R. H. (1964). A study of fine structure of a mammalian ovary (*Felis domestica*). *Anat. Rec.* **148**, 385.

Lobel, B. L., and Levy, E. (1968). Enzymic correlates of development, secretory function and regression of follicles and corpora lutea in the bovine ovary. *Acta Endocrinol.* (*Copenhagen*), *Suppl.* **132,** 1–63.

Lobel, B. L., Rosenbaum, R. M., and Deane, H. W. (1961). Enzymatic correlates of physiological regression of follicles and corpora lutea in ovaries of normal rats. *Endocrinology* **68,** 232–247.

Long, J. A. (1973). Corpus luteum of pregnancy in the rat—ultrastructural and cytochemical observations. *Biol. Reprod.* **8,** 87–99.

Longenecker, D. E., Waite, A. B., and Day, B. N. (1968). Similarity in the number of corpora lutea during two stages of pregnancy in swine. *J. Anim. Sci.* **27,** 466–467.

Loubet, R., and Loubet, A. (1961). Les cellules du hile de l'ovaire et leurs rapports avec les autres éléments endocrines de la glande. *Ann. Anatomie Pathol.* [N.S.] **6,** 189–212.

Lyngset, O. (1968a). Reproduction in the goat. 1. Normal genital organs of the non-pregnant goat. *Acta Vet. Scand.* **9,** 208–222.

Lyngset, O. (1968b). Reproduction in the goat. 2. Genital organs of the pregnant goat. *Acta Vet. Scand.* **9,** 242–252.

Lyngset, O. (1968c). Reproduction in the goat. 3. Functional activity of the ovaries. *Acta Vet. Scand.* **9,** 268–276.

McDonald, D. M., and Goldfien, A. (1965). The relation of lutein cell Golgi apparatus morphology to plasma progesterone concentration during the estrous cycle. *Anat. Rec.* **151,** 385.

McDonald, D. M., Seiki, K., Prizant, M., and Goldfien, A. (1969). Ovarian secretion of progesterone in relation to the Golgi apparatus in lutein cells during the estrous cycle of the rat. *Endocrinology* **85,** 236–243.

McGill, T. E., and Coughlin, R. C. (1970). Ejaculatory reflex and luteal activity induction in *Mus musculus*. *J. Reprod. Fertil.* **21,** 215–220.

McKay, D. G., and Robinson, D. (1947). Fluorescence, birefringence and histochemistry of the human ovary during the menstrual cycle. *Endocrinology* **41,** 378–394.

McKay, D. G., Pinkerton, J. H. M., Hertig, A. T., and Danziger, S. (1961). The adult human ovary: a histochemical study. *Obstet. Gynecol.* **18,** 13–39.

McKeever, S. (1966). Reproduction in *Citellus beldingi* and *Citellus lateralis* in North Eastern California. *Symp. Zool. Soc. London* **15,** 365–385.

Macaluso, M. C. (1965). The fine structure of the corpus luteum of the Swiss mouse after parturition. *Anat. Rec.* **151,** 463.

Mackintosh, N. A. (1946). The natural history of whalebone whales. *Biol. Rev. Cambridge Philos. Soc.* **21,** 60–74.

Mackintosh, N. A. (1965). "The Stocks of Whales." Fishing News (Books) Ltd., London.

Mackintosh, N. A., and Wheeler, J. F. G. (1929). Southern blue and fin whales. *'Discovery' Rep.* **1,** 257–540.

MacRae, D. J., Willmott, M. P., Ismail, A. A. A., and Love, D. N. (1971). Androgen patterns in a case of hilar cell hyperplasia of the ovary. *J. Obstet. Gynaecol. Br. Commonw.* **78,** 741–745.

Malven, P. V. (1969). Hypophysial regulation of luteolysis in the rat. *In* "The Gonads" (K. W. McKerns, ed.), pp. 367–382. Appleton, New York.

Mandl, A. M. (1963). Preovulatory changes in the oocyte of the adult rat. *Proc. R. Soc. London, Ser. B* **158,** 105–118.

Manotaya, T., and Potter, E. L. (1963). Oocytes in prophase of meiosis from squash preparations of human fetal ovaries. *Fertil. Steril.* **14**, 378–392.

Mansfield, A. W. (1958). The breeding behaviour and reproductive cycle of the Weddell seal (*Leptonychotes weddelli* Lesson). *Falkland Isl. Depend. Surv., Sci. Rep.* **18**, 1–41.

Maqueo, M., and Goldzieher, J. W. (1966). Hormone-induced alterations of ovarian morphology. *Fertil. Steril.* **17**, 676–683.

Marcotty, A. (1914). Ueber das Corpus luteum menstruationis und das Corpus luteum graviditatis. *Arch. Gynaekol.* **103**, 63–106.

Marion, G. B., Gier, H. T., and Choudary, J. B. (1968). Micromorphology of the bovine ovarian follicular system. *J. Anim. Sci.* **27**, 451–465.

Martinez-Esteve, P. (1942). Observations on the histology of the opossum ovary. *Contrib. Embryol. Carnegie Inst.* **30**, 17–26.

Matthews, L. H. (1935). The oestrous cycle and intersexuality of the female mole (*Talpa europaea* Linn.). *Proc. Zool. Soc. London* 347–383.

Matthews, L. H. (1947). A note on the female reproductive tract in the tree kangaroo (*Dendrolagus*). *Proc. Zool. Soc. London* **117**, 313–333.

Mattner, P. E., and Thorburn, G. D. (1969). Ovarian blood flow in sheep during the oestrous cycle. *J. Reprod. Fertil.* **19**, 547–549.

Mead, R. A. (1968a). Reproduction in eastern forms of the spotted skunk (genus *Spilogale*). *J. Zool.* **156**, 119–136.

Mead, R. A. (1968b). Reproduction in western forms of the spotted skunk (genus *Spilogale*). *J. Mammal.* **49**, 373–390.

Merker, H. J. (1961). Elektronen mikroskopische Untersuchungen über die Bildung der Zona Pellucida in den Follikeln des Kaninchenovars. *Z. Zellforsch. Mikrosk. Anat.* **54**, 677–688.

Merker, H. J., and Diaz-Encinas, J. (1969). Das elecktronmikroskopische Bild des ovars juveniler Ratten und Kaninchen nach Stimulierung mit PMS und HCG. 1. Theka und Stroma (Interstitielle Drüsse). *Z. Zellforsch. Mikrosk. Anat.* **94**, 605–623.

Meyer, R. (1911). Über corpus Bildung beim Menschen. *Arch. Gynaekol.* **93**, 354–404.

Mirecka, J. (1969). Characters of granulo-lutein and theca-lutein cells as shown by the lipid metabolism of corpus luteum. *Folia Histochem. Cytochem.* **7**, 117–132.

Modak, S. P., and Kamat, D. N. (1968). Study of periodic acid Schiff material in the ovarian follicles of tropical bats. *Cytologia* **33**, 54–59.

Møller, O. M. (1973). The fine structure of the ovarian interstitial gland cells in the mink, *Mustela vison*. *J. Reprod. Fertil.* **34**, 171–174.

Montemagno, U., and Caramia, F. G. (1965). Effect of the PMS hormone stimulation on the ultrastructure of the oocyte in the rabbit. *Folia Endocrinol.* **18**, 77–93.

Moor, R. M., and Rowson, L. E. A. (1966). Local uterine mechanisms affecting luteal function in the sheep. *J. Reprod. Fertil.* **11**, 307–310.

Mori, H. (1970). Fine structure of interstitial gland cells in rabbit ovaries. *Med. J. Osaka Univ.* **20**, 215–233.

Mori, H., and Matsumoto, K. (1970). On the histogenesis of the ovarian interstitial gland in rabbits. 1. Primary interstitial gland. *Am. J. Anat.* **129**, 289–306.

Mori, H., and Matsumoto, K. (1973). Development of the secondary interstitial gland in the rabbit ovary. *J. Anat.* **116**, 417–429.

Moricard, R. (1958). Fonction méiogène et fonction oestrogène du follicule ovarien des mammifères (cytologie Golgienne, traceurs, microscopie électronique). *Ann. Endocrinol.* **19,** 943–967.
Moricard, R., and Sallusto, A. (1968). Gonadotrophines et cytophysiologie ultrastructural de la méiose et de l'ovulation chez la lapine. *Proc. Int. Congr. Anim. Reprod. Artif. Insem., 6th, 1968* 1, 165–168.
Moricard, R., Gothie, S., and Traub, A. (1969). Modifications ultrastructurales des cellules periovocytaires apparaissant apres coït avant l'ovulation. *Gynecol. Obstet.* **68,** 79–92.
Morris, B., and Sass, M. B. (1966). The formation of lymph in the ovary. *Proc. R. Soc. London, Ser. B* **164,** 577–591.
Morrison, J. A. (1960). Ovarian characteristics in elk of known breeding history. *J. Wildl. Manage.* **24,** 297–307.
Morrison, J. A. (1971). Morphology of corpora lutea in the Uganda kob antelope, *Adenota kob thomasi* (Neumann). *J. Reprod. Fertil.* **26,** 297–305.
Morrison, J. A., and Buechner, N. A. (1971). Reproductive phenomena during the *post partum*-preconception interval in the Uganda kob. *J. Reprod. Fertil.* **26,** 307–317.
Moss, S., Wrenn, T. R., and Sykes, J. F. (1954). Some histological and histochemical observations of the bovine ovary during the estrous cycle. *Anat. Rec.* **120,** 409–434.
Mossman, H. W. (1937). The thecal gland and its relation to the reproductive cycle. A study of the cyclic changes in the ovary of the pocket gopher *Geomys bursarius* (Shaw). *Am. J. Anat.* **61,** 289–319.
Mossman, H. W. (1938). The homology of the vesicular ovarian follicles of the mammalian ovary with the coelom. *Anat. Rec.* **70,** 643–655.
Mossman, H. W. (1966). The rodent ovary. *Symp. Zool. Soc. London* **15,** 455–470.
Mossman, H. W., and Duke, K. L. (1973). "Comparative Morphology of the Mammalian Ovary." Univ. of Wisconsin Press, Madison.
Mossman, H. W., and Judas, I. (1949). Accessory corpora lutea, lutein cell origin and the ovarian cycle in the Canadian porcupine. *Am. J. Anat.* **85,** 1–39.
Mossman, H. W., Koering, M. J., and Ferry, D. (1964). Cyclic changes of interstitial gland tissue of the human ovary. *Am. J. Anat.* **115,** 235–255.
Motta, P. (1966). Osservazioni sulla fine struttura delle cellule interstiziali dell'ovaio (ricerche in *Lepus cuniculus* Linn.). *Biol. Lat.* **19,** 107–133.
Motta, P. (1967). Sui corpi de call ed exner in diverse specie di mammiferi. *Biol. Lat.* **20,** 303–318.
Motta, P. (1969). Electron microscope study of the human lutein cell with special reference to its secretory activity. *Z. Zellforsch. Mikrosk. Anat.* **98,** 233–245.
Motta, P., and Nesci, E. (1969). "Call-Exner bodies" of mammalian ovaries with reference to the problem of rosette formation. *Arch. Anat. Microsc. Morphol. Exp.* **58,** 283–290.
Motta, P., and Takeva, Z. (1969). Electron microscopy and histochemical study of Δ^5-3β-hydroxysteroid dehydrogenase in the interstitial cells of the human ovary in different stages of fetal and adult life. *Arch. Ostet. Ginecol.* **74,** 415–424.
Motta, P., Takeva, Z., and Bourneva, V. (1970). A histochemical study of Δ^5-3β-hydroxysteroid dehydrogenase activity in the interstitial cells of the mammalian ovary. *Experientia* **26,** 1128–1129.
Motta, P., Takeva, Z., and Palermo, D. (1971). On the presence of cilia in different cells of the mammalian ovary. *Acta Anat.* **78,** 591–603.

Mozanska, T. (1968). Histologic studies on the structure and origin of tubular structures and epithelial crypts in the ovaries of white rats. *Folia Morphol.* **27,** 249–258.

Mullen, D. A. (1968). Reproduction in brown lemmings (*Lemmus trimucronatus*) and its relevance to their cycle of abundance. *Univ. Calif., Berkeley, Publ. Zool.* **85,** 1–24.

Mulligan, R. N. (1942). Histological studies on the canine female genital tract. *J. Morphol.* **71,** 431–448.

Muta, T. (1958). The fine structure of the interstitial cell in the mouse ovary studied with electron microscope. *Kurume Med. J.* **5,** 167–185.

Mutere, F. (1965). Delayed implantation in an equatorial fruit bat. *Nature* (*London*) **207,** 780.

Nalbandov, A. (1970). Comparative aspects of corpus luteum function. *Biol. Reprod.* **2,** 7–13.

Neal, E. G., and Harrison, R. J. (1958). Reproduction in the European Badger (*Meles meles* L.). *Trans. Zool. Soc. London* **29,** 67–131.

Neilson, D., Jones, G. S., Woodruff, J. D., and Goldberg, B. (1970). The innervation of the ovary. *Obstet. & Gynecol. Surv.* **25,** 889–904.

Nekola, M. V., and Nalbandov, A. V. (1971). Morphological changes of rat follicular cells as influenced by oocytes. *Biol. Reprod.* **4,** 154–160.

Nelson, W. W., and Greene, R. R. (1953). The human ovary in pregnancy. *Surg., Gynecol. Obstet.* **97,** 1–22.

Nelson, W. W., and Greene, R. R. (1958). Some observations on the histology of the human ovary during pregnancy. *Am. J. Obstet. Gynecol.* **76,** 66–90.

Newson, J. (1964). Reproduction and prenatal mortality of snowshoe hares on Manitoulin Island, Ontario. *Can. J. Zool.* **42,** 988–1005.

Nishizuka, Y. (1954). Histological studies on ovaries: mainly histochemical observations on the ovarian endocrine functions. *Acta Sch. Med. Univ. Kioto* **31,** 215–257.

Novak, E. R., Goldberg, B., Jones, G. S., and O'Toole, R. V. (1965). Enzyme histochemistry of the menopausal ovary associated with normal and abnormal endometrium. *Am. J. Obstet. Gynecol.* **93,** 669–682.

Novoa, C. (1970). Reproduction in Camelidae. *J. Reprod. Fertil.* **22,** 3–20.

O'Donoghue, C. H. (1912). The corpus luteum in the non-pregnant *Dasyurus* and polyovular follicles in *Dasyurus*. *Anat. Anz.* **41,** 353–368.

O'Donoghue, C. H. (1916). On the corpora lutea and interstitial tissue of the ovary in the Marsupialia. *Q. J. Microsc. Sci.* [N.S.] **61,** 433–473.

O'Donoghue, P. N. (1963). Reproduction in the female hyrax (*Dendrohyrax arborea ruwenzorii*). *Proc. Zool. Soc. London* **141,** 207–238.

Odor, D. L. (1960). Electron microscopic studies on ovarian oocytes and unfertilized tubal ova in the rat. *J. Biophys. Biochem. Cytol.* **7,** 567–576.

Odor, D. L. (1965). The ultrastructure of unilaminar follicles of the hamster ovary. *Am. J. Anat.* **116,** 493–522.

Odor, D. L. (1967). Electron microscopic study of the histogenesis of the fetal and early postnatal mouse ovary. *Anat. Rec.* **157,** 293–294.

Odor, D. L., and Blandau, R. J. (1969). Ultrastructural studies on fetal and early postnatal mouse ovaries. 1. Histogenesis and organogenesis. *Am. J. Anat.* **124,** 163–186.

Odor, D. L., and Renninger, D. F. (1960). Polar body formation in the rat oocyte as observed with the electron microscope. *Anat. Rec.* **137,** 13–23.

O'Gara, B. W. (1969). Unique aspects of reproduction in the female pronghorn (*Antilocapra americana* Ord.). *Am. J. Anat.* **125,** 217–232.
Ogle, T. F., Braach, H. H., and Buss, T. O. (1973). Fine structure and progesterone concentration in the corpus luteum of the African elephant. *Anat. Rec.* **175,** 707–724.
Ohno, S., and Smith, J. B. (1964). Role of fetal follicular cells in meiosis of mammalian oocytes. *Cytogenetics* **3,** 324–333.
Ohno, S., Makino, S., Kaplan, W. D., and Kinosita, R. (1961). Female germ cells in man. *Exp. Cell Res.* **24,** 106–110.
Ohno, S., Klinger, H. P., and Atkin, N. B. (1962). Human oogenesis. *Cytogenetics* **1,** 42–51.
Ohsumi, S. (1964). Comparison of maturity and accumulation rate of corpora albicantia between the left and right ovaries in Cetacea. *Sci. Rep. Whales Res. Inst.* **18,** 123–148.
Okamoto, T. (1970). Studies on the uteri and ovaries in fur seals. *J. Fac. Fish Anim. Husb., Hiroshima Univ.* **9,** 161–175.
O'Shea, J. D. (1966). Histochemical observations on mucin secretion by subsurface epithelial structures in the canine ovary. *J. Morphol.* **120,** 347–358.
O'Shea, J. D. (1968). Histological study of non-follicular cysts in the ovulation fossa region of the equine ovary. *J. Morphol.* **124,** 313–320.
O'Shea, J. D. (1970). Ultrastructural study of smooth muscle-like cells in the theca externa of ovarian follicles in the rat. *Anat. Rec.* **167,** 127–139.
O'Shea, J. D. (1971a). Smooth muscle-like cells in the theca externa of ovarian follicles in the sheep. *J. Reprod. Fertil.* **24,** 283–285.
O'Shea, J. D. (1971b). Ovarian nerve supply in rats. *J. Anat.* **108,** 205.
Osvaldo-Decima, L. (1970). Smooth muscle in the ovary of the rat and monkey. *J. Ultrastruct. Res.* **29,** 218–237.
Owman, C., and Sjöberg, N.-O. (1966). Adrenergic nerves in the female genital tract of the rabbit. With remarks on cholinesterase-containing structures. *Z. Zellforsch. Mikrosk. Anat.* **74,** 182–197.
Owman, C., Rosengren, E., and Sjöberg, N.-O. (1967). Adrenergic innervation of the human female reproductive organs: a histochemical and chemical investigation. *Obstet. Gynecol.* **30,** 763–773.
Pace, E. (1966). The biochemical constituents of horse embryonic gonads. *Ital. J. Biochem.* **15,** 30–42.
Parkes, A. S. (1931). Reproductive processes of certain mammals. I. Oestrous cycle of the Chinese hamster (*Cricetulus griseus* Milne-Edwards). *Proc. R. Soc. London, Ser. B* **108,** 138–149.
Pearson, A. K., and Enders, R. K. (1951). Further observations on the reproduction of the Alaskan fur seal. *Anat. Rec.* **111,** 695–712.
Pearson, O. P. (1944). Reproduction in the shrew (*Blarina brevicauda* Say). *Am. J. Anat.* **75,** 39–93.
Pearson, O. P. (1949). Reproduction of a South American rodent, the mountain viscacha. *Am. J. Anat.* **84,** 143–174.
Pearson, O. P., and Baldwin, P. H. (1953). Reproduction and age structure of a Mongoose population in Hawaii. *J. Mammal.* **34,** 436–447.
Pearson, O. P., and Enders, R. K. (1943). Ovulation, maturation and fertilization in the fox. *Anat. Rec.* **85,** 69–83.
Pearson, O. P., Koford, M. R., and Pearson, A. K. (1952). Reproduction of the lump-nosed bat (*Corinorhynus rafinesquei*) in California. *J. Mammal.* **33,** 273–320.

Pedersen, P. H., and Larsen, J. F. (1968). The ultrastructure of the human granulosal lutein cell of the first trimester of gestation. *Acta Endocrinol. (Copenhagen)* **58,** 481–496.

Pedersen, T. (1970a). Determination of follicle growth rate in the ovary of the immature mouse. *J. Reprod. Fertil.* **21,** 81–93.

Pedersen, T. (1970b). Follicle kinetics in the ovary of the cyclic mouse. *Acta Endocrinol. (Copenhagen)* **64,** 304–323.

Pedersen, T., and Peters, H. (1968). Proposal for a classification of oocytes and follicles in the mouse ovary. *J. Reprod. Fertil.* **17,** 555–557.

Pedersen, T., and Peters, H. (1971). Follicle growth and cell dynamics in the mouse ovary during pregnancy. *Fertil. Steril.* **22,** 42–52.

Perry, J. S. (1953). The reproduction of the African elephant, *Loxodonta africana. Phil. Trans. R. Soc. London, Ser. B* **237,** 93–149.

Perry, J. S. (1964). The structure and development of the reproductive organs of the female African elephant. *Phil. Trans. R. Soc. London, Ser. B* **248,** 33–51.

Perry, J. S., and Rowlands, I. W. (1962). Early pregnancy in the pig. *J. Reprod. Fertil.* **4,** 175–188.

Perry, J. S., and Rowlands, I. W. (1963). Hypophysectomy of the immature guinea-pig and the ovarian response to gonadotrophins. *J. Reprod. Fertil.* **6,** 393–404.

Peters, H. (1969). The development of the mouse ovary from birth to maturity. *Acta Endocrinol. (Copenhagen)* **62,** 98–116.

Peters, H., and Clarke, J. (1974). The development of the ovary from birth to maturity in the bank vole (*Clethrionomys glareolus*) and the vole (*Microtus agrestis*). *Anat. Rec.* **179,** 241–252.

Peters, H., and Levy, E. (1966). Cell dynamics of the ovarian cycle. *J. Reprod. Fertil.* **11,** 227–236.

Peters, N. (1939). Über Grosse, Wachstum und Alter des Blauwales (*Balaenoptera musculus* L.) und Finwales (*Balaenoptera physalus* L.). *Zool. Anz.* **127,** 193–204.

Petter-Rousseaux, A. (1962). Recherches sur la biologie de la reproduction des primates inférieurs. *Mammalia* **26,** 1–87.

Petter-Rousseaux, A., and Bourlière, F. (1965). Persistance des phénomènes d'ovogénèse chez l'adulte de *Daubentonia madagascariensis* (Prosimii, Lemuriformes). *Folia Primatol.* **3,** 241–244.

Peyre, A. (1962). Recherches sur l'intersexualité specifique chez *Galemys pyrenaicus,* G. (Mammifère, Insectivore). *Arch. Biol.* **73,** 1–74.

Peyre, A., and Herlant, M. (1963). Ovo-implantation différée et corrélations hypophyso-génitales chez la femelle du Minioptère (*Miniopterus schrebersii* B.). *C. R. Acad. Sci. (Paris)* **257,** 524–526.

Pilton, P. E., and Sharman, G. B. (1962). Reproduction in the marsupial *Trichosurus vulpecula. J. Endocrinol.* **25,** 119–136.

Pinkerton, J. H. M., McKay, D. G., Adams, E. C., and Hertig, A. T. (1961). Development of the human ovary—a study using histochemical techniques. *Obstet. Gynecol.* **18,** 152–181.

Poole, W. E., and Pilton, P. E. (1964). Reproduction in the grey kangaroo, *Macropus canguru* in captivity. *CSIRO Wildl. Res.* **9,** 218–234.

Prasad, M. R. N. (1961). Reproduction in the female Indian gerbille *Tatera indica cuvieri. Acta Zool.* **42,** 1–12.

Pratt, J. P. (1935). The human corpus luteum. *Arch. Pathol. Lab. Med.* **19,** 380–545.

Price, H. (1953). The reproductive cycle of the water shrew, *Neomys sodiens bicolor* Shaw. *Proc. Zool. Soc. London* **123,** 599–621.

Priedkalns, J., and Weber, A. F. (1968a). Ultrastructural studies of the bovine Graafian follicle and corpus luteum. *Z. Zellforsch. Mikrosk. Anat.* **91,** 554–573.

Priedkalns, J., and Weber, A. F. (1968b). Quantitative ultrastructural analysis of the follicular and luteal cells of the bovine ovary. *Z. Zellforsch. Mikrosk. Anat.* **91,** 574–585.

Priedkalns, J., and Weber, A. F. (1968c). The succinic dehydrogenase and lipid content of follicular and luteal cells of the bovine ovary. *Acta Anat.* **71,** 542–564.

Priedkalns, J., Weber, A. F., and Zemjanis, R. (1968). Qualitative and quantitative morphological studies of the cells of the membrana granulosa, theca interna and corpus luteum of the bovine ovary. *Z. Zellforsch. Mikrosk. Anat.* **85,** 501–520.

Provost, E. E. (1962). Morphological characteristics of the beaver ovary. *J. Wildl. Manage.* **26,** 272–278.

Quay, W. B. (1960). The reproductive organs of the collared lemming under diverse temperature and light conditions. *J. Mammal.* **41,** 74–89.

Raczynski, J. (1964). Studies on the European hare. V. Reproduction. *Acta Theriol.* **19,** 305–353.

Rajakoski, E. (1960). The ovarian follicular system in sexually mature heifers with special reference to seasonal, cyclical and left-right variations. *Acta Endocrinol.* (*Copenhagen*), *Suppl.* **53,** 1–68.

Rajakoski, E., and Hafez, E. S. E. (1963). Derivatives of cortical cords in adult freemartin gonads of bovine quintuplets. *Anat. Rec.* **147,** 457–467.

Ramaswamy, K. R. (1961). Studies on the sex-cycle of the Indian vampire bat, *Megaderma* (*Lyroderma*) *lyra lyra* (Geoffrey). I. Breeding habits. *Proc. Natl. Inst. Sci. India, Part B* **27,** 287–307.

Rand, R. W. (1955). Reproduction in the female Cape Fur Seal, *Arctocephalus pusillus* (Schreber). *Proc. Zool. Soc. London* **124,** 717–740.

Rankin, J. J. (1961). The bursa ovarica of the beaked whale, *Mesoplodon gervaisi* Deslongchamps. *Anat. Rec.* **139,** 379–386.

Raps, G. (1948). The development of the dog ovary from birth to six months of age. *Am. J. Vet. Res.* **9,** 61–64.

Rathmacher, R. P., and Anderson, L. L. (1968). Blood flow and progesterone levels in the ovary of cycling and pregnant pigs. *Am. J. Phyiol.* **214,** 1014–1018.

Reitmeyer, J. C., and Sorenson, A. M. (1965). Accessory corpora lutea in swine. *J. Anim. Sci.* **24,** 928.

Rennels, E. G. (1966). Observations on the ultrastructure of luteal cells from PMS and PMS-HCG treated immature rats. *Endocrinology* **79,** 373–386.

Rennie, P., Davenport, G. R., and Welborn, W. (1965). Cellular changes, protein and glycogen content in developing rabbit corpora lutea. *Anat. Rec.* **153,** 289–294.

Restall, B. J. (1964). The growth and retrogression of the corpus luteum in the ewe. *Aust. J. Exp. Agric. Anim. Husb.* **4,** 274–276.

Reynolds, S. R. M. (1950). The vasculature of the ovary and ovarian function. *Recent Prog. Horm. Res.* **5,** 65–100.

Richmond, M., and Conaway, C. H. (1969). Induced ovulation and oestrus in *Microtus ochrogaster. J. Reprod. Fertil., Suppl.* **6,** 357–376.

Robertson, H. A. (1969). Endogenous control of estrus and ovulation in sheep, cattle and swine. *Vitam. Horm.* (*N.Y.*) **27,** 91–130.

Robins, J. P. (1954). Ovulation and pregnancy corpora lutea in the ovaries of the humpback whale. *Nature* (*London*) **173,** 201–203.

Robinson, A. (1918). The formation, rupture and closure of ovarian follicles in ferrets and ferret-polecat hybrids and some associated phenomena. *Trans. R. Soc. Edinburgh* **52,** 302–362.

Rocereto, T., Jacobowitz, D., and Wallach, E. E. (1969). Observations of spontaneous contractions of the cat ovary *in vitro. Endocrinology* **84,** 1336–1341.

Rosati, P., and Pelagalli, G. V. (1968). La dinamica vascolare sanguigna dell'ovario nell'evoluzione del follicolo ooforo e nel corpo luteo. Ricerche in *Bos taurus. Acta Med. Vet.* **14,** 273–303.

Rosengren, E., and Sjöberg, N. O. (1967). The adrenergic nerve supply to the female reproductive tract of the cat. *Am. J. Anat.* **121,** 271–283.

Rossman, I. (1942). On the lipin and pigment in the corpus luteum of the rhesus monkey. *Contrib. Embryol. Carnegie Inst.* **30,** 97–109.

Rowlands, I. W. (1956). The corpus luteum of the guinea-pig. *Ciba Found. Colloq. Ageing* **2,** 69–83.

Rowlands, I. W. (1961). Effect of hysterectomy at different stages in the life cycle of the corpus luteum in the guinea-pig. *J. Reprod. Fertil.* **2,** 341–360.

Rowlands, I. W., and Heap, R. B. (1966). Histological observations on the ovary and progesterone levels in the coypu, *Myocastor coypus. Symp. Zool. Soc. London* **15,** 335–352.

Rowlands, I. W., and Sadleir, R. M. F. S. (1968). Induction of ovulation in the lion, *Panthera leo. J. Reprod. Fertil.* **16,** 105–111.

Rowlands, I. W., Tam, W. H., and Kleiman, D. G. (1970). Histological and biochemical studies on the ovary and progesterone levels in the systemic blood of the green acouchi (*Myoprocta pratti*). *J. Reprod. Fertil.* **22,** 533–545.

Roy, D. J., and Mullick, D. N. (1964). Endocrine function of corpus luteum of buffaloes during estrous cycle. *Endocrinology* **75,** 284–287.

Ruby, J. R., Dyer, R. F., and Skalko, R. G. (1969). The occurrence of intercellular bridges during oogenesis in the mouse. *J. Morphol.* **127,** 307–340.

Ryan, K. J., and Petro, Z. (1966). Steroid biosynthesis by human ovarian granulosa and thecal cells. *J. Clin. Endocrinol. Metab.* **26,** 46–52.

Ryan, K. J., and Short, R. V. (1965). Formation of estradiol by granulosa and theca cells of the equine ovarian follicle. *Endocrinology* **76,** 108–114.

Ryan, K. J., and Short, R. V. (1966). Cholesterol formation by granulosa and thecal cells of the equine ovary. *Endocrinology* **78,** 214–216.

Ryan, K. J., and Smith, O. W. (1965). Biogenesis of steroid hormones in the human ovary. *Recent Prog. Horm. Res.* **21,** 367–402.

Săglik, S. (1938). Ovaries of gorilla, chimpanzee, orangutan and gibbon. *Contrib. Embryol. Carnegie Inst.* **27,** 179–189.

Saint Girons, H., Brosset, A., and Saint Girons, M. C. (1969). Contribution à la connaissance du cycle annuel de la chauve-souris *Rhinolophus ferrumequinum* (Schreber, 1774). *Mammalia* **33,** 357–370.

Sammelwitz, P. H., Aldred, J. P., and Nalbandov, A. V. (1961). Mechanism of formation of corpora lutea in guinea-pigs. *J. Reprod. Fertil.* **2,** 394–399.

Sandes, E. P. (1903). The corpus luteum of *Dasyurus viverrinus* with observations on the growth and atrophy of the Graafian follicle. *Proc. Linn. Soc. N.S.W.* **28,** 364–405.

Santoro, A. (1964). L'ultrastruttura delle cellule costitutive la teca interna del follicolo oophoro in *Lepus cuniculus* (Linn). *Biol. Lat.* **17,** 391–409.
Sato, H. (1955). A histological study of the afferent innervation of the ovary of the dog. *Archiv. Japanische Chir.* **24,** 456–469.
Sato, S. (1965). An electron microscope study on the fine structure of the ovary in normal mature rats. *Arch. Histol. Jpn.* (*Okayama, Jpn.*) **24,** 115–149.
Sauramo, H. (1954a). The histology and function of the ovary from embryonic period to fertile age. Development, occurrence, function and pathology of the rete ovarii. *Acta Obstet. Gynecol. Scand.* **33,** Suppl. 2, 29–46.
Sauramo, H. (1954b). The histology and function of the ovary from embryonic period to fertile age. Development, occurrence and pathology of aberrant adreno-cortical tissue in the region of the ovary. *Acta Obstet. Gynecol. Scand.* **33,** Suppl. 2, 49–58.
Sauramo, H. (1954c). The histology and function of the ovary from embryonic period to fertile age. Occurrence, function and pathology of ovarian sympathico-tropic cells, with special reference to their differentiation from interstitial or hilus cells. *Acta Obstet. Gynecol. Scand.* **33,** Suppl. 2, 69–81.
Sauramo, H. (1954d). The histology and function of the ovary from embryonic period to fertile age. The anatomy, histology, histopathology and function of the ovarian vascular system. *Acta Obstet. Gynecol. Scand.* **33,** Suppl. 2, 111–124.
Savard, K., Marsh, J. M., and Rice, B. F. (1965). Gonadotropins and ovarian steroidogenesis. *Recent Prog. Horm. Res.* **21,** 285–365.
Schoenbaum, M., and Klopper, U. (1969). Cystic changes in the ovaries of guinea-pigs. *Refuah Vet.* **26,** 118–121.
Scott, R. S., and Rennie, P. I. C. (1970). Factors controlling the life-span of the corpora lutea in the pseudopregnant rabbit. *J. Reprod. Fertil.* **23,** 415–422.
Scully, R. E., and Cohen, R. B. (1964). Oxidative enzyme activity in normal and pathologic human ovaries. *Obstet. Gynecol.* **24,** 667–681.
Sergeant, D. E. (1962). The biology of the pilot or pothead whale *Globicephala melaena* (Traill) in Newfoundland waters. *Bull. Fish. Res. Board Can.* **132,** 1–84.
Seth, P., and Prasad, M. R. N. (1967). Effect of bilateral adrenalectomy on the ovary of the five-striped Indian palm squirrel, *Funambulus pennanti. Gen. Comp. Endocrinol.* **8,** 152–162.
Shalash, M. R., and Salama, A. (1962). Seasonal variation in the ovarian activity of the buffalo-cow. *Proc. Int. Congr. Anim. Reprod., Artif. Insem., 4th, 1961* Vol. 2, pp. 190–191.
Sharman, G. B. (1955a). Studies on marsupial reproduction. II. The oestrous cycle of *Setonix brachyurus. Aust. J. Zool.* **3,** 44–55.
Sharman, G. B. (1955b). Studies on marsupial reproduction. III. Normal and delayed pregnancy in *Setonix brachyurus. Aust. J. Zool.* **3,** 56–70.
Sharman, G. B. (1959). Marsupial reproduction. *Monogr. Biol.* **8,** 322–368.
Sharman, G. B. (1964). The female reproductive system of the red kangaroo *Megaleia rufa. CSIRO Wildl. Res.* **9,** 50–57.
Sharman, G. B., and Clark, M. J. (1967). Inhibition of ovulation by the corpus luteum in the red kangaroo, *Megaleia rufa. J. Reprod. Fertil.* **14,** 129–137.
Sharman, G. B., and Pilton, P. E. (1964). The life history and reproduction of the red kangaroo (*Megaleia rufa*). *Proc. Zool. Soc. London* **142,** 29–48.

Sharman, G. B., Calaby, J. H., and Poole, W. E. (1966). Patterns of reproduction in female diprotodont marsupials. *Symp. Zool. Soc. London* **15,** 205–232.

Shehata, R. (1974). Polyovular Graafian follicles in a newborn kitten with a study of polyovuly in the cat. *Acta Anat.* **89,** 21–30.

Shippel, S. (1950). The ovarian theca cell. *J. Obstet. Gynaecol. Br. Emp.* **57,** 362–387.

Shivers, C. A., Metz, C. B., and Lutwak-Mann, C. (1964). Some properties of pig follicular fluid. *J. Reprod. Fertil.* **8,** 115–120.

Short, R. V. (1962). Steroids in the follicular fluids and the corpus luteum of the mare. A "two-cell type" theory of ovarian steroid synthesis. *J. Endocrinol.* **24,** 59–63.

Short, R. V. (1966). Oestrous behaviour, ovulation and the formation of the corpus luteum in the African elephant (*Loxodonta africana*). *East Afr. Wildl. J.* **4,** 56–68.

Short, R. V., and Buss, I. O. (1965). Biochemical and histological observations on the corpora lutea of the African elephant, *Loxodonta africana. J. Reprod. Fertil.* **9,** 61–67.

Short, R. V., and Hay, M. F. (1966). Delayed implantation in the roe deer, *Capreolus capreolus. Symp. Zool. Soc. London* **15,** 173–194.

Simić, V., and Gadev, H. (1968). Recherches anatomo-radiologiques sur les artères et leur rôle fonctionnel dans la vascularisation des ovaires, des oviductes et parties craniales des cornes utérines chez quelques équides domestiques. *Acta Vet.* (*Belgrade*) **18,** 101–118.

Simkins, C. S. (1932). Development of the human ovary from birth to sexual maturity. *Am. J. Anat.* **51,** 465–505.

Sinha, A. A., and Conaway, C. H. (1968). The ovary of the sea otter. *Anat. Rec.* **160,** 795–805.

Sinha, A. A., Conaway, C. H., and Kenyon, K. W. (1966). Reproduction in the female sea otter. *J. Wildl. Manage.* **30,** 121–130.

Sinha, A. A., Seal, U. S., and Doe, R. P. (1971a). Fine structure of the corpus luteum of the raccoon during pregnancy. *Z. Zellforsch. Mikrosk. Anat.* **117,** 35–45.

Sinha, A. A., Seal, U. S., and Doe, R. P. (1971b). Ultrastructure of the corpus luteum of the white-tailed deer during pregnancy. *Am. J. Anat.* **132,** 189–206.

Skalko, R. G., Ruby, J. R., and Dyer, R. (1968). Demonstration of mast cells in the post natal mouse ovary. *Anat. Rec.* **161,** 459–464.

Sleptsov, M. M. (1940). [On the biology of reproduction of the Black Sea dolphin, *Delphinus*.] *Zool. Zh.* **20,** 632–653.

Slijper, E. J. (1966). Functional morphology of the reproductive system in Cetacea. *In* "Whales, Dolphins and Porpoises" (K. S. Norris, ed.), Chapter 15, pp. 277–319. Univ. of California Press, Berkeley.

Sluiter, J. W., and Bels, L. (1951). Follicular growth and spontaneous ovulation in captive bats during the hibernation period. *Proc. K. Ned. Akad. Wet., Ser. C* **54,** 585–593.

Smeaton, T. C., and Robertson, H. A. (1971). Studies on the growth and atresia of Graafian follicles in the ovary of the sheep. *J. Reprod. Fertil.* **25,** 243–252.

Smith, H. C., and Waites, G. M. (1969). Blood flow in the brain and reproductive organs of the conscious rabbit before and during ovulation. *Aust. J. Exp. Biol. Med. Sci.* **47,** P-25.

Smith, J. G., Hanks, J., and Short, R. V. (1969). Biochemical observations on the corpora lutea of the African elephant, *Loxodonta africana*. *J. Reprod. Fertil.* **20,** 111–117.

Smith, N. S. (1969). The persistence and functional life of the corpus luteum in the African elephant. *Diss. Abstr. Int. B* **30,** 1416–1417.

Solomons, B., and Gatenby, J. W. B. (1924). Notes on the formation, structure and physiology of the corpus luteum of man, the pig and the duck-billed platypus. *J. Obstet. Gynaecol. Br. Emp.* **31,** 580–594.

Sotelo, J. R. (1959). An electron microscope study on the cytoplasmic and nuclear components of rat primary oocytes. *Z. Zellforsch. Mikrosk. Anat.* **50,** 749–765.

Sotelo, J. R., and Porter, K. R. (1959). An electron microscope study of rat ovum. *J. Biophys. Biochem. Cytol.* **5,** 327–342.

Spinage, C. (1969). Reproduction in the Uganda defassa waterbuck, *Kobus defassa ugandae* (Neumann). *J. Reprod. Fertil.* **18,** 445–457.

Stafford, W. T., and Mossman, H. W. (1945). Ovarian interstitial tissue and its relation to the pregnancy cycle in the guinea-pig. *Anat. Rec.* **93,** 97–107.

Stangel, J. J., Richart, R. M., Okagari, T., and Cottral, G. (1970). Nuclear DNA content of luteinized cells of the human ovary. *Am. J. Obstet. Gynecol.* **108,** 543–549.

Stanley, H. P., and Hilleman, H. H. (1960). Histology of the reproductive organs of nutria *Myocastor coypus* (Molina). *J. Morphol.* **106,** 277–299.

Stegner, N. E. (1967). Die elektronmikroskopische Struktur der Eizelle. *Ergeb. Anat. Entwicklungsgesch.* **39,** 5–113.

Stegner, H. E., and Wartenberg, H. H. (1963). Elektronmikroskopische Untersuchungen an Eizellen des Menschen in verschiedenem Stadien der Oogenese. *Arch. Gynaekol.* **199,** 151–172.

Sternberg, W. H. (1949). Morphology, androgenic function, hyperplasia and tumors of the human ovarian hilus cells. *Am. J. Pathol.* **25,** 493–521.

Stieve, H. (1950). Anatomisch-biologische Untersuchungen ueber die Fortpflanzungstatigkeit des europaischen Rehes (*Capreolus capreolus capreolus* L.). *Z. Zellforsch. Mikrosk. Anat.* **55,** 427–530.

Strassman, E. O. (1941). The theca cone and its tropism toward the ovarian surface, a typical feature of growing human and mammalian follicles. *Am. J. Obstet. Gynecol.* **41,** 363–378.

Strauss, F. (1938). Die Befruchtung und der Vorgang der ovulation bei *Ericulus* aus der Familie der Centetiden. *Biomorphosis* **1,** 281–312.

Strauss, F. (1939). Die Bildung der Corpus luteum bei Centetiden. *Biomorphosis* **1,** 489–544.

Strauss, F. (1966). Weibliche Geschlechtsorgane. *Handb. Zool.* **9,** 97–202.

Strauss, F., and Bracher, F. (1954). Das Epoophoron des Goldhamsters. *Rev. Suisse Zool.* **61,** 494–503.

Sydow, H. (1968). Elektronmikroskopische Untersuchung über zytosomale Lamellenkörper in den Eizellen des Ovars vom Igel (*Erinaceus europaeus,* L). *Z. Zellforsch. Mikrosk. Anat.* **88,** 387–407.

Täher, E. S. (1964). Das innere Lymphgefässsystem des Eierstocks. *Tieraerztl. Umsch.* **19,** 194–197.

Talmage, R. V., and Buchanan, G. D. (1954). The armadillo (*Dasypus novemcinctus*). A review of its natural history, ecology, anatomy and reproductive physiology. *Rice Inst. Pam.* **41,** 1–135.

Tam, W. H. (1970). The function of the accessory corpora lutea in the hystricomorph rodents. *J. Endocrinol.* **48,** 54–55.

Tam, W. H. (1971). The production of hormonal steroids by ovarian tissues of the chinchilla (*Chinchilla langier*). *J. Endocrinol.* **50,** 267–279.

Tardini, A., Vitali-Mazza, L., and Mansani, F. E. (1960). Ultrastructura dell'ovocita umano maturo. I. Rapporti fra cellule della corona radiata, pellucida ed ovoplasma. *Arch. Vecchi* **33,** 281–305.

Tardini, A., Vitali-Mazza, L., and Mansani, F. E. (1961). Ultrastruttura dell'ovocita umano maturo. II. Nucleo e citoplasma ovulane. *Arch. Vecchi* **35,** 25–71.

Tayes, M. A. F. (1948). Studies on the anatomy of the ovary and the corpus luteum of the camel. *Vet. J.* **104,** 179–186.

Taylor, J. M., and Horner, B. E. (1970). Observations on reproduction in *Leggadina* (Rodentia, Muridae). *J. Mammal.* **51,** 9–17.

Thomas, D. C. (1970). "The Ovary, Reproduction and Productivity of Female Columbian Black-tailed Deer." Ph.D. Thesis, University of British Columbia.

Thwaites, C. J., and Edey, T. N. (1970). Histology of the corpus luteum in the ewe: changes during the estrous cycle and early pregnancy, and in response to some experimental treatments. *Am. J. Anat.* **129,** 439–448.

Tokida, A. (1965). Electron microscopic studies on the corpora lutea obtained from normal human ovaries. *Mie Med. J.* **15,** 27–76.

Toyoda, Y., and Chang, M. C. (1969). Delayed ovulation and embryonic development in the rat treated with pentobarbitone sodium. *Endocrinology* **84,** 1456–1460.

Trauger, D. L., and Haugen, A. O. (1965). Corpora lutea variations of white-tailed deer. *J. Wildl. Manage.* **29,** 487–492.

Tripp, H. R. H. (1970). "Reproduction in the Macroscelididae with Special Reference to Ovulation." Ph.D. Thesis, University of London.

Tripp, H. R. H. (1971). Reproduction in elephant shrews (*Macroscelididae*) with special reference to ovulation. *J. Reprod. Fertil.* **26,** 149–159.

Trujillo-Cenoz, O., and Sotelo, J. R. (1959). Relationships of the ovular surface with follicle cells and the origin of the zona pellucida in rabbit oocytes. *J. Biophys. Biochem. Cytol.* **5,** 347–348.

Tuohimaa, P., and Niemi, M. (1969). Cell renewal in the ovarian follicles of the rat during the oestrous cycle and in persisting oestrus. *Acta Endocrinol. (Copenhagen)* **62,** 306–314.

Tyndale-Biscoe, C. H. (1955). Observations on the reproduction and development of the brush-tailed possum *Trichosurus vulpecula* Kerr (Marsupialia) in New Zealand. *Aust. J. Zool.* **3,** 162–184.

Tyndale-Biscoe, C. H. (1963). The role of the corpus luteum in the delayed implantation of marsupials. *In* "Delayed Implanation" (A. C. Enders, ed.), pp. 15–32. Univ. of Chicago Press, Chicago, Illinois.

Tyndale-Biscoe, C. H. (1965). The female urino-genital system and reproduction of the marsupial *Lagostrophus fasciatus*. *Aust. J. Zool.* **13,** 255–267.

Tyndale-Biscoe, C. H. (1968). Reproduction and post-natal development in the marsupial *Bettongia lesueur* (Quay and Gaimard). *Aust. J. Zool.* **16,** 577–602.

Umbaugh, R. E. (1949). Superovulation and ovum transfer in cattle. *Am. J. Vet. Res.* **10,** 295–305.

Unsicker, K. (1970). Über den Feinban der Hiluszischenzellen im Ovar des Schweins (*Sus scrofa* L.). Mit bemerkungen zur Frage ihrer Innervation. *Z. Zellforsch. Mikrosk. Anat.* **109,** 495–516.

Unsicker, K. (1971). Über den Feinban von Marksträngen und Markschlänchen im Ovar juveniler und geschlechtsreifer Schweine (*Sus scrofa* L.). *Z. Zellforsch. Mikrosk. Anat.* **114,** 344–364.

Vaček, Z. (1967). Ultrastructure and enzyme histochemistry of the corpus luteum graviditatis and its correlation to the decidual transformation of the endometrium. *Folia Morphol.* **15,** 375–383.

Valdes-Dapena, M. A. (1967). The normal ovary of childhood. *Ann. N.Y. Acad. Sci.* **142,** 597–614.

Valentine, G. L., and Kirkpatrick, R. L. (1970). Seasonal changes in reproductive and related organs in the pine vole, *Microtus pinetorum,* in South Western Virginia. *J. Mammal.* **51,** 553–560.

van der Horst, C. J., and Gillman, J. (1940). Ovulation and corpus luteum formation in *Elephantulus. S. Afr. J. Med. Sci.* **5,** 73–91.

van der Horst, C. J., and Gillman, J. (1942). The life history of the corpus luteum of menstruation in *Elephantulus. S. Afr. J. Med. Sci.* **7,** 21–41.

van der Horst, C. J., and Gillman, J. (1946). The corpus luteum of *Elephantulus* during pregnancy, its form and function. *S. Afr. J. Med. Sci.* **11,** Biol. Suppl., 87–102.

van Lennep, E. W. (1950). Histology of the corpora lutea in blue and fin whale ovaries. *Proc. K. Ned. Akad. Wet.* **53,** 593–599.

van Lennep, E. W., and Madden, L. M. (1965). Electron microscopic observations on the involution of the human corpus luteum of menstruation. *Z. Zellforsch. Mikrosk. Anat.* **66,** 365–380.

van Rensburg, S. J., and van Niekerk, C. H. (1968). Ovarian function, follicular oestradiol-17β, and luteal progesterone and 20α-hydroxy-pregn-4-en-3-one in cycling and pregnant equines. *Onderstepoort J. Vet. Res.* **35,** 301–317.

van Wagenen, G., and Simpson, M. E. (1965). "Embryology of the Ovary and Testis in *Homo sapiens* and *Macaca mulatta.*" Yale Univ. Press, New Haven, Connecticut.

Vongpayabal, P., and Callahan, W. P. (1970). Oocyte numbers and follicular development in the rat ovary. *Anat. Rec.* **166,** 393.

von Kaulla, K. N., and Shettles, L. B. (1956). Thromboplastic activity of human cervical mucus and ovarian follicular and seminal fluids. *Fertil. Steril.* **7,** 166–169.

Waring, H., Sharman, G. B., Lovat, D., and Kahn, M. (1955). Studies on marsupial reproduction. I. General features and techniques. *Aust. J. Zool.* **3,** 34–43.

Wartenberg, H., and Stegner, H. E. (1960). Über die elektronmikroskopische feinstruktur des menschlichen ovarialeies. *Z. Zellforsch. Mikrosk. Anat.* **52,** 450–474.

Watzka, M. (1940). Mikroskopisch anatomische Untersuchungen ueber die Ranzzeit und Tragdaner des Hermelins (*Putorius ermineus*). *Z. Mikrosk. Anat. Forsch.* **48,** 359–374.

Watzka, M. (1957). Weibliche Genitalorgane. Das Ovarium. *Handb. Mikrosk. Anat.* **7,** 1–178.

Weakley, B. S. (1966). Electron microscopy of the oocyte and granulosa cells in the developing ovarian follicles of the golden hamster (*Mesocricetus auratus*). *J. Anat.* **100,** 503–534.

Weakley, B. S. (1967). Light and electron microscopy of developing germ cells in the ovary of the golden hamster: twenty-four hours before birth to eight days *post partum. J. Anat.* **101,** 435–459.

Weakley, B. S. (1968). Comparison of cytoplasmic lamellae and membranous elements in the oocytes of five mammalian species. *Z. Zellforsch. Mikrosk. Anat.* **85,** 109–123.

Weakley, B. S. (1969). Electron microscope study of the surface epithelium of the hamster ovary from the perinatal period to maturity. *J. Anat.* **104,** 198.

Weakley, B. S. (1971). Basic protein and RNA in the cytoplasm of the ovarian oocyte in the golden hamster. *Z. Zellforsch. Mikrosk. Anat.* **112,** 69–84.

Weir, B. J. (1966). Aspects of reproduction in the chinchilla. *J. Reprod. Fertil.* **12,** 410–441.

Weir, B. J. (1971a). Observations on reproduction in the female green acouchi, *Myoprocta pratti. J. Reprod. Fertil.* **24,** 193–201.

Weir, B. J. (1971b). Observations on reproduction in the female agouti, *Dasyprocta aguti. J. Reprod. Fertil.* **24,** 203–211.

Weir, B. J. (1971c). The reproductive organs of the female plains viscacha, *Lagostomus maximus. J. Reprod. Fertil.* **25,** 365–373.

Weir, B. J. (1971d). Some notes on reproduction in the Patagonian mountain viscacha, *Lagidium boxi* (Mammalia: Rodentia). *J. Zool.* **164,** 463–467.

Weir, B. J., and Rowlands, I. W. (1974). Functional anatomy of the hystricomorph ovary. *Symp. Zool. Soc. London* **34,** 303–332.

Wenzel, J. (1966). Untersuchungen über die Lymphgefässsystem juveniler Kaninchenovarien. *Z. Mikrosk. Anat. Forsch.* **74,** 471–481.

Wenzel, J., and Staudt, J. (1966). Neue Befunde über das Lymphgefässsystem des Kaninchenovariums durch differazierte Darstellung von Lymph- und Blutgefässen. *Z. Mikrosk. Anat. Forsch.* **74,** 457–470.

White, R. F., Hertig, A. T., Rock, J., and Adams, E. I. (1951). Histological and histochemical observations on the corpus luteum of human pregnancy with special reference to corpora lutea associated with early normal and abnormal ova. *Contrib. Embryol. Carnegie Inst.* **34,** 55–74.

Wilcox, D. E., and Mossman, H. W. (1945). The common occurrence of "testis" cords in the ovaries of a shrew (*Sorex vagrans* Baird). *Anat. Rec.* **92,** 183–195.

Wiltbank, J. N., and Casida, L. E. (1956). Alteration of ovarian activity by hysterectomy. *J. Anim. Sci.* **15,** 134–140.

Wimsatt, W. A. (1944). Growth of the ovarian follicle and ovulation in *Myotis lucifugus lucifugus. Am. J. Anat.* **74,** 129–173.

Wimsatt, W. A. (1963). Delayed implantation in the Ursidae with particular reference to the black bear (*Ursus americanus* Pallas). *In* "Delayed Implantation" (A. C. Enders, ed.), pp. 49–76. Univ. of Chicago Press, Chicago, Illinois.

Wimsatt, W. A., and Kallen, F. C. (1957). The unique maturation response of the Graafian follicles of hibernating vespertilionid bats and the question of its significance. *Anat. Rec.* **129,** 115–131.

Wimsatt, W. A., and Parks, H. F. (1966). Ultrastructure of the surviving follicle of hibernation and of the ovum-follicle cell relationship in the vespertilionid bat, *Myotis lucifugus. Symp. Zool. Soc. London* **15,** 419–454.

Wimsatt, W. A., and Trapido, H. (1952). Reproduction and the female reproductive cycle in the tropical American vampire bat, *Desmodus rotundus murinus. Am. J. Anat.* **91,** 415–446.

Wimsatt, W. A., and Waldo, L. M. (1945). The normal occurrence of a peritoneal opening in the bursa ovarii of the mouse. *Anat. Rec.* **93,** 47–57.

Winter, G. F. (1962). Follikel-Zählungen an den Eierstöcken gesunder, nicht schwangerer weiblicher Personen. *Z. Gynaekol.* **47,** 1824–1829.

Wischnitzer, S. (1964). Phase-contrast microscope observations on the germinal epithelium of the mouse ovary. *Z. Mikrosk. Anat. Forsch.* **71,** 618–622.
Wischnitzer, S. (1965). Ultrastructure of the germinal epithelium of the mouse ovary. *J. Morphol.* **17,** 387–400.
Wislocki, G. B. (1931). Notes on the female reproductive tract (ovaries, uterus, and placenta) of the collared peccary (*Pecari angulatus bangsi* Goldman). *J. Manmal.* **12,** 143–149.
Witherspoon, D. M., and Talbot, R. B. (1970). Ovulation site in the mare. *J. Am. Vet. Assoc.* **157,** 1452–1459.
Wotton, R. M., and Village, P. A. (1951). The transfer function of certain cells in the wall of the Graafian follicle as revealed by their reaction to previously stained fat in the cat. *Anat. Rec.* **110,** 121–127.
Wright, P. L. (1942). Delayed implantation in the long-tailed weasel (*Mustela frenata*), the short-tailed weasel (*Mustela cicognani*) and the marten (*Martes americana*). *Anat. Rec.* **83,** 341–353.
Wright, P. L. (1963). Variations in reproductive cycles in North American mustelids. *In* "Delayed Implantation" (A. C. Enders, ed.), pp. 77–97. Univ. of Chicago Press, Chicago, Illinois.
Wright, P. L. (1966). Observations on the reproductive cycle of the American badger, *Taxidea taxus. Symp. Zool. Soc. London* **15,** 26–45.
Wurtman, R. J. (1964). An effect of luteinizing hormones on the fractional perfusion of the rat ovary. *Endocrinology* **75,** 927–933.
Yamada, E., and Ishikawa, T. M. (1960). The fine structure of the corpus luteum in the mouse ovary as revealed by electron microscopy. *Kyushu J. Med. Sci.* **11,** 235–259.
Yamada, E., Muta, T., Motomura, A., and Koga, H. (1957). The fine structure of the oocyte in the mouse ovary studied with electron microscope. *Kurume Med. J.* **4,** 148–171.
Yamashita, T. (1961). Histological studies on the ovaries of sows. IV. Stereographical study of the vascular arrangement in the various structures of the ovaries by use of neoprene latex casting specimens. *Jpn. J. vet. Res.* **9,** 31–45.
Yamashita, T. (1962). Histological studies of the ovaries of sows. V. Histological observations of the various corpora lutea in the ovaries of sows which have definite histories of parturition. *Jpn. J. vet. Res.* **10,** 1–18.
Yasuda, T., and Doi, K. (1953). Histochemical studies on histogenesis of lutein tissue in the rabbit ovary. *Med. J. Osaka Univ.* **4,** 241–253.
Young, W. C. (1961). The mammalian ovary. *In* "Sex and Internal Secretions" (W. C. Young, ed.), 3rd ed., Vol. I, Chapter 7, pp. 449–496. Williams & Wilkins, Baltimore, Maryland.
YoungLai, E. V., and Short, R. V. (1970). Pathways of steroid biosynthesis in the intact Graafian follicle of mares in oestrus. *J. Endocrinol.* **47,** 321–331.
Zachariae, F. (1957). Studies on the mechanism of ovulation, autoradiographic investigation on the uptake of radioactive sulphate (^{35}S) into the ovarian follicular mucopolysaccharides. *Acta Endocrinol.* (*Copenhagen*) **26,** 215–224.
Zamboni, L., and Gondos, B. (1968). Intercellular bridges and synchronization of germ cell differentiation during oogenesis in the rabbit. *J. Cell Biol.* **36,** 376–382.
Zamboni, L., and Mastroianni, L. (1966). Electron microscope studies on rabbit ova. 1. The follicular oocyte. *J. Ultrastruct. Res.* **14,** 95–117.

Zawistowska, H. (1956). Participation of blood vessel endothelium in the secretion processes of the ovary. *Folia Morphol.* **7,** 175–183.

Zuckerman, S. (1951). The number of oocytes in the mature ovary. *Recent Prog. Horm. Res.* **6,** 63–109.

Zuckerman, S., and Parkes, A. S. (1932). The menstrual cycle of the primates. V. The cycle of the baboon. *Proc. Zool. Soc. London* 138–191.

5

The Structure of the Ovary of Nonmammalian Vertebrates

J. M. Dodd

I. INTRODUCTION

The purpose of this chapter is to provide a comparative account of ovarian morphology, histology, and fine structure in the nonmammalian vertebrates. A detailed description of the mammalian ovary is given in Volume I, Chapter 4, and the present account should be read in conjunction with this and Volume I, Chapter 2 and Volume II, Chapter 6, which deal with ovarian development and ovarian cycles, respectively.

The ovary of nonmammalian vertebrates, unlike that of mammals, is highly variable in appearance, and its size, relative to the size of the body, is larger. The final form of the ovary will clearly depend on a number of variables among which are mode of origin, amount of stroma, size and number of eggs, whether the ovary is paired or single, whether the ovarian tissue is solid or hollow, and whether or not ovarian gestation takes place. Furthermore, the appearance of the ovary in these animals, which breed seasonally, will depend to a great extent on whether it is developing, mature, or spent. It follows from such considerations that it is impossible to generalize regarding ovarian morphology in the nonmammalian groups though the basic constituents of the ovary are in all cases similar even if strict homology between the various components has not always been established.

The ovaries of all nonmammalian vertebrates consist of oogonia, oocytes and their surrounding follicle cells, supporting tissue or stroma, and vascular and nervous tissue. But whereas in the testis there is usually a separate population of endocrine cells in the form of Leydig or interstitial cells and possibly Sertoli cells, in the ovary the endocrine function is met by follicular cells or their derivatives. Even when interstitial cells or glands are present in lower vertebrate ovaries they appear to originate from follicle cells. In viviparous teleosts, in which the ovary has a nutritive role as well as a gametogenic and endocrine one, other tissues may be present though these appear to be merely derivatives of the follicular and peritoneal epithelia exercising their nutritive and phagocytic functions.

The outer limiting epithelium of the ovary is derived from the peritoneum and in the bony fishes a similar epithelium lines the ovarian cavities. These epithelia are often misleadingly termed "germinal" though there is no good evidence that the germinal elements in the ovary ever originate from them (see Volume I, Chapters 2 and 4).

II. COMPARATIVE MORPHOLOGY

The ovaries of all lower vertebrates are intraperitoneal and they are suspended from the dorsal body wall by the mesovarium which has a much more extended attachment than in mammals. The embryological origins of the ovarian constituents are discussed in Volume I, Chapter 2 and it is shown that the fate of cortex and medulla are variable in the different groups though the cortical elements and their derivatives always predominate. A true medullary component is probably absent from the ovaries of cyclostomes, teleosts, and ganoid fishes and in the other groups it either gives rise to a loose, spongy, highly vascularized tissue or, as in the Amphibia, it is replaced by lymph-filled spaces.

These introductory remarks provide a basis for the more detailed discussion of ovarian structure in each of the nonmammalian classes which follows.

A. Stroma

The cortex and its investment of peritoneal cells give rise to the ovarian stroma and to the follicular cells which invest the developing oocytes. In the nonmammalian groups the stroma is relatively sparse and the large follicles may become pedunculate and project into lymph-filled cavities in the central regions of the ovary. The stroma consists of connective tissue fibers, collagen fibers, fibroblasts, and a variety of other cell types and its main function appears to be to provide a supporting skeleton for the follicles and their associated blood vessels and nerves though, with few exceptions, the vasculature and innervation of the ovary of lower vertebrates have received little detailed attention.

B. Follicles

When a secondary oogonium enters the prophase of meiosis it becomes an oocyte and acquires an investment of epithelial cells. Such units are called follicles and as the oocyte grows the follicular investment increases in complexity. In mammals two basic regions are recognized in the follicular wall, a granulosa or follicular epithelium and a theca which can usually be separated into theca externa and theca interna. Granulosa and theca are separated from each other by the membrana propria. In nonmammalian vertebrates a follicular epithelium and an outer coat of cells, which may or may not be two-layered, can always be recognized and these are often called granulosa and theca. Whether they are embryologically or physiologically homologous with these layers in mammals is doubtful though these terms are used here for convenience. The region of the follicle lying between oocyte and granulosa is of particular interest and importance. It separates cell types of very different origins and functions and all the various substances entering the oocyte must pass across it. It has now been studied by electron microscopy in all the major vertebrate groups, except elasmobranchs. In all follicles examined in this way the region lying between follicular cells and the oocyte cell membrane is highly complex. It contains an extracellular perivitelline substance, equivalent to the zona pellucida of mammals, across which pass cytoplasmic processes from the follicle cells that approach the oocyte surface. The situation is rendered still more complex by the abundant microvilli which arise from the oocyte cell membrane and penetrate the buffer zone to form the zona radiata.

In most lower vertebrates the follicular epithelium or granulosa consists of a single layer of uniform cells, though in elasmobranchs, reptiles, and possibly in birds, it may be stratified, and in the reptiles several different cell types can be recognized, one of which is believed to be nutritive. In elasmobranchs and reptiles fine protoplasmic fibrils were believed to traverse the zona pellucida and fuse with the cytoplasm of the oocyte (Wallace, 1903; Loyez, 1906; Retzius, 1912; Kemp, 1958), but this question has recently been reexamined by electron microscopy in reptiles and the presence of intercellular bridges between follicle cells and oocyte has not been substantiated, except in the early previtellogenic follicles (Neaves, 1971).

Less is known about the theca and, although two zones can be recognized in birds (Hett, 1923a; Cunningham and Smart, 1934; Bellairs, 1965; Wyburn *et al.*, 1965) and reptiles (Hett, 1924; Weekes, 1934), it is doubtful whether these are homologous with the theca externa and interna of mammals (Brambell, 1956). The theca interna consists of fibroblasts, connective tissue fibers, collagen fibers and capillaries. Smooth muscle fibers have been identified in the theca of the frog (Rugh, 1935) and in the fowl (Phillips and Warren, 1937) and elastic fibers are said to develop in the theca externa of reptiles at a certain point in follicular development. The theca, whether single- or double-layered, is richly supplied with blood vessels and lymphatics and it may become stretched and extremely thin in follicles containing heavily yolked eggs. In teleosts the theca is said to be merely a simple fibroelastic layer which never shows sudanophilia (Hoar, 1969), but more recent work has shown that, in some species at least, it contains Δ^5-3β-hydroxysteroid dehydrogenase (3β-HSD), an enzyme essential for steroidogenesis (Bara, 1965a,b).

"Interstitial" tissue has been identified in certain lower vertebrate ovaries (Brambell, 1956) and it appears to be derived from thecal cells after ovulation.

C. Corpora Atretica and Corpora Lutea

A great deal of attention has been given to certain structures, almost universally present in the ovaries of lower vertebrates, which are often collectively called "corpora lutea" even though they are of two quite dissimilar origins and there is no direct evidence that they have an endocrine function. However, it seems desirable to make a distinction between the two structures, reserving the term "corpora lutea" for postovulatory structures, without implying any strict homology with the corpora lutea of mammals, and calling the others, which arise by atresia of the follicle, "corpora atretica."

Corpora atretica may arise from follicles of any size in lower vertebrates though vitellogenic oocytes are more frequently affected than previtellogenic ones. The stimulus leading to their formation is not known in intact animals though hypophysectomy invariably results in atresia of yolky oocytes. They usually arise by invasion of the oocyte by the follicular cells (granulosa) and sometimes by the theca, the follicular cells indulging their digestive and absorptive functions, and occasionally atresia occurs by bursting of the follicles. The question of follicular atresia has been frequently reviewed (Brambell, 1956; Hoar, 1957, 1965, 1969; Dodd, 1960, 1972; Chieffi, 1962; Volume I, Chapter 6).

Postovulatory follicles in all lower vertebrates, other than lampreys, undergo a "tidying-up" operation and in elasmobranchs and reptiles structures reminiscent of mammalian corpora lutea are formed, especially in viviparous species. Postovulatory corpora lutea are also found in the other lower vertebrate groups but they are poorly developed and transitory (Harrison, 1948; Brambell, 1956; Dodd, 1960, 1972; Hoar, 1965, 1969). The epithelial cells of the ruptured follicles hypertrophy though there is some doubt whether they divide; they invade the follicular cavity and usually become luteinized. The granulosa cells appear to become secretory and, in elasmobranchs, reptiles and birds, connective tissue originating from the theca gives support to the luteal cells and carries blood to them.

III. PATTERNS OF OVARIAN STRUCTURE

A. Cyclostomes

The class Cyclostomata contains only two orders: the Petromyzontia, or lampreys, and the Myxinoidea, or hagfishes, both of which have an extended larval phase in which the gonads start to develop though they remain immature and may show signs of intersexuality. After metamorphosis the gonads, of both orders, mature, giving rise in genetic females to ovaries which are strikingly different in the two groups, both in structure and development. The differences are correlated with two contrasting types of life history; the lampreys mature and spawn only once, whereas the hagfishes breed over several seasons.

1. The Petromyzontia

The lamprey ovary is elongate, lobate, and unpaired. It extends the entire length of the body cavity and, when mature, distends and fills the latter, since by this time the gut has virtually disappeared (Fig. 1a). The follicles mature synchronously during the anadromous migration of the common European lamprey, *Lampetra fluviatilis,* and Lanzing (1959) reports that the gonadosomatic index of females increases from 3.9 to 13.9 during this

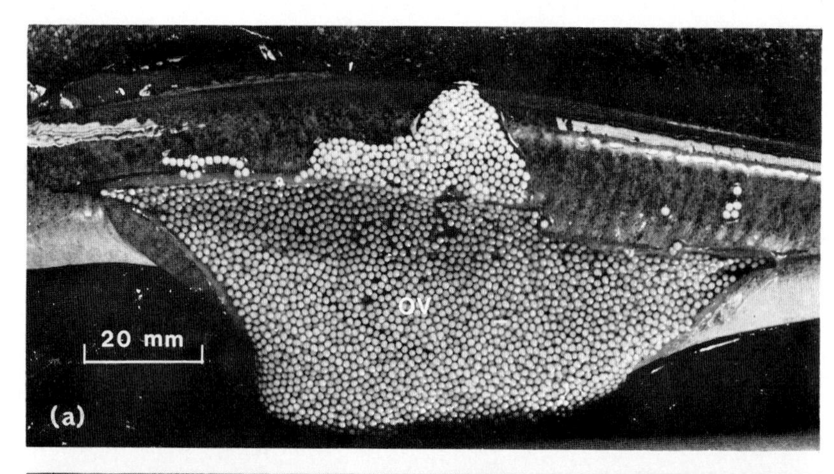
OV
20 mm
(a)

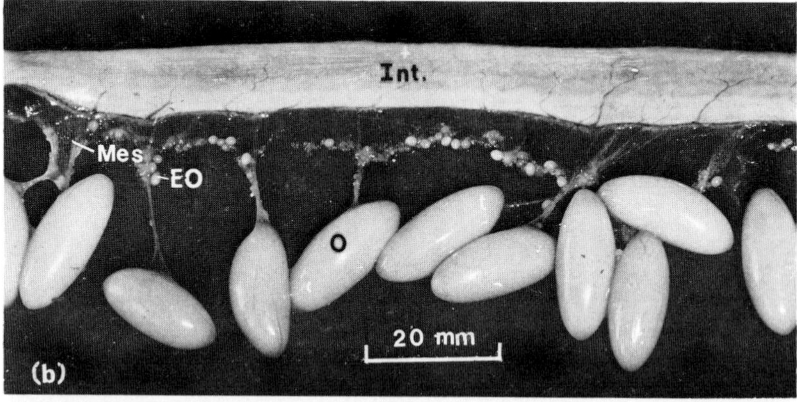
Int.
Mes
EO
O
20 mm
(b)

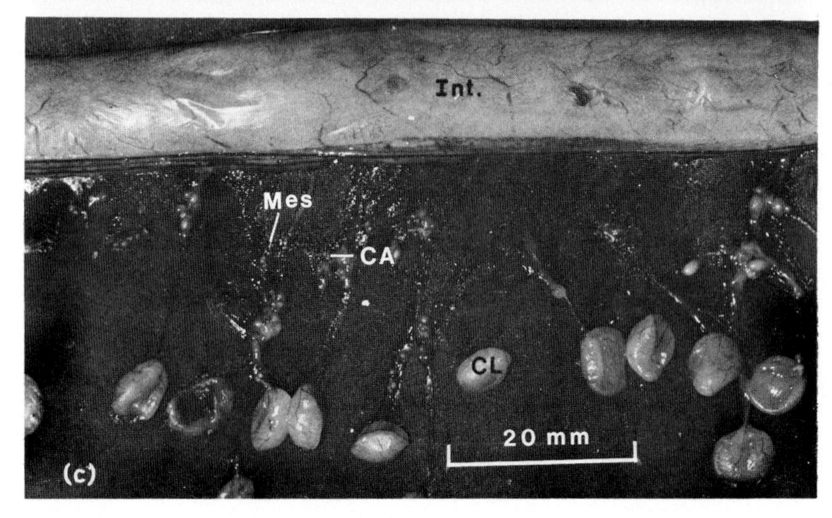
Int.
Mes
CA
CL
20 mm
(c)

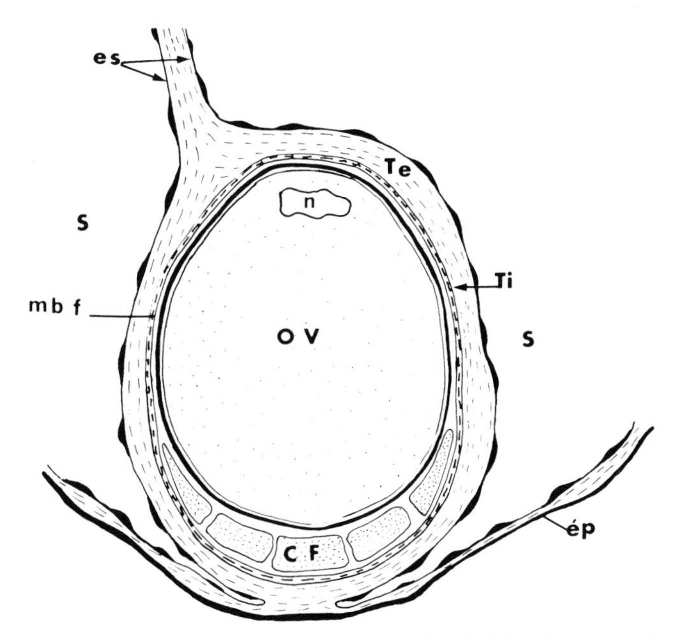

Fig. 2. Diagrammatic representation of an ovarian follicle of *L. planeri.* Note that follicular cells are associated with only one pole of the oocyte. CF, follicular cells (granulosa); ép, peritoneal epithelium; es, epithelium lining ovarian sacs; mbf, basement membrane; n, germinal vesicle; OV, oocyte; S, interfollicular space; Te, theca externa; Ti, theca interna. (Reproduced by permission from Busson-Mabillot, 1967.)

period in lampreys from the river Meuse. There are no gonoducts; eggs are ovulated into the body cavity and they reach the exterior via abdominal pores that become patent only at the time of oviposition (Knowles, 1939).

Ovarian structure has been described in *Lampetra fluviatilis* (Lanzing, 1959; Evennett, 1963), *Petromyzon marinus* (Lewis and McMillan, 1965), and *L. planeri* (Busson-Mabillot, 1966, 1967a,b). The stroma of the mature ovary is sparse, it consists of fibroblasts and bundles of collagen fibers forming thin strands of tissue which carry blood vessels and nerves to the follicles. The latter are stalked and separated from each other by large interfollicular spaces (Fig. 2). The follicles in the maturing ovary are irregular in shape, due to compression, though the oocytes, when freed from

---

Fig. 1. Cyclostome ovaries. (a) Ripe ovary of *Petromyzon fluviatilis.* (b) Ripe ovary of *Myxine glutinosa* showing oocytes of different sizes. (c) Postovulatory ovary of *M. glutinosa* showing corpora lutea and corpora atretica. CA, corpus atreticum; CL, corpus luteum; EO, early oocytes; Int., intestine; Mes., mesovarium; O, late oocyte; OV, ovary. [(a) Reproduced by courtesy of Dr. P. J. Evennett and by permission from Dodd, 1972; (b), (c) Reproduced by courtesy of Dr. Finn Walvig.]

the follicles, are ellipsoidal. In *L. fluviatilis* their major diameter increases from approximately 0.6 mm in October to 1.1 mm in April (Evennett, 1963).

Busson-Mabillot (1967a,b) has described and illustrated the structure of the ovarian follicle in *L. planeri* (Figs. 2 and 3) and shown that it consists of a theca of two distinct layers, and a granulosa (follicular epithelium) which is unique among vertebrates in forming an incomplete investment; it provides a shallow basal cup of cells in which one pole of the oocyte lies and it is separated from the theca interna by a double membrane. However, in spite of this unique feature, the fine structure of the ovarian follicle in these animals, which have been a separate group for some 400,000,000 years, is virtually indistinguishable in basic structure from that of other vertebrate groups of much more recent origin such as the birds (see Fig. 14). The outer thecal layer consists of well-vascularized connective tissue while the theca interna is glandular and believed to secrete estrogens. Each oocyte is surrounded by a zona pellucida (radiata) which consists of regularly arranged microvilli arising from the oocyte's surface, embedded in a matrix of two layers, the inner being finely filamentous, the outer being granular.

No atretic follicles are found in the maturing gonads of *L. fluviatilis* or *L. planeri* though atresia has been reported in immature ovaries of these species (Okkelberg, 1921). However, in *P. marinus* Lewis and McMillan (1965) have noted that atresia occurs throughout the adult stages and after spawning. Coalescence of yolk particles precedes atresia and the degenerating oocytes are then invaded by phagocytes said to be derived from the follicular epithelium, and ultimately all that remains is a ball of cuboidal and ovoid follicular and stromal cells; the atresia that occurs in unovulated follicles after the vast majority of oocytes have been shed is identical. There is no attempt at any kind of reorganization of the ovulated follicles and no corpora lutea are found. Lampreys appear to be unique in this respect and this is no doubt correlated with the synchronous development of the follicles and the absence of any requirement to "tidy up" the empty follicles or to produce postovulatory secretory structures in animals which spawn only once and then die. Thus, except in *P. marinus*, the granulosa of the lamprey follicles does not appear to have acquired the phagocytic role it has in other vertebrates.

2. The Myxinoidea

The situation in the hagfishes, as exemplified by *Myxine glutinosa*, is different. The unpaired ovary lies on the right side of the gut and stretches from gall bladder to anus (Fig. 1b). It arises from the right gonad rudiment, the left one atrophying at an early stage of development (Okkelberg, 1921). The mature ovary contains only 1 to 21 vitellogenic oocytes whereas that of

Petromyzon marinus produces between 24,000 and 236,000 (Walvig, 1963). Furthermore, follicles develop serially in the hagfishes and ovulation continues throughout the lives of mature females.

The morphology of the myxinoid ovary has been described in considerable detail by Cunningham (1886), Conel (1917), Lyngnes (1931, 1936), Schreiner (1955), and Walvig (1963). The stroma consists of little more than a continuation of the mesovarium and its caudal region may contain nonfunctional islands of testicular tissue (Schreiner, 1955). The largest and oldest of the eggs lie along the proximal border of the mesovarium, parallel to the intestine (Fig. 1b); each is suspended by a ribbon-shaped stalk and enclosed in a tough shell with micropyle and anchoring filaments at each pole. Secretion by the follicle of such a sophisticated shell, is unique among vertebrates, shell secretion being usually a function of the oviducts, but these are absent in *Myxine*. Oocytes of up to 10 mm are invested by a single layer of follicle cells. When they reach a size of about 17 mm stratification and thickening occurs, first at the poles and then round the entire oocyte, and mature follicles are four layered. The inner layer forms a simple follicular epithelium; this is surrounded by two layers of connective tissue and an outer squamous covering of epithelial cells (Lyngnes, 1936).

In addition to normal eggs the mature hagfish ovary may contain corpora atretica and corpora lutea (Fig. 1c). Nothing is known of the physiology of these structures, or to what extent the pituitary is involved in their formation, but Lyngnes (1936) reported that in July, 51.3% of eggs between 2 mm and 23 mm in the *Myxine* ovary may be atretic; he also described the genesis and structure of the corpora atretica in some detail. These arise by both open involution and closed involution, processes reminiscent of other lower vertebrates with yolky eggs. In the former case the shell and follicle wall atrophy and burst at a particular point and the oocyte contents escape to collect in pockets in the stroma from which they are removed, presumably by phagocytes. Closed involution involves migration of cells from the theca into the oocyte in which they form a porous network with phagocytic and absorptive functions.

Postovulatory corpora lutea again are of two kinds. Some are solid masses of tissue derived from the follicle cells while the others are described as fluid-filled cysts (Lyngnes, 1936). The functional significance of these structures and of the corpora atretica is unknown.

B. Elasmobranchs

Elasmobranchs have one or two ovaries and their highly variable structure is correlated with the reproductive habits of the species. Extreme differences are illustrated by the ovary of the oviparous *Scyliorhinus canicula*,

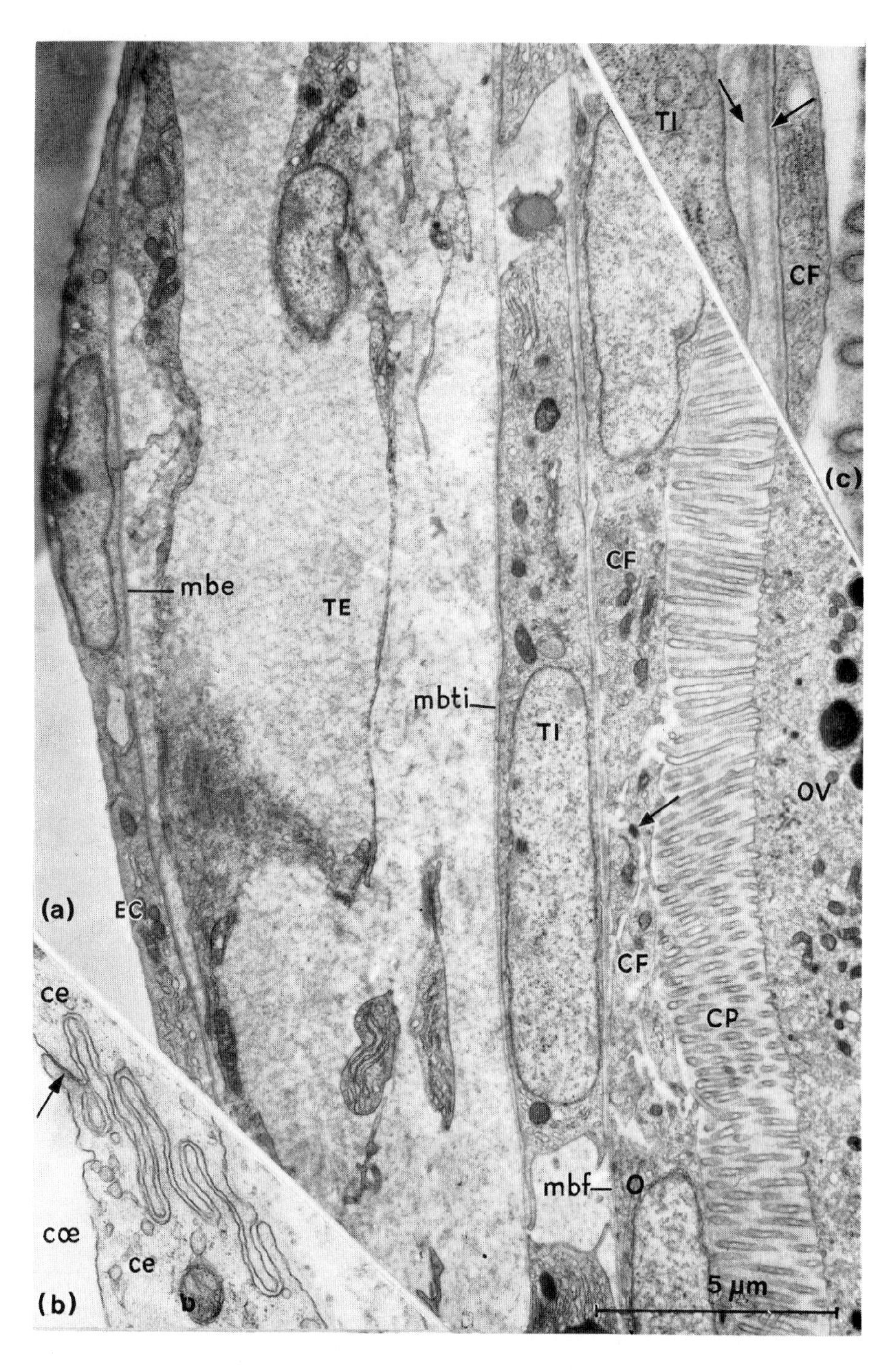
TI
CF
(c)
mbe
TE
CF
mbti
TI
OV
(a)
EC
CF
CP
ce
cœ
ce
(b)
mbf
O
5 µm

which is probably typical of oviparous species, and by that of *Cetorhinus maximus* which is believed to be viviparous (Matthews, 1950). The ovary of *S. canicula* (Figs. 4a and b) is large and unpaired and develops from the right primordium. It is suspended by a conspicuous mesovarium carrying blood vessels and nerves and retains traces of a sac-like structure though the ova are ovulated into the body cavity. During development a distinct medullary region appears but this becomes replaced by hematopoietic cells derived from the kidney. These give rise to the so-called epigonal organ (Figs. 4a and c) which is prominent in the elasmobranch testis as well as in the ovary, and is believed to be lympho-myeloid though its function has not been definitely established (Chieffi, 1949; Matthews, 1950). The epigonal organ greatly augments the stroma and the small oocytes appear to be embedded in it.

It seems probable that the mature ovary (Fig. 4b) contains no oogonia since the meiotic prophase is believed to be precocious in elasmobranchs (see Volume I, Chapter 2). Oocytes, varying in color from the whitish, translucent, yolk-free oocytes to the large, yellow, or green, heavily yolked oocytes are present throughout the year in the mature ovary (Metten, 1939), as are corpora atretica and corpora lutea, though their relative proportions show seasonal differences. The oocytes are enclosed in characteristic follicles consisting of a granulosa and a two-layered theca (Fig. 5a) but the cells are so stretched in the largest follicle that the layers are difficult to identify.

The histology of the corpora atretica of *S. canicula* is illustrated in Fig. 5b and detailed descriptions exist for the corpora atretica and corpora lutea of other elasmobranch species (Samuel, 1943; Hisaw and Hisaw, 1959; Chieffi, 1961, 1962; Botte, 1963); Brambell (1956) has reviewed the early literature. Hisaw and Hisaw (1959) described corpora atretica and corpora lutea in the ovaries of oviparous (*Raja binoculata, R. erinacea*), ovoviviparous (*Squalus suckleyi, S. acanthias*), and viviparous (*Mustelus canis*) elasmobranchs and showed that their modes of formation and structure are virtually identical.

The ovary of *Cetorhinus maximus*, the basking shark, has been described in some detail by Matthews (1950) who states that it is conspicuously different from that of any other elasmobranch. The differences are believed to indicate that the basking shark is viviparous though all the evidence for this

---

Fig. 3. Fine structure of the basal pole (see Fig. 2) of an early vitellogenic follicle of *L. planeri.* (a) General view. (b) Peritoneal (coelomic) epithelium. (c) Region between follicular epithelium and theca interna. ce, cells of the coelomic epithelium; coe, coelomic cavity; CP, zona pellucida (radiata); EC, coelomic epithelium; mbe, basement membrane between coelomic epithelium and theca externa; mbf, basement membrane between follicular epithelium (granulosa) and theca interna; mbti, basement membrane separating theca interna from theca externa; TE, theca externa; TI, theca interna, CF, follicular cells; OV, oocyte. (Reproduced by permission from Busson-Mabillot, 1967b.)

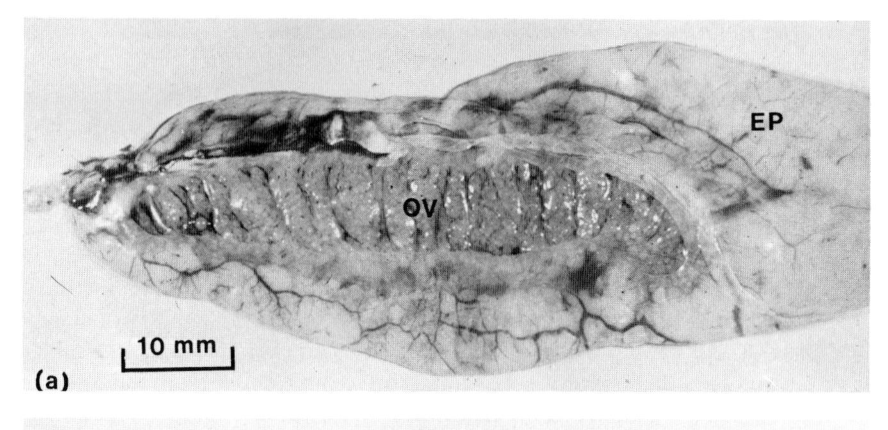

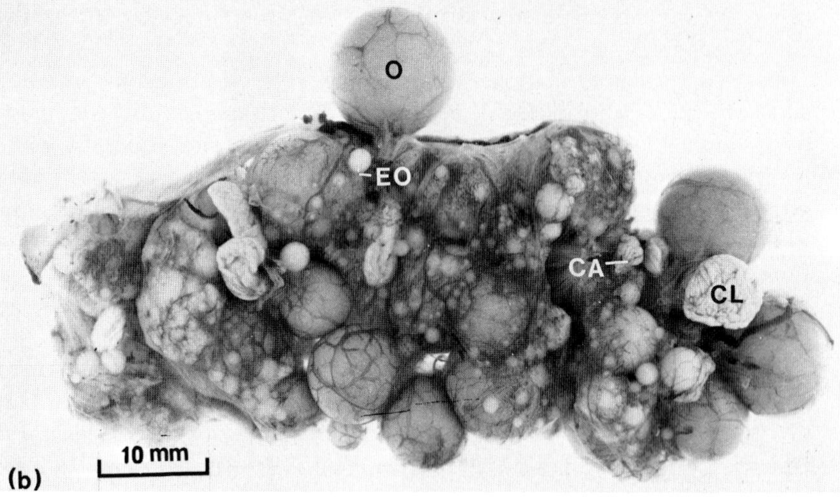

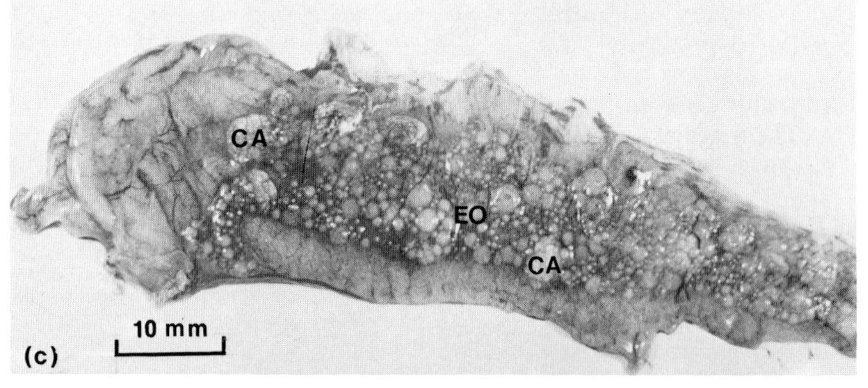

Fig. 4. The ovary of the spotted dogfish, *Scyliorhinus canicula*. (a) Immature ovary (OV) showing early oocytes embedded in the epigonal organ (EP). (b) Mature ovary with mature oocytes (O), early oocytes (EO), corpora lutea (CL), and corpora atretica (CA). (c) Ovary after hypophysectomy containing only previtellogenic oocytes (EO) and corpora atretica (CA). ((b) Photograph by Dr. T. J. Nicholls; (c) from Dodd, 1960.)

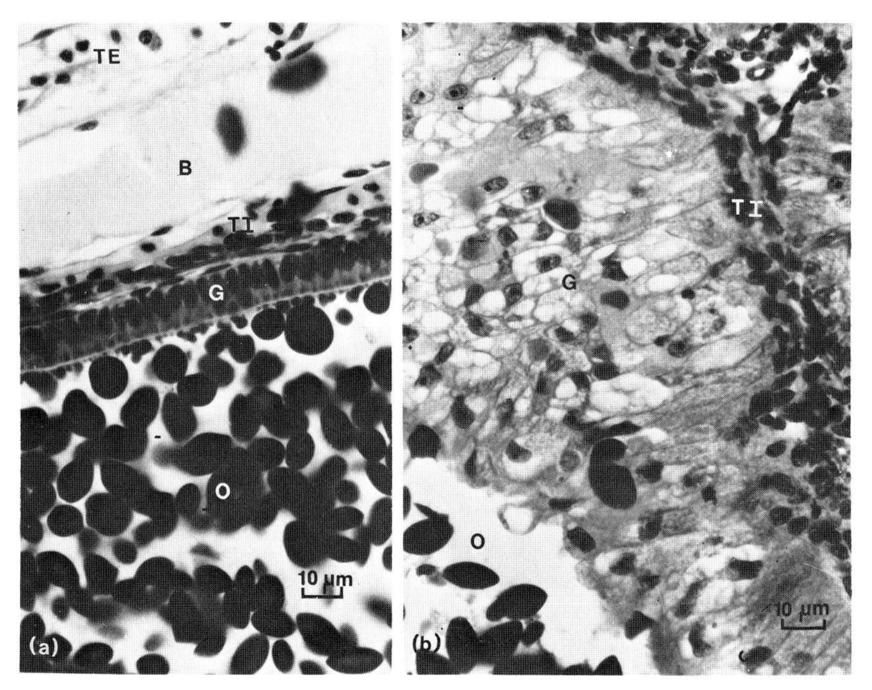

Fig. 5. Photomicrographs of the ovary of the spotted dogfish, *Scyliorhinus canicula* (a) Transverse section of vitellogenic follicle. (b) Transverse section of early corpus atreticum. B, blood vessels; G, granulosa; O, oocyte with yolk platelets; TE, theca externa; TI, theca interna.

is indirect. The ovary contains very little stroma and is ramified by fine tubules which are extensions of the peritoneal cavity and which lead into wider tubules that finally open into a conspicuous pocket on the right side of the ovary into which, presumably, the eggs are ultimately shed. The bulk of the ovary consists of follicles at various stages of development together with corpora lutea and blood vessels. Matthews (1950) has calculated that the ovary contains at least 6 million ova of 0.5 mm or more in diameter, which already have some yolk, the largest being not more than 5.0 mm in diameter. The medium-sized oocytes are contained in follicles which are said to consist of a syncytial layer in contact with the oocyte and a theca consisting of a single layer of extremely flattened cells. In the larger follicles the follicular epithelium is two or three cells thick and from these cells fine protoplasmic strands arise to constitute a zona radiata. The theca consists of four or five layers of cells which are separated from the follicular epithelium by an extremely thin basement membrane.

Corpus luteum-like structures of two kinds, distinguished by size and origin, are prominent features of the *Cetorhinus* ovary. The numerous small ones are believed to be formed by atresia of follicles that have reached a diameter of approximately 1.0 mm. They consist of peripheral cells, some of which are highly vacuolated, surrounding a central cavity. These corpora atretica are well vascularized and appear to persist in this form since no stages in degeneration or resorption are seen. The larger corpora lutea are 4 to 5 mm in diameter; these also have a central cavity which is filled by loose cells, lymphocytes and cellular debris. They gradually shrink in size, the cavity becoming obliterated, and cellular degeneration occurs. Matthews (1950) believed that these corpora lutea are formed from the empty follicles after ovulation though the evidence for this view is circumstantial.

Ovarian structure in the Holocephali, a related group of cartilaginous fishes, is basically similar to that in the elasmobranchs though the ovary is not associated with an epigonal organ (Dean, 1906; Stanley, 1963).

C. Bony Fishes

Bony fishes, especially the teleosts, show a greater degree of diversity in reproduction than any other vertebrates (see Volume II, Chapter 6) and this is reflected by their range of ovarian structure. Most are cyclical breeders and the ovary varies greatly in appearance at different times in the cycle. Gokhale (1957), on the basis of seasonal changes, recognized seven different stages in ovarian development, based on morphological criteria, in the whiting and Norwegian pout and eight stages have been recognized by Polder (1961) in *Clupea harengus*. Yet, in spite of the striking differences in gross morphology between the ovaries of different species and between one and the same ovary at different times of the year, fundamental structure is basically similar within the group.

The ovarian primordia of ganoids and teleosts appear to lack a medulla (see Volume I, Chapter 2). Early in development, longitudinal ridges arise on the ventrolateral surface of the developing gonads, these fuse and enclose a cavity which is clearly coelomic and is lined by peritoneal epithelium, unlike the ovarian cavities of other vertebrates which are lined by mesenchyme. In Teleostei and Lepidostei the cavity persists, giving rise to the cystovarian condition (Figs. 6b and c) whereas in Dipnoi, Chondrostei and *Amia*, among Holostei, the ovary is solid and naked (Fig. 6a), the gymnovarian condition. Cystovarian ovaries usually have posterior extensions of their wall and cavities which form oviducts and these open into the cloaca (Fig. 6b). But in some teleosts they degenerate and ova are ovulated into the body cavity, not into the ovarian cavity (Fig. 6c).

The so-called "germinal epithelium" of the teleost gonad is probably peritoneal in origin, there being no evidence that the germ cells originate from anything other than the primordial gonia. The stroma has received lit-

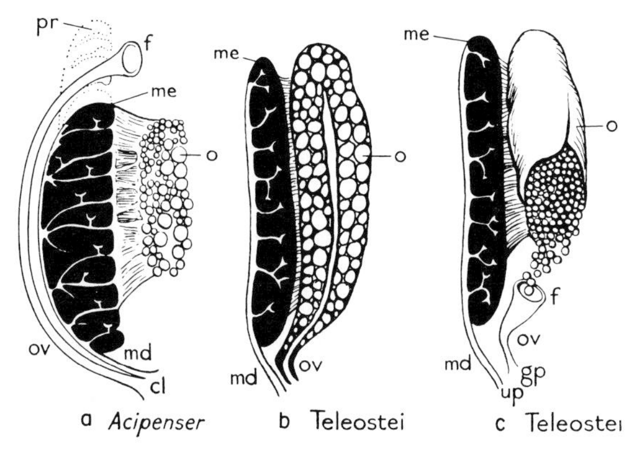

Fig. 6. Diagrammatic representations of gymnovarian (a) and cystovarian (b and c) ovaries in bony fishes. cl, cloaca; f, ostium of oviduct; gp, genital papilla; md, mesonephric duct; me, mesonephros; o, ovary; ov, oviduct; pr, pronephros; up, urinary pore. (Reproduced by permission from Hoar, 1957.)

tle consideration though it appears to be relatively sparse and to consist mainly of connective tissue which supports the ovigerous lamellae and carries the appreciable vasculature and nerve supply to the follicles.

The follicles of the fish ovary have been frequently studied both by optical microscopy and electron microscopy (Flügel, 1964a,b,c; Götting, 1967; Moser, 1967); and for the purpose of considering structure, it is convenient to recognize developing follicles, mature follicles, atretic follicles and postovulatory follicles. Prior to folliculogenesis the germ cells exist as oogonia, usually in small nests in which mitotic divisions can sometimes be seen (Hann, 1927, *Cottus bairdii*; Yamamoto and Shirai, 1962, *Rhodeus ocellatus*; Yamazaki, 1965, *Carassius auratus*). The question of the origin of seasonal crops of oogonia is fully considered in Volume I, Chapter 2.

The structure and growth of fish oocytes, especially vitellogenesis (Droller and Roth, 1966; Rastogi, 1969) have received a great deal of attention. Yamamoto and Yamazaki (1961) have described the morphological changes that occur during the maturation of oocytes. They recognize three previtellogenic phases, mainly identifiable by size, nuclear and nucleolar characteristics, and six stages in vitellogenesis ending with the ripe-egg stage. The development of yolk vesicles and yolk globules and the various nuclear changes during vitellogenesis are described in some detail. The so-called yolk nucleus has been described in eight species of bony fishes by Nayyar (1964) who has also reviewed the literature and shown that although such a structure is probably universal in fish oocytes, ideas concerning its origin, role, and fate are contradictory. In *Anabas scandens*, Dutt (1964) has described the yolk nucleus as a spherical structure with three concentric zones in contact with the nucleus in the early oocyte and containing RNA, lipid, sulfhydryl groups, carbohydrates, and certain amino acids. It appears to consist of a mass of lipid granules and mitochondria which migrate from their situation in contact with the nucleus, at a particular stage in vitellogenesis, to the periphery of the oocyte where they become dispersed in the cytoplasm (Nayyar, 1964).

The more primitive groups, Dipnoi, Chondrostei, and Holostei are oviparous as are most of the teleosts; but some teleosts have evolved viviparity and this necessitates ovarian gestation since oviducts of renal origin, which serve for gestation in all other vertebrates, are absent. Ovarian gestation, whether or not it involves simultaneous gestation of several broods of different ages (superfetation), has a profound effect on ovarian structure.

1. The Ovary in Oviparous Species

Egg production in some species of oviparous teleosts reaches quite remarkable proportions as Norman and Greenwood (1963) indicated: "A single Ling (*Molva*), 61 inches long and weighing 54 lbs. was found to have 28,361,000 eggs in the ovaries, a Turbot (*Scophthalmus*) of 17 lbs. weight, more than 9,000,000, and a Cod (*Gadus*) weighing 21½ lbs., 6,652,000; a Flounder (*Platichthys*), however, produces a mere million ova on the average, and a Sole (*Solea*) only 570,000." These figures refer, of course, to a single year's production by a mature female and most of the species mentioned breed over several seasons. Thus, the ovary in these fishes may account for a high proportion of their total weight though its size and appearance vary greatly throughout the year, being most variable in those species in which the ripening and spawning periods are partial and protracted. Detailed accounts of ovarian structure are available for a number of common species of oviparous teleosts.

Craig-Bennett (1931) has described ovarian morphology in the three-spined stickleback, *Gasterosteus aculeatus*, a typical oviparous species in which the ovaries are paired and continuous with short oviducts. When mature the ovaries measure 10 mm by 4 mm and each contains 70 to 150 mature eggs having a diameter of 1.0 to 1.5 mm. The "germinal epithelium" is applied to connective tissue lamellae which originate from the dorsal wall and invade the central cavity. Stroma is meager in amount and there are no interstitial cells. No mention is made of corpora atretica but structures "analogous to mammalian corpora lutea," though much simpler and more transient, are said to be produced in the empty follicles after ovulation. There are many such descriptions of ovarian structure in fishes dating back to those of Brock (1878) who examined fifty-seven species of bony fish and identified eight different morphological types. Among the oviparous species studied more recently are *Carassius auratus* (Beach, 1959;

Yamamoto and Yamazaki, 1961; Yamazaki, 1965), *Clupea harengus* (Bowers and Holliday, 1961; Polder, 1961), *Mugil capito, M. cephalus* (Stahl and Leray, 1961), *Salvelinus fontinalis* (Hurley and Fisher, 1966), *Oryzias latipes* (Yamamoto, 1963a,b), *Eucalia inconstans* (Braekevelt and McMillan, 1967), *Hippocampus erectus, Syngnathus fuscus* (Anderson, 1967), *Clarias batrachus* (Lehri, 1968), and *Pseudopleuronectes americanis* (Dunn and Tyler, 1969). Flügel (1964a,b, 1967) has described the fine structure of the follicles of *Fundulus heteroclitus* and a number of salmonids with special reference to the region of contact between the oocyte and the follicle cells, the zona radiata. In all the species studied this region consists of two zones, the outer being more electron dense than the inner, both being traversed by canals which give the radiata its striated appearance in the light microscope. Each canal comes to contain a microvillus from the oocyte and a process from a follicular cell; these are withdrawn from the canals shortly before ovulation. During follicular development the cells of the granulosa are separated by irregular spaces into which processes from the zona radiata extend. As the follicles mature these spaces disappear and the granulosa becomes once more a continuous cell layer.

2. The Ovary in Viviparous Species

Ovoviviparity is commonly encountered in lower vertebrates though it appears to be rare in amphibians and absent in cyclostomes and birds; true viviparity is much less common. The former has little effect on the basic structure of the ovary though it is usually associated with small numbers of relatively yolky eggs, whereas in true viviparity the ovary produces numerous small follicles and the eggs are virtually yolk free. However, in teleosts, in which viviparity has arisen independently, and repeatedly, in at least eight of the large groups, ovarian structure may become markedly modified since viviparity invariably involves the ovary in a gestational role. Some grade of viviparity commonly occurs in eight families of teleost fishes belonging to two orders: Cyprinodontes (families Poeciliidae, Anablepidae, Jenynsiidae and Goodeidae) and Acanthopteri (families Embiotocidae, Scorpaenidae, Bleniidae and Zoarcidae), and in these the ovaries of the viviparous species are atypical, often grossly so.

In the Scorpaenidae, in which the embryos are very numerous and are born while still immature, the ovary is only slightly modified on the oviparous pattern. The Poeciliidae are intermediate in the grade of viviparity they achieve; the embryos have a large yolk sac but they are not released until they have reached an active free swimming and feeding stage though they are sexually immature; here the ovaries are modified for gestation. The embiotocids show the most extreme, not to say bizarre, forms of viviparity in which the ovary is strikingly modified to permit follicular gestation and superfetation (Turner, 1937). In those teleosts in which viviparity is well established, and in which gestation occurs in the ovary, the epithelium lining the ovarian cavity becomes nutritive when it contains a brood, and spermatozoa become embedded in some of the cells to be kept alive for varying lengths of time, up to 10 months in *Heterandria formosa* (Turner, 1947). In other vertebrates it is the oviduct which assumes a sperm-storage function, never the ovary.

The ovary of *Goodea bilineata,* which is typical of those species of viviparous teleosts in which gestation is wholly or mainly in the ovarian cavity, has been described by Turner (1933). The virgin ovary in this species is an oval sac suspended between swim bladder and large intestine by dorsal and ventral mesenteries. The anterior four-fifths of the cavity is subdivided by a median septum and each half contains ovigerous folds. The ovarian wall consists of a thin peritoneal coat, two layers of muscle, and a thin layer of connective tissue on which the epithelium of the intraovarian cavity rests. This epithelium is cuboidal, or columnar, and glandular. The central region of the ovigerous folds consists of well-vascularized connective tissue and stroma in which primordial germ cells and oocytes are embedded. Each oocyte lies in a follicle delimited by one layer of cuboidal cells, and about sixty oocytes are ready for fertilization at any one time. Full-grown oocytes degenerate and are invaded by follicle cells if they are not fertilized. After fertilization the ovary becomes a nutritive organ as well as a gonad; vascularity increases, the internal epithelium becomes more glandular and folded, and the stroma develops fluid-filled spaces. The epithelium discharges food into the ovarian cavity and this is supplemented by degenerating embryos and oocytes. In the Jenynsiidae the ovarian epithelium is drawn out into club-like processes; these penetrate beneath the opercula of the developing embryos and enter their branchial and mouth cavities.

In all vertebrates, except certain teleost fishes, whether oviparous or viviparous, reproduction proper starts with ovulation. But some viviparous teleosts, notably the Hemiramphidae and the cyprinodont families Anablepidae and Poeciliidae, adopt intrafollicular gestation and in these cases the structure of the follicles is considerably modified and there is no ovulation. For example, in *Anableps* the ovisac is pear-shaped, the anterior four-fifths being ovary, the rest forming the ovarian duct. Under the peritoneum there is a thick muscular coat of two layers. At the time of fertilization the almost yolkless ovum fills the follicle but a fluid-filled space soon develops and the follicle cells degenerate to be replaced by a fibrous hypervascular capsule which develops branched and irregular villi. These are especially numerous adjacent to the yolk-sac of the embryo where they form a thick bed of filaments in which the yolk-sac bulbs of the embryo rest. This structure has been called a follicular pseudoplacenta. Another species which shows modifications associated with follicular gestation is *Poecilia reticulata,* and Lambert (1970a,b) has recently described its ovary (Fig. 7). It is hollow and saccular, the walls containing follicles at different stages of development prior to fertilization and also atretic follicles and follicles containing embryos at different stages of gestation. Lambert (1970a), using the eight stages recognized by Takano (1964), has described the development of oocytes and the follicular envelope in some detail. A theca and granulosa can be recognized once the oocytes reach a diameter of 0.06 to 0.15 mm. The theca shows few changes during the developmental process but the granulosa becomes hyperactive and folded during the early stages of yolk deposition, regressing as the follicle matures.

Another feature of the ovary of some viviparous teleosts, unique among vertebrates, is the phenomenon of superfetation (Turner, 1937). This occurs when two or more broods of embryos at different stages of development are present within the follicles of the ovary at the same time. It is incipient or actual in the families Poeciliidae, Goodeidae, and Jenynsiidae and it is seen in its most extreme manifestations in *Heterandria formosa* and several species of *Poeciliopsis* and *Aulophallus* in which as many as nine small broods, at different stages of development, are present simultaneously in an ovary. Superfetation results from a combination of factors of which the following are the most important: differentiation and maturation of oocytes in an already gravid ovary; intrafollicular fertilization rendering ovulation

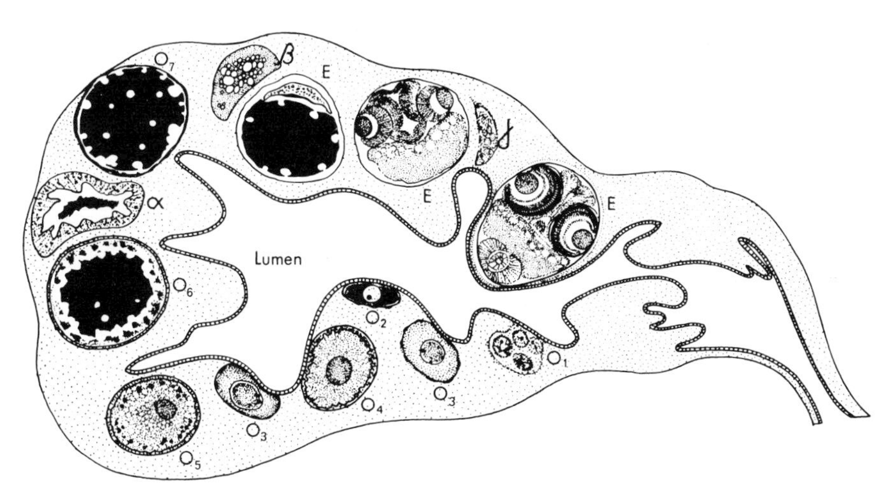

Fig. 7. Diagrammatic representation of T.S. of the ovary of *Poecilia reticulata.* All stages of follicular growth and atresia are shown though these are not normally present simultaneously. O_1–O_7, oocytes in various stages of development; E, embryos; α, β, γ, stages in follicular atresia. (Reproduced by permission from Lambert, 1970a.)

unnecessary; and short successive reproductive cycles in which maturation of oocytes overlaps gestation periods (Turner, 1940).

Early work on developing and mature follicles in viviparous species was based on optical microscopy; this has more recently been effectively augmented by fine-structural studies (Fig. 8). Kemp and Allen (1956) have described the development of the chorion and Jollie and Jollie (1964a,b) and Jollie (1964) have described the structure and development of the follicle and pseudoplacenta in the ovoviviparous *Lebistes reticulatus*. The smallest follicles have two layers of squamous cells surrounding the oocyte. As development proceeds, a basement membrane appears between them; the inner layer (granulosa equivalent) remains simple but the outer layer stratifies and appears to become fibrous and cornify. The outer region of the oocyte produces villus-like microprojections which become intimately associated with the inner layer of the follicle wall. Between these projections the bilaminar vitelline membrane is deposited and constitutes the corona radiata. The outer ends of the villi coalesce immediately below the

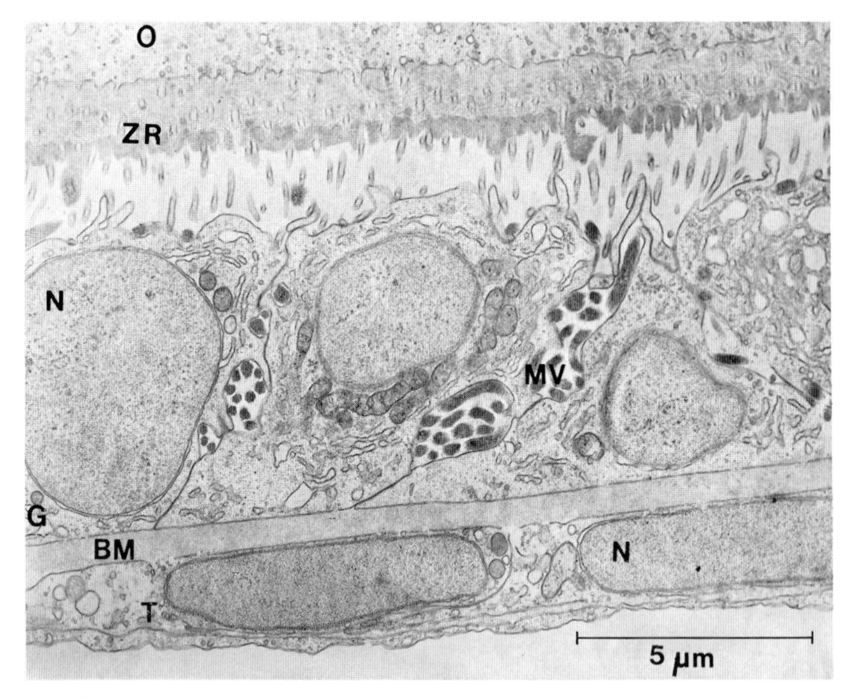

Fig. 8. Fine structure of the follicular epithelium surrounding an early vitellogenic oocyte of *Cichlasoma nigrofasciatum*. Granulosa cells (G) are separated from thecal cells (T) by a distinct basememt membrane (BM). Microvilli from granulosa and oocyte intermingle to form the zona radiata (ZR). N, nucleus; MV, intergranulosal microvilli; O, oocyte. (Reproduced by permission from Nicholls and Maple, 1972.)

presumed granulosa and form the chorion. After fertilization these various layers undergo modifications and together with certain membranes of the interfollicular embryo form the pseudoplacenta.

Other viviparous species in which ovaries and ovarian cycles have been studied are *Neotoca bilineata* (Mendoza, 1940), *Cymatogaster aggregata* (Wiebe, 1968), and *Cichlasoma nigrofasciatum* (Nicholls and Maple, 1972).

It has recently been shown that the coelacanth, *Latimeria chalumnae*, is ovoviviparous (Lavett Smith *et al.*, 1975), the uniquely large yolky eggs being retained in the single oviduct throughout development. Anthony and Millot (1972) have summarized what little is known of the morphology of the ovary in this fish; nothing seems to be known of its histology or fine structure. Twenty females have now been examined, most of which were immature, though in three of them the single ovary, the right one as in elasmobranchs, contained a number of eggs. In one specimen these had a diameter of 1.0 cm, in another they were 2.0 cm, and in the third a single egg, 7.0 cm in diameter, was found in the oviduct while the ovary contained about a dozen much smaller eggs. However, in early 1972, a sexually mature female was caught in which 19 spherical, or subspherical eggs, 8.5 to 9.0 cm in diameter, were found in the body cavity, apparently newly ovulated. They ranged between 300 and 334 gm in weight and are the largest fish eggs ever recorded.

3. Corpora Atretica and Corpora Lutea

Interest in the structure and significance of corpora atretica and postovulatory corpora lutea in bony fish dates back to the work of Bretschneider and Duyvené de Wit (1947) who described the development and mature organization of both types of structure in great detail in the bitterling, *Rhodeus amarus*. They have subsequently been described in a number of other species (*Salvelinus fontinalis*: Henderson, 1963; *Oryzias latipes*: Yamamoto, 1963b; *Rhodeus amarus*: Polder, 1964; *Scomber scomber*: Bara, 1965a,b; *Gobius giuris*: Rajalakshmi, 1966; *Tor* (*Barbus*) *tor*: Rai, 1966; *Eucalia inconstans*: Braekvelt and McMillan, 1967; *Poecilia reticulata*: Lambert, 1969, 1970a,b). Lambert (1970b) has shown that the atretic follicle in *Poecilia reticulata* arises when cells of the granulosa and theca invade mature and developing oocytes, the granulosa resorbing the yolk. When all the yolk has gone there remains a compact well-vascularized structure, consisting mainly of granulosa, with a supporting framework of thecal cells. Regression now occurs, the cells being resorbed by phagocytic action. The final stage of regression is a small group of cells containing yellow–brown granules. Unlike Stolk (1951), Lambert (1970b) was unable to find any correlation between particular stages of gestation and the occurrence of corpora atretica. Several recent investigations have been concerned with attempts to demonstrate enzymes, cofactors, and precursors, known to be essential for steroidogenesis, in the corpora atretica but the results are discordant and beyond the scope of this chapter.

Postovulatory follicles are, of course, present in the fish ovary after spawning, but unlike the situation in some other vertebrates the ruptured follicles usually show no hypertrophy and no signs of physiological activity (Barr, 1963; Rajalakshmi, 1966; Rai, 1966). Some yolk may continue to be secreted after ovulation but after minor reorganization this yolk and the ruptured follicular membranes are merely resorbed.

4. Hermaphrodite Gonads in Teleosts

The occurrence of hermaphroditism in teleosts, mainly in the families Labridae, Sparidae, and Maenidae, is well documented (Chan, 1970; Remacle, 1970) and its manifestation spans the entire range of possibilities. Some species are simultaneous (synchronous) hermaphrodites (Reinboth, 1962; Atz, 1964) and in these both eggs and sperms are mature at the same time (*Serranus scriba, S. hepatus, S. phoebe,* and *Hypoplectrus unicolor*); others are protogynous (*Epinephelus guttatus, E. striatus, Pagellus erythrinus, Maena smaris, M. chryselis,* and *M. maena*) or protandrous (*Sparus auratus, Boops salpa, Pagellus acarne,* and *P. mormyrus*), and yet others show rudimentary hermaphroditism (*Diplodus sargus, D. annularis,* and *Pagellus mormyrus*).

The gonad of the protogynous *Monopterus albus* (Synbranchidae) the ricefield eel, during natural sex reversal has been described by Chan and Phillips (1967). It is an elongated tubular structure lying to the right in the body cavity. The internal sinus is invaded by two hollow structures which constitute the gonadal lamellae. Thus, the gonad contains two inner cavities, which fuse posteriorly, and an outer cavity. Germinal elements in the gonadal lamellae are seen as isolated clusters or cords of small undifferentiated cells, sometimes in mitosis. There is some indication of segregation, the gonocytes destined to become oogonia lying in the outer regions of the lamellae, the others being more centrally situated. In young animals of 20 to 30 cm the gonad is filled with primary oocytes undergoing vitellogenesis. This typical ovarian structure matures, ovulation takes place and this is followed by an hermaphrodite phase which lasts between 2 and 4 months and, in all but a few specimens, which may develop and function again as females, it is characterized by atretic follicles, developing testicular lobules, and interstitial (Leydig) cells. Thus, sex change in *Monopterus* appears to follow a functional female phase and it is suggested by Chan and Phillips (1967) that the ovary already contains presumptive male gonia which are quiescent until the testis begins to develop.

Chan *et al.* (1967) have described the atretic structures that occur in *Monopterus* and shown that five types of atretic follicles can be recognized. Two of these occur only during the female phase, the others are associated with gonads undergoing sex change. All appear to be typical corpora atretica and are associated with the breakdown and resorption of developing or mature follicles; there is no evidence that they have a function other than this. *In vitro* studies have shown that the *Monopterus* gonad is capable of synthesizing progesterone, androgens, and estrogens in all the developmental stages tested though androgen production increases markedly at the time of sex reversal.

D. Amphibians

Amphibian ovaries are paired and sac-like, their internal cavities arising during early development when the medullary regions of the primordial gonads become gelatinous, their few remaining cells forming the squamous internal lining of the ovary. The internal cavities become subdivided by septa to form numerous fluid-filled chambers to which the "germinal epithelium" is applied; their outer surfaces are covered by peritoneal epithelium which may or may not be ciliated. The ovarian stroma is restricted to the thin connective tissue septa which support the walls of the ovarian sacs and carry vasculature and nerves to the follicles. Oocytes arise seasonally from persistent nests of oogonia and in some species they are divisible into discrete populations on the basis of their developmental stage (Smith, 1955).

The ovary and oogenesis in the oviparous anuran *Xenopus laevis* have recently been described by Dumont (1972). Each ovary consists of about twenty-four lobes and all stages in the development of oocytes are present throughout adult life, though this is exceptional in amphibians. Six stages in follicular development are recognized by Dumont (1972) and they have been characterized in some detail (Fig. 9). Stage I is previtellogenic and the oocytes are between 50 and 300 μm in diameter. During vitellogenesis, which covers four developmental stages, the oocyte diameter increases from 300 μm to 1200 μm. A final postvitellogenic stage, Stage VI, is characterized by a relatively unpigmented equatorial band and a diameter of 1300 μm. The follicular wall consists of a squamous layer of peritoneal cells, an undivided theca containing collagen, fibrocytes, and small blood vessels, and a follicular epithelium (granulosa) the cells of which become separated from the surface of the oocyte and form macrovilli while the oocyte surface becomes microvillous.

Hope *et al.* (1963, 1964a,b) have described the relationship between oocyte and follicle in the urodele *Triturus viridescens* in some detail and

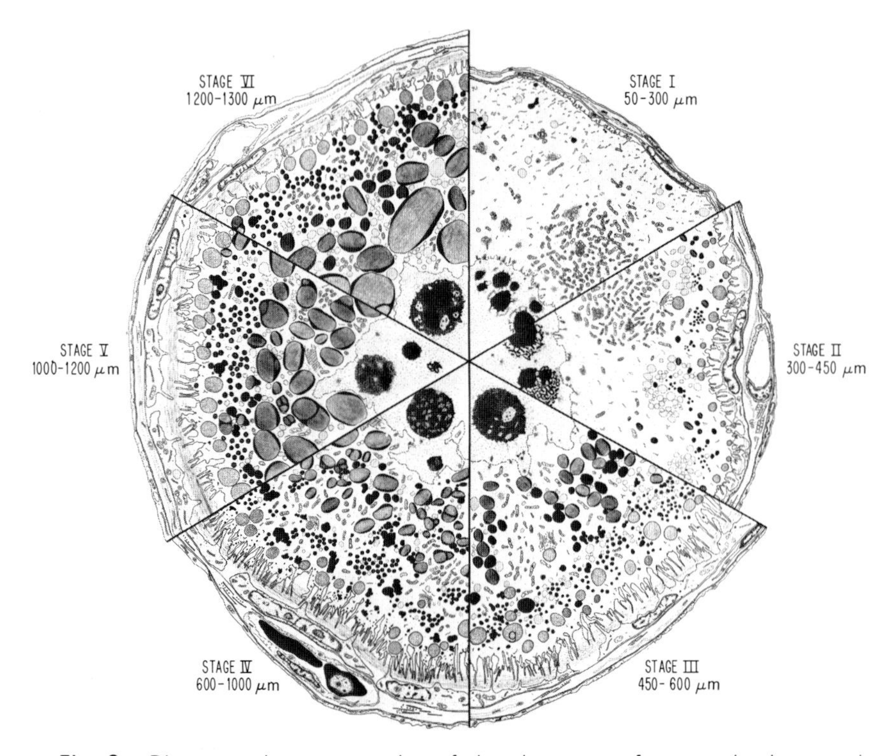

Fig. 9. Diagrammatic representation of the six stages of oocyte development in *Xenopus laevis*. Stage I oocytes are characterized by a very thin covering of follicle cells, a large mitochondrial mass, a few lipid droplets (light spheres), and small Golgi complexes. During Stage II the follicle cells increase in thickness and arch over the oocyte surface which now has microvilli. The vitelline envelope forms beneath the arches of the follicle cells. Cortical granules, premelanosomes (small dark spheres), and some yolk appear. Vitellogenesis begins in Stage III, when pigment and cortical granules increase and the vitelline envelope forms a continuous layer over the oocyte. Irregularly shaped yolk platelets are present in the peripheral cytoplasm of Stage IV oocytes; microvilli are large and numerous. During Stage V the accumulation of yolk gradually ceases and the microvilli become shorter and less numerous. During Stage VI many of the microvilli are lost (retracted?) and the follicle cells decrease in thickness. (Reproduced by permission from Dumont, 1972.)

shown that in oocytes between 100 and 200 μm in diameter the plasma membrane of the oocyte is in continuous contact with the inner surface of the follicular epithelium, though prior to the development of the microvilli by the oocyte spaces appear between contiguous cells of the follicular epithelium and between epithelium and oocyte. These spaces soon become occupied by the microvilli, by processes from the follicular cells, and by an electron-dense homogeneous material which together form the zona radiata.

The microvilli reach a maximum length of 2.0 to 2.2 μm in oocytes with a diameter of 650 to 750 μm. When they and the follicular cell processes are withdrawn the remaining material forms the vitelline membrane. Fine-structural studies such as these have added appreciably to our understanding of follicular structure in the Amphibia and to knowledge concerning cytoplasmic changes in oocyte and follicle during development. Other species studied have included the following: *Rana esculenta* (Wartenberg and Gusek, 1960; Wartenberg, 1962), *R. temporaria, Triton alpestris, Ambystoma mexicanum* (Wartenberg, 1962), *Rana pipiens* (Kemp, 1953; Kessel, 1969), and *Xenopus laevis* (Wartenberg, 1962; Dumont, 1972). Other features of ovarian structure, including the oocyte chromosomes, especially in their lampbrush stages, have been studied in some detail (Callan and Lloyd, 1960), as has the nuclear membrane (Callan and Tomlin, 1950; Merriam, 1961; Cole, 1969), yolk synthesis (Ward, 1962; Wartenberg, 1962; Hope *et al.*, 1964a), and pigment formation (Hope *et al.*, 1964b).

Most amphibian species are oviparous; however, a few are viviparous or ovoviviparous and the ovaries of these show certain superficial modifications, mainly associated with the number and size of oocytes and the presence of apparently functional corpora lutea. Ovarian structure and function have recently been described in the viviparous anuran *Nectophrynoides occidentalis* (Xavier *et al.*, 1970) and in the ovoviviparous urodele *Salamandra salamandra* (Joly and Picheral, 1972). In the former, at the end of September, immediately before ovulation, three oogenetic stages are present in the ovaries: oogonia, numerous small follicles having a diameter of 300 μm or less, and ripe follicles of approximately 550 μm. Only some of the latter ovulate, the others become atretic. In *Salamandra salamandra*, the ovaries are paired and at the time of ovulation contain ten to twenty yolky oocytes, which are creamy white in color and approximately 5 mm in diameter, and also smaller yolked oocytes and large numbers of previtellogenic oocytes less than 100 μm in diameter. Joly and Picheral (1972) have described the structure and ultrastructure of the ripe follicles. An outer epithelium, consisting of flattened cells with dense cytoplasm is separated by a basement membrane from the theca, which is undivided and consists mainly of fibroblasts containing liposomes. Bundles of collagen fibers lie between the fibroblasts and the outer and inner basement membranes. The follicular epithelium consists of a regular layer of cells that show pronounced intercellular spaces between scattered desmosomes and whose cytoplasm contains smooth endoplasmic reticulum and mitochondria with tubular cristae. A zona pellucida, into which project macrovilli from the inner border of the follicular epithelium and microvilli from the oocyte, lies between the oocyte and the follicular epithelium.

Corpora Atretica and Corpora Lutea

Corpora atretica and corpora lutea may also be present in the amphibian ovary, the former arising only from yolky oocytes (Dumont, 1972). Guraya (1968, 1969a,b) has described the corpora atretica in certain anuran amphibians (*Rana catesbeiana, R. pipiens, Bufo stomatictus*) and shown that they appear to be concerned solely with the digestion and absorption of yolk. They contain triglycerides derived from yolk but not cholesterol or its esters. In the viviparous *Nectophrynoides occidentalis* also, corpora atretica are said to have the histological characteristics of degeneration rather than secretion (Xavier *et al.*, 1970). However, Guraya (1969b) reports that in *R. pipiens* and *B. stomatictus* some thecal cells of the corpus atreticum, after absorption of the oocyte is complete, become thickened to form conspicuous nodules of so-called interstitial tissue and there is histochemical evidence suggesting that these cells may secrete steroids. The granulosa cells disappear and only the pigment of the oocyte remains when absorption is complete. Humeau and Sentein (1968) have described fine-structural aspects of the oocyte and follicle cells at the start of atresia in *Triturus helveticus*.

It has long been known (Brambell, 1956) that ovulated follicles in oviparous amphibians are transitory. Hett (1923b) has described postovulatory events in *Triturus vulgaris* and shown that yolk continues to be secreted into the empty follicle for a time; the theca then shrinks and becomes multilayered and may show an inner zone and a well vascularized outer zone. In *Xenopus laevis*, however, absorption of the relict follicles seems to commence at the time of ovulation and it is complete $6\frac{1}{2}$ days later (Cunningham and Smart, 1934), but in *Nectophrynoides occidentalis* (Lamotte and Rey, 1954; Vilter and Lugand, 1959) and in *Salamandra salamandra* (Joly and Picheral, 1972) postovulatory corpora lutea are present. In the former species they are maintained throughout most of the 9-month gestation period and show signs of active secretion during the first 7 months. In *Salamandra salamandra* they are hollow structures with walls two or three layers thick, derived from the follicular epithelium, and a fluid-filled cavity containing free cells. A month after ovulation the corpora lutea are smaller in size than when they are first formed and they continue to decrease in size. Δ^5-3β-Hydroxysteroid dehydrogenase activity is present in all of the cells but especially in the luteinized ones which are said by Joly and Picheral (1972) to show also the main histological characters of steroid-secreting cells.

In both sexes of bufonid anurans (toads), in addition to the definitive gonads, paired rudimentary ovaries are present. These structures, the so-called Bidder's organs (Fig. 10), lie at the anterior pole of the gonads and consist of early oocytes which may, in certain circumstances, develop yolk

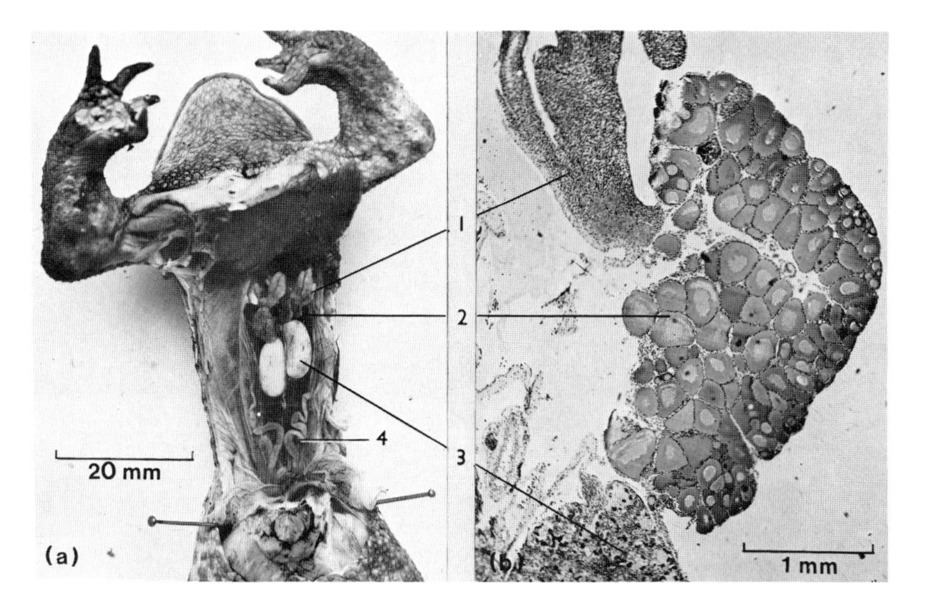

Fig. 10. Bidder's organ in a male toad. (a) Anatomical relationships. (b) Sagittal section through fat body, Bidder's organ, and rostral end of testis. 1, fat body; 2, Bidder's organ; 3, mature testis; 4, rudimentary oviduct. (Reproduced by permission from Ponse, 1949.)

(Ponse, 1924, 1949). They persist throughout the life of male toads but tend to disappear in mature females. However, if male toads are castrated the ovarian potentialities of the Bidder's organs are realized and they become transformed into fully functional ovaries. Anderson and Chomyn (1964) have described the fine structure of the Bidder's organ in *Bufo americanus* and *B. marinus* and report that, except for the absence of yolk bodies, their oocytes show fine-structural features identical to those of developing ovarian oocytes.

E. Reptiles

Ovarian morphology has been described in several species of reptiles (Boyd, 1940; Samuel, 1943; Bragdon, 1952; Panigel, 1956; Betz, 1963a,b; Eyeson, 1971; Neaves, 1971) and most of the differences between the ovaries of different species appear to be superficial and due mainly to differences in the numbers of large vitellogenic oocytes present in the ovary. In lizards, for example, few are present, in *Agama agama*, two to four (Fig. 11a), while in geckos it is only one per ovary (Boyd, 1940).

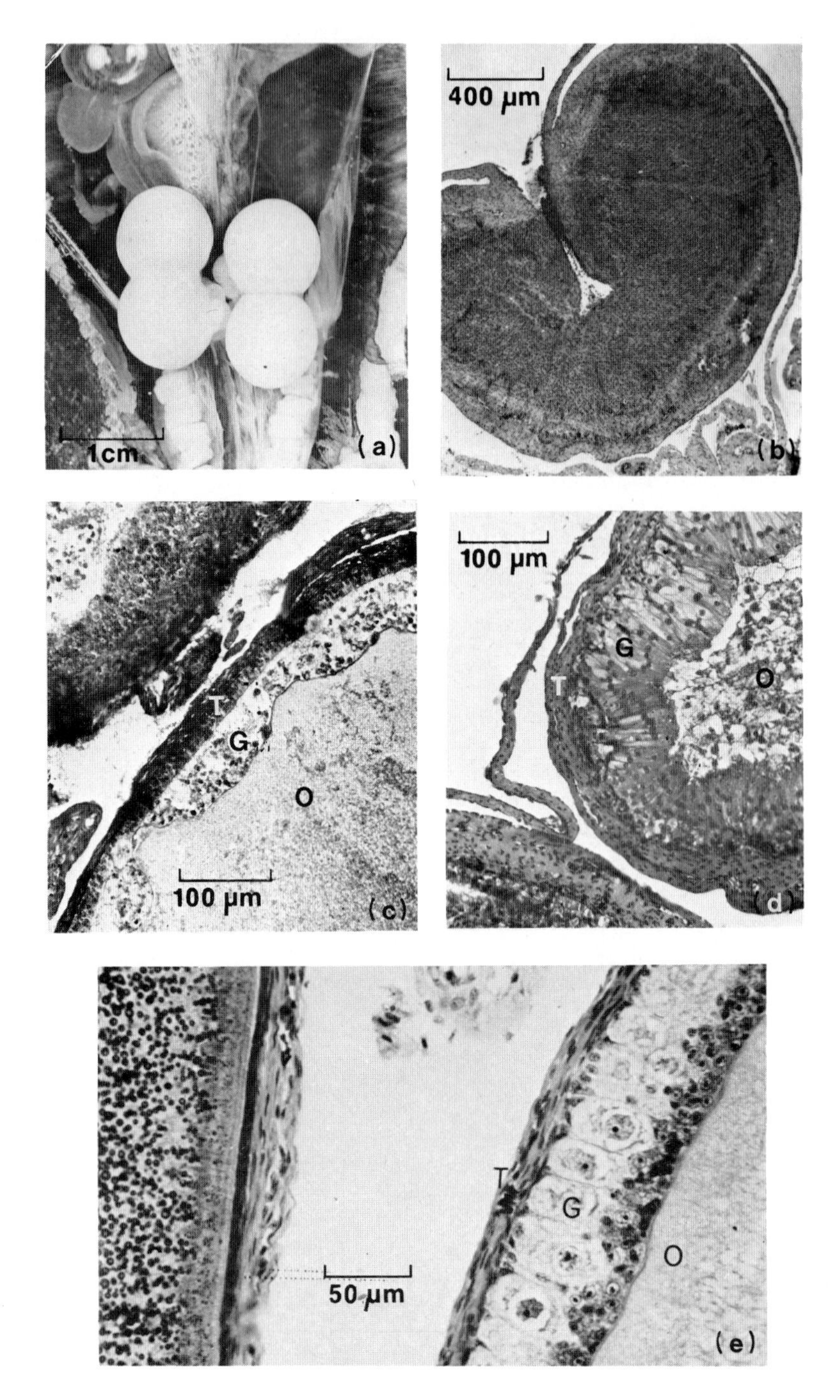
1cm
(a)
400 μm
(b)
T
G
O
100 μm
(c)
100 μm
G
T
O
(d)
T
G
O
50 μm
(e)

Betz (1963a,b) has described the ovaries of *Natrix rhombifera*, a viviparous colubrid snake, in which each ovary is an elongate, thin-walled, saccular structure with an irregular, lymph-filled, central cavity. This cavity is a characteristic feature of the reptile ovary; it is derived from the medullary cords of the ovarian primordium and is, thus, homologous with the cavities of the amphibian ovary though not with those of teleosts. The ovaries are asymmetric, the right being more anterior than the left and each is suspended by a mesovarium which extends from the anterior region of the oviducts to the posterior region of the kidneys. The stroma is sparse and semitransparent, the bulk of the ovary being contributed by the follicles which, once vitellogenesis has started, are creamy-white and relatively avascular. Oogonia are randomly scattered in small groups in the stroma while the smallest oocytes are restricted to the lateral aspects of the ovary and become surrounded by one layer of undifferentiated cells. Corpora atretica and corpora lutea are also present at certain times of the year.

The description by Neaves (1971) of the ovary of the lizard *Anolis carolinensis* illustrates the magnitude of the seasonal changes in the ovaries of a cyclically breeding species. During autumn and winter the largest follicles have a diameter of 1 mm and there is only one of these in each ovary. There are also a number of smaller follicles ranging in size from 0.8 mm to less than 0.1 mm. But in spring and summer the reproductively active ovary contains follicles ranging from less than 0.1 mm to 8.0 mm.

The theca of the ovarian follicles in many reptiles appears to be divisible into theca externa and interna (Guraya, 1965) though, in lizards, Weekes (1934) could only recognize the distinction after ovulation. Both layers contain collagen fibers (Hett, 1924) and the theca externa also contains elastic fibers. Eyeson (1971) has described the structure of "vitellogenic" and "previtellogenic" follicles in *Agama agama* (Fig. 11e) in which the theca, although several-layered is not obviously subdivided into interna and externa.

The granulosa of small follicles is more differentiated in lizards and snakes than in any other vertebrate group (Figs. 11e, 12a, and b) and the early literature on this subject has been summarized by Brambell (1956). The granulosa consists of distinct cell types called "small," "intermediate or transitional" and "large" by Loyez (1906) and Bhattacharya (1925). The largest cells, also called "pyriform" or "nutritive" were believed to establish

Fig. 11. Ovarian structure of the lizard, *Agama agama.* (a) Ovaries in an advanced stage of vitellogenesis. (b) Transverse section of postovulatory corpus luteum. (c) Transverse section of atretic previtellogenic follicle. (d) Transverse section of atretic vitellogenic follicle, granulosa cells invading the oocyte. (e) Transverse section of vitellogenic follicle (left) and previtellogenic follicle (right). G, granulosa; O, oocyte; T, theca. (Reproduced by permission from Eyeson, 1971.)

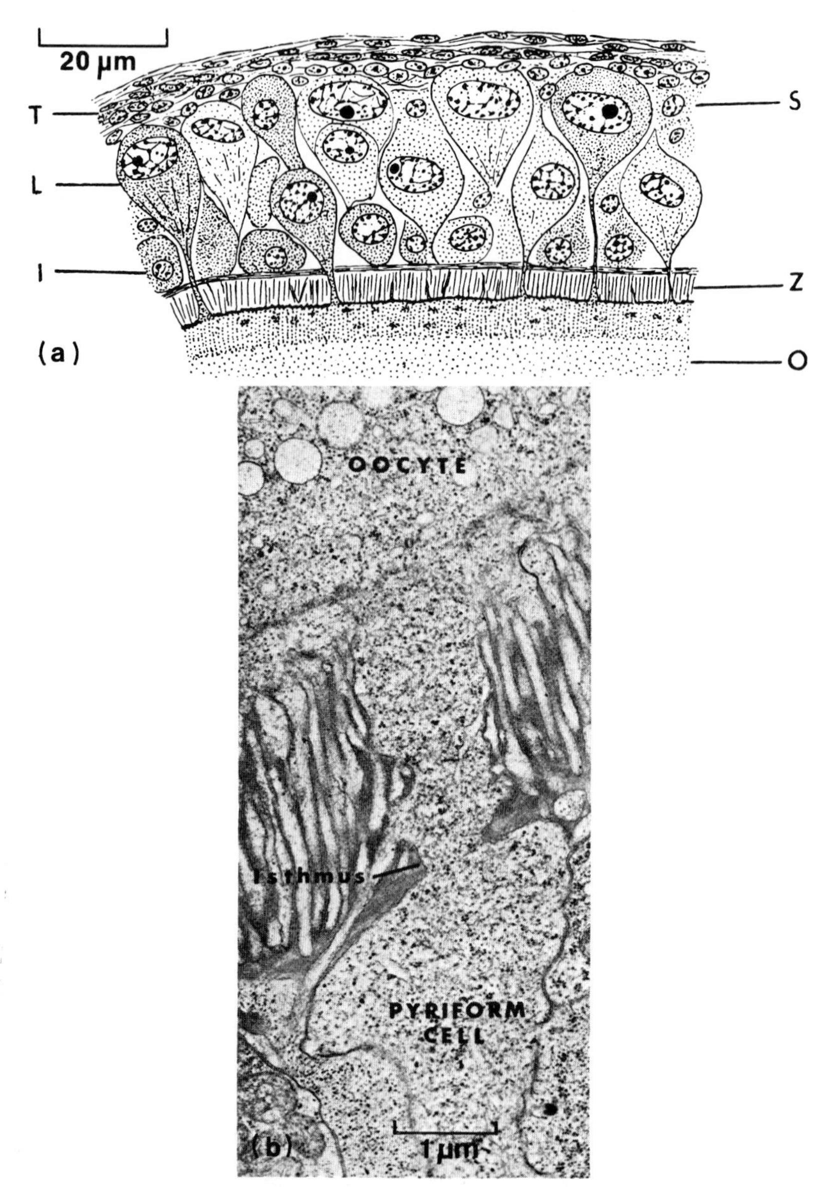

Fig. 12. Structure of the follicle wall in lizards. (a) Part of the follicle wall of *Lacerta muralis* showing the theca (T), and the small (S), intermediate (I), and large (L) cells of the granulosa. The large (pyriform) cells make contact with the oocyte (O) through the zona radiata (Z). (b) Follicular wall in *Anolis carolinensis* at the time of its maximum complexity (November, follicle diameter 1.0 mm) with the pyriform cell extending toward the oocyte. (Reproduced by permission from Neaves, 1971.)

a unique relationship with the oocyte in that they achieved cytoplasmic fusion with it. Neaves (1971) has reexamined, by electron microscopy, the claim that intercellular bridges occur between certain follicle cells and the oocyte and shown it to be well-founded in *Anolis carolinensis*, though such bridges are restricted to early developmental stages. He found that the follicular epithelium in follicles of 1 mm always contains pyriform, basal, intermediate, and apical cells. The pyriform cells, not found in other vertebrates, produce narrow cytoplasmic bridges which traverse the zona pellucida and fuse with the cytoplasm of the oocyte. However, they have nothing to do with vitellogenesis; indeed, when the follicles enlarge beyond a diameter of 1.0 mm and vitellogenesis starts the bridges undergo fatty degeneration and they have disappeared by the time the follicles have reached a diameter of 2.0 mm. Furthermore, although the follicular epithelium in early follicles is stratified and consists of cells of different kinds and sizes, it eventually becomes single layered.

Corpora Atretica and Corpora Lutea

Follicular atresia in the reptile ovary, reviewed by Betz (1963b), may occur in follicles of any size but is most frequently seen in those with a polymorphic granulosa and in mature follicles that have failed to ovulate (Boyd, 1940; Bragdon, 1951, 1952; Betz, 1963a,b; Guraya, 1965). Typical early corpora atretica in *Agama agama* (Fig. 11d) are hyperemic and flaccid, the yolk is more fluid and paler in color and the follicular walls are thin and fragile. When digestion and absorption of the oocyte contents are complete all that remains is a small scar of residual tissue which may persist for some time (Betz, 1963a). A peculiar kind of atresia, by bursting, has been described (Raynaud and Raynaud, 1961; Betz, 1963a) in which yolk is found in the ovarian stroma and cavity, and in the body cavity, and is presumably removed by phagocytosis. Raynaud and Raynaud (1961) have shown that in *Anguis fragilis* the epithelium lining the ovarian cavity becomes phagocytic and removes this ectopic yolk.

Postovulatory corpora lutea (Fig. 11b) are better developed and more persistent in reptiles than in any other nonmammalian vertebrates and, especially in viviparous species, there is circumstantial evidence that they may be functional. Betz (1963a) has described the corpora lutea of the viviparous snake *Natrix rhombifera*. They are usually oval bodies, embedded in the stroma, pale yellow in color, and hyperemic during the first month of gestation. In the second month they become dark yellow and smaller and in the last month they are deep yellow and less than 5 mm in diameter. After parturition they degenerate rapidly and become small orange patches in the stroma, which disappear by the following spring. Cunningham and Smart (1934) and Weekes (1934) found that corpora lutea in viviparous species were more persistent and better developed than those in oviparous species, a finding that is consistent with the view that these structures may have a function during pregnancy. Other descriptions of the development and structure of reptilian corpora lutea are given by Hett (1924), Rahn (1939), Boyd (1940), Altland (1951), Bragdon (1951, 1952), and Badir (1968).

F. Birds

In birds there are two ovarian primordia but in most species only that of the left side matures, though there are exceptions. For example, in some birds of prey the right ovary is frequently as large as the left one and it may be functional. Stanley (1937), studying three species of American hawks, found nearly symmetric paired ovaries, and Romanoff and Romanoff (1949) have reported that a well-developed right ovary is present in more than 50% of mature hawks and about 24% of ring doves; one is also occasionally present in chickens and pigeons (Stanley and Witschi, 1940). The structural basis for this variability lies in the relative development of cortex and medulla in the primordial gonads, and Brode (1928) has shown that in 60% of hens the right ovary consists entirely of medullary tissue whereas in the remaining 40% a rudimentary cortex is present. These differences affect the results when the left gonad is removed (Benoit, 1923; Domm, 1924, 1939; Kornfeld, 1960) or its function is impaired (Crew, 1923; Brambell and Marrian, 1929); the right gonad then develops into a testis or an ovotestis depending upon its cellular constitution.

General features of ovarian structure in birds (Fig. 13) are well established and have been frequently reviewed (Benoit, 1950; Sturkie, 1954; Breneman, 1955; Wyburn *et al.*, 1965; Gilbert, 1971). The morphology of a mature bird ovary at a particular point in time obviously depends on whether the species in question breeds cyclically, rearing one, two, or several broods per year or breeds more or less continuously. It contains no oogonia, since birds belong to the group of vertebrates in which oogenesis is precocious (Volume I, Chapter 2), though oocytes are present and are either previtellogenic, vitellogenic, or atretic. The ovarian follicles are embedded in stroma which, in the fowl, consists of an extensive network of connective tissue containing large lacunae and supporting a rich supply of blood vessels and nerves. Pearl and Schoppe (1921) counted 1906 oocytes visible to the naked eye in the chick though Hutt (1949) estimated that the number of oocytes in chicks may run into millions, the few that mature enlarging to about 40 mm before being ovulated. Their rapid gross enlargement, due to

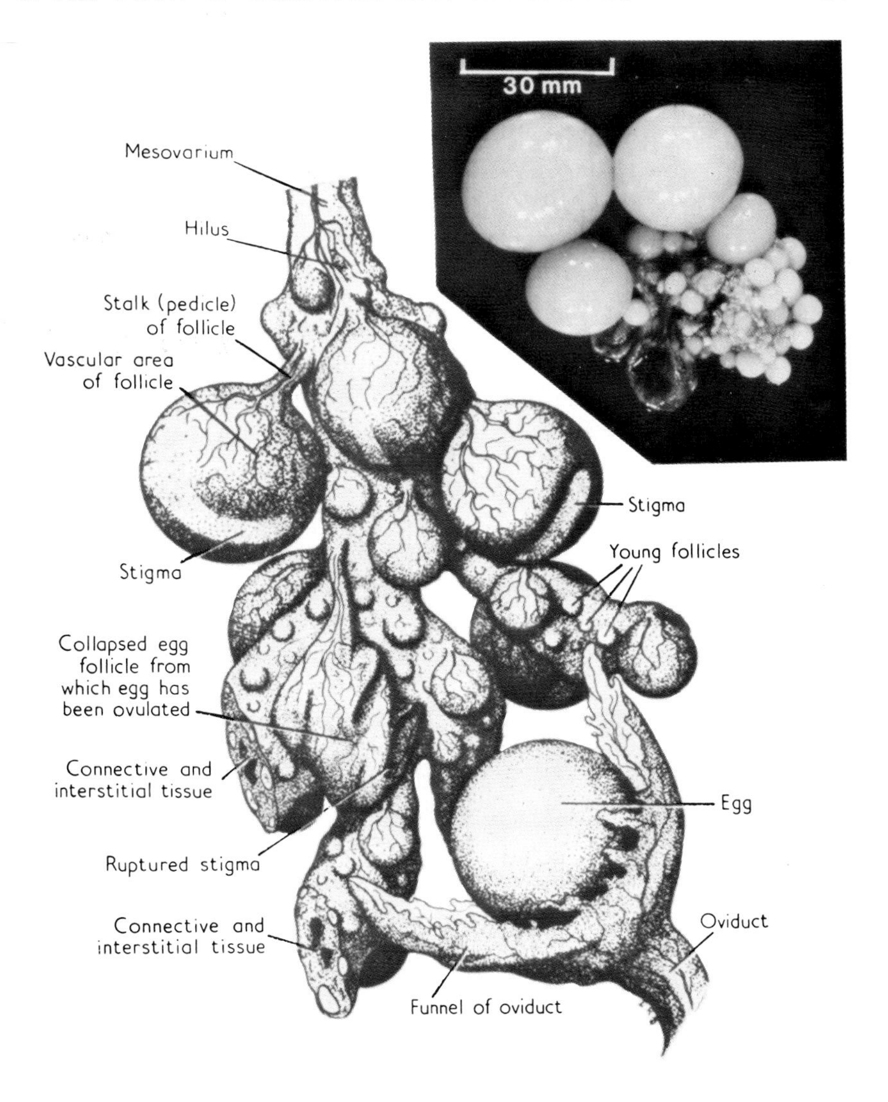

Fig. 13. Diagram showing external structure of mature ovary of a bird. (Reproduced by permission from Nelsen, 1953.) Inset, photograph of mature fowl ovary. Note hierarchy of follicles and postovulatory follicles. (Reproduced by permission from Lorenz, 1969.)

accumulation of yolk, causes the oocytes to project from the surface of the ovary and ultimately to become pedunculate (Fig. 13.).

Early studies on the follicular wall, critically reviewed by Brambell (1956), have established that it consists of a granulosa and a theca of two zones. More recent work on follicular development and histology has been mainly on the fine structure (Fig. 14) and virtually restricted to the fowl (Bellairs *et al.*, 1963; Bellairs, 1965, 1967; Wyburn *et al.*, 1965; Greenfield, 1966; Gilbert, 1971). The outer layer of the oocyte is the vitelline membrane, which is a coarse network of fibers (Bellairs *et al.*, 1963). This is surrounded by a zona radiata, a narrow zone, about 5 μm wide (Wyburn *et al.*, 1965) which with the perivitelline layer or zona pellucida, probably secreted by the granulosa, constitutes the region of contact between oocyte and follicle. A layer of electron-dense rods is present in the region of the perivitelline layer adjacent to the zona radiata. In a follicle of 7-mm diameter the latter is formed by primary and secondary processes from the oocyte cytoplasm and a few villi arising from granulosa cells. The theca interna is a compact cellular layer derived from the connective tissue of the stroma and three regions can be identified in it; an inner layer of collagen fibers, a middle layer of fibroblasts, and an outer layer of vacuolated cells (Gilbert, 1971); the theca externa consists of an extensive zone of fibroblasts interspersed with collagen fibers. Each follicle has, in addition, an outer connective tissue sheath which contains smooth muscle and is covered by the so-called germinal epithelium (Gilbert, 1971). Two to four arteries enter the follicular stalk and provide an extensive vascular system. The follicle at the opposite pole to the stalk is differentiated to form the stigma (Fig. 13), the region along which the follicle bursts at ovulation. It is thinner than the rest of the follicle since the connective tissue layer is absent from it, and it is relatively avascular.

Gilbert (1965, 1969) has shown that the ovarian follicle of the domestic fowl is extremely well innervated. Nerve fibers traverse the follicular stalk and end in the walls; many are efferent and end on the blood vessels and smooth muscles of the follicles, but afferent fibers may also be present since structures resembling mammalian receptors have been identified. Dahl (1970b) also has studied the innervation of the follicle in the fowl, by electron microscopy, and has found that the theca interna is extremely well innervated, in particular this applies to the so-called thecal glands. Myelinated and nonmyelinated fibers are seen, some of which are said to be in "membranous contact" with the presumed steroidogenic cells of the thecal gland.

These thecal glands have recently been described in considerable detail by Dahl (1970a). They consist of lipid-rich cells and "enclosing" cells constituted to form spherical, oval, or elongated islets located mainly in the theca interna but also in the interfollicular stroma. The lipidic cells are said to have the fine-structural characteristics of steroid-secreting cells, to be organized like endocrine glands, and to be extremely well innervated (Dahl, 1970b).

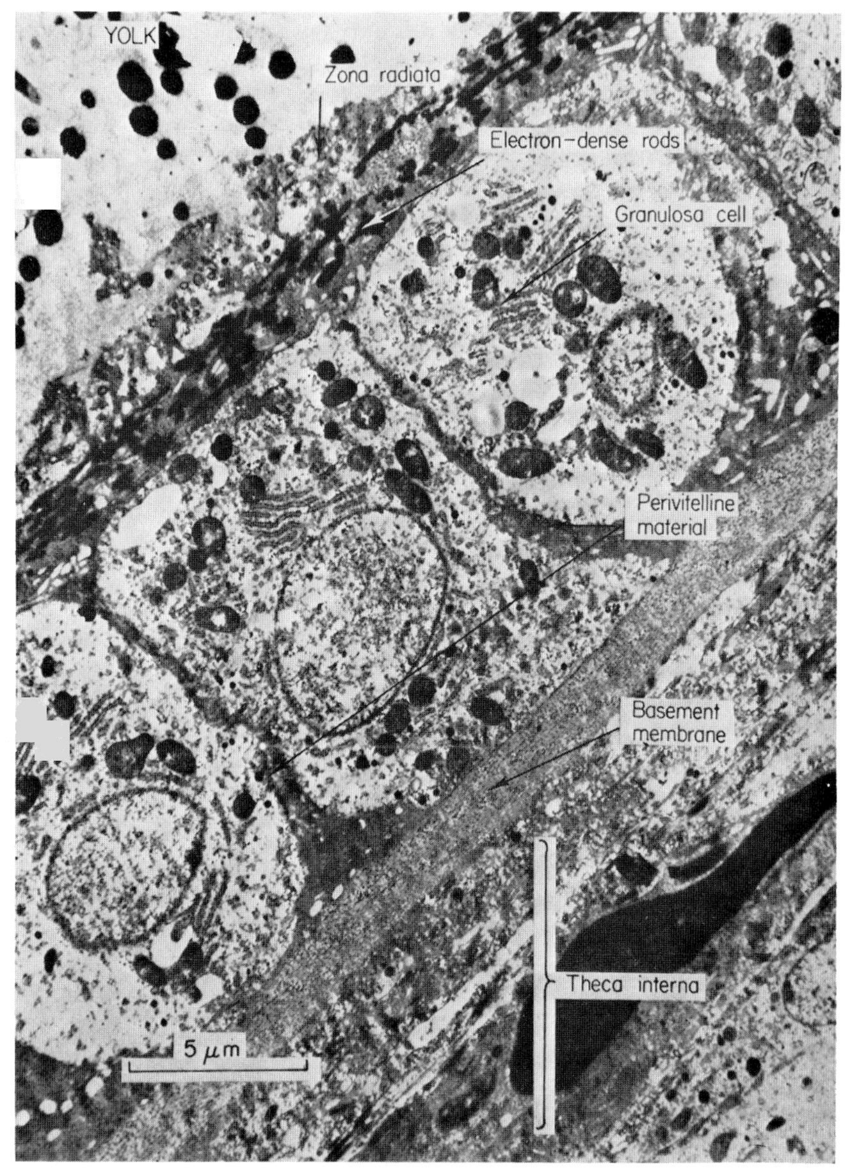

Fig. 14. Fine structure of the follicular wall of the fowl oocyte. (Reproduced by permission from Gilbert, 1971.)

Corpora Atretica and Corpora Lutea

In addition to oocytes which have a future there are many in birds as in most other vertebrates which become atretic. Brambell (1925) found that most previtellogenic follicles in the fowl become atretic before they reach a diameter of 400 μm, and Stieve (1919) estimated that not more than 0.01% of the oocytes of the jackdaw *Colaeus monedula* are ovulated. Brambell (1956) has reviewed the early literature on follicular atresia in birds and shown that the course it takes depends largely on the size of the follicle at the start of atresia. The small ones are resorbed *in situ*, some of the larger ones may rupture into the stromal lacunae, the resulting debris being removed by phagocytes and leucocytes, and others, the true corpora atretica, appear to be formed by invasion and phagocytosis of the developing egg by the follicular epithelium.

Postovulatory structures of doubtful status are found in the bird ovary but are transient. The literature on their formation and structure has been reviewed by Brambell (1956), and Stieve (1918) has described the corpora lutea of the hen as being cup shaped and stalked with a slit-like opening opposite the stalk; he states that it survives ovulation by only 10 to 14 days and appears to arise by contraction of the follicular capsule after ovulation, not by the genesis of new cells. More recently Floquet and Grignon (1964) have reported histochemical studies on the postovulatory follicle in the hen. Lipids appear in it, though they are transitory, and it is suggested that these and other signs of secretory activity may indicate that the corpora lutea play a part in the transit of the eggs through the oviducts. However, Wyburn *et al.* (1966) state that there is no corpus luteum in the hen though cells of the granulosa and theca interna become vacuolated and can be recognized up to 72 hours after ovulation.

IV. CONCLUDING REMARKS

This review illustrates the remarkable range of ovarian morphology in lower vertebrates, compared with that of mammals, but it also illustrates a consonance of basic structures that is equally remarkable. The only important differences between the ovaries of nonmammalian vertebrates are due to embryological causes, as when the medullary component appears to be absent from the gonadal primordia and, to a lesser extent, to the fate of this component when present. All the other differences arise from the amount of yolk present in the eggs. Differences between the homologies and significance of corpora atretica and corpora lutea in mammalian and non-mammalian animals are more difficult to resolve, but atresia appears to be an almost universal feature in vertebrate ovaries and any differences in the products of atresia may be due to the amount of yolk encountered during the process; the end result is the same. Corpora lutea, on the other hand, pose a more difficult problem. In mammals their formation is obviously under pituitary control and they have an important endocrine function. In the nonmammalian vertebrates neither pituitary involvement in their formation nor an endocrine function have as yet been directly demonstrated.

REFERENCES

Altland, P. D. (1951). Observations on the structure of the reproductive organs of the box turtle. *J. Morphol.* **89,** 599–621.

Anderson, E. (1967). The formation of the primary envelope during oocyte differentiation in teleosts. *J. Cell Biol.* **35,** 193–212.

Anderson, E., and Chomyn, E. (1964). Fine structure of Bidder's organ and young oocytes from two species of toads, *Bufo americanus* and *B. marinus*. *Anat. Rec.* **148,** 254–255.

Anthony, J., and Millot, J. (1972). Première capture d'une femelle de Coelacanthe en état de maturité sexuelle. *C. R. Acad. Sci. Ser. D* **274,** 1925–1926.

Atz, J. W. (1964). Intersexuality in fishes. *In* "Intersexuality in Vertebrates Including Man" (C. N. Armstrong and A. J. Marshall, eds.), pp. 145–232. Academic Press, New York.

Badir, N. (1968). Structure and function of corpus luteum during gestation in the viviparous lizard *Chalcides ocellatus*. *Anat. Anz.* **122,** 1–10.

Bara, G. (1965a). Histochemical localization of Δ^5-3β-hydroxysteroid dehydrogenase in the ovaries of a teleost fish *Scomber scomber* L. *Gen. Comp. Endocrinol.* **5,** 284–296.

Bara, G. (1965b). Glucose-6-phosphate dehydrogenase activity in the ovaries of *Scomber scomber* L. *Experientia* **21,** 638–640.

Barr, W. A. (1963). Endocrine control of the sexual cycle in the plaice *Pleuronectes platessa*. 1. Cyclical changes in the normal ovary. *Gen. Comp. Endocrinol.* **3,** 197–204.

Beach, A. W. (1959). Seasonal changes in the cytology of the ovary and of the pituitary gland of the goldfish. *Can. J. Zool.* **37,** 615–626.

Bellairs, R. (1965). The relationship between oocyte and follicle in the hen's ovary as shown by electron microscopy. *J. Embryol. Exp. Morphol.* **13,** 215–233.

Bellairs, R. (1967). Aspects of the development of yolk spheres in the hen's oocyte, studied by electron microscopy. *J. Embryol. Exp. Morphol.* **17,** 267–281.

Bellairs, R., Harkness, M., and Harkness, R. D. (1963). The vitelline membrane of the hen's egg: a chemical and electron microscopical study. *J. Ultrastruct. Res.* **8,** 339–359.

Benoit, J. (1923). A-propos du changement expérimental de sexe par ovariotomie chez la poule. (Présentation d'animaux et de preparations microscopiques.) *C. R. Soc. Biol.* **89,** 1326–1328.

Benoit, J. (1950). Organes uro-genitaux. *In* "Traité de Zoologie" (P.-P. Grassé, ed.), Vol. XV, pp. 341–377. Masson, Paris.

Betz, T. W. (1963a). The gross ovarian morphology of the diamond-backed water snake, *Natrix rhombifera,* during the reproductive cycle. *Copeia* pp. 692–697.
Betz, T. W. (1963b). The ovarian histology of the diamond-backed water snake, *Natrix rhombifera* during the reproductive cycle. *J. Morphol.* **113,** 245–260.
Bhattacharya, D. R. (1925). "Les inclusions cytoplasmiques dans l'oogénèse de certains reptiles. Ph.D.-Thesis, Faculty of Science, University of Paris.
Botte, V. (1963). Osservazioni istologiche ed istochimiche sui follicoli post-ovulatore e atresici di *Raja spp. Acta med. Romana* **1,** 117–123.
Bowers, A. B., and Holliday, F. G. T. (1961). Histological changes in the gonad associated with the reproductive cycle of the herring *Clupea harengus* L. *Mar. Res. Ser. Scott. Home Dep.* **5,** 1–16.
Boyd, M. M. (1940). The structure of the ovary and the formation of the corpus luteum in *Hoplodactylus maculatus* Gray. *Q. J. Microsc. Sci.* [N.S.] **82,** 337–376.
Braekevelt, C. R., and McMillan, D. B. (1967). Cyclic changes in the ovary of the brook stickleback *Eucalia inconstans* (Kirtland). *J. Morphol.* **123,** 373–396.
Bragdon, D. E. (1951). The non-essentiality of the corpora lutea for the maintenance of gestation in certain live-bearing snakes. *J. Exp. Zool.* **118,** 419–435.
Bragdon, D. E. (1952). Corpus luteum formation and follicular atresia in the common garter snake, *Thamnophis sirtalis. J. Morphol.* **91,** 413–445.
Brambell, F. W. R. (1925). The oogenesis of the fowl (*Gallus bankiva*). *Philos. Trans. R. Soc. London, Ser. B* **214,** 113–151.
Brambell, F. W. R. (1956). Ovarian changes. *In* "Marshall's Physiology of Reproduction" (A. S. Parkes, ed.), 3rd ed, Vol. 1, Part I, pp. 397–542. Longmans Green, London and New York.
Brambell, F. W. R., and Marrian, G. F. (1929). Sex reversal in a pigeon (*Columba livia*). *Proc. R. Soc. London, Ser. B* **104,** 459–470.
Breneman, W. R. (1955). Reproduction in birds: the female. *Mem. Soc. Endocrinol.* **4,** 94–113.
Bretschneider, L. H., and Duyvené de Wit, J. J. (1947). "Sexual Endocrinology of Non-Mammalian Vertebrates." Elsevier, Amsterdam.
Brock, J. (1878). Beiträge zur Anatomie und Histologie der Geschlechtsorgane der Knochenfische. *Morphol. Jahrb.* **4,** 505–572.
Brode, M. D. (1928). The significance of the asymmetry of the ovaries of the fowl. *J. Morphol.* **46,** 1–57.
Busson-Mabillot, S. (1966). Présence d'une thèque interne, glandulaire dans le follicule primaire de la lamproie de Planer (*Lampetra planeri,* Bloch) vertébré cyclostome. *C. R. Acad. Sci. Ser. D* **262,** 117–118.
Busson-Mabillot, S. (1967a). Structure ovarienne de la Lamproie adulte (*Lampetra planeri* Bloch). I. Zone pellucide, Morphogenèse et constitution chimique. *J. Microsc.* (*Paris*) **6,** 577–598.
Busson-Mabillot, S. (1967b). Structure ovarienne de la lamproie adulte (*Lampetra planeri* Bloch). *J. Microsc.* (*Paris*) **6,** 807–838.
Callan, H. G., and Lloyd L. (1960). Lampbrush chromosomes of crested newts *Triturus cristatus* (Laurenti). *Philos. Trans. R. Soc. London, Ser. B* **243,** 135–219.
Callan, H. G., and Tomlin, S. G. (1950). Experimental studies on amphibian oocyte nuclei. I. Investigation of the structure of the nuclear membrane by means of the electron microscope. *Proc. R. Soc. London, Ser. B* **137,** 367–378.
Chan, S. T. H. (1970). Natural sex reversal in vertebrates. *Philos. Trans. R. Soc. London, Ser. B* **259,** 59–71.

Chan, S. T. H., and Phillips, J. G. (1967). The structure of the gonad during natural sex reversal in *Monopterus albus* (Pisces; Teleostei). *J. Zool.* **151,** 129–141.

Chan, S. T. H., Wright, A., and Phillips, J. G. (1967). The atretic structures in the gonads of the rice-field eel (*Monopterus albus*) during natural sex-reversal. *J. Zool.* **153,** 527–539.

Chieffi, G. (1949). Richerche sul differenziamento dei sessi negli embrioni di *Torpedo ocellata. Pubbl. Stn. Zool. Napoli* **22,** 57–78.

Chieffi, G. (1961). La luteogenesi nei Selaci ovovivipari. I. Richerche istologische e istochimiche in *Torpedo marmorata* e *Torpedo ocellata. Pubbl. Stn. Zool. Napoli* **32,** 145–166.

Chieffi, G. (1962). Endocrine aspects of reproduction in elasmobranch fishes. *Gen. Comp. Endocrinol., Suppl. 1,* 275–285.

Cole, M. B., Jr. (1969). Ultrastructural cytochemistry of granules associated with the nuclear pores of frog oocytes. *J. Cell Biol.* **43,** 24a.

Conel, J. Le Roy (1917). The urogenital system of Myxinoids. *J. Morphol.* **29,** 75–163.

Craig-Bennett, A. (1931). The reproductive cycle of the three-spined stickleback *Gasterosteus aculeatus* L. *Philos. Trans. R. Soc. London, Ser. B* **219,** 197–279.

Crew, F. A. E. (1923). Studies in intersexuality. II. Sex-reversal in the fowl. *Proc. R. Soc. London, Ser. B* **95,** 256–278.

Cunningham, J. T. (1886). On the structure and development of the reproductive elements in *Myxine glutinosa* L. *Q. J. Microsc. Sci.* [N.S.] **27,** 49–76.

Cunningham, J. T., and Smart, W. A. M. (1934). The structure and origin of the corpora lutea in some of the lower vertebrata. *Proc. R. Soc. London, Ser. B* **116,** 258–281.

Dahl, E. (1970a). Studies of the fine structure of ovarian interstitial tissue. 2. The ultrastructure of the thecal gland of the domestic fowl. *Z. Zellforsch. Mikrosk. Anat.* **109,** 195–211.

Dahl, E. (1970b). Studies on the fine structure of ovarian interstitial tissue. 3. The innervation of the thecal gland of the domestic fowl. *Z. Zellforsch. Mikrosk. Anat.* **109,** 212–226.

Dean, B. (1906). Chimaeroid fishes and their development. *Carnegie Inst. Washington Publ.* **32,** 195p.

Dodd, J. M. (1960). Gonadal and gonadotrophic hormones in lower vertebrates. *In* "Marshall's Physiology of Reproduction" (A. S. Parkes, ed.), 3rd ed., Vol. 1, Part 2, pp. 417–582. Longmans Green, London and New York.

Dodd, J. M. (1972). Ovarian control in cyclostomes and elasmobranchs. *Am. Nat.* **12,** 325–339.

Domm, L. V. (1924). Sex-reversal following ovariotomy in the fowl. *Proc. Soc. Exp. Biol. Med.* **22,** 28–35.

Domm, L. V. (1939). Modifications in sex and secondary sexual characters in birds. *In* "Sex and Internal Secretions" (E. Allen, C. H. Danforth, and E. A. Doisy ed.), 2nd ed., p. 227–327. Williams and Wilkins, Baltimore, Maryland.

Droller, M. J., and Roth, T. F. (1966). An electron microscope study of yolk formation during oogenesis in *Lebistes reticulatus* (Guppy). *J. Cell Biol.* **28,** 209–232.

Dumont, J. N. (1972). Oogenesis in *Xenopus laevis* (Daudin) 1. Stages of oocyte development in laboratory maintained animals. *J. Morphol.* **136,** 153–180.

Dunn, R. S., and Tyler, A. V. (1969). Aspects of the anatomy of the winter flounder ovary with hypotheses on oocyte maturation time. *J. Fish. Res. Board Can.* **26,** 1943–1947.

Dutt, N. H. G. (1964). Yolk-nucleus in the oocytes of *Anabas scandens* (Teleostei). *Q. J. Microsc. Sci.* [N.S.] **105,** 349–352.
Evennett, P. J. (1963). "The endocrine control of reproduction in the river lamprey *Lampetra fluviatilis* L." Ph.D. Thesis, University of St. Andrews.
Eyeson, K. N. (1971). Pituitary control of ovarian activity in the lizard, *Agama agama. J. Zool.* **165,** 367–372.
Floquet, A., and Grignon, G. (1964). Etude histologique de follicule post-ovulatoire chez la poule. *C. R. Soc. Biol.* **158,** 132–135.
Flügel, H. (1964a). On the fine structure of the zona radiata of growing trout oocytes. *Naturwissenschaften* **51,** 542.
Flügel, H. (1964b). Electron microscopic investigations on the fine structure of the follicular cells and the zona radiata of trout oocytes during and after ovulation. *Naturwissenschaften* **51,** 564–565.
Flügel, H. (1964c). Desmosomes in the follicular epithelium of growing oocytes of the eastern brook trout *Salvelinus fontinalis* (electron microscopic investigations). *Naturwissenschaften* **51,** 566.
Flügel, H. (1967). Licht und elecktronenmikroskopische Untersuchungen an Oozyten und Eiern einiger Knochenfische. *Z. Zellforsch. Mikrosk. Anat.* **83,** 82–116.
Gilbert, A. B. (1965). Innervation of the ovarian follicle of the domestic hen. *Q. J. Exp. Physiol. Cogn. Med. Sci.* **50,** 437–445.
Gilbert, A. (1969). Innervation of the ovary of the domestic hen. *Q. J. Exp. Physiol. Cogn. Med. Sci.* **54,** 404–411.
Gilbert, A. B. (1971). The ovary. *In* "Physiology and Biochemistry of the Domestic Fowl" (D. J. Bell and B. M. Freeman, eds.), Vol. 3, pp. 1163–1208. Academic Press, New York.
Gokhale, S. V. (1957). Seasonal histological changes in the gonads of the whiting *Gadus merlangus L.* and the Norway pout *G. esmarkii* Nilsson. *Indian J. Fish.* **4,** 92–112.
Götting, K. J. (1967). Der Follikel und die peripheren strukturen der Oocyten der Teleosteer und Amphibien Eine vergleichende Betrachtung auf der Grundlage elektronenmikroskopischer Untersuchungen. *Z. Zellforsch. Mikrosk. Anat.* **79,** 481–491.
Greenfield, M. L. (1966). Oocyte of the domestic chicken shortly after hatching, studied by electron microscopy. *J. Embryol. Exp. Morphol.* **15,** 297–316.
Guraya, S. S. (1965). A histochemical study of follicular atresia in the snake ovary. *J. Morphol.* **117,** 151–170.
Guraya, S. S. (1968). Histochemical study of granulosa (follicular) cells in the preovulatory and postovulatory follicles of amphibian ovary. *Gen. Comp. Endocrinol.* **10,** 138–146.
Guraya, S. S. (1969a). Histochemical observations on the corpora atretica of amphibian ovary. *Gen. Comp. Endocrinol.* **12,** 165–167.
Guraya, S. S. (1969b). Histochemical observations on the interstitial gland cells of amphibian ovary (frog, toad). *Gen. Comp. Endocrinol.* **13,** 173–178.
Hann, H. W. (1927). The history of the germ cells of *Cottus bairdii,* Girard. *J. Morphol.* **43,** 427–497.
Harrison, R. J. (1948). The development and fate of the corpus luteum in the vertebrate series. *Biol. Rev. Cambridge Philos. Soc.* **23,** 269–331.
Henderson, N. E. (1963). Extent of atresia in mature ovaries of the Eastern brook trout, *Salvelinus fontinalis. J. Fish. Res. Board Can.* **20,** 899–908.

Hett, J. (1923a). Das Corpus luteum der Dohle (*Colaeus monedula*). *Arch. Mikrosk. Anat. Entwicklungs Mech.* **97,** 718–838.

Hett, J. (1923b). Das Corpus luteum des Molches (*Triton vulgaris*). *Z. Anat. Entwicklungsgesch.* **68,** 243–272.

Hett, J. (1924). Das Corpus luteum der Zauneidechse (*Lacerta agilis*). *Z. Mikrosk. Anat. Forsch.* **1,** 41–48.

Hisaw, F. L., and Hisaw, F. L., Jr. (1959). Corpora lutea of elasmobranch fishes. *Anat. Rec.* **135,** 269–277.

Hoar, W. S. (1957). The gonads and reproduction. *In* "The Physiology of Fishes" (M. E. Brown, ed.), Vol. 1, pp. 287–321. Academic Press, New York.

Hoar, W. S. (1965). Hormones and reproduction in fishes. *Annu. Rev. Physiol.* **27,** 51–70.

Hoar, W. S. (1969). Reproduction. *In* "Fish Physiology" (W. S. Hoar and D. J. Randall, eds.), Vol. 3, pp. 1–72. Academic Press, New York.

Hope, J., Humphries, A. A., Jr., and Bourne, G. H. (1963). Ultrastructural studies on developing oocytes of the salamander *Triturus viridescens*. 1. The relationship between follicle cells and developing oocytes. *J. Ultrastruct. Res.* **9,** 302–324.

Hope, J., Humphries, A. A., Jr., and Bourne, G. H. (1964a). Ultrastructural studies on developing oocytes of the salamander *Triturus viridescens*. II. The formation of yolk. *J. Ultrastruct. Res.* **10,** 547–556.

Hope, J., Humphries, A. A., Jr., and Bourne, G. H. (1964b). Ultrastructural studies on developing oocytes of the salamander *Triturus viridescens*. III. Early cytoplasmic changes and the formation of pigment. *J. Ultrastruct. Res.* **10,** 557–566.

Humeau, C., and Sentein, P. (1968). Rapports entre les cellules folliculeuses et l'ovocyte en début d'atresie chez *Triturus helveticus* Raz. *C. R. Seances Soc. Biol. Ses Fil.* **162,** 2181–2184.

Hurley, D. A., and Fisher, K. C. (1966). The structure and development of the external membrane in young eggs of the brook trout, *Salvelinus fontinalis* (Mitchill). *Can. J. Zool.* **44,** 173–190.

Hutt, F. B. (1949). "Genetics of the Fowl." McGraw-Hill, New York.

Jollie, W. P. (1964). Fine structure of the ovarian pseudoplacenta in *Lebistes reticulatus*. *Anat. Rec.* **148,** 295–296.

Jollie, W. P., and Jollie, L. G. (1964a). The fine structure of the ovarian follicle of the ovoviviparous poeciliid fish *Lebistes reticulatus*. I. Maturation of follicular epithelium. *J. Morphol.* **114,** 479–502.

Jollie, W. P., and Jollie, L. G. (1964b). The fine structure of the ovarian follicle of the ovoviviparous poeciliid fish *Lebistes reticulatus*. II. Formation of follicular pseudoplacenta. *J. Morphol.* **114,** 503–526.

Joly, J., and Picheral, B. (1972). Ultrastructure, histochimie et physiologie du follicule pre-ovulatoire et du corps jaune de l'urodèle ovo-vivipare *Salamandra salamandra* (L.). *Gen. Comp. Endocrinol.* **18,** 235–259.

Kemp, N. E. (1953). Synthesis of yolk in oocytes of *Rana pipiens* after induced ovulation. *J. Morphol.* **92,** 487–511.

Kemp, N. E. (1958). Protoplasmic bridges between oocytes and follicle cells in vertebrates. *Anat. Rec.* **130,** 324.

Kemp, N. E., and Allen, M. D. (1956). Electron microscopic observations on the development of the chorion of *Fundulus*. *Biol. Bull.* (*Woods Hole, Mass.*) **111,** 293.

Kessel, R. G. (1969). Cytodifferentiation in the *Rana pipiens* oocyte. 1. Association between mitochondria and nucleolus-like bodies in young oocytes. *J. Ultrastruct. Res.* **28,** 61–77.

Knowles, F. G. W. (1939). The influence of anterior-pituitary and testicular hormones on the sexual maturation of lampreys. *J. Exp. Biol.* **16,** 535–547.

Kornfeld, W. (1960). Experimentally induced proliferation of the rudimentary gonad of an intact domestic fowl. *Nature* (*London*) **185,** 320.

Lambert, J. G. D. (1969). "Steroidproductie in het ovarium van *Poecilia reticulata.*" Thesis, University of Utrecht.

Lambert, J. G. D. (1970a). The ovary of the guppy *Poecilia reticulata.* The granulosa cells as sites of steroid biosynthesis. *Gen. Comp. Endocrinol.* **15,** 464–476.

Lambert, J. G. D. (1970b). The ovary of the guppy *Poecilia reticulata.* The atretic follicle, a *corpus atreticum* or a *corpus luteum praeovulationis. Z. Zellforsch. Mikrosk. Anat.* **107,** 54–67.

Lamotte, M., and Rey, P. (1954). Existence de corpora lutea chez un Batracien anoure vivipare. *Nectophrynoides occidentalis* Angel; leur évolution morphologique, *C. R. Acad. Sci.* (*Paris*), **238,** 393–395.

Lanzing, W. J. R. (1959). "Studies on the River Lamprey, *Lampetra fluviatilis* during its Anadromous Migration." Thesis, University of Utrecht.

Lavett Smith, C., Rand, C. S., Schaeffer, B., and Atz, J. W. (1975). *Latimeria,* the living Coelacanth, is ovoviviparous. *Science* **190,** 1105–1106.

Lehri, G. K. (1968). Cyclical changes in the ovary of the catfish *Clarias batrachus. Acta Anat.* **69,** 105–124.

Lewis, J. C., and McMillan, D. B. (1965). The development of the ovary of the sea lamprey (*Petromyzon marinus* L.). *J. Morphol.* **117,** 425–466.

Lorenz, F. W. (1969). Reproduction in domestic fowl. *In* "Reproduction in Domestic Animals" (H. H. Cole and P. T. Cupps, eds.), 2nd ed., pp. 569–608. Academic Press, New York.

Loyez, M. (1906). Recherches sur le développement ovarien des oeufs méroblastiques à vitellus nutritif abondant. *Arch. Anat. Microsc. Morphol. Exp.* **8,** 69–397.

Lyngnes, R. (1931). Über atretische und hypertrophische Gebilde im ovarium der *Myxine glutinosa. Z. Morphol. Öekol. Tiere* **19,** 591–608.

Lyngnes, R. (1936). Rückbildung der ovuliertung und nicht ovulierten Follikel im ovarium der *Myxine glutinosa* L. *Skr. Nor. Vidensk-Akad. Oslo,* **4,** 1–116.

Matthews, L. H. (1950). Reproduction in the basking shark *Cetorhinus maximus* (Gunner). *Philos. Trans. R. Soc. London, Ser. B* **234,** 247–316.

Mendoza, G. (1940). The reproductive cycle of the viviparous teleost, *Neotoca bilineata,* a member of the family Goodeidae. II. The cyclic changes in the ovarian soma during gestation. *Biol. Bull.* (*Woods Hole, Mass.*) **84,** 87–97.

Merriam, R. W. (1961). On the fine structure and composition of the nuclear envelope. *J. Biophys. Biochem. Cytol.* **11,** 559–570.

Metten, H. (1939). Studies on the reproduction of the dogfish. *Philos. Trans. R. Soc. London, Ser. B* **230,** 217–238.

Moser, H. G. (1967). Seasonal histological changes in the gonads of *Sebastodes paucispinis,* an ovoviviparous teleost. *J. Morphol.* **123,** 329–354.

Nayyar, R. P. (1964). The yolk nucleus of fish oocytes. *Q. J. Microsc. Sci.* [N.S.] **105,** 353–358.

Neaves, W. B. (1971). Intercellular bridges between follicle cells and oocyte in the lizard *Anolis carolinensis*. *Anat. Rec.* **170**, 285–302.
Nelsen, O. E. (1953). "Comparative Embryology of the Vertebrates," Chapter 2. McGraw-Hill (Blakiston), New York.
Nicholls, T. J., and Maple, G. (1972). Ultrastructural observations on possible sites of steroid biosynthesis in the ovarian follicular epithelium of two species of cichlid fish, *Cichlasoma nigrofasciatum* and *Haplochromis multicolor*. *Z. Zellforsch. Mikrosk. Anat.* **128**, 317–335.
Norman, J. R., and Greenwood, P. H. (1963). "A History of Fishes." Ernest Benn Ltd., London.
Okkelberg, P. (1921). The early history of the germ cells in the brook lamprey *Entosphenus wilderi* (Gage), up to and including the period of sexual differentiation. *J. Morphol.* **35**, 1–151.
Panigel, M. (1956). Contribution à l'étude de l'ovoviviparité chez les reptiles; gestation et parturition chez le lézard vivipare *Zootoca vivipara*. *Ann. Sci. Nat., Zool. Biol. Anim.* [11] **18**, 569–668.
Pearl, R., and Schoppe, W. F. (1921). Studies on the physiology of reproduction in the domestic fowl. XVIII. Further observations on the anatomical basis of fecundity. *J. Exp. Zool.* **34**, 101–118.
Phillips, R. E., and Warren, D. C. (1937). Observations concerning the mechanics of ovulation in the fowl. *J. Exp. Zool.* **76**, 117–136.
Polder, J. J. W. (1961). Cyclical changes in testis and ovary related to maturity stages in the North sea herring *Clupea harengus L. Arch. Neerl. Zool.* **14**, 45–60.
Polder, J. J. W. (1964). Occurrence and significance of atretic follicles (pre-ovulatory corpora lutea) in ovaries of the bitterling, *Rhodeus amarus* (Bloch). *Proc. K. Ned. Akad. Wet., Ser. C* **67**, 218–222.
Ponse, K. (1924). L'organe de Bidder et le déterminisme des caractères sexuels secondaires du crapaud (*Bufo vulgaris* L.). *Rev. Suisse Zool.* **31**, 177–336.
Ponse, K. (1949). "La Différenciation du Sexe et l'Intersexualité Chez les Vertébrés." Lausanne.
Rahn, H. (1939). Structure and function of placenta and corpus luteum in viviparous snakes. *Proc. Soc. Exp. Biol. Med.* **40**, 381–382.
Rai, B. P. (1966). Corpora atretica and the so-called corpora lutea in the ovary of *Tor* (*Barbus*) *tor*. *Anat. Anz.* **119**, 459–465.
Rajalakshmi, M. (1966). Atresia of oocytes and ruptured follicles in *Gobius giuris*. *Gen. Comp. Endocrinol.* **6**, 378–385.
Rastogi, R. K. (1969). Studies on the fish oogenesis. 3. Vitellogenesis in some freshwater teleosts. *Anat. Anz.* **125**, 24–36.
Raynaud, A., and Raynaud, J. (1961). Modifications de structure présentées par des ovaires d'Orvet (*Anguis fragilis* L.) au cours de la gestation; leur signification. *Bull. Soc. Zool. Fr.* **86**, 29–38.
Reinboth, R. (1962). Morphologische und funktionelle zweigeschlechtlichkeit bei marinen Teleostiern (Serranidae, Sparidae, Centracanthidae, Labridae). *Zool. Jahrb., Abt. All. Zool. Physiol. Tiere* **69**, 405–480.
Remacle, C. (1970). Contribution à l'étude de la sexualité chez certains labridae et sparidae (téléostéens perciformes). *Bull. Inst. R. Sci. Nat. Belg.* **46**, 1–13.
Retzius, G. (1912). Zur kenntnis der Hüllen und besonders des Follikelepithels an den Eiern der Wirbeltiere. *Biol. Untersuch.* **17**, 1–52.

Romanoff, A. L., and Romanoff, A. J. (1949). "The Avian Egg." Wiley, New York.

Rugh, R. (1935). Ovulation in the frog. 1. Pituitary relations in induced ovulation. *J. Exp. Zool.* **71,** 149–162.

Samuel, M. (1943). Studies on the corpus luteum of *Rhinobatus granulatus* Cuv. *Proc. Indian Acad. Sci. Sect. B* **18,** 133–162.

Schreiner, K. E. (1955). Studies on the gonad of *Myxine glutinosa* L. *Univ. Bergen Arbok Naturvitensk. Rekke* **8,** 1–36.

Smith, C. L. (1955). Reproduction in female amphibia. *Mem. Soc. Endocrinol.* **4,** 39–56.

Stahl, A., and Leray, C. (1961). L'ovogénèse chez les poissons téléostéens. I. Origine et signification de la zona radiata et de ses annexes. *Arch. Anat. Microsc. Morphol. Exp.* **50,** 251–268.

Stanley, A. J. (1937). Sexual dimorphism in North American hawks. 1. Sex organs. *J. Morphol.* **61,** 321–349.

Stanley, A. J., and Witschi, E. (1940). Germ cell migration in relation to asymmetry in the sex glands of hawks. *Anat. Rec.* **76,** 329–342.

Stanley, H. P. (1963). Urogenital morphology in the chimaeroid fish *Hydrolagus colliei* (Lay and Bennett). *J. Morphol.* **112,** 99–128.

Stieve, H. (1918). Über experimentell durch veränderte äussere Bedingungen hervorgerufene Rückbildungsvorgänge am Eierstock das Haushuhnes (*Gallus domesticus*). *Arch. Entwicklungsmech. Org.* **44,** 531–588.

Stieve, H. (1919). Die Entwicklung des Eierstockseies der Dohle (*Colaeus monedula*). *Arch. Mikrosk. Anat. Entwicklungsmech.* **92,** 137–288.

Stolk, A. (1951). Histo-endocrinological analysis of gestation phenomena in the cyprinodont *Lebistes reticulatus* Peters. IV. The oocyte cycle during pregnancy. *Proc. K. Ned. Akad. Wet. Ser. C* **54,** 574–578.

Sturkie, P. D. (1954). "Avian Physiology." Cornell Univ. Press (Comstock), Ithaca, New York.

Takano, K. (1964). On the egg formation and the follicular changes in *Lebistes reticulatus*. *Bull. Fac. Fish. Hokkaido Univ.* **15,** 147–155.

Turner, C. L. (1933). Viviparity superimposed upon ovo-viviparity in the Goodeidae, a family of cyprinodont teleost fishes of the mexican plateau. *J. Morphol.* **55,** 207–251.

Turner, C. L. (1937). Reproductive cycles and superfetation in poeciliid fishes. *Biol. Bull* (*Woods Hole, Mass.*) **72,** 145–164.

Turner, C. L. (1940). Superfetation in viviparous cyprinodont fishes. *Copeia* pp. 88–91.

Turner, C. L. (1947). Viviparity in teleost fishes. *Sci. Mon.* **65,** 508–518.

Vilter, V., and Lugand, A. (1959). Recherches sur le déterminisme interne et externe du corps jaune gestatif chez le crapaud vivipare du Mont Nimba, le *Nectophrynoides occidentalis* Ang., de la Haute Guinée. *C. R. Soc. Biol.* **153,** 294–297.

Wallace, W. (1903). Observations on ovarian ova and follicles in certain teleostean and elasmobranch fishes. *Q. J. Microsc. Sci.* [N.S.] **47,** 161–213.

Walvig, F. (1963). The gonads and the formation of the sexual cells. *In* "The Biology of *Myxine*" (A. Brodal and R. Fänge, eds.), pp. 530–580. Oslo Univ. Press, Oslo.

Ward, R. T. (1962). The origin of protein and fatty yolk in *Rana pipiens*. II. Electron microscopic and cytochemical observations of young and mature oocytes. *J. Cell Biol.* **14,** 309–341.

Wartenberg, H. (1962). Elektronenmikroskopische und histochemische studien über die Oogenese der Amphibieneizelle. *Z. Zellforsch. Mikrosk. Anat.* **58,** 427–486.

Wartenberg, H., and Gusek, W. (1960). Electronenoptische Untersuchungen über die Feinstruktur des Ovarialeies und des Follikelepithels von Amphibien. *Exp. Cell Res.* **19,** 199–209.

Weekes, H. C. (1934). The corpus luteum in certain oviparous and viviparous reptiles. *Proc. Linn. Soc. N.S.W.* **59,** 380–391.

Wiebe, J. P. (1968). The reproductive cycle of the viviparous sea perch *Cymatogaster aggregata* Gibbons. *Can. J. Zool.* **46,** 1221–1234.

Wyburn, G. M., Aitken, R. N. C., and Johnston, H. S. (1965). The ultrastructure of the zona radiata of the ovarian follicle of the domestic fowl. *J. Anat.* **99,** 469–484.

Wyburn, G. M., Johnston, H. S., and Aitken, R. N. C. (1966). Fate of the granulosa cells in the hen's follicle. *Z. Zellforsch. Mikrosk. Anat.* **72,** 53–65.

Xavier, F., Zuber-Vogeli, M., and Le Quang Trong, Y. (1970). Recherches sur l'activité endocrine de l'ovaire de *Nectophrynoides occidentalis* Angel (Amphibien Anoure vivipare). 1. Etude histochimique. *Gen. Comp. Endocrinol.* **15,** 425–431.

Yamamoto, M. (1963a). Electron microscopy of fish development. II. Oocyte-follicle cell relationship and formation of chorion in *Oryzias latipes. J. Fac. Sci., Univ. Tokyo, Sect. 4* **10,** 123–127.

Yamamoto, K. (1963b). Cyclical changes in the wall of the ovarian lumen in the medaka, *Oryzias latipes. Annot. Zool. Jpn.* **36,** 179–186.

Yammamoto, K., and Shirai, K. (1962). Origin of the yearly crop of eggs in the bitterling, *Rhodeus ocellatus. Annot. Zool. Jpn.* **35,** 218–222.

Yamamoto, K. and Yamazaki, F. (1961). Rhythm of development in the oocyte of the goldfish *Carassius auratus. Bull. Fac. Fish. Hokkaido Univ.* **12,** 93–110.

Yamazaki, F. (1965). Endocrinological studies on the reproduction of the female goldfish *Carassius auratus* L. with special reference to the function of the pituitary gland. *Mem. Fac. Fish., Hokkaido Univ.* **13,** 1–64.

6

Ovulation and Atresia

Barbara J. Weir and I. W. Rowlands

I. INTRODUCTION

Ovulation and atresia were reviewed separately in the first edition of this book. These two topics are presented together in this chapter to emphasize views that have been elaborated during the past decade on the fate of the ovarian follicle in vertebrates. Although ovulation, the release of an egg by the follicle, has been extensively studied and, in mammals at least, marks the transition between the follicular and luteal phases of the ovarian cycle, only a relatively small number of follicles reach this stage: "en un mot, l'atrésie phenoméne anormal est la règle; la ponte, processus physiologique, l'exception" (Branca, 1925). In many vertebrates most follicles degenerate or undergo atresia rather than ovulate, but in mammals and some nonmammals the atretic follicles may still serve important functions relating to the endocrinology of the ovary and the well-being of the embryo. Even after 70 years (see Heape, 1905; Greenwald, 1972) it is not yet known what mechanisms determine which oocytes will become associated with incipient granulosa cells to form primary follicles (see Volume I, Chapter 2) and which follicles will ovulate.

Three main types of study have been carried out during the past 10–15 years to answer these problems: (1) the examination of the structure and morphology of follicles by electron microscopy and laparoscopy (see Volume I, Chapters 4 and 5); (2) the detection of changes at the cell level by cytochemical (see Volume I, Chapter 7) and *in vitro* (see Volume III, Chapter 6) methods; (3) the exploration of neural and hormonal mechanisms (see Volume II, Chapter 7 and Volume III, Chapters 3 and 4). Many parts of the pattern have been clarified but a theory which adequately explains all the facts has not yet been proposed. As in other reproductive processes (e.g., control of breeding, cyclicity, and maintenance of gestation—Volume II, Chapters 6 and 8; see also Brambell, 1956), it is possible that there is no single formula which applies to all members of a class and that the variations will be legion. Detailed reviews of the process of atresia (Velloso de Pinho, 1923; Marshall, 1943; Aron and Aron, 1953; Ingram, 1962) are far fewer than those of ovulation (see Hisaw, 1947; Brambell, 1956; Fraps, 1961; Nalbandov, 1961; Asdell, 1962; Blandau, 1966, 1967, 1970; Donovan, 1967; Gorski, 1968; Harris, 1969; Gilbert, 1969, 1971; Mauleon, 1969; Schwartz and Hoffmann, 1972; Espey, 1974; Bahr *et al.*, 1974; Rondell, 1974; Parr, 1975), and the factors that influence one event must inevitably be associated with those affecting the other (see Sturgis, 1961).

II. OVULATION

A. Background

Rupture of the follicle and release of the egg are controlled by gonadotropic hormones (see Young, 1961; Mauleon, 1969) and involve maturation of the oocyte, growth of the follicle, and timing in relation to estrus, which itself is brought about by equally complex hormonal changes associated with follicular activity. The structure of follicles has been discussed in Volume I, Chapters 4 and 5 and will be referred to in this chapter only when relevant to the various theories of the mechanism of ovulation (see Section II,B).

1. Oocyte Maturation

The first meiotic division of the mammalian oocyte begins during embryonic life but is arrested in late prophase and the oocytes remain in the dictyate or germinal vesicle stage until some signal is given for the oocyte to resume meiosis, leading to the extrusion of the first polar body (see Franchi *et al.*, 1962; Baker, 1972; Volume I, Chapter 2). Donahue (1972) gives a list of twenty-one mammalian species in which oocyte maturation has been investigated *in vivo* or *in vitro* to determine structural changes and timing (see also Thibault, 1972), control by gonadotropins (see also Baker and Neal, 1972), influence of follicular cells (see also Nalbandov, 1972), and metabolism.

The oocyte responds to a gonadotropic stimulus, e.g., FSH or LH, only if the follicle is of the right size (Baker and Neal, 1972; Neal and Baker, 1975), although the stimulus itself may be needed only for a short time (Ayalon *et al.,* 1972). In general, the maturation sequence to metaphase II and ovulation after such a stimulus takes between 10 and 48 hours (see Donahue, 1972). Maturation and ovulation are not, however, always associated since (1) meiosis will resume if oocytes are mechanically released from follicles (see Edwards, 1962); (2) mature oocytes may occur in follicles which fail to ovulate (Engle, 1927); and (3) apart from the occasional release of immature oocytes in many species, ovulation of oocytes in the germinal vesicle stage has been reported for dogs (Evans and Cole, 1931) and foxes (van der Stricht, 1923), and can be induced in pigs (R. H. F. Hunter, personal communication). Fertilization precedes ovulation in the centetine insectivores (Strauss, 1938) and short-tailed shrew (Pearson, 1944).

2. Follicular Growth

The relationship between the size of the oocyte and the follicle has been discussed in Volume I, Chapter 4 and has been the subject of many studies (see Brambell, 1956). The final size of a mature follicle is generally related to body size in mammals. Although there may be many well-developed follicles in the ovaries at any one time, only a characteristic number will respond to gonadotropins in each succeeding ovarian cycle (see Volume II, Chapter 6). Relatively little is known, however, of the recruitment of follicles into the various size categories or follicular growth rates. Histological studies, in which assumptions have to be made for follicle populations based on the largest follicles present at any one time (Rajakoski, 1960; Hutchinson and Robertson, 1966), have been superseded by studies with tritiated thymidine for marking populations of follicles (see Peters, 1969; Pedersen, 1972).

B. Mechanism of Rupture

The ovulatory follicle is usually, but not always, one of the largest in the ovary at the time of rupture. The changes indicating that a follicle is about to ovulate (see Walton and Hammond, 1928; Blandau, 1966) include a sudden rapid increase in size due to the secretion of secondary liquor folliculi; the dissociation, first, of the cells of the cumulus oophorus and then of the membrana granulosa; the formation of the macula pellucida—the translucent avascular area where the follicle begins to bulge at the surface of the ovary and where the stigma will form; and the loosening of the cells of the follicle wall and the tunica albuginea and surface epithelium at the impending point of rupture.

Many hypotheses have been advanced to explain the mechanism of ovulation and several excellent reviews are available (Hisaw, 1947; Blandau and Rumery, 1963; Rondell, 1970). The present account indicates the changes that occur in the mammalian follicle as it approaches ovulation and summarizes the various theories which have been related to each event.

1. Stimulus for Ovulation

Mammalian species are often distinguished as being spontaneous (e.g., rat, hamster, pig, sheep, man) or reflex (e.g., cat, rabbit, ferret) ovulators. There is, however, mounting recognition (see Aron *et al.,* 1966; Conaway, 1971; Weir and Rowlands, 1973) that many mammals cannot be placed categorically into one or other of these groups, because although members of any one species usually ovulate in one fashion, they may sometimes ovulate in the other. For example, rabbits and other reflex ovulators may ovulate without the stimulus of coitus (Fee and Parkes, 1930) and rats (Aron, 1965; Everett, 1967; Zarrow and Clark, 1968; Ying and Meyer, 1969), chinchilla (Weir, 1973), pigs (Signoret *et al.,* 1972), and probably many other spontaneous ovulators may ovulate in response to coitus if this stimulus precedes that which normally triggers the luteinizing hormone (LH) surge. The evidence for the overlap between reflex and spontaneous ovulation in various species is discussed by Jöchle (1975). Zarrow *et al.* (1970) have reported that immature female mice treated with gonadotropins will ovulate more readily if they are exposed to males, and have postulated pheromonal facilitation of ovulation in this and other spontaneous ovulators.

Investigations of the timing of the trigger and the subsequent hormonal changes have given rise to an extensive literature, mostly relating to the rat, hamster, rabbit, and man. The critical factor appears to be the release from the pituitary of the ovulation-inducing hormone which, if follicle-stimulating hormone (FSH) and LH are believed to be distinct substances (see Volume III, Chapter 4), is LH (Schwartz and McCormack, 1972), or a synergism between FSH and LH since the plasma concentrations of both hormones rise at the time of ovulation (Labhsetwar, 1974). In the rabbit, ovulation occurs 10 to 14 hours after coitus or injection of LH (Walton and Hammond, 1928; Harper, 1963). In spontaneously ovulating animals, the preovulatory surge of LH takes place during a "critical phase" and can be blocked by sedation with barbiturates (rat: Everett and Sawyer, 1950; Everett, 1956, 1972; hamster: Greenwald, 1971; Norman *et al.*, 1972; mouse: Bingel and Schwartz, 1969). Further investigations into the route whereby the pituitary is influenced with respect to the ovulatory discharge have involved electrical stimulation of different parts of the brain (Gorski, 1966; Quinn, 1969; Kawakami and Teresawa, 1970; Holsinger and Everett, 1970; Przekop and Domanski, 1970; Koves and Halasz, 1970; see also Volume III, Chapter 4), inhibition of neural transmitters (Szego, 1965; Coppola *et al.*, 1966; Harrington *et al.*, 1970; Frith and Hooper, 1971; Labhsetwar, 1972) and feedback effects of ovarian steroids (Dutt, 1953; Zeilmaker, 1966; Döcke and Dörner, 1969; Labhsetwar, 1970; Stevens *et al.*, 1970; McDonald and Gilmore, 1971). Diurnal sensitivity of the pituitary to release of LH has also been shown in the domestic hen (Fraps, 1954, 1965; Nelson and Nalbandov, 1966; Tanaka, 1968; Gilbert, 1971).

Although FSH has been shown to induce ovulation independently of LH, its role is considered to be mainly a priming one on the follicle. The extent of synergism between LH and FSH probably varies from species to species.

2. Structural Changes in Response to LH

The great precision with which the rabbit responds to LH, or human chorionic gonadotropin (HCG) has led to the extensive use of this animal for studies of the sequence of changes before ovulation, although most investigations have been limited to certain facets of the process. One of the first noticeable effects of LH is hyperemia of the ovary (Zondek *et al.*, 1945; Burr and Davies, 1951). It has not yet been established whether there is an increased blood flow to the ovulating follicle. Muscle cells are present in the larger blood vessels and their associated and/or the general innervation of the ovary has been suggested as a determining factor in the selection of follicles for maturation or atresia (Jacobowitz and Wallach, 1967; Fink and Schofield, 1970). Little is known about the role of nerves in ovarian processes despite numerous experiments involving denervation of the ovaries of mammals and birds (see Bahr *et al.*, 1974).

Edema and extravasation of blood are associated with ovarian hyperemia and are prominent in the theca interna of the ovulating follicle (see Espey and Lipner, 1963; Bjersing and Cajander, 1974a,e). These changes are believed to be associated with the breakdown of the collagen fibers of the blood vessels as part of the general dissolution of follicular collagen (see Section II,B,5). There are also conspicuous changes in the vasculature of the follicle wall where it bulges at the ovarian surface: at first the area is covered by an extensive network of capillaries which gradually becomes more diffuse and then one area becomes avascular. This represents the site of the potential stigma. In birds and amphibians the avascularity of the stigma is believed to lead to necrosis and thus to weakening of the area (see Nalbandov, 1961; Gilbert, 1971).

The histological changes in the follicle as it approaches ovulation have been frequently described for different mammals (see Brambell, 1956; Blandau, 1967). The use of the electron microscope has led to several detailed reports. About 8 hours after mating in the rabbit, as the time of ovulation approaches, the cells of the follicular layers at the apex of the follicle become dissociated, first within the tunica albuginea and then in the theca externa and theca interna (Espey, 1967; Bjersing and Cajander, 1974c,d). This is believed to be due to enzymic action (see Section II,B,5) and the source of the enzymes has been postulated as the multivesicular bodies of thecal fibroblasts by Espey (1971) and the dense bodies in the cells of the surface (germinal) epithelium by Bjersing and Cajander (1974b). Desquamation and disintegration of the surface epithelium was also noted in mice (Byskov, 1969), rats (Parr, 1974) and in several other species (Motta *et al.,* 1971). Parr also claims that there are no granulosa cells beneath the stigma.

The changes occurring in the thecal and granulosa cells, as observed in the light and electron microscopes at the time of ovulation, have often been described with reference to steroid secretion (see Volume I, Chapter 4).

3. Intrafollicular Pressure

An increase in intrafollicular pressure to bursting point has for long been assumed to be the cause of the extrusion of the egg (Rouget, 1858; Gothie, 1967). Several observers, who have studied the process cinematographically, have stated that ovulation is explosive, but most accounts of direct observation of ovulation describe a steady exudation of the follicular fluid containing the egg and its associated cumulus cells (rat: Blandau, 1955; rabbit: Walton and Hammond, 1928; Kelley, 1931; Smith, 1934; Hill *et al.,* 1935; Markee and Hinsey, 1936; Fujimoto *et al.,* 1974; cow: Umbaugh, 1949; horse: Heinze, 1975; monkey: Rawson and Dukelow, 1973; woman: Decker, 1951; Doyle, 1954). Moreover, the stigma through which the cumulus passes has a smooth edge rather than the ragged tear that would be expected if ovulation were an explosive event. The increased pressure has been variously ascribed to increased hyperemia of the ovary and especially the follicle wall (Heape, 1905; Pearson, 1944), the contractions of smooth muscle cells (see below), and the secretion of osmotically active substances into the antrum (Zachariae, 1957, 1958; Zachariae and Jensen, 1958; see also Larsen, 1970).

Although Guttmacher and Guttmacher (1921) were unable to induce rupture in pig follicles by increasing the pressure in the ovaries, the "increased pressure" theory was not challenged until the 1960's when it was shown by direct measurement that the intrafollicular pressure does not increase before ovulation and may in fact decrease slightly (Blandau and Rumery, 1963; Espey and Lipner, 1963; Rondell, 1964). Veterinarians have long known that they can feel an appreciable drop in the turgidity of the large follicle (5–6 cm in diameter) in the mare during a 48-hour period preceding ovulation.

Many authors have claimed that smooth muscle cells are involved in the process of ovulation. The evidence put forward is (1) the demonstration of cells, particularly in the stroma and theca externa, that have the characteristics, usually ultrastructural, of smooth muscle cells (von Kölliker, 1849; Guttmacher and Guttmacher, 1921; O'Shea, 1970, 1971; Osvaldo-Decima, 1970; Burden, 1972; Okamura *et al.*, 1972; Bjersing and Cajander, 1974d), and (2) the observations of spontaneous contractions *in vitro* of the ovaries of cats (Rocereto *et al.*, 1969) and women (Palti and Freund, 1972). Such contractions, however, do not appear to be related to the hormonal state, and could not be induced in the follicles of the sow by a variety of smooth muscle stimulants (Espey, 1970). Claesson (1947) was unable to demonstrate with the polarizing microscope any muscle cells in the follicle walls of cows, pigs, rabbits, or guinea pigs. Lipner and Maxwell (1960) discounted any role for smooth muscle cells, and Espey (1974) considers that none of the ultrastructural findings demonstrates the existence of thecal muscle cells.

4. Hormonal Changes

The changes in the concentrations of steroid and pituitary hormones about the time of ovulation have been studied extensively, particularly those in blood (see Volume III, Chapter 5), and have become important aids in the prediction of the time of ovulation in women (Johansson and Wide, 1969; Miyata *et al.*, 1970), monkeys (Mahoney, 1970), and domestic farm animals (Hansel and Snook, 1970; Bjersing *et al.*, 1972).

In the rabbit, 20α-dihydroprogesterone is secreted by the interstitial tissue and reaches a peak in the ovarian venous plasma within 2 hours of mating (Hilliard *et al.*, 1967). The first change in most mammals is an increase in progesterone secretion, often before luteinization becomes recognizable. Since progesterone is known to be capable of causing ovulation in mammals (Döcke and Dörner, 1969) and amphibians (Wright, 1961a), its secretion is believed by some workers (see Rondell, 1974) to be an important step in the ovulatory process, although progesterone (and estrogen) levels are known to be low in rabbit follicular fluid (YoungLai, 1972) and sheep plasma (Bjersing *et al.*, 1972) before ovulation.

5. Biochemical Changes

Hormones are known to combine at their target site with a receptor protein and much work has been done on the binding of gonadotropins, particularly HCG (see Edwards and Steptoe, 1975). Channing and Kammerman (1974) have found that large follicles in the pig contain 10- to 1000-fold more HCG receptors than do small or medium-sized follicles. The receptors are found in the thecal and granulosa cells, indicating that the arrival of gonadotropin initiates cellular changes in both follicular layers.

The next step after formation of the gonadotropin–receptor complex is uncertain. Prostaglandins have been postulated as mediators between LH and the stimulation of adenyl cyclase (Armstrong *et al.*, 1974; Labhsetwar, 1974; LeMaire and Marsh, 1975). It is possible that the effects of prostaglandins which simulate those of LH are reached through a different pathway (see Rondell, 1974). The involvement of adenyl cyclase (Dorrington and Baggett, 1969), leading to stimulation of cyclic AMP (Marsh and Savard, 1964; Robison, 1970; Major and Kilpatrick, 1972; Ellsworth and Armstrong, 1973), is generally accepted (but see LeMaire *et al.*, 1972).

The increased amounts of cyclic AMP in turn lead to increased protein synthesis (Pool and Lipner, 1964, 1966, 1969; Civen *et al.*, 1966; Barros and Austin, 1968) which is essential for the structural changes needed to enable the granulosa cells to secrete progesterone after ovulation and for the elaboration of the factor(s) which brings about the dissociation of the cells of the various follicular layers. These factors are believed to be proteolytic enzymes (Reichert, 1962a,b; Lipner, 1965; Unbehaan *et al.*, 1965; Espey, 1970) or possibly only one, a collagenase (Espey and Lipner, 1965; Espey and Rondell, 1966, 1968; Espey, 1974), which attacks the cement substance between the cells and the collagen fibers of the tunica albuginea and blood vessels. Proteolytic enzymes are known to be involved in the ovulatory process in the domestic fowl (Nakajo *et al.*, 1973). One of the main effects of collagenolytic activity is increased distensibility of the follicle wall as ovulation approaches (Rondell, 1974). The follicle can increase in size, without any rise in its internal pressure, to accommodate the increased volume of follicular fluid caused by transudation from the "leaky" blood vessels and hyperosmotic change in the follicular fluid itself (Zachariae, 1958; Edwards, 1974).

6. Conclusions

The final stimulus for the release of the egg is not yet known. It is probable that the event is the culmination of a complex sequence of reactions. Attempts have been made to represent the various steps and their controlling factors as mathematical models (Rodbard, 1968; Schwartz, 1969; Vande Wiele *et al.*, 1970), but these have not generally been felt to provide a useful approach.

As pointed out earlier, the various works referred to above cover only a few of the many species of mammals. As more mammalian species are investigated in detail, it is likely that more variations will be found, as is already apparent in the structural diversity of the gonads (see Volume I, Chapter 4 and Volume II, Chapter 6), and it is doubtful whether further studies would simplify or complicate interpretations of the basic mechanisms of ovulation. For example, the ovulatory sequence in centetids (Strauss, 1938) and plains viscacha (Weir, 1971b), animals which have little or no antrum, would be interesting to examine in detail; there is no preovulatory swelling of the follicles, although the granulosa cells of *Centetes* hypertrophy and obliterate the incipient antrum.

C. Induction of Ovulation

1. Mammals

The essential feature of mammalian reproduction is the release of a fertile egg and many attempts have been made to induce ovulation by treatment with various exogenous gonadotropins. The capacity of these substances to cause a normal or a superovulatory response has been used to clarify some of the events associated with ovulation. Many techniques have been used, with conflicting results, some of which may be attributable to differences in strains of animals or reproductive states (Hafez, 1969). Ideally, animals should be hypophysectomized to eliminate the complicating effects of circulating endogenous hormones before attempting to mimic the normal sequence of ovulation by administration of exogenous gonadotropins or steroid hormones (see Rowlands and Williams, 1942–1944; Lostroh and Johnson, 1966; Rowlands and Parkes, 1966). Usually, however, the aim is to achieve ovulation in animals (1) that are not breeding because of infertility or anestrus or (2) at times convenient for the collection of eggs for other purposes such as fertilization tests *in vitro* or for artificial insemination. The most commonly used substances are gonadotropins—pituitary extracts (FSH and LH), urinary extracts (HMG), or nonpituitary hormones such as pregnant mares' serum gonadotropin (PMSG), which has FSH-like properties, and human chorionic gonadotropin (HCG), which resembles LH in its action. Standard pituitary extracts are available for a few species such as the horse, sheep, pig, rat, or man, but the commercial preparations of PMSG and HCG are more widely used. Whichever hormones are used, a suitable regimen involves stimulation of the ovary so that enough follicles develop to ovulable size. The size, duration, and route of the dose are important considerations for the avoidance of superovulation or cystic follicle formation (Rowlands, 1944; Kennelly and Foote, 1965). Administration of the ovulatory hormone (LH or HCG) then induces ovulation. This basic formula has been applied to animals of many species—laboratory, domestic, and exotic—and responses have generally been good (for example, marsupials: Nelsen and White, 1941; Smith and Godfrey, 1970; sheep: Hammond *et al.*, 1942; rabbits: Kennelly and Foote, 1965; Maurer *et al.*, 1968; rodents: Rowlands, 1944; Fowler and Edwards, 1957; Seth and Prasad, 1967; Greenwald and Choudary, 1969; Weir, 1969, 1973; primates: Bennett, 1967; Wan and Balin, 1969; Dukelow, 1970; Harrison and Dukelow, 1970; Breckwoldt *et al.*, 1971; man: Gemzell, 1965; Lunenfeld *et al.*, 1969; Edwards and Steptoe, 1975). Although release of eggs can be obtained within a range of doses, the eggs may not all be fertilizable (see Jainudeen *et al.*, 1966). A regimen which simulates more closely the normal situation by also inducing estrus is harder to achieve (see Weir, 1973). The follicles of the nonpregnant guinea pig (Hamburger and Pedersen-Bjergaard, 1946) and mare (W. R. Allen, personal communication) rarely ovulate in response to exogenous gonadotropins. Rabbits are normally reflex ovulators and ovulation can therefore be induced by LH or HCG alone, as it can in alpaca (Fernandez-Baca *et al.*, 1970) and voles (Cross, 1972). In such animals, stimulation of the cervix with a rod (Greulich, 1934; Milligan, 1975) may be sufficient to cause release of LH.

In many animals, e.g., sheep, cows, and guinea pigs, ovulation will not take place in the presence of a functional corpus luteum. Prostaglandins and their analogues have been shown to be potent luteolysins in cows (Rowson *et al.*, 1972) and mares (Allen and Rowson, 1973), and estrus and ovulation usually follow within a few days of treatment.

A natural estrus and concomitant ovulation may be induced as a rebound phenomenon when substances such as methallibure or clomiphene, inhibitors of pituitary secretions, are withdrawn after a course of treatment (Rosenberg, 1973). This form of therapy is frequently used, sometimes with additional HCG, to treat infertile women (Gemzell and Johansson, 1971). Progesterone may also be used to suppress the gonadotropic activity of the pituitary. Early tests involved oral administration of substances such as 6α-

methyl-17α-acetoxyprogesterone (Graves and Dziuk, 1968), but it was difficult to administer known doses to individual animals when they were fed in groups, as in normal husbandry. The intravaginal insertion of progesterone-impregnated sponges has been used successfully to inhibit estrus in sheep (Robinson, 1965) and cattle (Sreenan, 1975). Although the first estrus after withdrawal of the sponge may not be pronounced or fertile, the next is associated with consistent ovulation, estrus, and conception. Subcutaneous implants of progesterone are also effective in this way (Roche, 1974).

Ovarian steroids may induce ovulation (Döcke and Dörner, 1965, 1966; Brown-Grant, 1969), as may copper and cadmium salts in rabbits (Emmens, 1940–41; Suzuki and Bialy, 1964). Metallic copper has, however, been shown to be effective as an intrauterine contraceptive (see Webb, 1973).

Since the development of theories of pituitary regulation by hypothalamic factors (see Volume III, Chapter 4) and the chemical identification of these factors, they and analogues have been manufactured and shown to be potent inducers of ovulation (McCann and Porter, 1969).

2. Other Vertebrates

The process of ovulation has been studied in some nonmammalian vertebrates, especially anurans and poultry, to determine the specific mechanisms involved and to relate them to a scheme applicable to all vertebrates.

The effects of heterologous and homologous pure or crude pituitary extracts and mammalian hormone preparations are discussed in Volume III, Chapter 4 (Section III,B). In fish, folliculogenesis and vitellogenesis are generally easier to produce than ovulation. Rugh (1935) induced ovulation in the frog with homozoic whole pituitaries. Amphibians are more sensitive to mammalian pituitary preparations than to those of fish (Greep, 1961), and the ability of gonadotropins to induce ovulation in *Xenopus laevis* (Hogben, 1930) was an early and reliable test for pregnancy in women (see Bellerby, 1934; Shapiro and Zwarenstein, 1934). In reptiles too, mammalian and avian pituitary preparations appear to be more effective in inducing ovulation than substances of fish origin (Licht and Stockell Hartree, 1971); and ovine FSH is more potent than ovine LH (Licht, 1970). The hormonal control of ovulation in birds seems to be complex. Purified preparations of chicken pituitary extracts are less able to maintain a normal follicular sequence in hypophysectomized hens than are crude extracts which presumably contain FSH and LH acting synergistically (Mitchell, 1967, 1970).

Steroid hormones are also known to affect ovulation in amphibians (Chang and Witschi, 1957; Bergers and Li, 1960; Wright, 1961a,b; Edgren and Carter, 1963; Schuetz, 1971). Wright (1971) has postulated that in *Rana pipiens* gonadotropins induce secretion of a 3-keto-Δ^4-steroid which is the mediator of the ovulatory sequence of events, as 20α-dihydroprogesterone is in the rabbit (Hilliard *et al.*, 1967).

D. Factors Affecting Ovulation Number

The number of follicles which ovulate at each estrus is considered to be characteristic of the species. Strains or breeds of species have also been shown to have different ovulation rates (pigs: Squiers *et al.*, 1952; Bhalla *et al.*, 1969; sheep: Packham and Triffitt, 1966; Land, 1970a; Bradford, 1972; rats: Land *et al.*, 1974; mice: Bradford, 1969; Land and Falconer, 1969). Selection for prolificacy has been significant in farming practice, but there have been few detailed studies on the way in which the genetic influence on ovulation number is effected. Land and Falconer (1969) have shown in mice that the ovulation rate depends on FSH output and the sensitivity of the ovary to gonadotropins: a high ovulation rate is due to an increased output of FSH, and a low ovulation rate to increased ovarian sensitivity. In sheep breeds, a positive relationship exists between the ovulation number (indicated by litter size) and the duration of estrus (Land, 1970b). Land and Carr (1975) have found that the plasma testosterone levels in ram lambs can be correlated with the ovulation number of ewes of the same breed.

Improved nutrition before the start of the breeding season ("flushing") has a beneficial effect on the numbers of eggs ovulated and is good farming practice (Self *et al.*, 1955; Lutwak-Mann, 1962; Bellows *et al.*, 1963; Brooks *et al.*, 1972; Gunn and Doney, 1973; see also Sadleir, 1969).

Falconer *et al.* (1961) and McLaren (1963) showed that in mice a random number of eggs was shed from each ovary. When one ovary is removed the number of eggs shed from the remaining ovary is similar to that which would have been shed had both ovaries been present (rat: Mandl and Zuckerman, 1951; Peppler and Greenwald, 1970a,b; Peppler, 1971; Chatterjee and Greenwald, 1972; mouse: Jones and Krohn, 1960; Biggers *et al.*, 1962; hamster: Greenwald, 1961; Chatterjee and Greenwald, 1972; guinea pig: Hermreck and Greenwald, 1964; sow: Short *et al.*, 1965; Dailey *et al.*, 1970; cow: Saiduddin *et al.*, 1970; ewe: Mallampati and Casida, 1970). The compensation in ovulation number was first shown by the experiments of Hunter (1787) who compared the number of piglets produced by two sows, one of which was unilaterally ovariectomized, until they ceased to breed. Although the sows gave birth to almost the same number of young after eight litters, the unilaterally ovariectomized sow then stopped breeding and the other continued to farrow. Hunter's conclusion that hemi-spaying curtailed the reproductive life span, although based on only two animals, has been confirmed by Jones and Krohn (1960) and Biggers *et al.* (1962) who found that unilaterally castrated mice produced only half the number of young that their intact littermate controls did. Several studies have shown that the effect of unilateral ovariectomy becomes clear very soon after the operation and that the increased ovulation number is due to recruitment of follicles to the pool of follicles responsive to gonadotropins rather than to ovulation of already existing large follicles (see Greenwald, 1961; Peppler and Greenwald, 1970a,b).

An extreme example of the remaining ovary taking over the function of the other after unilateral ovariectomy is seen in the mountain viscacha, *Lagidium*. These animals usually ovulate only from the right ovary, but if this ovary is removed the left then becomes functional (Pearson, 1949; Weir, 1971c). Whether mountain viscacha that are thus forced to rely on the left ovary have a shorter reproductive life span is not known. It would also be interesting to examine the results of unilateral ovariectomy of the plains viscacha, *Lagostomus maximus*, and the elephant shrew, *Elephantulus myurus*, which ovulate about 800 (Weir, 1971b) and 120 (van der Horst and Gillman, 1940) eggs, respectively, at each estrus.

III. ATRESIA

The word atresia is derived from the Greek (*a* = not; *tretos* = perforated) and is correctly used as a medical term to describe the closure of a natural channel (Shorter Oxford English Dictionary, 1973). For example, Grosser (1903) described as "atresia" the obliteration of the vagina of certain bats by cornified detritus which accumulates during hibernation. It is clear, therefore, that when referring to the ovary the term atresia can properly be applied only to vesicular follicles, and should not be used as a synonym for degeneration. Degenerating follicles are, however, commonly called atretic, as are oocytes which are being lost from the ovary for any reason other than ovulation (Ingram, 1962). Indeed, many dictionary definitions are now including the degeneration of germ cells as an example of atresia (see Chambers' Technical Dictionary, 1958; Webster's Fourth International Dictionary of the English Language, Unabridged, 1971). Many oocytes and some follicles do degenerate but a large number of the other follicles fail to ovulate and then become atretic. The conversion of follicles to the atretic state is considered by us to be an integral part of ovarian function whose importance has only recently been recognized. According to the definition given above, it would be incorrect to refer to the changes in the unovulated follicles of nonmammalian vertebrates (i.e., animals having macrolecithal eggs and no follicular antrum) as atresia. However, the homology with mammals is good in that the whole follicle is involved, and the use of the word atresia is considered justified.

A. Oocytes

The germ cell population in the ovary increases by mitotic divisions to a maximum at a point varying with the species (see Baker, 1972), and then decreases because of the cessation of mitosis and waves of degeneration which occur at most stages of the first meiotic division (mouse: Jones and Krohn, 1959; Borum, 1961; man: Baker, 1963; rat: Beaumont and Mandl, 1962; guinea-pig: Ioannou, 1964; monkey: Baker, 1966; pig: Black and Erickson, 1968; cow: Erickson, 1966). The loss of over a 1000 oocytes at each estrus throughout adult life in the plains viscacha (Weir, 1971b) poses the question of whether any significant loss of germ cells occurs during fetal life as it does in other mammals. Some selection of oocytes for ovulation must, however, take place at each estrus because an egg is released only from about 70% of the apparently mature follicles. The number of oocytes in the mammalian ovary at birth still exceeds by far the number that will actually be ovulated during the reproductive lifetime of an individual. For example, of the 2,000,000 oocytes said to be present in the human ovary at birth, only about 400 could possibly be ovulated (see Baker, 1963). Degeneration of oocytes of nonmammalian vertebrates has also been reported (see Brambell, 1956), but exact oocyte counts have not been made.

The changes in the appearance of the degenerating oocyte have been investigated with the light microscope (see Burkl, 1962, 1965; Ingram, 1962) and with the electron microscope (Franchi and Mandl, 1962; Vazquez-Nin and Sotelo, 1963–1964). About 50–70% of the oocytes in a mammalian ovary appear, in the light microscope, to be degenerating, but the percentage cannot be determined precisely because of the subjective element involved in the recognition of degeneration (Mandl and Zuckerman, 1950). The numbers of oocytes with recognizable degenerative changes are found to be low, however, with the electron microscope (Baker and Franchi, 1972), indicating that fixation may be an important factor. Changes in the oocytes of X-irradiated ovaries are generally similar to those in oocytes degenerating spontaneously and are discussed in Volume III, Chapter 1.

The number of oocytes in an ovary falls steadily after birth (Zuckerman, 1951). After an oocyte is surrounded by follicular cells to form a primordial follicle, it becomes difficult to determine whether failure of the oocyte or membrana granulosa cells is the primary cause of the loss. Not all oocytes become invested with a layer of granulosa cells, and those that remain naked usually degenerate (Gillman, 1948; Ohno and Smith, 1964). Groups of germ cells seem to degenerate at the same time, although they may be at different stages of development, and this synchrony may be due to the presence of intercellular bridges (see Zamboni, 1972).

The rate of loss of oocytes throughout reproductive life is to some extent genetically determined. Differences in the numbers of oocytes in the ovaries and in the rate of germ cell degeneration at particular reproductive stages have been described for various strains of mouse (Jones and Krohn, 1961a) and breeds of sheep (Land, 1970a).

B. Follicles

Atresia occurs throughout the vertebrate classes, and many descriptions have been given of the resulting structural changes. In recent years, advances in biochemical techniques for measuring steroids in small amounts of body tissues have led to the belief that atresia is a functional rather than a degenerative process. In mammals, follicular atresia may give rise to interstitial tissue or to corpora lutea accessoria (see Volume I, Chapter 4), both of which may be hormonally active (see below). The term "corpus luteum" is frequently applied to certain structures in nonmammalian ovaries, but Dodd (Volume I, Chapter 5) has pointed out that they are not all homologous with the corpora lutea of mammals since they are not always derived from the ovulated follicle and their endocrine functions are unknown. Dodd does, however, suggest that the postovulatory structures should continue to be called corpora lutea while the bodies resulting from follicles that do not undergo the normal ovulatory processes should be termed corpora atretica.

1. Cyclostomes

Atresia is more common in the hagfishes than in the lampreys and has been described by Lyngnes (1936; see also Volume I, Chapter 5). Corpora atretica are formed by the escape of the yolky follicle contents through atrophied parts of the follicular wall to the stroma where they are removed by phagocytosis, or by invasion of the yolk by phagocytic thecal cells.

2. Fishes

a. Elasmobranchs. The early literature on atretic structures in the ovaries of cartilaginous fishes has been discussed by Brambell (1956) and Barr (1968). Atresia occurs in follicles of all sizes and its progress can be determined by the amount of yolk accumulated in the oocyte. The atretic process in yolky follicles involves hypertrophy, vacuolization, and folding of the granulosa cell layer. The yolk granules are released during degeneration of the oocyte and are taken up by the granulosa cells. Eventually a solid body is produced which resembles that formed in follicles after ovulation. Several workers (Chieffi, 1967; Lance and Callard, 1969; TeWinkel, 1972) have tried to show that large numbers of corpora atretica are associated with the occurrence of internal fertilization which is the predominant mode in these fishes. Hisaw and Hisaw (1959) demonstrated that the formation and structure of the corpora lutea and corpora atretica are very similar in oviparous, ovoviviparous, and viviparous species. Histochemical studies have suggested that steroidogenesis may occur in both types of atretic follicles (Lance and Callard, 1969; TeWinkel, 1972), and Chieffi (1967) and Simpson *et al.* (1968) have claimed that secretion of estrogens and progesterone has been demonstrated.

b. Teleosts. There have been many studies (see Volume I, Chapter 5 and Volume II, Chapter 6) of the fate of the unruptured follicle in bony fishes since the first account by Bretschneider and Duyvené de Wit (1947). The process of atresia is similar to that in elasmobranchs, i.e., phagocytosis of the degenerating oocyte by the granulosa cells with the eventual formation of a solid body. The possibility of secretion of steroids by the corpora atretica has been investigated by several workers (Sathyanesan, 1961; Nair, 1963; Polder, 1964; Rai, 1966; Rastogi, 1966, 1969; Chan *et al.*, 1967; Pant, 1968; Lambert, 1970). Much of the evidence is indirect and any steroids found in the ovary are most probably breakdown products associated with the removal of the yolk, rather than essential hormones (Hoar, 1955; Ball, 1961).

3. Amphibians

Perry and Rowlands (1962) suggested that, although the corpora lutea of amphibians are probably not secretory in spite of being "luteinized," the corpora atretica (formed as in fish by phagocytosis of the yolk by granulosa cells) may secrete a substance which is involved in the production of mucus by the oviduct. Recent reports indicate that the corpus luteum of some viviparous anurans is endocrinologically important in the maintenance of gestation (Vilter and Vilter, 1964; Kessel and Panje, 1968; Xavier, 1969, 1970; Joly and Picherel, 1972).

4. Reptiles

Betz (1963) provides a good review of the literature relating to atresia in reptiles, and other details are given in Volume I, Chapter 5. As in the vertebrate classes discussed above, the granulosa cells may remove the yolk, but a type of "bursting" atresia whereby the follicle liberates the yolk into the body cavity, whence it is removed, has also been described (Bragdon, 1952). The corpus luteum in reptiles appears to be more definitely related to the maintenance of pregnancy than it is in fishes or amphibians, and the evidence is discussed in Volume I, Chapter 5 and Volume II, Chapters 6 and 8.

5. Birds

At least five types of atresia of avian follicles have been described (see Brambell, 1956; Kern, 1972). "Bursting" atresia, as found in reptiles, is common, and so is lipoglandular atresia in which lipid accumulates in the central cells of the atretic follicle and the thecal cells hypertrophy. Sometimes all the cells of the corpus atreticum contain lipid—the result of lipoidal atresia. Yolky atresia is the form common in other nonmammalian vertebrates, i.e., phagocytosis of the yolk by the granulosa cells without the participation of the thecal cells. Liquefaction atresia occurs in medium to large follicles; the yolk contents break down and the follicles may burst. Several authors consider that the type of atresia undergone depends on the size of the follicle and the amount of accumulated yolk (Dominic, 1961; Kern, 1972; Erpino, 1973), both of which are related to the stage of the ovarian cycle (see Marshall and Coombs, 1957; also Volume II, Chapter 6). Gilbert (1970) has suggested that in the fowl atresia may be reversible and may be a feature related to the intensive egg-laying associated with domestication. As for other vertebrates, controversy exists about a possible function of the postovulatory follicle. Wyburn *et al.* (1966) consider that a corpus luteum is not formed in the hen, but disruption of the timing mechanism for oviposition and disturbed nesting behavior occur if the postovulatory follicle is removed (Rothchild and Fraps, 1944; Wood-Gush and Gilbert, 1964).

6. Mammals

Only two types of follicular atresia, as distinct from degeneration, occur in mammals. The most common form is "thecal" atresia whereby the granulosa cells become pyknotic and eventually disappear or are restricted to a small mass in the center of the body which is formed by hypertrophy of the cells of the theca interna. The hypertrophied thecal cells may take the form of interstitial cells or become "luteinized" (a corpus luteum atreticum: see Volume I, Chapter 4). The latter process is the rarer, but the former is generally agreed to be the source of the secondary interstitial tissue in all mammals (Velloso de Pinho, 1923; Guthrie and Jeffers, 1938; Kingsbury, 1939; Aron and Aron, 1953; Brambell, 1956; Sturgis, 1961; Mossman *et al.*, 1964; Dalmane, 1967; Deanesly, 1972; Mori and Matsumoto, 1973; Mossman and Duke, 1973; Weir and Rowlands, 1974). Recent ultrastructural, histochemical and endocrinological studies (see Volume I, Chapter 4) have shown that the interstitial tissue in those animals in which it is found secretes hormones. In the rabbit, for example, the interstitial cells secrete a progestin in response to the LH surge caused by coitus, and the progestin has a feed-back action on LH release that leads to ovulation (Hilliard *et al.*, 1967).

The second form of atresia in mammals takes place when the follicle behaves as it would have done had ovulation occurred, but the egg is not released. The resulting structure is termed a corpus luteum accessorium, although it is often erroneously called a corpus luteum atreticum (see Volume I, Chapter 4). The accessory corpora lutea may be found occasionally in any species (see Volume I, Chapter 4), but they are characteristic of the ovaries of a few groups, the most notable being the rodent suborder Hystricomorpha (see Weir and Rowlands, 1974). In animals such as the coypu, chinchilla, and acouchi, "thecal" luteinization also occurs but there is as yet no indication of the factor(s) that determines which process a particular follicle will follow. There is some evidence in the agouti that the luteinized granulosa cells of an accessory corpus luteum may become converted into interstitial cells (Weir, 1971a). Follicles of all sizes may become accessory corpora lutea; in the North American porcupine (Mossman and Judas, 1949) and mountain viscacha (Pearson, 1949) they remain smaller than the primary corpus luteum, but in the chinchilla (Weir, 1966), agouti (Weir, 1971a) and plains viscacha (Weir, 1971b) they may become indistinguishable from the primary corpora lutea. Apart from the presence of an entrapped egg, the accessory corpora lutea resemble true corpora lutea in structure and in their ability to secrete progesterone (Tam, 1970). The function of accessory corpora lutea has not been established; chinchilla are able to maintain normal gestation without such corpora lutea, and hystricomorph rodents do, in any case, have a progesterone-conserving mechanism which operates during pregnancy (Heap and Illingworth, 1974). Corpora lutea accessoria may be found in ovaries of other mammals and their formation appears to be related to an imbalance in hormonal stimulation caused, for example, by treatment with exogenous hormones or by grafting the ovary under the kidney capsule or cornea (Welschen, 1971; Jones and Nalbandov, 1972). El-Fouly *et al.* (1970) have suggested that the ovum itself has an antiluteotropic action, and that when its influence is removed by ovulation or degeneration the granulosa cells will automatically luteinize. Such a theory does not appear to explain why, in most mammals, "thecal" atresia occurs rather than granulosa cell luteinization when the ovum degenerates. Richards (1975) has suggested that in the hypophysectomized rat the response of the follicle is related to its stage of differentiation and its previous stimulation by FSH.

Studies on atresia are generally of two types: (1) those relating to the recognition of atresia and (2) those concerning the degree of atresia at specific stages of the reproductive cycle.

Studies in rodents designed to determine the onset of and reasons for atresia have been primarily histochemical (see Strassman, 1945; Deane, 1952; Kar *et al.*, 1962; Guraya and Greenwald, 1964; Zerbian, 1966; Zerbian and Goslar, 1968; Mozanska, 1969). It is clear that enzyme changes in follicles can be detected before karyorhexis begins in the granulosa cells; e.g., in an atretic follicle, the granulosa cells contain alkaline phosphatase, glycogen, lipids, ascorbic acid, and esterase, all of which would be absent in a normal follicle. Moor *et al.* (1971) have shown that although several large follicles may be present in the ewe a few days before ovulation, only one will be "activated" in that the steroid-converting enzyme Δ^5-3β-hydroxysteroid dehydrogenase appears in the thecal cells.

There are many useful accounts, with good illustrations, of follicular atresia during the reproductive cycle (platypus: Garde, 1930; marsupials: Sandes, 1903; Martínez-Esteve, 1942; rat: Mandl and Zuckerman, 1950; bats: van der Stricht, 1901; mouse: Engle, 1927; Nakamura, 1957; Byskov, 1974; hamster: Knigge and Leathem, 1956; Horowitz, 1967; guinea pig: Harman and Kirgis, 1938; Adams *et al.*, 1966; rabbit: Asami, 1920; ewe: Smeaton and Robertson, 1971; cow: Rajakoski, 1960; monkey: Sturgis, 1949; women: Allen *et al.*, 1930; Block *et al.*, 1953; Watzka, 1957; Mossman *et al.*, 1964; general: Brambell, 1956; Mossman and Duke, 1973).

C. Conclusions

In spite of an increased interest in the study of atresia in recent years, there has been little advance in our understanding of the factors affecting the process since the reviews of Ingram (1962) and Krohn (1967). One of the reasons for this is the practical difficulty of distinguishing oocyte degeneration from follicular atresia. For example, hypophysectomy leads to a reduction in the numbers of antral follicles and yet to an increase in the numbers of normal oocytes (Jones and Krohn, 1961b). When estrogen or PMSG is given to hypophysectomized rats fewer follicles become atretic (Ingram, 1959). Follicles may actually grow when estrogen is administered after hypophysectomy (Williams, 1956), although atresia is often considered to be most intense at estrus when estrogen levels are high. High levels of atresia were found in the badger during the period of delayed implantation (Canivenc *et al.*, 1965) and in the hibernating hedgehog (Balboni, 1969). It has been suggested (Sturgis, 1961) that the surge of LH that initiates the maturation of the oocyte also affects other ovulable follicles to make them atretic. Estrogen and progesterone promote the growth of medium-sized follicles in the rabbit, but the affected follicles become atretic when they enlarge (Wallach and Noriega, 1970). Peters *et al.* (1973) have suggested that follicular growth in the mouse is influenced by a factor in the follicular fluid of atretic follicles.

It is clear that follicular atresia is not necessarily a degenerative process and that the metamorphosis of the follicular wall into a different kind of, probably functional, tissue occurs without ovulation as a normal and essential event in the ovarian cycle of many vertebrates. It is also obvious that there is an intricate relationship between the pituitary and ovarian hormones in their control of atresia and of ovulation. But we do not yet know the answers to the following questions: Why do so many oocytes degenerate in the fetal ovary? What determines which oocytes become enclosed by granulosa cells? (It has been suggested that the rete ovarii plays an important role in this step: Byskov and Lintern-Moore, 1973). Which follicles start to grow when? Since early follicular growth is independent of hypophysial hormones (see Perry and Rowlands, 1963; also Challoner, 1975), why do some follicles become atretic before they even become vesicular? The most difficult and perhaps most important question to answer, however, is: What determines which follicles escape the fate of the majority to become the few to ovulate?

REFERENCES

Adams, E. C., Hertig, A. T., and Foster, S. (1966). Studies on guinea pig oocytes. 2. Histochemical observations on some phosphatases and lipid in developing and in atretic oocytes and follicles. *Am. J. Anat.* **119,** 303–339.

Allen, E., Pratt, J. P., Newell, Q. U., and Bland, L. J. (1930). Human ova from large follicles including a search for maturations and observations on atresia. *Am. J. Anat.* **46,** 1–54.

Allen, W. R., and Rowson, L. E. A. (1973). Control of the mare's oestrous cycle by prostaglandins. *J. Reprod. Fertil.* **33,** 539–543.

Armstrong, D. T., Grinwich, D. L., Moon, Y. S., and Zamecnik, J. (1974). Inhibition of ovulation in rabbits by intrafollicular injection of indomethacin and prostaglandin F antiserum. *Life Sci.* **14,** 129–140.

Aron, C. (1965). Données nouvelles sur les mécanismes de la ponte chez des femelles dites ≪ à la ponte spontanée ≫. Mise en évidence, chez la ratte, de l'action ovulatoire du coit au cours du cycle oestral. *Rev. Roum. Endocrinol.* **2,** 221–227.

Aron, C., Asch, G., and Roos, J. (1966). Triggering of ovulation by coitus in the rat. *Int. Rev. Cytol.* **20,** 139–172.

Aron, M., and Aron, C. (1953). L'atrésie folliculaire: déterminisme et signification. *Arch. Anat. Histol. Embryol.* **36,** 69–86.

Asami, G. (1920). Observations on the follicular atresia in the rabbit ovary. *Anat. Rec.* **18,** 323–343.

Asdell, S. A. (1962). The mechanism of ovulation. *In* "The Ovary" (S. Zuckerman, ed.), 1st ed., Vol. 1, Chapter 8, pp. 435–449. Academic Press, New York.

Ayalon, D., Tsafriri, A., Lindner, H. R., Cordova, T., and Harell, A. (1972). Serum gonadotrophin levels in pro-oestrous rats in relation to the resumption of meiosis by the oocytes. *J. Reprod. Fertil.* **31,** 51–58.

Bahr, J., Kao, L., and Nalbandov, A. V. (1974). The role of catecholamines and nerves in ovulation. *Biol. Reprod.* **10,** 273–290.

Baker, T. G. (1963). A quantitative and cytological study of germ cells in human ovaries. *Proc. R. Soc. London, Ser. B* **158,** 417–433.

Baker, T. G. (1966). A quantitative and cytological study of oogenesis in the rhesus monkey. *J. Anat.* **100,** 761–776.

Baker, T. G. (1972). Oogenesis and ovarian development. *In* "Reproductive Biology" (H. A. Balin and S. R. Glasser, eds.), pp. 398–437. Excerpta Med. Found., Amsterdam.

Baker, T. G., and Franchi, L. L. (1972). The fine structure of oogonia and oocytes in the rhesus monkey (*Macaca mulatta*). *Z. Zellforsch. Mikrosk. Anat.* **126,** 53–74.

Baker, T. G., and Neal, P. (1972). Gonadotrophin-induced maturation of mouse Graafian follicles in organ culture. *In* "Oogenesis" (J. D. Biggers and A. W. Schuetz, eds.), pp. 377–396. Univ. Park Press, Baltimore, Maryland.

Balboni, G. C. (1969). Observations on the ovary of the hedgehog (*Erinaceus europaeus*) in physiological and experimental awakening. *Acta Anat.* **73,** Suppl. 56, 60–67.

Ball, J. N. (1961). Reproduction in female bony fishes. *Symp. Zool. Soc. London* **1,** 105–135.

Barr, W. A. (1968). Patterns of ovarian activity. *In* "Perspectives in Endocrinology" (E. J. W. Barrington and C. Barker-Jørgensen, eds.), pp. 163–238. Academic Press, New York.

Barros, C., and Austin, C. R. (1968). Inhibition of ovulation by systemically administered actinomycin D in the hamster. *Endocrinology* **83,** 177–179.

Beaumont, H. M., and Mandl, A. M. (1962). A quantitative and cytological study of oogonia and oocytes in the foetal and neonatal rat. *Proc. R. Soc. London, Ser. B* **155,** 557–579.

Bellerby, C. W. (1934). A rapid test for the diagnosis of pregnancy. *Nature (London)* **133,** 494–495.

Bellows, R. A., Pope, A. L., Meyer, R. K., Chapman, A. B., and Casida, L. E. (1963). Physiological mechanisms in nutritionally induced differences in ovarian activity of mature ewes. *J. Anim. Sci.* **22,** 93–100.

Bennett, J. P. (1967). Induction of ovulation in the squirrel monkey (*Saimiri sciureus*) with pregnant mares serum (PMS) and human chorionic gonadotrophin (HCG). *J. Reprod. Fertil.* **13,** 357–359.

Bergers, A. C. J., and Li, C. H. (1960). Amphibian ovulation *in vitro* induced by mammalian pituitary hormones and progesterone. *Endocrinology* **66,** 255–259.

Betz, T. W. (1963). Ovarian histology of the diamond-backed water snake, *Natrix rhombifera,* during the reproductive cycle. *J. Morphol.* **113,** 245–260.

Bhalla, R. C., First, N. L., Chapman, A. B., and Casida, L. E. (1969). Quantitative variation in ovarian and follicular development in four genetic groups of pigs at different ages. *J. Anim. Sci.* **28,** 780–784.

Biggers, J. D., Finn, C. A., and McLaren, A. (1962). Long-term reproductive performance of female mice. 1. Effect of removing one ovary. *J. Reprod. Fertil.* **3,** 303–312.

Bingel, A. S., and Schwartz, N. B. (1969). Timing of LH release and ovulation in the cyclic mouse. *J. Reprod. Fertil.* **19,** 223–229.

Bjersing, L., and Cajander, S. (1974a). Ovulation and the mechanism of follicle rupture. I. Light microscopic changes in rabbit ovarian follicles prior to induced ovulation. *Cell Tissue Res.* **149,** 287–300.

Bjersing, L., and Cajander, S. (1974b). Ovulation and the mechanism of follicle rupture. II. Scanning electron microscopy of rabbit germinal epithelium prior to induced ovulation. *Cell Tissue Res.* **149,** 301–312.

Bjersing, L., and Cajander, S. (1974c). Ovulation and the mechanism of follicle rupture. IV. Ultrastructure of membrana granulosa of rabbit Graafian follicles prior to induced ovulation. *Cell Tissue Res.* **149,** 1–14.

Bjersing, L., and Cajander, S. (1974d). Ovulation and the mechanism of follicle rupture. V. Ultrastructure of tunica albuginea and theca externa of rabbit Graafian follicles prior to induced ovulation. *Cell Tissue Res.* **153,** 15–30.

Bjersing, L., and Cajander, S. (1974e). Ovulation and the mechanism of follicle rupture. VI. Ultrastructure of theca interna and the inner vascular network surrounding rabbit Graafian follicles prior to induced ovulation. *Cell Tissue Res.* **153,** 31–44.

Bjersing, L., Hay, M. F., Kann, G., Moor, R. M., Naftolin, F., Scaramuzzi, R. J., Short, R. V., and YoungLai, E. V. (1972). Changes in gonadotrophins, ovarian steroids and follicular morphology in sheep at oestrus. *J. Endocrinol.* **52,** 465–479.

Black, J. L., and Erickson, B. H. (1968). Oogenesis and ovarian development in the prenatal pig. *Anat. Rec.* **161,** 45–56.

Blandau, R. J. (1955). Ovulation in the living albino rat. *Fertil. Steril.* **6,** 391–404.

Blandau, R. J. (1966). The mechanism of ovulation. *In* "Ovulation" (R. Greenblatt, ed.), pp. 1–15. Lippincott, Philadelphia, Pennsylvania.

Blandau, R. J. (1967). Anatomy of ovulation. *Clin. Obstet. Gynecol.* **10,** 347–360.

Blandau, R. J. (1970). Growth of the ovarian follicle and ovulation. *Prog. Gynecol.* **5,** 58–76.

Blandau, R. J., and Rumery, R. E. (1963). Measurements of intrafollicular pressure in ovulatory and preovulatory follicles of the rat. *Fertil. Steril.* **14,** 330–341.

Block, E., Magnusson, G., and Odeblad, E. (1953). A study of normal and atretic follicles with autoradiography. *Acta Obstet. Gynecol. Scand.* **32,** 1–6.

Borum, K. (1961). Oogenesis in the mouse; study of the meiotic prophase. *Exp. Cell Res.* **24,** 495–507.

Bradford, G. E. (1969). Genetic control of ovulation rate and embryo survival of mice. I. Response to selection. *Genetics* **61,** 905–921.

Bradford, G. E. (1972). Genetic control of litter size in sheep. *J. Reprod. Fertil., Suppl.* **15,** 23–41.

Bragdon, D. E. (1952). Corpus luteum formation and atresia in the common garter snake, *Thamnophis sirtalis. J. Morphol.* **91,** 413–445.

Brambell, F. W. R. (1956). Ovarian changes. *In* "Marshall's Physiology of Reproduction" (A. S. Parkes, ed.), 3rd ed., Vol. 1, pp. 397–542. Longmans Green, London and New York.

Branca, A. (1925). L'ovocyte atrésique et son involution. *Arch. Biol.* **35,** 325–440.

Breckwoldt, M., Bettendorf, G., and García, R. (1971). Induction of ovulation in noncycling and hypophysectomized rhesus monkeys with various human gonadotrophins. *Fertil. Steril.* **22,** 451–455.

Bretschneider, L. H., and Duyvené de Wit, J. J. (1947). "Sexual Endocrinology of Non-Mammalian Vertebrates." Elsevier, Amsterdam.

Brooks, P. H., Cooper, K. J., Lamming, G. E., and Cole, D. J. A. (1972). The effect

of feed level during oestrus on ovulation rate in the gilt. *J. Reprod. Fertil.* **30,** 45–53.
Brown-Grant, K. (1969). The induction of ovulation by ovarian steroids in the adult rat. *J. Endocrinol.* **43,** 553–562.
Burden, H. W. (1972). Ultrastructural observations on ovarian perifollicular smooth muscle in the cat, guinea pig, and rabbit. *Am. J. Anat.* **133,** 125–142.
Burkl, W. (1962). Segmentation und Fragmentation der Eizellen in atretischen Eierstockfollikeln. *Z. Zellforsch. Mikrosk. Anat.* **58,** 369–386.
Burkl, W. (1965). Zur kauselen Genese der Follikelatresie. *Arch. Gynaekol.* **200,** 689–698.
Burr, J. H., Jr., and Davies, J. I. (1951). The vascular system of the rabbit ovary and its relationship to ovulation. *Anat. Rec.* **111,** 273–297.
Byskov, A.-G., and Lintern-Moore, S. (1973). Follicle formation in the immature mouse ovary: the role of the rete ovarii. *J. Anat.* **116,** 207–217.
Byskov, A. G. S. (1969). Ultrastructural studies on the preovulatory follicle in the mouse ovary. *Z. Zellforsch. Mikrosk. Anat.* **100,** 285–299.
Byskov, A. G. S. (1974). Cell kinetics of follicular atresia in the mouse ovary. *J. Reprod. Fertil.* **37,** 277–285.
Canivenc, R., Bonnin-Laffargue, M., and Lajus, M. (1965). Action du nitrate d'argent sur l'atrésie folliculaire chez le Blaireau européen (*Meles meles* L.). *C. R. Soc. Biol.* (*Paris*) **159,** 1953–1955.
Challoner, S. (1975). Studies on oogenesis and follicular development in the golden hamster. 3. The initiation of follicular growth. *J. Anat.* **119,** 157–162.
Chan, S. T. H., Wright, A., and Phillips, J. (1967). The atretic structures in the gonads of the rice-field eel, *Monopterus albus*, during natural sex-reversal. *J. Zool.* **153,** 527–539.
Chang, C., and Witschi, E. (1957). Cortisone effect on ovulation in the frog. *Endocrinology* **61,** 514–519.
Channing, C. P., and Kammerman, S. (1974). Binding of gonadotropins to ovarian cells. *Biol. Reprod.* **10,** 179–198.
Chatterjee, A., and Greenwald, G. S. (1972). The long term effects of unilateral ovariectomy of the cycling hamster and rat. *Biol. Reprod.* **7,** 238–246.
Chieffi, G. (1967). The reproductive system of elasmobranchs; developmental and endocrinological aspects. *In* "Sharks, Skates and Rays" (P. W. Gilbert, R. F. Mathewson, and D. P. Ralls, eds.), pp. 553–580. Johns Hopkins Press, Baltimore, Maryland.
Civen, M., Brown, C. B., and Hilliard, J. (1966). Ribonucleic acid and protein synthesis in ovary. *Biochim. Biophys. Acta* **114,** 127–134.
Claesson, L. (1947). Is there any smooth musculature in the wall of the Graafian follicle? *Acta Anat.* **3,** 295–311.
Conaway, C. H. (1971). Ecological adaptation and mammalian reproduction. *Biol. Reprod.* **4,** 239–247.
Coppola, J. A., Leonard, R. G., and Lippman, W. (1966). Ovulatory failure in rats after treatment with brain norepinephrine depletors. *Endocrinology* **78,** 225–228.
Cross, P. C. (1972). Observations on the induction of ovulation in *Microtus montanus*. *J. Mammal.* **53,** 210–212.
Dailey, R. A., Cloud, J. G., First, N. L., Chapman, A. B., and Casida, L. E. (1970). Response of 170-day-old prepuberal Poland China, Yorkshire and crossbred gilts to unilateral ovariectomy. *J. Anim. Sci.* **31,** 937–939.

Dalmane, A. R. (1967). [Atresia of mammalian ovarian follicles.] *Arkh. Anat., Gistol. Embriol.* **53,** 73–84. (In Russian.)
Deane, H. W. (1952). Histochemical observations on the ovary and oviduct of the albino rat during the estrous cycle. *Am. J. Anat.* **91,** 363–413.
Deanesly, R. (1972). Origins and development of interstitial tissue in ovaries of rabbit and guinea-pig. *J. Anat.* **113,** 251–260.
Decker, A. (1951). Culdoscopic observation on the tubo-ovarian mechanism of ovum reception. *Fertil. Steril.* **2,** 253–259.
Döcke, F., and Dörner, G. (1965). The mechanism of the induction of ovulation by oestrogens. *J. Endocrinol.* **33,** 491–499.
Döcke, F., and Dörner, G. (1966). Facilitative action of progesterone in the induction of ovulation by oestrogen. *J. Endocrinol.* **36,** 209–210.
Döcke, F., and Dörner, G. (1969). A possible mechanism by which progesterone facilitates ovulation in the rat. *Neuroendocrinology* **4,** 139–149.
Dominic, C. J. (1961). Study of the atretic follicle in the ovary of the domestic pigeon. *Proc. Natl. Acad. Sci., India Sect. B* **31,** 273–286.
Donahue, R. P. (1972). The relation of oocyte maturation to ovulation in mammals. *In* "Oogenesis" (J. D. Biggers and A. W. Schuetz, eds.), pp. 413–438. Univ. Park Press, Baltimore, Maryland.
Donovan, B. T. (1967). Control of follicular growth and ovulation. *In* "Reproduction in the Female Mammal" (G. E. Lamming and E. C. Amoroso, eds.), pp. 3–27. Butterworth, London.
Dorrington, J. H., and Baggett, B. (1969). Adenyl cyclase activity in the rabbit ovary. *Endocrinology* **84,** 989–996.
Doyle, J. B. (1954). Ovulation and the effects of selective uterotubal denervation; direct observation by culdotomy. *Fertil. Steril.* **5,** 105–130.
Dukelow, W. R. (1970). Induction and timing of single and multiple ovulations in the squirrel monkey (*Saimiri sciureus*). *J. Reprod. Fertil.* **22,** 303–309.
Dutt, R. H. (1953). The role of estrogens and progesterone in ovulation. *Iowa State Coll. J. Sci.* **28,** 55–66.
Edgren, R. A., and Carter, D. L. (1963). Studies on progesterone-induced *in vitro* ovulation in *Rana pipiens*. *Gen. Comp. Endocrinol.* **3,** 526–528.
Edwards, R. G. (1962). Meiosis in ovarian oocytes of adult mammals. *Nature* (*London*) **196,** 446–450.
Edwards, R. G. (1974). Follicular fluid. *J. Reprod. Fertil.* **37,** 189–219.
Edwards, R. G., and Steptoe, P. C. (1975). Induction of follicular growth, ovulation and luteinization in the human ovary. *J. Reprod. Fertil., Suppl.* **22,** 121–163.
El-Fouly, M. A., Cook, B., Nekola, M., and Nalbandov, A. V. (1970). The role of the ovum in follicular luteinization. *Endocrinology* **87,** 288–293.
Ellsworth, L. R., and Armstrong, D. T. (1973). Luteinization of transplanted ovarian follicles in the rat induced by dibutyryl cyclic AMP. *Endocrinology* **92,** 840–846.
Emmens, C. W. (1940–1941). The production of ovulation in the rabbit by the intravenous injection of salts of copper and cadmium. *J. Endocrinol.* **2,** 63–69.
Engle, E. T. (1927). A quantitative study of follicular atresia in the mouse. *Am. J. Anat.* **39,** 187–203.
Erickson, B. H. (1966). Development and radio-response of the prenatal bovine ovary. *J. Reprod. Fertil.* **11,** 97–105.
Erpino, M. J. (1973). Histogenesis of atretic ovarian follicles in a seasonally breeding bird. *J. Morphol.* **139,** 239–250.

Espey, L. L. (1967). Ultrastructure of the apex of the rabbit's Graafian follicle during the ovulatory process. *Endocrinology* **81,** 267–277.
Espey, L. L. (1970). Effect of various substances on tensile strength of sow ovarian follicles. *Am. J. Physiol.* **219,** 230–233.
Espey, L. L. (1971). Decomposition of connective tissue in rabbit ovarian follicles by multivesicular structures of thecal fibroblasts. *Endocrinology* **88,** 437–444.
Espey, L. L. (1974). Ovarian proteolytic enzymes and ovulation. *Biol. Reprod.* **10,** 216–235.
Espey, L. L., and Lipner, H. (1963). Measurements of intrafollicular pressures in the rabbit ovary. *Am. J. Physiol.* **205,** 1067–1072.
Espey, L. L., and Lipner, H. (1965). Enzyme-induced rupture of the rabbit Graafian follicle. *Am. J. Physiol.* **208,** 208–213.
Espey, L. L., and Rondell, P. (1966). Deterioration of the connective tissue of the Graafian follicle as ovulation approaches. *Physiologist* **9,** 177.
Espey, L. L., and Rondell, P. (1968). Collagenolytic activity in the rabbit and sow Graafian follicle during ovulation. *Am. J. Physiol.* **214,** 326–329.
Evans, H. M., and Cole, H. H. (1931). An introduction to the study of the oestrous cycle in the dog. *Mem. Univ. Calif.* **9,** 65–225.
Everett, J. W. (1956). The time of release of ovulating hormone from the rat hypophysis. *Endocrinology* **59,** 580–585.
Everett, J. W. (1967). Provoked ovulation of long-delayed pseudopregnancy from coital stimuli in barbiturate-blocked rats. *Endocrinology* **80,** 145–154.
Everett, J. W. (1972). Brain, pituitary gland, and the ovarian cycle. *Biol. Reprod.* **6,** 3–12.
Everett, J. W., and Sawyer, C. H. (1950). A 24 hour periodicity in the "LH release apparatus" of female rats, disclosed by barbiturate sedation. *Endocrinology* **47,** 198–218.
Falconer, D. S., Edwards, R. G., Fowler, R. E., and Roberts, R. C. (1961). Analyses of differences in the numbers of eggs shed by the two ovaries of mice during natural oestrus and after superovulation. *J. Reprod. Fertil.* **2,** 418–437.
Fee, A. R., and Parkes, A. S. (1930). Studies on ovulation. Effect of vaginal anaesthesia on ovulation in the rabbit. *J. Physiol.* (*London*) **70,** 385–388.
Fernandez-Baca, S., Hansel, W., and Novoa, C. (1970). Corpus luteum function in the alpaca. *Biol. Reprod.* **3,** 251–261.
Fink, G., and Schofield, G. C. (1970). Innervation of ovary in cats. *J. Anat.* **106,** 191.
Fowler, R. E., and Edwards, R. G. (1957). Induction of superovulation and pregnancy in mature mice by gonadotrophins. *J. Endocrinol.* **15,** 374–384.
Franchi, L. L., and Mandl, A. M. (1962). Ultrastructure of oogonia and oocytes in the foetal and neonatal rat. *Proc. R. Soc. London, Ser. B* **157,** 99–114.
Franchi, L. L., Mandl, A. M., and Zuckerman, S. (1962). The development of the ovary and the process of oogenesis. *In* "The Ovary" (S. Zuckerman, ed.), 1st ed., Vol. I, Chapter 1, pp. 1–88. Academic Press, New York.
Fraps, R. M. (1954). Neural basis of diurnal periodicity in release of ovulation-inducing hormone in fowl. *Proc. Natl. Acad. Sci. U.S.A.* **40,** 348–356.
Fraps, R. M. (1961). Ovulation in the domestic fowl. *In* "Control of Ovulation" (C. A. Villee, ed.), pp. 133–162. Pergamon, Oxford.
Fraps, R. M. (1965). Twenty-four hour periodicity in the mechanism of pituitary gonadotrophin release for follicular maturation and ovulation in the chicken. *Endocrinology* **77,** 5–18.

Frith, D. A., and Hooper, K. C. (1971). Action of some ovulation inhibitors on the rabbit hypothalamus. *Acta Endocrinol.* (*Copenhagen*) **66,** 221–228.

Fujimoto, S., Rawson, J. M. R., and Dukelow, W. R. (1974). Hormonal influences on the time of ovulation in the rabbit as determined by laparoscopy. *J. Reprod. Fertil.* **38,** 97–103.

Garde, M. L. (1930). The ovary of *Ornithorhynchus* with special reference to follicular atresia. *J. Anat.* **64,** 422–453.

Gemzell, C. (1965). Induction of ovulation with human gonadotropins. *Recent Prog. Horm. Res.* **21,** 179–204.

Gemzell, C. A., and Johansson, E. D. B. (1971). Human gonadotrophins. Factors stimulating ovarian function. *Control Hum. Fertil., Proc. Nobel Symp., 15th, 1970* pp. 241–262.

Gilbert, A. B. (1969). A reassessment of certain factors which affect egg production in the domestic fowl. *World's Poult. Sci. J.* **25,** 239–258.

Gilbert, A. B. (1970). The possible reactivation of atretic follicles in the domestic hen. *J. Reprod. Fertil.* **22,** 184–185.

Gilbert, A. B. (1971). The control of ovulation. *In* "Physiology and Biochemistry of the Domestic Fowl" (D. J. Bell and B. M. Freeman, eds.), Vol. 3, pp. 1225–1236. Academic Press, New York.

Gillman, J. (1948). The development of the gonads in man, with consideration of the role of fetal endocrines and the histogenesis of ovarian tumors. *Contrib. Embryol. Carnegie Inst.* **32,** 81–131.

Gorski, R. A. (1966). Localization and sexual differentiation of the nervous structures which regulate ovulation. *J. Reprod. Fertil., Suppl.* **1,** 67–88.

Gorski, R. A. (1968). The neural control of ovulation. *In* "Biology of Gestation" (N. Assali, ed.), Vol. 1, pp. 1–66. Academic Press, New York.

Gothie, S. (1967). Contribution à l'étude cinétique de l'ovulation chez la Souris prépubère. *C. R. Soc. Biol.* **161,** 554–556.

Graves, C. N., and Dziuk, P. J. (1968). Control of ovulation in dairy cattle with human chorionic gonadotrophin after treatment with 6α-methyl-17α-acetoxyprogesterone. *J. Reprod. Fertil.* **17,** 169–172.

Greenwald, G. S. (1961). Quantitative study of follicular development in the ovary of the intact or unilaterally ovariectomized hamster. *J. Reprod. Fertil.* **2,** 351–361.

Greenwald, G. S. (1971). Preovulatory changes in ovulating hormone in the cyclic hamster. *Endocrinology* **88,** 671–677.

Greenwald, G. S. (1972). Of eggs and follicles. *Am. J. Anat.* **135,** 1–3.

Greenwald, G. S., and Choudary, J. B. (1969). Follicular development and induction of ovulation in the pregnant mouse. *Endocrinology* **84,** 1512–1520.

Greep, R. O. (1961). Physiology of the anterior hypophysis in relation to reproduction. *In* "Sex and Internal Secretions" (W. C. Young, ed.), 3rd ed., Vol. 1, pp. 240–301. Williams & Wilkins, Baltimore, Maryland.

Greulich, W. W. (1934). Artificially induced ovulation in the cat (*Felis domesticus*). *Anat. Rec.* **58,** 217–224.

Grosser, O. (1903). Der physiologische bindegewebige Atresia der Genitokanales von Vesperugo nach erfoger Cohabitation. *Verh. Anat. Ges., Jena* **23,** 129–132.

Gunn, R. G., and Doney, J. M. (1973). The effects of nutrition and rainfall at the time of mating on the reproductive performance of ewes. *J. Reprod. Fertil., Suppl.* **19,** 253–258.

Guraya, S. S., and Greenwald, G. S. (1964). A comparative histochemical study of the interstitial tissue and follicular atresia in the mammalian ovary. *Anat. Rec.* **149,** 411–434.

Guthrie, M. J., and Jeffers, K. R. (1938). A cytological study of the ovaries of the bats *Myotis lucifugus lucifugus* and *Myotis grisescens*. *J. Morphol.* **62,** 523–558.

Guttmacher, M. S., and Guttmacher, A. F. (1921). Morphological and physiological studies of the musculature of the mature Graafian follicle of the sow. *Johns Hopkins Hosp., Bull.* **32,** 394–399.

Hafez, E. S. E. (1969). Superovulation and preservation of mammalian ova. *Acta Endocrinol.* (*Copenhagen*), *Suppl.* **40,** 1–44.

Hamburger, C., and Pedersen-Bjergaard, K. (1946). On the effect of gonadotrophins in normal infantile female guinea-pigs. *Acta Pathol.* **23,** 84–102.

Hammond, J., Jr., Hammond, J., and Parkes, A. S. (1942). Hormonal augmentation of fertility in sheep. 1. Induction of ovulation, superovulation and heat in sheep. *J. Agric. Sci.* **32,** 308–323.

Hansel, W., and Snook, R. B. (1970). Pituitary-ovarian relationships in the cow. *J. Dairy Sci.* **53,** 945–961.

Harman, M. T., and Kirgis, H. D. (1938). The development and atresia of the Graafian follicle and the division of intraovarian ova in the guinea-pig. *Am. J. Anat.* **63,** 79–99.

Harper, M. J. K. (1963). Ovulation in the rabbit: the time of follicular rupture and expulsion of the eggs in relation to injection of LH. *J. Endocrinol.* **26,** 307–316.

Harrington, F. E., Bex, F. J., Elton, R. L., and Roach, J. B. (1970). The ovulatory effects of follicle stimulating hormone treated with chymotrypsin in chlorpromazine blocked rats. *Acta Endocrinol.* (*Copenhagen*) **65,** 222–228.

Harris, G. W. (1969). Ovulation. *Am. J. Obstet. Gynecol.* **105,** 659–669.

Harrison, R. M., and Dukelow, W. R. (1970). Induced ovulation in the non-human primate. *Fed. Proc., Fed. Am. Soc. Exp. Biol.* **29,** 643.

Heap, R. B., and Illingworth, D. V. (1974). The maintenance of gestation in the guinea-pig and other hystricomorph rodents: changes in the dynamics of progesterone metabolism and the occurrence of progesterone-binding globulin (PBG). *Symp. Zool. Soc. London* **34,** 385–415.

Heape, W. (1905). Ovulation and the degenerating ova in the rabbit. *Proc. R. Soc. London, Ser. B* **76,** 260–268.

Heinze, H. (1975). Pelviscopy in the mare. *J. Reprod. Fertil., Suppl.* **23,** 319–321.

Hermreck, A. S., and Greenwald, G. S. (1964). The effects of unilateral ovariectomy on follicular maturation in the guinea-pig. *Anat. Rec.* **148,** 171–176.

Hill, J. P., Allen, E., and Kramer, T. C. (1935). Cinematographic studies of rabbit ovulation. *Anat. Rec.* **63,** 239–245.

Hilliard, J., Penardi, R., and Sawyer, C. H. (1967). A functional role for 20α-hydroxypregn-4-en-3-one in the rabbit. *Endocrinology* **80,** 901–909.

Hisaw, F. L. (1947). The development of the Graafian follicle and ovulation. *Physiol. Rev.* **27,** 95–119.

Hisaw, F. L., Jr., and Hisaw, F. L. (1959). The corpora lutea of elasmobranch fishes. *Anat. Rec.* **135,** 269–277.

Hoar, W. S. (1955). Comparative physiology: hormones and reproduction in fishes. *Annu. Rev. Physiol.* **27,** 51–70.

Hogben, L. (1930). Some remarks on the relation of the pituitary gland to ovulation and skin secretion in *Xenopus laevis*. *Proc. R. Soc. S. Afr.* [see *Trans. R. Soc. S. Afr.* **22,** xvii–xviii (1934)].
Holsinger, J. W., and Everett, J. W. (1970). Thresholds to pre-optic stimulation at varying times in the rat estrous cycle. *Endocrinology* **86,** 251–256.
Horowitz, M. (1967). Anovular corpus luteum in the golden hamster (*Mesocricetus auratus* Waterhouse) and comparisons with the normal corpus luteum and follicle of atresia. *Acta Anat.* **66,** 199–225.
Hunter, J. (1787). An experiment to determine the effect of extirpating one ovarium upon the number of young produced. *Philos. Trans. R. Soc. London, Ser. B* **77,** 233–239.
Hutchinson, J. S. M., and Robertson, H. A. (1966). Growth of the follicle and corpus luteum in the ovary of the sheep. *Res. Vet. Sci.* **7,** 17–24.
Ingram, D. C. (1959). The effect of gonadotrophins and oestrogens on ovarian atresia in the immature rat. *J. Endocrinol.* **19,** 117–122.
Ingram, D. C. (1962). Atresia. *In* "The Ovary" (S. Zuckerman, ed.), 1st ed., Vol. 1, Chapter 4, pp. 247–273. Academic Press, New York.
Ioannou, J. M. (1964). Oogenesis in the guinea-pig. *J. Embryol. Exp. Morphol.* **12,** 673–691.
Jacobowitz, D., and Wallach, E. E. (1967). Histochemical and chemical studies of the autonomic innervation of the ovary. *Endocrinology* **81,** 1132–1139.
Jainudeen, M. R., Hafez, E. S. E., and Lineweaver, J. A. (1966). Superovulation in the calf. *J. Reprod. Fertil.* **12,** 149–153.
Jöchle, W. (1975). Current research in coitus-induced ovulation: a review. *J. Reprod. Fertil., Suppl.* **22,** 165–207.
Johansson, E. D. B., and Wide, L. (1969). Periovulatory levels of plasma progesterone and LH in women. *Acta Endocrinol.* (*Copenhagen*) **62,** 82–88.
Joly, J., and Picherel, B. (1972). Ultrastructure, histochemie et physiologie du follicule pré-ovulatoire et du corps jaune de l'urodèle ovo-vivipare, *Salamandra salamandra* (L.). *Gen. Comp. Endocrinol.* **18,** 235–257.
Jones, E. C., and Krohn, P. L. (1959). Influence of the anterior pituitary on the ageing process in the ovary. *Nature* (*London*) **183,** 1155–1158.
Jones, E. C., and Krohn, P. L. (1960). The effect of unilateral ovariectomy on the reproductive lifespan of mice. *J. Endocrinol.* **20,** 129–134.
Jones, E. C., and Krohn, P. L. (1961a). The relationship between age, numbers of oocytes and fertility in virgin and multiparous mice. *J. Endocrinol.* **21,** 469–495.
Jones, E. C., and Krohn, P. L. (1961b). The effect of hypophysectomy on age changes in the ovaries of mice. *J. Endocrinol.* **21,** 497–509.
Jones, E. E., and Nalbandov, A. V. (1972). Effects of intrafollicular injection of gonadotrophins on ovulation or luteinization of ovarian follicles. *Biol. Reprod.* **7,** 87–93.
Kar, A. B., Kamboj, V. P., and Das, R. D. (1962). Incipient follicular atresia and acid phosphatase activity in ovary and serum of rats. *J. Sci. Ind. Res., Sect. C* **21,** 231–237.
Kawakama, M., and Teresawa, E. (1970). Effect of electrical stimulation of the brain on ovulation during the estrous cycle in the rat. *Endocrinol. Jpn.* **17,** 7–14.
Kelley, G. L. (1931). Direct observations of rupture of Graafian follicles in the mammal. *J. Fla. Med. Assoc.* **17,** 422–423.

Kennelly, J. J., and Foote, R. H. (1965). Superovulatory response of pre- and post-pubertal rabbits to commercially available gonadotrophins. *J. Reprod. Fertil.* **9,** 177–188.

Kern, M. D. (1972). Seasonal changes in the reproductive system of the female white-crowned sparrow, *Zonotrichia leucophrys gambelii,* in captivity and in the field. 1. The ovary. *Z. Zellforsch. Mikrosk. Anat.* **126,** 297–319.

Kessel, R. C., and Panje, W. R. (1968). Organisation and activity in the pre- and post-ovulatory follicle of *Necturus maculosus. J. Cell Biol.* **39,** 1–34.

Kingsbury, B. F. (1939). Atresia and the interstitial cells of the ovary. *Am. J. Anat.* **65,** 309–331.

Knigge, K. M., and Leathem, J. H. (1956). Growth and atresia of follicles in the ovary of the hamster. *Anat. Rec.* **124,** 679–707.

Koves, K., and Halasz, B. (1970). Location of the neural structures triggering ovulation in the rat. *Neuroendocrinology* **6,** 180–193.

Krohn, P. L. (1967). Factors influencing the number of oocytes in the ovary. *Arch. Anat. Microsc.* **56,** Suppl. 3–4, 151–159.

Labhsetwar, A. P. (1970). The role of oestrogens in spontaneous ovulation: evidence for positive oestrogen feedback in the 4-day oestrous cycle. *J. Endocrinol.* **47,** 481–493.

Labhsetwar, A. P. (1972). Role of monoamines in ovulation: evidence for a serotoninergic pathway for inhibition of spontaneous ovulation. *J. Endocrinol.* **54,** 269–275.

Labhsetwar, A. P. (1974). Prostaglandins and the reproductive cycle. *Fed. Proc., Fed. Am. Soc. Exp. Biol.* **33,** 61–77.

Lambert, J. G. D. (1970). The ovary of the guppy (*Poecilia reticulata*). The atretic follicle a *corpus atreticum* or a *corpus luteum praeovulationis*? *Z. Zellforsch. Mikrosk. Anat.* **107,** 54–67.

Lance, V., and Callard, I. P. (1969). A histochemical study of ovarian function in the ovoviviparous elasmobranch, *Squalus acanthis. Gen. Comp. Endocrinol.* **13,** 255–267.

Land, R. B. (1970a). Number of oocytes present at birth in the ovaries of pure and Finnish Landrace cross Blackface and Welsh sheep. *J. Reprod. Fertil.* **21,** 517–521.

Land, R. B. (1970b). A relationship between the duration of oestrus, ovulation rate and litter size of sheep. *J. Reprod. Fertil.* **23,** 49–53.

Land, R. B., and Carr, W. R. (1975). Testis growth and LH concentration following hemicastration and its relation with female prolificacy in sheep. *J. Reprod. Fertil.* **45,** 495–501.

Land, R. B., and Falconer, D. S. (1969). Genetic studies of ovulation rate in the mouse. *Genet. Res.* **13,** 25–46.

Land, R. B., de Reviers, M.-M., Thompson, R., and Mauléon, P. (1974). Quantitative physiological studies of genetic variation in the ovarian activity of the rat. *J. Reprod. Fertil.* **38,** 29–39.

Larsen, L. O. (1970). The lamprey egg at ovulation (*Lampetra fluviatilis*). *Biol. Reprod.* **2,** 37–47.

LeMaire, W. J., and Marsh, J. M. (1975). Interrelationships between prostaglandins, cyclic AMP and steroids in ovulation. *J. Reprod. Fertil., Suppl.* **22,** 53–74.

LeMaire, W. J., Mills, T., Ito, Y., and Marsh, J. M. (1972). Inhibition by 3′,5′-

cyclic AMP of luteinization induced by intrafollicular injection of LH. *Biol. Reprod.* **6,** 109–116.
Licht, P. (1970). Effects of mammalian gonadotrophins (ovine FSH and LH) in female lizards. *Gen. Comp. Endocrinol.* **14,** 98–106.
Licht, P., and Stockell Hartree, A. (1971). Actions of mammalian, avian and piscine gonadotrophins in the lizard. *J. Endocrinol.* **49,** 113–124.
Lipner, H. (1965). Induction of Graafian follicle proteolytic enzyme activity by HCG in rat Graafian follicle. *Am. Zool.* **5,** 167.
Lipner, H., and Maxwell, B. (1960). Hypothesis concerning the role of follicular contractions in ovulation. *Science* **131,** 1737–1738.
Lostroh, A. J., and Johnson, R. E. (1966). Amounts of ICSH and FSH required for follicular development, uterine growth and ovulation in the hypophysectomized rat. *Endocrinology* **79,** 991–996.
Lunenfeld, B., Insler, V., and Rabau, E. (1969). Induction de l'ovulation par les gonadotrophines. *In* "L'ovulation, méiose et ouverture folliculaire. Traitements de l'anovulation" (R. Moricard and J. Ferrin, eds.), pp. 291–340. Masson, Paris.
Lutwak-Mann, C. (1962). The influence of nutrition on the ovary. *In* "The Ovary" (S. Zuckerman, ed.), 1st ed., Vol. 2, Chapter 18, pp. 291–315. Academic Press, New York.
Lyngnes, R. (1936). Ruckbildung der ovuliertung und nicht ovulierten Follikel im Ovarium der *Myxine glutinosa* L. *Skr. Nor. Vidensk.-Akad. Oslo* **4,** 1–116.
McCann, S. M., and Porter, J. C. (1969). Hypothalamic pituitary-stimulating and inhibiting hormones. *Physiol. Rev.* **49,** 240–284.
McDonald, P. G., and Gilmore, D. R. (1971). Effect of ovarian steroids on hypothalamic thresholds for ovulation in the female rat. *J. Endocrinol.* **49,** 421–429.
McLaren, A. (1963). The distribution of eggs and embryos between sides in the mouse. *J. Endocrinol.* **27,** 157–181.
Mahoney, C. J. (1970). A study of the menstrual cycle in *Macaca irus* with special reference to the detection of ovulation. *J. Reprod. Fertil.* **21,** 153–163.
Major, P. W., and Kilpatrick, R. (1972). Cyclic AMP and hormone action. *J. Endocrinol.* **52,** 593–630.
Mallampati, R. S., and Casida, L. E. (1970). Ovarian compensatory hypertrophy following unilateral ovariectomy during the breeding season in the ewe. *Biol. Reprod.* **3,** 43–46.
Mandl, A. M., and Zuckerman, S. (1950). The numbers of normal and atretic ova in the mature rat. *J. Endocrinol.* **6,** 426–435.
Mandl, A. M., and Zuckerman, S. (1951). Numbers of normal and atretic oocytes in unilaterally spayed rats. *J. Endocrinol.* **7,** 112–119.
Markee, J. E., and Hinsey, J. C. (1936). Observations on ovulation in the rabbit. *Anat. Rec.* **64,** 309–319.
Marsh, J. M., and Savard, K. (1964). The activation of luteal phosphorylase by luteinizing hormone. *J. Biol. Chem.* **239,** 1–7.
Marshall, A. J., and Coombs, C. J. F. (1957). The interaction of environmental, internal and behavioural factors in the rook, *Corvus frugilegus* Linnaeus. *Proc. Zool. Soc. London* **128,** 545–589.
Marshall, F. H. A. (1943). The undischarged follicle. *In* "Essays in Biology in Honor of Herbert M. Evans," pp. 381–385. Univ. of California Press, Berkeley.

Martínez-Esteve, P. (1942). Observations on the histology of the opossum ovary. *Contrib. Embryol. Carnegie Inst.* **30,** 17–26.

Mauleon, P. (1969). Oogenesis and folliculogenesis. *In* "Reproduction in Domestic Animals" (H. H. Cole and P. T. Cupps, eds.), 2nd ed., pp. 187–225. Academic Press, New York.

Maurer, R. R., Hunt, W. L., and Foote, R. H. (1968). Repeated superovulation following administration of exogenous gonadotrophins in Dutch-belted rabbits. *J. Reprod. Fertil.* **15,** 93–102.

Milligan, S. R. (1975). Mating, ovulation and corpus luteum formation in the vole, *Microtus agrestis. J. Reprod. Fertil.* **42,** 35–44.

Mitchell, M. E. (1967). Stimulation of the ovary in hypophysectomized hens by an avian pituitary preparation. *J. Reprod. Fertil.* **14,** 249–256.

Mitchell, M. E. (1970). Treatment of hypophysectomized hens with partially purified avian FSH. *J. Reprod. Fertil.* **22,** 233–241.

Miyata, J., Taymor, M. L., Levesque, L., and Lymeburner, N. (1970). Timing of ovulation by a rapid luteinizing hormone assay. *Fertil. Steril.* **21,** 784–790.

Moor, R. M., Hay, M. F., and Caldwell, B. V. (1971). The sheep follicle: relationship between sites of steroid dehydrogenase activity, gonadotrophic stimulation and steroid production. *J. Reprod. Fertil.* **27,** 464–485.

Mori, H., and Matsumoto, K. (1973). Development of the secondary interstitial gland in the rabbit ovary. *J. Anat.* **116,** 417–430.

Mossman, H. W., and Duke, K. L. (1973). "Comparative Morphology of the Mammalian Ovary." Univ. of Wisconsin Press, Madison.

Mossman, H. W., and Judas, I. (1949). Accessory corpora lutea, lutein cell origin and the ovarian cycle in the Canadian porcupine. *Am. J. Anat.* **85,** 1–39.

Mossman, H. W., Koering, M. J., and Ferry, D. (1964). Cyclic changes of interstitial gland tissue of the human ovary. *Am. J. Anat.* **115,** 235–256.

Motta, P., Cherney, D. D., and Didio, L. J. A. (1971). Scanning and transmission electron microscopy of the ovarian surface in mammals with special reference to ovulation. *J. Submicrosc. Cytol.* **3,** 85–100.

Mozanska, T. (1969). Morphological studies of the atresia of Graafian follicles in the ovaries of the albino rat. *Folia Histochem. Cytochem.* **7,** 86 (abstr.).

Nair, P. V. (1963). Ovular atresia and the formation of the so-called 'corpus luteum' in the ovary of the Indian catfish, *Heteropneustes fossilis. Proc. Zool. Soc. Bengal* **16,** 51–61.

Nakajo, S., Zakaria, A. H., and Imai, K. (1973). Effect of the local administration of proteolytic enzymes on the rupture of the ovarian follicles in the domestic fowl, *Gallus domesticus. J. Reprod. Fertil.* **34,** 235–240.

Nakamura, T. (1957). Cytological studies on abnormal ova in immature ovaries of mice observed at different phases of the cycle. *J. Fac. Fish. Anim. Husb., Hiroshima Univ.* **1,** 343–356.

Nalbandov, A. V. (1961). Mechanisms controlling ovulation of avian and mammalian follicles. *In* "Control of Ovulation" (C. A. Villee, ed.), pp. 122–132. Pergamon, Oxford.

Nalbandov, A. V. (1972). Interaction between oocytes and follicular cells. *In* "Oogenesis" (J. D. Biggers and A. W. Schuetz, eds.), pp. 513–522. Univ. Park Press, Baltimore, Maryland.

Neal, P., and Baker, T. G. (1975). Response of mouse Graafian follicles in organ culture to various doses of follicle-stimulating hormone and luteinizing hormone. *J. Endocrinol.* **65,** 27–32.

Nelsen, O. E., and White, E. L. (1941). A method for inducing ovulation in the anoestrous opossum (*Didelphys virginiana*). *Anat. Rec.* **81,** 529–535.
Nelson, D. M., and Nalbandov, A. V. (1966). Hormone control of ovulation. *In* "Physiology of the Domestic Fowl" (C. Horton-Smith and E. C. Amoroso, eds.), pp. 3–10. Oliver & Boyd, Edinburgh.
Norman, R. L., Blake, C. A., and Sawyer, C. H. (1972). Delay of the proestrous ovulatory surge of LH in the hamster by pentobarbital or ether. *Endocrinology* **91,** 1025–1029.
Ohno, S., and Smith, J. B. (1964). Role of foetal follicular cells in meiosis of mammalian oocytes. *Cytogenetics* **3,** 324–333.
Okamura, H., Virtumasen, P., Wright, K. H., and Wallach, E. (1972). Ovarian smooth muscle in the human being, rabbit and cat. *Am. J. Obstet. Gynecol.* **112,** 183–191.
O'Shea, J. D. (1970). Ultrastructural study of smooth muscle-like cells in the theca externa of ovarian follicles in the rat. *Anat. Rec.* **167,** 127–139.
O'Shea, J. D. (1971). Smooth muscle-like cells in the theca externa of ovarian follicles in the sheep. *J. Reprod. Fertil.* **24,** 283–285.
Osvaldo-Decima, L. (1970). Smooth muscle in the ovary of the rat and monkey. *J. Ultrastruct. Res.* **29,** 218–237.
Packham, A., and Triffitt, L. K. (1966). Association between ovulation rate and twinning in Merino sheep. *Aust. J. Agric. Res.* **17,** 515–520.
Palti, Z., and Freund, M. (1972). Spontaneous contractions of the human ovary *in vitro*. *J. Reprod. Fertil.* **28,** 113–115.
Pant, M. C. (1968). Process of atresia and the fate of the discharged follicle in the ovary of *Glyptothorax pectinopterus*. *Zool. Anz.* **181,** 153–160.
Parr, E. L. (1974). Histological examination of the rat ovarian follicle wall before ovulation. *Biol. Reprod.* **11,** 483–503.
Parr, E. L. (1975). Rupture of ovarian follicles at ovulation. *J. Reprod. Fertil., Suppl.* **22,** 1–17.
Pearson, O. P. (1944). Reproduction in the shrew (*Blarina brevicauda* Say). *Am. J. Anat.* **75,** 39–93.
Pearson, O. P. (1949). Reproduction of a South American rodent, the mountain viscacha. *Am. J. Anat.* **84,** 143–174.
Pedersen, T. (1972). Follicle growth in the mouse. *In* "Oogenesis" (J. D. Biggers and A. W. Schuetz, eds.), pp. 361–376. Univ. Park Press, Baltimore, Maryland.
Peppler, R. D. (1971). Effect of unilateral ovariectomy on follicular development and ovulation in cycling, aged rats. *Am. J. Anat.* **132,** 423–427.
Peppler, R. D., and Greenwald, G. S. (1970a). Effects of unilateral ovariectomy on ovulation and cycle length in 4- and 5-day cycling rats. *Am. J. Anat.* **127,** 1–7.
Peppler, R. D., and Greenwald, G. S. (1970b). Influence of unilateral ovariectomy on follicular development in cycling rats. *Am. J. Anat.* **127,** 9–14.
Perry, J. S., and Rowlands, I. W. (1962). The ovarian cycle in vertebrates. *In* "The Ovary" (S. Zuckerman, ed.), 1st ed., Vol. 1, Chapter 5, pp. 275–309. Academic Press, New York.
Perry, J. S., and Rowlands, I. W. (1963). Hypophysectomy of the immature guinea-pig and the ovarian response to gonadotrophins. *J. Reprod. Fertil.* **6,** 393–404.
Peters, H. (1969). The development of the mouse ovary from birth to maturity. *Acta Endocrinol.* (*Copenhagen*) **62,** 68–116.
Peters, H., Byskov, A. G., and Faber, M. (1973). Intraovarian regulation of follicle growth in the immature mouse. *In* "The Development and Maturation of the

Ovary and its Functions" (H. Peters, ed.), Int. Congr. Ser. No. 267, pp. 20–25. Excerpta Med. Found., Amsterdam.

Polder, J. J. W. (1964). Occurrence and significance of atretic follicles ("preovulatory corpora lutea") in ovaries of the bitterling, *Rhodeus amarus*. *Proc. K. Ned. Akad. Wet., Ser. C* **67**, 218–222.

Pool, W. R., and Lipner, H. (1964). Inhibition of ovulation in the rabbit by actinomycin D. *Nature (London)* **203**, 1385–1387.

Pool, W. R., and Lipner, H. (1966). Inhibition of ovulation by antibiotics. *Endocrinology* **79**, 858–864.

Pool, W. R., and Lipner, H. (1969). Radioautography of newly synthesized RNA and protein in pre-ovulatory follicles. *Endocrinology* **84**, 711–717.

Przekop, F., and Domanski, E. (1970). Induction of ovulation in sheep by electrical stimulation of hypothalamic regions. *J. Endocrinol.* **46**, 305–311.

Quinn, D. L. (1969). Neural activation of gonadotropic hormone release by electrical stimulation in the hypothalamus of the guinea-pig and the rat. *Neuroendocrinology* **4**, 254–263.

Rai, B. P. (1966). The corpora atretica and the so-called corpora lutea in the ovary of *Tor (Barbus) tor* (Ham.). *Anat. Anz.* **119**, 459–465.

Rajakoski, E. (1960). The ovarian follicular system in sexually mature heifers with special reference to seasonal, cyclical and left–right variations. *Acta Endocrinol. (Copenhagen), Suppl.* **52**, 1–68.

Rastogi, R. K. (1966). Study of the follicular atresia and evacuated follicles in the Indian teleost, *Xenentodon cancila* (Ham). *Acta Biol. Acad. Sci. Hung.* **17**, 52–63.

Rastogi, R. K. (1969). Occurrence and significance of ovular atresia in the freshwater mud-eel, *Amphipnous cuchia* (Ham). *Acta Anat.* **73**, 148–159.

Rawson, J. M. R., and Dukelow, W. R. (1973). Observation of ovulation in *Macaca fascicularis*. *J. Reprod. Fertil.* **34**, 187–190.

Reichert, L. E. (1962a). Endocrine influences on rat ovarian proteinase activity. *Endocrinology* **70**, 697–700.

Reichert, L. E. (1962b). Further studies on proteinases of the rat ovary. *Endocrinology* **71**, 838–839.

Richards, J. S. (1975). Estradiol receptor content in rat granulosa cells during follicular development: Modification by estradiol and gonadotropins. *Endocrinology* **97**, 1174–1184.

Robinson, T. J. (1965). The practical control of ovulation and oestrus in domestic animals using new synthetic progestogens. *In* "Recent Advances in Ovarian and Synthetic Steroids and the Control of Ovarian Function" (R. P. Shearman, ed.), pp. 12–24. Searle, High Wycombe, Bucks, England.

Robison, G. A. (1970). Cyclic AMP as a second messenger. *J. Reprod. Fertil., Suppl.* **10**, 55–74.

Rocereto, T., Jacobowitz, D., and Wallach, E. E. (1969). Observations on spontaneous contractions of the cat ovary *in vitro*. *Endocrinology* **84**, 1335–1341.

Roche, J. F. (1974). Effect of short-term progesterone treatment on oestrous response and fertility in heifers. *J. Reprod. Fertil.* **40**, 433–440.

Rodbard, D. (1968). Mechanics of ovulation. *J. Clin. Endocrinol. Metab.* **28**, 849–961.

Rondell, P. (1964). Follicular pressure and distensibility in ovulation. *Am. J. Physiol.* **207**, 590–594.

Rondell, P. (1970). Biophysical aspects of ovulation. *Biol. Reprod., Suppl.* **2**, 64–89.

Rondell, P. (1974). Role of steroid synthesis in the process of ovulation. *Biol. Reprod.* **10,** 199–215.
Rosemberg, E. (1973). Induction of ovulation. *In* "Human Reproduction, Conception and Contraception" (E. S. E. Hafez and T. N. Evans, eds.), Chapter 30. Harper, New York.
Rothchild, I., and Fraps, R. M. (1944). On the function of the ruptured ovarian follicle. *Proc. Soc. Exp. Biol. Med.* **56,** 79–82.
Rouget, C. I. (1858). Recherches sur les organes erectiles de la femme et sur l'appareil musculaire tubo-ovarian dans leur rapports avec l'ovulation et la menstruation. *J. Physiol.* (*Paris*) **1,** 320–343.
Rowlands, I. W. (1944). The production of ovulation in the immature rat. *J. Endocrinol.* **3,** 384–391.
Rowlands, I. W., and Parkes, A. S. (1966). Hypophysectomy and the gonadotrophins. *In* "Marshall's Physiology of Reproduction" (A. S. Parkes, ed.), 3rd ed., Vol. 3, Chapter 25, pp. 26–146. Longmans Green, London and New York.
Rowlands, I. W., and Williams, P. C. (1942–1944). Production of ovulation in hypophysectomized rats. *J. Endocrinol.* **3,** 310–315.
Rowson, L. E. A., Tervit, H. R., and Brand, A. (1972). The use of prostaglandins for synchronization of oestrus in cattle. *J. Reprod. Fertil.* **29,** 145–148.
Rugh, R. (1935). Ovulation in the frog. II. Follicular rupture to fertilization. *J. Exp. Zool.* **71,** 163–193.
Sadleir, R. M. F. S. (1969). "The Ecology of Reproduction in Wild and Domestic Mammals." Methuen, London.
Saiduddin, S., Rowe, R. F., and Casida, L. E. (1970). Ovarian follicular changes following unilateral ovariectomy in the cow. *Biol. Reprod.* **2,** 408–412.
Sandes, E. P. (1903). The corpus luteum of *Dasyurus viverrinus* with observations on the growth and atrophy of the Graafian follicle. *Proc. Linn. Soc. N. S. W.* **28,** 364–405.
Sathyanesan, A. G. (1961). A histological study of the ovular atresia in the catfish, *Mystus seenghala* (Sykes). *Rec. Indian Mus.* **59,** 75–82.
Schuetz, A. W. (1971). *In vitro* induction of ovulation and oocyte maturation in *Rana pipiens* ovarian follicles: effects of steroidal and non-steroidal hormones. *J. Exp. Zool.* **178,** 377–386.
Schwartz, N. B. (1969). Model for the regulation of ovulation in the rat. *Recent Prog. Horm. Res.* **25,** 1–56.
Schwartz, N. B., and Hoffmann, J. C. (1972). Ovulation: basic aspects. *In* "Reproductive Biology" (H. Balin and S. Glasser, eds.), pp. 438–476. Excerpta Med. Found., Amsterdam.
Schwartz, N. B., and McCormack, C. E. (1972). Reproduction: gonadal function and its regulation. *Annu. Rev. Physiol.* **34,** 425–472.
Self, H. L., Grummer, R. H., and Casida, L. E. (1955). The effects of various sequences of full and limited feeding on the reproductive phenomena in Chester White and Poland China gilts. *J. Anim. Sci.* **14,** 573–592.
Seth, P., and Prasad, M. R. N. (1967). Induction of ovulation by gonadotrophins in the Indian five-striped palm squirrel, *Funambulus pennanti* (Wroughton). *J. Endocrinol.* **39,** 369–378.
Shapiro, H. A., and Zwarenstein, H. (1934). A rapid test for pregnancy. *Nature* (*London*) **133,** 762.
Short, R. E., Peters, J. B., First, N. L., and Casida, L. E. (1965). Some effects of unilateral ovariectomy in gilts. *J. Anim. Sci.* **24,** 929–930.

Signoret, J. P., du Mesnil du Buisson, F., and Mauléon, P. (1972). Effect of mating on the onset and duration of ovulation in the sow. *J. Reprod. Fertil.* **31**, 327–330.
Simpson, T. H., Wright, R. S., and Hunt, S. V. (1968). Steroid biosynthesis *in vitro* by the component tissues of the ovary of the dogfish (*Scyliorhinus caniculus* L.). *J. Endocrinol.* **42**, 519–527.
Smeaton, T. C., and Robertson, H. A. (1971). Studies on the growth and atresia of Graafian follicles in the ovary of the sheep. *J. Reprod. Fertil.* **25**, 243–252.
Smith, J. T. (1934). Some observations on rupture of Graafian follicles in rabbits. *Am. J. Obstet. Gynecol.* **27**, 728–730.
Smith, M. J., and Godfrey, G. K. (1970). Ovulation induced by gonadotrophins in the marsupial, *Sminthopsis crassicaudata. J. Reprod. Fertil.* **22**, 41–47.
Squiers, C. D., Dickerson, G., and Mayer, D. T. (1952). Influence of inbreeding, age and growth rate of sows on sexual maturity, rate of ovulation, fertilization and embryonic survival. *Mo., Agric. Exp. Stn., Res. Bull.* **494.**
Sreenan, J. M. (1975). Effect of long- and short-term intravaginal progestagen treatments on synchronization of oestrus and fertility in heifers. *J. Reprod. Fertil.* **45**, 479–485.
Stevens, K. R., Spies, H. G., Hilliard, J., and Sawyer, C. H. (1970). Site(s) of action of progesterone in blocking ovulation in the rat. *Endocrinology* **86**, 970–975.
Strassmann, E. O. (1945). Development and degeneration of ovum and follicle as observed by intravital staining. *Am. J. Obstet. Gynecol.* **49**, 343–355.
Strauss, F. (1938). Die Befruchtung und der Vorgang der Ovulation bei *Ericulus* aus der Familie der Centetiden. *Biomorphosis* **1**, 281–312.
Sturgis, S. H. (1949). Rate and significance of atresia of the ovarian follicle of the rhesus monkey. *Contrib. Embryol. Carnegie Inst.* **33**, 67–80.
Sturgis, S. H. (1961). Factors influencing ovulation and atresia of ovarian follicles. *In* "Control of Ovulation" (C. A. Villee, ed.), pp. 213–218. Pergamon, Oxford.
Suzuki, M., and Bialy, G. (1964). Fertilizability of copper-ovulated rabbit ova. *Endocrinology* **75**, 288–289.
Szego, C. M. (1965). Role of histamine in mediation of hormone action. *Fed. Proc., Fed. Am. Soc. Exp. Biol.* **24**, 1343–1352.
Tam, W. H. (1970). The function of accessory corpora lutea in the hystricomorph rodents. *J. Endocrinol.* **48**, liv–lv.
Tanaka, K. (1968). Fluctuation of pituitary LH levels during the egg-laying cycle of the hen. *Jpn. J. Zootech. Sci.* **39**, 377–385.
TeWinkel, L. E. (1972). Histological and histochemical studies of post-ovulatory and preovulatory atretic follicles in *Mustelus canis. J. Morphol.* **136**, 433–458.
Thibault, C. G. (1972). Final stages of mammalian oocyte maturation. *In* "Oogenesis" (J. D. Biggers and A. W. Schuetz, eds.), pp. 397–411. Univ. Park Press, Baltimore, Maryland.
Umbaugh, R. E. (1949). Superovulation and ovum transfer in cattle. *Am. J. Vet. Res.* **10**, 295–305.
Unbehaan, V., Jung, G., and Kidess, E. (1965). Enzymuntersuchungen im Liquor folliculi. *Arch. Gynaekol.* **202**, 225–228.
van der Horst, C. J., and Gillman, J. (1940). Ovulation and corpus luteum formation in *Elephantulus. S. Afr. J. Med. Sci.* **5**, 73–91.
van der Stricht, O. (1901). L'atrésie ovulaire et l'atrésie folliculaire du follicule de de Graaf, dans l'ovaire de Chauve-souris. *Anat. Anz.* **19**, 108–121.
van der Stricht, O. (1923). Etude comparée des ovules des mammifères aux dif-

ferentes periodes de l'ovogenèse d'après les travaux du laboratoire d'histologie et d'embryologie de l'université de Gand. *Arch. Biol.* **33,** 229–300.
Vande Wiele, R. L., Bogumil, J., Dyenfurth, I., Ferin, M., Jewelewicz, R., Warren, M., Rizkallah, T., and Mikhail, G. (1970). Mechanisms regulating the menstrual cycle in women. *Recent Prog. Horm. Res.* **26,** 63–95.
Vazquez-Nin, G. H., and Sotelo, J. R. (1963–1964). Electron microscope study of the atretic oocytes of the rat. *Z. Zellforsch. Mikrosk. Anat.* **80,** 518–533.
Velloso de Pinho, A. (1923). Atrésie de l'epithélium ovarique chez les mammifères. *C. R. Soc. Biol.* **88,** 830–833.
Vilter, V., and Vilter, A. (1964). Sur l'évolution des corps jaunes ovariens chez *Salamandra atra* Laur. des Alpes vaudoises. *C. R. Soc. Biol.* (*Paris*) **158,** 457–461.
von Kölliker, A. (1849). Beitrage zur Kenntnis der glatten Muskeln. *Z. Wiss. Zool.* **1,** 30–44.
Wallach, E. E., and Noriega, C. (1970). Effects of local steroids on follicular development and atresia in the rabbit. *Fertil. Steril.* **21,** 253–267.
Walton, A., and Hammond, J. (1928). Observations on ovulation in the rabbit. *Br. J. Exp. Biol.* **6,** 190–204.
Wan, L. S., and Balin, H. (1969). Induction of ovulation in rhesus monkeys: a comparative study. *Fertil. Steril.* **20,** 111–126.
Watzka, M. (1957). Weibliche Genitalorgane. Das Ovarium. *In* "Handbuch der Mikroskopische Anatomie der Menschen" (W. Von Möllendorf and W. Bargamann, eds.), Vol. 7, pp. 1–178. Springer-Verlag, Berlin and New York.
Webb, F. T. G. (1973). The contraceptive action of the copper IUD in the rat. *J. Reprod. Fertil.* **32,** 429–439.
Weir, B. J. (1966). Aspects of reproduction in chinchilla. *J. Reprod. Fertil.* **12,** 410–411.
Weir, B. J. (1969). The induction of ovulation in the chinchilla. *J. Endocrinol.* **43,** 55–60.
Weir, B. J. (1971a). Observations on reproduction in the female agouti, *Dasyprocta aguti. J. Reprod. Fertil.* **24,** 203–211.
Weir, B. J. (1971b). The reproductive organs of the female plains viscacha, *Lagostomus maximus. J. Reprod. Fertil.* **25,** 365–373.
Weir, B. J. (1971c). Some notes on reproduction in the Patagonian Mountain viscacha, *Lagidium boxi* (Mammalia: Rodentia). *J. Zool.* **164,** 463–467.
Weir, B. J. (1973). The induction of ovulation and oestrus in the chinchilla. *J. Reprod. Fertil.* **33,** 61–68.
Weir, B. J., and Rowlands, I. W. (1973). Reproductive strategies of mammals. *Annu. Rev. Ecol. Syst.* **4,** 139–163.
Weir, B. J., and Rowlands, I. W. (1974). Functional anatomy of the hystricomorph ovary. *Symp. Zool. Soc. London* **34,** 303–332.
Welschen, R. (1971). Corpora lutea atretica in ovarian grafts. *J. Endocrinol.* **49,** 693–694.
Williams, P. C. (1956). The history and fate of redundant follicles. *Ciba Found. Colloq. Ageing* **2,** 59–66.
Wood-Gush, D. G. M., and Gilbert, A. B. (1964). The control of the nesting behaviour of the domestic hen. II. The role of the ovary. *Anim. Behav.* **12,** 451–453.
Wright, P. A. (1961a). Induction of ovulation *in vitro* in *Rana pipiens* with steroids. *Gen. Comp. Endocrinol.* **1,** 20–23.
Wright, P. A. (1961b). Influence of estrogens on induction of ovulation *in vitro* in *Rana pipiens. Gen. Comp. Endocrinol.* **1,** 381–385.

Wright, P. A. (1971). 3-Keto-Δ^4 steroid: requirement for ovulation in *Rana pipiens*. *Gen. Comp. Endocrinol.* **16,** 511–515.

Wyburn, G. M., Johnston, H. S., and Aitken, R. N. C. (1966). Fate of the granulosa cells in the hen's follicle. *Z. Zellforsch. Mikrosk. Anat.* **72,** 53–65.

Xavier, F. (1969). Corps jaunes de post-ovulations actifs chez les femelles non fecondées de *Nectophrynoides occidentalis* (Amphibien anoure vivipare). *Gen. Comp. Endocrinol.* **13,** 542 (abstr.).

Xavier, F. (1970). Analyse du rôle des corpora lutea dans le mantien de la gestation chez *Nectophrynoides occidentalis* Angel. *C. R. Acad. Sci. Ser. D* **270,** 2018–2020.

Ying, S.-Y., and Meyer, R. K. (1969). Effect of coitus on barbiturate-blocked ovulation in immature rats. *Fertil. Steril.* **20,** 772–778.

Young, W. C. (1961). The mammalian ovary. *In* "Sex and Internal Secretions" (W. C. Young, ed.), 2nd ed., Vol. 1, pp. 449–496. Williams & Wilkins, Baltimore, Maryland.

YoungLai, E. V. (1972). Effect of mating on follicular fluid steroids in the rabbit. *J. Reprod. Fertil.* **30,** 157–159.

Zachariae, F. (1957). Studies on the mechanism of ovulation. Autoradiographic investigation on the uptake of radioactive sulphate (^{35}S) into the ovarian follicular mucopolysaccharides. *Acta Endocrinol.* (*Copenhagen*) **26,** 215–224.

Zachariae, F. (1958). Studies on the mechanism of ovulation. Permeability of the blood–liquor barrier. *Acta Endocrinol.* (*Copenhagen*) **27,** 339–342.

Zachariae, F., and Jensen, C. E. (1958). Studies on the mechanism of ovulation. Histochemical and physico-chemical investigations on genuine follicular fluids. *Acta Endocrinol.* (*Copenhagen*) **27,** 343–355.

Zamboni, L. (1972). Comparative studies on the ultrastructure of mammalian oocytes. *In* "Oogenesis" (J. D. Biggers and A. W. Schuetz, eds.), pp. 2–45. Univ. Park Press, Baltimore, Maryland.

Zarrow, M. X., and Clark, J. H. (1968). Ovulation following vaginal stimulation in a spontaneous ovulator and its complications. *J. Endocrinol.* **40,** 343–352.

Zarrow, M. X., Estes, S. A., Denenberg, V. H., and Clark, J. H. (1970). Pheromonal facilitation of ovulation in the immature mouse. *J. Reprod. Fertil.* **23,** 357–360.

Zeilmaker, G. H. (1966). The biphasic effect of progesterone on ovulation in the rat. *Acta Endocrinol.* (*Copenhagen*) **51,** 461–468.

Zerbian, K. (1966). Über das histochemische Verhalten einiger Enzyme bei der Follikelatresie im Ovarium des Meerschweinchens. *Acta Histochem. Cytochem.* **23,** 303–321.

Zerbian, K., and Goslar, H. G. (1968). Das histochemische Verhaltender Azolesterasen sowie weiterer Oxydoreduktasen bei der Follikelatresie im Ovarium einiger Nager. *Histochemie* **13,** 45–56.

Zondek, B., Sulman, F., and Black, R. (1945). The hyperaemia effect of gonadotropins on the ovary and its use in a rapid pregnancy test. *J. Am. Med. Assoc.* **128,** 939–944.

Zuckerman, S. (1951). The number of oocytes in the mature ovary. *Recent Prog. Horm. Res.* **6,** 63–107.

7

Ovarian Histochemistry

L. Bjersing

I. INTRODUCTION

As a rule, histochemistry lacks the accuracy and refinement of the biochemical approach, and it usually provides qualitative rather than quantitative information. However, its unique value and advantage lies in the capacity to locate substances and biochemical reactions in tissues, cells, and organelles without disturbing the normal structural organization. Therefore, histochemistry can supply information about physiological activities and pathological states that is difficult or impossible to obtain by other means. This special advantage is particularly obvious in an organ with such a complicated anatomy and physiology as the ovary. Most *in vivo* or *in vitro* methods fail to reveal which structural subunits are responsible for certain physiological events. With histochemical methods, however, it is possible to survey the activities of various enzymes of individual corpora lutea, follicles, and other subunits.

During the 10 years that have elapsed since the chapter on ovarian histochemistry was written for the first edition of this book, the field of histochemistry has undergone an almost explosive development. This is also true of the histochemistry of the ovary, and it is clearly impossible to cover all aspects. How, then, can the subject be restricted without losing too much of general interest and value?

First, it is clearly unnecessary to repeat what was so ably covered by Jacoby (1962) in the first edition. Since his essay, however, no review covering the whole field of ovarian histochemistry has appeared.

Second, it is of value to define what we mean by histochemistry and to determine which aspects are important to review. This question is dealt with in Section II.

Third, in order to facilitate reference to the mass of available information in its proper structural and functional context, the data have been organized, as much as possible, under different anatomical and physiological subheadings, as can be seen in the Table of Contents.

Germ cells and atresia will be dealt with in only the most cursory manner. These topics are discussed in Chapters 2 and 6 of this volume, and in earlier essays, e.g., by Velardo and Rosa (1963) and Rossi (1964).

This review is mainly concerned with nonneoplastic mammalian ovaries under normal and experimental conditions; however, the histochemistry of ovarian tumors, particularly steroid-producing tumors, was considered worthy of special consideration and has been covered under a separate heading. Finally, histochemistry at the electron microscopical level and future developments are briefly discussed.

II. METHODS WITH RELEVANCE TO OVARIAN HISTOCHEMISTRY

Histochemistry in a restricted sense can be defined as the application of chemical methods for characterization, localization, and visualization of substances or groups of substances in the cells and intercellular materials of a tissue. Most of these methods are merely qualitative.

In histology dyestuffs are used empirically only for improvement of contrast. Histochemistry goes beyond structure and can also provide some information about function.

Histochemistry can be divided into several branches (see Sections A to D below) and, since so many special techniques are now in use, the average reader cannot be expected to know them all. However, a general knowledge of the possibilities and limitations of the methods is necessary to put the histochemical results into perspective. Therefore, a short and critical evaluation of the main histochemical methods has been included; a brief compilation of all these techniques is not readily available elsewhere.

Besides histochemical methods in the stricter sense, there are several so-called histochemical methods based on physical techniques such as autoradiography, ultraviolet spectrophotometry, and electron-microprobe analysis. Consideration of these methods would exceed the scope of this chapter; however, when results of central interest have emerged from such techniques they have been referred to in the text.

A. Conventional Histochemistry

By conventional histochemistry we mean demonstration of carbohydrates, lipids, proteins, and nucleic acids by chromogenic agents which selectively bind to these substances or to some specific groups of these substances.

1. Carbohydrates

The histochemistry of carbohydrates is complex. It is therefore necessary to interpret histochemical results cautiously and to be aware of both the possibilities and limitations of the available methods. The technique most used, the periodic acid–Schiff reaction (PAS reaction), demonstrates carbohydrates with 1,2-glycol groupings: periodic acid oxidises adjacent hydroxyl groups with liberation of aldehydes which produce a red product with the Schiff reagent. However, according to Hotchkiss (1948), any substance with the groups CHOH·CHOH, $CHOH \cdot CHNH_2$, CHOH·CHNHR, or CHOH·CO will give a positive reaction if it yields an oxidation product which is not diffusible, and if it is present in sufficient concentration to cause a detectable final color. In addition,

$$\begin{array}{cc} | & | \\ -C & =C- \end{array}$$

groups in unsaturated fatty acids can be oxidised to aldehydes and give a positive reaction.

By several types of controls, it is possible to prove or disprove that a positive reaction is due to a carbohydrate. After acetylation (or benzoylation), 1,2-glycol and secondary 1,2-aminohydroxy groups yield negative PAS reactions (Thompson, 1966). Therefore a substance that reacts positively to the simple PAS reaction and negatively in a control section after acetylation is almost certainly a carbohydrate. To test for preexisting aldehydes, e.g., plasmals, one section may be placed in 1 *N* hydrochloric acid at room temperature for 15 minutes and then exposed to the Schiff reagent (Chayen *et al.*, 1969) and a serial section may be subjected directly to the Schiff reagent. Dimedone blockage, according to Bulmer (1959), blocks the majority of diastase-resistant PAS-positive materials but not the reaction of glycogen. When it has been made certain that the section contains a carbohydrate, its chemical structure should be determined as closely as possible. Disappearance or reduction of PAS-positivity after treatment with saliva or diastase may be taken as evidence for glycogen.

If the substance is diastase-resistant, what kind of carbohydrate can it be? The most probable diastase-resistant PAS-positive carbohydrate is glycoprotein. Evidence for its presence may be obtained by applying a test for protein. A PAS-positive substance which has been shown to contain carbohydrate and which also reacts positively with tests for lipid must be a glycolipid. (Note that unsaturated lipids and phospholipids also may give a positive PAS reaction; however, they are still positive after acetylation.) A disappearance or reduction of PAS-positivity after treatment with neuraminidase proves that the substance contains sialoglycan. This carbohydrate is a constituent of both glycoproteins and glycolipids.

A popular stain for acid mucopolysaccharides is Alcian blue. Its specificity has been the subject of considerable debate. On prolonged staining, almost every tissue component becomes colored. On brief staining, acid mucopolysaccharides of epithelial and connective tissue mucin become positive, but most mucoproteins do not. The staining of connective tissues by Alcian blue can be masked by strong counterstains, and then only the nuclei appear blue. Spicer (1960) and Lev and Spicer (1964) showed that the dye reacted with many uronic acid groups at pH 2; at pH 1, however, it could be made specific for sulfated mucopolysaccharides. Stoward (1963) suggested that hydrogen binding of the unionized carboxyl group to the copper and phthalocyanin nitrogen to form a stable six-membered structure could explain the mechanism of Alcian blue staining. Observations by Scott *et al.* (1964) suggested that Alcian blue combined with polyanionic substances by salt linkages. Quintarelli *et al.* (1964) found that the addition of electrolytes (HCl, NaCl, KCl) increased Alcian blue staining. They assumed that the extra anions blocked all but one positive charge per dye molecule and that more dye was therefore bound by tissue anionic sites.

Scott and associates (see Pearse, 1968) observed that both sulfated mucins and glycosaminoglucuronoglycans with carboxyl groups (e.g., hyaluronic acid) bound Alcian blue *in situ* at low electrolyte concentrations (e.g., $MgCl_2$ below 0.3 *M*), whereas only sulfated mucosubstances did so at higher concentrations (above 0.8 *M*).

The combination of Alcian blue staining and treatment with hyaluronidase should disclose whether hyaluronic acid is present.

2. Lipids

Lipids can be demonstrated by their uptake of certain fat-soluble stains. The red Sudan dyes and Oil red O accumulate in neutral lipid droplets, while Sudan black B also dissolves in phospholipids and cholesterol. Acidic lipids, such as fatty acids and phospholipids are detected by Nile blue sulfate. Treatment of a section with chromium salts followed by acid hematein can also be used as a test for phospholipids. Furthermore, fat droplets can be identified by osmium tetroxide, which reacts mainly with unsaturated fatty acids. Other lipid substances, e.g., long-chain aliphatic aldehydes of plasmalogens, such as palmitaldehyde and steraldehyde, which are naturally occurring "free" aldehydes, give the so-called (Feulgen) plasmal reaction when tested for aldehyde with Schiff's reagent. Since these substances are soluble in organic solvents, the reaction must be studied in frozen sections. The Schiff reagent (bis-*N*-aminosulfonic acid) is a colorless compound which, together with aldehydes, produces a magenta red product.

In general, fresh cryostat sections should be used, because most fixatives can modify lipid–protein and lipid–carbohydrate complexes. If a fixative is used, solutions of formaldehyde, or other aldehydes, as well as Bouin's fluid are better than those containing organic solvents, particularly for preservation of neutral lipids. Loss of the hydrophilic phospholipids may be counteracted by $CaCl_2$ in Baker's formol-calcium. The large extent to which lipids are lost during fixation and preparation for ordinary electron microscopy has been studied by Korn and Weisman (1966) and others. The lipid droplets that are easily detected in tissue sections represent only a small part of the amount of lipid present in the tissue. Lipids can be so finely dispersed that the minute droplets cannot be resolved by the light microscope; an increase in neutral lipids or a decrease in phospholipids is necessary before the neutral lipids can be demonstrated by ordinary red Sudan stains.

Lipids may also be bound to structural molecules of nonlipid nature, particularly protein but also carbohydrate. These bound lipids are often resistant towards ether and several other lipid solvents. However, ethanol, methanol, and acetone are usually effective in splitting bonds between lipids and proteins; the liberated ("unmasked") lipid may then appear as Sudan-positive droplets. Methanol–chloroform has proved especially efficient in unmasking. The role of the methanol seems to be twofold; first, the hydrophilic obstruction to the chloroform is overcome, and second, the lipid–protein bonds are split (Chayen, 1968). As soon as the bonds are ruptured, the liberated lipid can be dissolved in the fat solvent. Phospholipids with their hydrophilic and hydrophobic end groups often play a prominent role in lipid–protein complexes. Cholesterol is also important in this context, but triglycerides do not form lipid–protein complexes (see Clara, 1965). Triglycerides may be masked, however, by concentric layers of phospholipid and water molecules. Masked lipid may also be freed by acid hydrolysis or proteolytic enzymes; this probably corresponds to what may cause lipophanerosis *in vivo* (see Clara, 1965). A recent and lucid discussion of the histochemistry of bound and unbound phospholipids and its significance for the interpretation of cell structure is given by Chayen (1968).

The best histochemical procedure for demonstrating phospholipids is Baker's acid hematein method (Baker, 1946). How it works is not yet fully understood, but it appears (see Chayen, 1968) that the reaction depends primarily on the unsaturated bonds of the fatty acids. The phosphate may merely allow a hydrophilic entry for the dichromate to the molecule or permit a stronger metal–dye complex. The method has been criticized and said to give variable results. This is only a reflection of a more general problem of lipid histochemistry; a major reason for variable results may be the degree to which masked or bound lipids become unmasked before or during the histochemical procedure. In serial sections of rat liver, unfixed tissue gave a moderate positive reaction to the acid hematein method, e.g., mitochondria and similarly sized particles reacted (Chayen, 1968). After treatment with methanol–chloroform for short periods, the acid hematein reaction was greatly enhanced throughout the cell. Bromination was also performed before and after treatment with methanol–chloroform. It was concluded (Chayen, 1968) that the bromination of unfixed sections blocked all free phospholipids but that the acid hematein procedure itself unmasked some more unsaturated groups. After methanol–chloroform, however, all these groups were fully unmasked and then abolished by bromination; the acid hematein reaction was then completely negative, since there were no more unsaturated fatty acids that could be unmasked and stained. Phospholipids may also be unmasked to various degrees under different conditions *in vivo* (Niles *et al.*, 1964; Clara, 1965; Chayen, 1963; Chayen and Bitensky, 1968). A great deal would be achieved if such pathophysiological variations could be evaluated histochemically on a quantitative basis; a change in the binding of lipid to cell structures appears to be a significant marker of cell damage. Lipids linked to protein or to carbohydrate occur in all cell fractions (Chayen and Bitensky, 1968). Clara (1965) lists granules of eosinophilic and neutrophilic leucocytes, lipofuscins, and mitochondria among the lipid–protein complexes; Chayen and Bitensky (1968) add the surface sheath and internal structural matter of chromosomes, nuclear membranes, and nucleoli. Lipid–protein complexes may be more or less strongly bound together. The most intractable lipids can be isolated only after the associated protein has been degraded mainly to the constituent amino acids (Chayen, 1968). The fact that some lipids become extractable only after proteolytic or hydrolytic degradation suggests (Chayen, 1968) the possibility that protoplasm itself may not be solely protein but lipid–protein with the lipid tightly conjugated. From the histochemical point of view, drastic methods for unmasking lipids produce an unfortunate derangement of structure and allow little precision in the localization of liberated lipid. Lipid and protein may also be more loosely attached to one another; even the acid hematein procedure can split some bonds, and alcoholic solutions such as methanol and ethanol are particularly effective.

More work on the histochemistry of bound lipids is urgently needed; bound lipids are intimately involved in the structure and function of cells and organelles, and a better understanding of the physiological role of these substances would be of profound importance.

3. Proteins

Proteins, or rather different amino acids, can be demonstrated by various tests or reactions. A red color in a tissue section treated with an alkaline mixture of α-naphthol and sodium hypochlorite indicates that arginine is present; a high concentration of arginine in turn suggests that the tissue contains basic proteins, such as histones. Diazonium hydroxide reacts with tyrosine, tryptophan and histidine groups and forms a colored complex.

4. Nucleic Acids

Nucleic acids give reactions depending on the three components of the nucleotide: the carbohydrate, the phosphoric acid, and the purine and pyrimidine bases. The Feulgen (nuclear) reaction is a highly specific histochemical test for DNA. No other naturally occurring substance will yield aldehydes if the reaction is properly performed. Sections of fixed tisssue are subjected to mild acid hydrolysis. This is sufficient to remove RNA from the tissue and also to remove the purines from DNA. By these means the aldehyde groups of DNA are unmasked and can react with Schiff's reagent; the chromonemata and particularly the chromocenters are strongly positive, as is the perinucleolar chromatin, while the nucleolus itself is Feulgen-negative. The specificity of the reaction can be checked by removing DNA by deoxyribonuclease. The basic stain azure B gives a specific reaction with

DNA and RNA. The basophilic properties of DNA and RNA depend on phosphoric acid residues.

B. Enzyme Histochemistry

What do we hope to achieve by the use of enzyme histochemistry? The most useful information is clearly that which tells, with the greatest possible accuracy, what has been going on in the living cell. The enzymes under study may be present but not always working at full capacity. With biochemical methods the total amount of enzyme can be determined; with gentle histochemical methods the actual physiological activity of the enzyme may be reasonably well evaluated. After suitable disruptive methods, e.g., repeated freezing and thawing, the total (potential, latent, or reserve) enzyme activity may also be studied histochemically.

Enzyme histochemistry is a rapidly expanding field, especially at the ultrastructural level; several of the methods originally devised for light microscopy, that is, methods employing metals as trapping agents (Gomori, 1952), have been modified and used at the submicroscopic level (see Section VI).

Up to 1939, the oxidative enzymes (dehydrogenases, oxidases) were the main enzymes that could be studied. The histochemical demonstration of alkaline phosphatase by Gomori (1939) and Takamatsu (1939) paved the way for localization of other hydrolytic enzymes in the tissues. In 1961 the International Union of Biochemistry (IUB) recommended a new classification of enzymes based on reaction types and reaction mechanisms. The system is complex, but informative and precise.

The six major classes of enzymes with several subclasses are tabulated below.

Name of major class	Function
1. Oxidoreductases	Catalyze oxidoreductions between two substrates
2. Transferases	Catalyze transfer of a group between two substrates
3. Hydrolases	Catalyze hydrolysis of ester, ether, peptide, glycosyl, acid–anhydride, C–C, C–halide, or P–N bonds
4. Lyases	Catalyze removal of groups from substrates by mechanisms other than hydrolysis, leaving double bonds
5. Isomerases	Catalyze interconversion of optical, geometric, or positional isomers
6. Ligases	Catalyze the linking together of two compounds coupled with the breaking of a pyrophosphate bond in adenosine triphosphate (ATP) or a similar compound

Only about 20% of a total of about 400 enzymes in vertebrate tissues have been studied histochemically so far.

a. Oxidoreductases. In contrast to oxidases, dehydrogenases cannot transfer hydrogen from the substrate directly to atmospheric oxygen. The acceptor is usually a coenzyme like NAD or NADP, or a flavoprotein. The hydrogen may then be passed either to an enzymatic process, e.g., for biosynthesis, or through a hydrogen transport system to reach oxygen. In the latter case, hydrogen moves along the hydrogen transport system of the mitochondrion from a relatively high negative electrode potential to the relatively high positive electrode potential of oxygen; as the final event, water is formed. Since all the methods for demonstration of the dehydrogenases depend on the same principle, it seems appropriate to briefly discuss the theoretical background of formazan deposition—especially as misconceptions have arisen.

When a section is studied for the presence of a particular dehydrogenase, the incubation solution usually contains NAD, a substrate, AH_2, and tetrazole as hydrogen acceptor. The AH_2 and NAD become attached to the active surface of the tissue dehydrogenase, and hydrogen is transferred from the substrate to NAD. The substrate thus becomes oxidized and the NAD is reduced to $NADH_2$. The $NADH_2$ cannot directly reduce tetrazole; it is necessary for the hydrogen to reach an oxidation–reduction system with an electrode potential compatible with that of tetrazole. Such a system is present in the tissue and is called the NAD–diaphorase system. This natural route in the tissue can be short-circuited by nonenzymatically acting substances, e.g., phenazine methosulphate in the reaction medium. Several enzymes may act as diaphorases: e.g., lipoic dehydrogenase (in the pig heart), $NADH_2$–cytochrome c reductase, $NADPH_2$–cytochrome c reductase, microsomal $NADH_2$–cytochrome reductase (see Barman, 1969; Chayen *et al.*, 1969). The diaphorases are flavoprotein enzymes and contain flavin adenine dinucleotide or flavin mononucleotide as coenzyme. In the mitochondria, hydrogen from $NADPH_2$ can be transferred to NAD with NAD(P)-transhydrogenase.

From a practical point of view, it is important to know that in the histochemical reaction formazan is not formed from the oxidation of the supplied substrate unless a system is present which will oxidize the reduced coenzyme and simultaneously reduce the tetrazolium salt. It is meaningless, for example, to argue about the localization of a dehydrogenase, e.g., a steroid dehydrogenase, unless it is certain that both the steroid dehydrogenase and the hydrogen-transferring enzyme system are located at the same site; as discussed above, formazan is deposited by the hydrogen-transferring enzyme, the "diaphorase." The amount of formazan formed may depend both on the oxidative activity of the steroid dehydrogenase and the diaphorase. By exposing a serial section to the reduced coenzyme and the tetrazole, the actual activity of the diaphorase can be tested and, thus, an indirect estimate of the steroid dehydrogenase activity is obtained.

One must also be aware that various interactions can occur between the enzyme tested and related enzyme systems. For example, if a section is supplied with glucose-6-phosphate and NADP in order to test its glucose-6-phosphate dehydrogenase activity, the glucose-6-phosphate might be acted on so rapidly by glucose-6-phosphatase in the section that little would be available for dehydrogenation. Such a possibility can be tested by another histochemical reaction. Practical principles for dehydrogenase histochemistry and possible causes of falsely negative or positive reactions have been discussed by Baillie *et al.* (1966a), Bjersing (1967a), Chayen *et al.* (1969), Fahimi and Karnovsky (1966), Jones *et al.* (1968a), and Wolrab and Cossel (1964).

b. Hydrolases. There are several types of phosphatases: some are relatively nonspecific and hydrolyze most substrates with a certain type of bond, others are specific for one or a few substrates. Examples of the former type are the so-called acid and alkaline phosphatases, which split monoesters of orthophosphoric acid optimally at acid and alkaline pH respectively; they have little effect on pyrophosphates, metaphosphates or on phosphoric diesters and their proper name is orthophosphoric monoester phosphohydrolases (I.U.B.). The nonspecific acid phosphatases have been particularly used as marker enzymes for lysosomes. Phosphodiesterase, which hydrolyses phosphoric diester into a phosphoric monoester and an alcohol, also has a wide range of specificity. Among the more or less substrate-specific phosphatases, 5′-nucleotidase removes H_3PO_4 from ribo- and deoxyribonucleoside-5-phosphates and ATPases hydrolyze ATP. In principle, only phosphatases liberating orthophosphoric acid can be demonstrated histochemically; the single exception appears to be thiamine pyrophosphatase. Since phosphatases catalyze the hydrolysis of phosphoric esters, they are actually esterases. However, the term "esterase" is, at least in the older histochemical literature, more or less reserved for enzymes which hydrolyze esters of carboxylic acids.

In the Gomori–Takamatsu method for alkaline phosphatase, phosphate ions are liberated from the substrate (phosphoric esters of glycerol) by hydrolysis and are finally incorporated into an insoluble metal salt. The metal is in turn made visible by transformation into lead sulfide, cobalt sulfide, metallic silver, or other colored compound. In another method used for hydrolytic enzymes, hydrolysis liberates β-naphthol, which is coupled to a diazonium salt and gives a colored azo compound at the site of enzyme activity. By changing the substrate and incubation conditions, enzymes such as alkaline and acid phosphatase, sulfatase, esterase, lipase, and β-glucuronidase can be demonstrated.

c. Esterases. In a restricted sense esterases are enzymes which hydrolyze esters of carboxylic acids. They can be divided into (1) simple esterases, aliesterases, or carboxylesterases, which hydrolyze short-chain aliphatic esters, (2) lipases (glycerol ester hydrolases), which act on esters with long carbon chains, and (3) cholinesterases, which attack esters of choline and can be divided into acetylcholinesterase and nonspecific (pseudo-) cholinesterase. Another esterase is the arylesterase acting on many phenolic esters. Strict separation in subgroups is not entirely justifiable as regards substrate specificity: all the enzymes mentioned, except acetylcholinesterase, are capable of hydrolyzing both choline and noncholine esters. However, the esterases can be distinguished histochemically by their different sensitivities to different inhibitors.

Of the *sulfatases,* only arylsulfatase and steroid sulfatase (sterol sulfatase) appear to be present in higher animals. Arylsulfatases (A and B) are found in lysosomes.

β-Glucuronidase, a glycosidase which hydrolyses and synthesizes glucuronides but acts only on β-glycoside linkages, has also been found in lysosomes. It is markedly affected by gonadal hormones (Fishman, 1963). Its main function is believed to be hydrolysis of steroid glucosiduronic acids and acidic mucopolysaccharides, and it is also said to function in cellular proliferation.

d. Proteolytic Enzymes. These enzymes that hydrolyze -CO-NH linkages of proteins and peptides were earlier divided into one type thought to degrade large molecules into smaller fragments and another type that attacked these small fragments. But the critical factor seems rather to be the chemical group on either side of the peptide bond. Certain peptidases—the exopeptidases—act where there is a terminal carboxyl or amino group on one side of the peptide bond. Others—the endopeptidases—may even be inhibited by such bonds and preferentially hydrolyze interior bonds of proteins; they also act more efficiently if a particular amino acid is present. The endopeptidase trypsin, for example, acts on the peptide bond between the carboxyl group of arginine and lysine; it can be inhibited by the proximity of carboxyl radicals. Furthermore, pepsin is inhibited by the close proximity of a free amino group. It is noteworthy that certain proteolytic enzymes, e.g., trypsin, split amides and esters even more efficiently than they catalyze the hydrolysis of peptide bonds. Both chymotrypsin and cathepsin C can hydrolyze amides and esters, and show activity against the substrate used for the histochemical aminopeptidase reaction.

It is clear, therefore, that the proteolytic enzymes can have esterase activity in a histochemical reaction and that they may also influence amidase reactions such as the phosphoamidase reaction. It is also probable that several proteolytic enzymes give the aminopeptidase reaction, and that a particular aminopeptidase does not act specifically on one amino acid peptide; for example, it is not possible to demonstrate specific leucine aminopeptidase activity histochemically, since numerous enzymes other than leucine aminopeptidase split leucyl-β-naphthylamide (see Vanha-Pertula and Hopsu, 1966). Therefore, the term aminopeptidase should be used instead of leucine aminopeptidase.

C. Immunohistochemistry

The basic principle in immunohistochemistry is an antigen–antibody reaction. Any protein may act as an antigen, and if it is injected into a suitable animal a highly specific antibody is produced. Other macromolecules and even small molecules, provided they are bound to larger molecular species, may also act as antigens. The antibodies which are present in the γ-globulin fraction of the serum can retain their specificity even if they are coupled to a fluorescent dye or to a molecule opaque to the electron beam. Thus, the localization of an antigen–antibody complex can be determined by fluorescence or electron microscopy. In the so-called direct method, the tagged antibody conjugate is allowed to react and combine with the homologous antigen of the cells or tissues. In the more widely used indirect method, an unlabeled specific antibody is bound by the antigen of the tissue. Then labeled anti-γ-globulin is allowed to combine with the specific antibody which acts as an antigen in respect to the labeled anti-γ-globulin. The label may be fluorescein for the light microscope and ferritin for the electron microscope. The label may be radioactive, in which case autoradiography is used for localization, or the label may be an enzyme such as peroxidase. The enzyme can then be demonstrated histochemically at the light or electron microscope level. This immunoenzyme method is better than the immunofluorescence technique because of the great stability of the reaction products.

The immunohistochemical methods are theoretically highly specific but they require exhaustive controls and the results must still be interpreted with caution. Glauert (1965) gives a good account of immunohistochemistry and histochemistry combined with electron microscopy.

D. Histochemistry with Electron Microscopy

Only the enzyme histochemical methods will be discussed here. A recent and excellent review covering most methods of histochemistry at the electron microscope level is that of Geyer (1969).

The main basis for enzyme histochemistry is the specific action of an enzyme on a substrate. If the reaction product is insoluble and electron-opaque, direct enzyme localization at the ultrastructural level should be possible. In other cases, however, it is necessary to convert the reaction product into an opaque one. It is of course important that the reaction products are precipitated quickly and that they are insoluble in water and in organic solvents. Diffusion artifacts are not unusual, particularly when long incubation times are used. Furthermore, formazan may be reoxidized during embedding for electron microscopy (Wolrab and Fuchs, 1967), and some reaction products are unstable in the electron beam.

1. Oxidoreductases

a. Dehydrogenase. Since most dehydrogenases are soluble, fixation is recommended to prevent loss and dislocation of enzyme and also to improve structure; cold formaldehyde solution (from paraformaldehyde) is a suitable fixative (see Fahimi and Karnovsky, 1966; Geyer, 1969). As hydrogen acceptors, tetrazolium salts are used. Seligman *et al.* (1967) synthesized a new osmiophilic tetrazolium salt (TC-NBT) with good substantivity and low lipid solubility. Tetranitro blue tetrazolium (TNBT) is also suitable for histochemistry at the electron microscope level; its formazan is very finely granular and stable toward postfixation in osmium. Some embedding media reoxidise the formazan, but with some precaution it is possible to embed in Epon or Araldite (Ogawa and Barnett, 1965; Wolrab and Fuchs, 1967).

Since the hydrogen from the NAD- or NADP-dependent dehydrogenase normally passes a diaphorase, the reaction product actually shows the localization of diaphorase and not the dehydrogenase under study. To eliminate the diaphorase dependence, phenazine methosulfate has sometimes been used both at the light and electron microscope levels (Fahimi and Karnovsky, 1966). This substance can transfer hydrogen directly from the pyridine nucleotide to the tetrazolium salt, and thus formazan is formed at the site of the dehydrogenase. However, in cells with a well-functioning cytochrome system the hydrogen may be transferred from phenazine methosulfate to the cytochrome system and not to the tetrazolium salt and the net result is an inhibitory action of phenazine methosulfate on the histochemical reaction. This inhibition can be prevented by blocking the cytochrome oxidase by cyanide ions. It is important that the phenazine methosulfate is chemically pure, and that the reaction is performed in complete darkness.

b. Peroxidase. Peroxidases prevent the accumulation of H_2O_2 in the cells; with 3,3′-diaminobenzidine as substrate, homogeneous and electron-opaque reaction products of peroxidation can be obtained (Graham and Karnovsky, 1966). The localization is sharp and the products are not attacked or dissolved during subsequent embedding and staining.

2. Hydrolases

Acid and alkaline phosphatases have been intensely studied at the ultrastructural level (Geyer, 1969), as has glucose-6-phosphatase. Several nucleoside phosphatases have been demonstrated with ultrahistochemistry. They attack the energy-rich pyrophosphate bonds and remove the terminal phosphate group. The substrate specificity is low; since nucleoside triphosphatase usually splits ATP preferentially, and since this has considerable importance, the enzyme is commonly termed ATPase. It is possible, however, to characterize the enzyme in question more closely by adding different substrates, activators, and inhibitors. Thiamine pyrophosphatase, which removes the terminal phosphate group of thiamine pyrophosphate, has also been demonstrated.

Other hydrolases which have been studied at the ultrastructural level include nonspecific cholinesterase and acetylcholinesterase, arylsulfatase and β-glucuronidase.

III. CARBOHYDRATES, LIPIDS, AND ENZYMES IN NORMAL OVARIAN TISSUE

A. Follicles

1. Carbohydrates

a. Man. McKay *et al.* (1961) recognized Call–Exner bodies in preantral follicles with only three cell layers. Centrally, the bodies contained saliva-resistant PAS-positive material ("glycoprotein"). No glycogen could be demonstrated in the granulosa cells up to this stage; the theca was not yet differentiated. The central PAS-positive material in the Call–Exner bodies was granular and appeared to be formed by the coalescence of small globules of glycoprotein, which were widely distributed between the granulosa cells. Intracellular glycoprotein was not detected, but a few glycogen granules were seen in granulosa cells of some preantral follicles. In follicles with a diameter of 0.4 mm or more, an antrum filled with liquor folliculi was found, and the thecal layers developed rapidly. In the large Graafian follicle, 0.5 cm or more in diameter, glycogen granules were present in granulosa and theca cells, though less abundant in the latter. The liquor folliculi contained globular PAS-positive material (apparently glycoprotein) adjacent to the inner surface of the granulosa layer; it closely resembled the material in the Call–Exner bodies. Glycogen granules could also be found in granulosa cells of early atretic follicles. Call–Exner bodies in various mammals, including man, have recently been studied by Motta (1965, 1967).

Ghosh and Sengupta (1970) detected glycogen in the membrana granulosa and theca interna. They also claimed that glycogen occurred in follicular fluid of normal mature follicles; in atretic follicles and follicular cysts the theca interna was negative, while the follicular contents was reported to give a positive reaction for glycogen. PAS-positive material resistant to diastase digestion (called glycoprotein by Ghosh and Sengupta) followed the glycogen pattern except in the follicles, where none was found. Alcian blue-positive material (presumably acid mucopolysaccharide) was found in granulosa cells and follicular fluid in five of eleven normal follicles. The reaction in the follicular fluid was totally abolished by hyaluronidase, and that in the granulosa cells partially abolished, indicating that both hyaluronic acid and other acid mucopolysaccharides were present in these cells. The theca interna gave a negative reaction with Alcian blue. Garcia-Bunuel and Monis (1964) reported the same pattern.

Csaba *et al.* (1970) also studied glycogen; none was found in granulosa cells of small and early developing follicles. However, in maturing healthy follicles the membrana granulosa was always positive and showed stronger reactions than the theca interna. In atretic follicles of normal ovaries the theca layer reacted negatively but the granulosa sometimes reacted positively to tests for glycogen. In contrast to Garcia-Bunuel and Monis (1964) and Ghosh and Sengupta (1970), Csaba *et al.* (1970) did not detect acid mucopolysaccharides in normal follicles with Alcian blue staining.

b. Other Mammals. Lobel and Levy (1968) found glycogen in the oocyte, its covering, granulosa cells, and also in thecal cells of the vesicular follicle of the cow. O'Shea (1966) showed that canine liquor folliculi contained both an acid and a neutral mucopolysaccharide or mucoprotein. In the acid component at least two separate substances were found, one removable by hyaluronidase and the other by neuraminidase. Carbohydrates, including mucopolysaccharides and glycogen, were also studied in the rat follicle by Ejsmont (1965), Georgieva *et al.* (1967), and Jirásek *et al.* (1970), and in the developing mouse ovary by Ben-Or (1963). Blanchette (1966), in a light and electron microscopic study on glycogen in rabbit ovarian follicles, suggested that accumulation of glycogen was "an early morphological indicator of lutein cell differentiation in a normal, nonatretic follicle."

Adams *et al.* (1966) found glycogen prominent in the cumulus oophorus in mature follicles of guinea pigs; during estrus, glycogen was seen throughout the coagulum containing the oocyte. These authors discussed the possibility that this might be evidence of cessation of the transport to the oocyte of glycogen precursors from the cells of the cumulus and the consequent deposition of glycogen in these cells. Modak and Kamat (1968) elucidated the distribution and nature of PAS-positive material in the growing Graafian follicle of tropical bats. They concluded that theca, granulosa, and cumulus cells contained mainly acid mucopolysaccharides and a small amount of glycogen.

2. Lipids

The histochemistry of ovarian lipids was reviewed by Wolman in 1964.

a. Man. Guraya (1967a, 1968b) reported that the membrana granulosa of developing ovarian follicles contained some irregular, scattered lipid bodies (L_2) consisting of phospholipids; according to Guraya such lipid bodies invariably occur in the oocyte of developing follicles in mammals and have been thought to supply energy-rich substances to the follicle. Diffuse lipids did not show any conspicuous development in the granulosa layer, but in theca interna cells all of the cytoplasm became stained with Sudan black B; the diffuse lipid component in the theca cells was interpreted as lipoprotein (Guraya, 1968b). The granular and rod-like mitochondria of the theca cells stained moderately with Sudan black B before, but not after, extraction with hot pyridine, and Guraya (1968b) thought that the positive reaction was due to lipoprotein. In addition, the theca interna cells showed some deeply sudanophilic lipid bodies. These were moderately positive with the acid hematein method, indicating the presence of phospholipids. According to Guraya (1968b), the theca interna cells of normal follicles did not show triglycerides or cholesterol or its esters, but these substances were stored in atretic follicles (Guraya, 1967a). The granulosa cells developed some fine lipid granules during ovulation, but otherwise changed very little. Theca interna cells of newly ruptured follicles were filled with coarse sudanophilic lipids, which stained moderately with acid hematein; they were colored pink in Nile blue, deep red in red Sudan dyes, and reacted positively to the Schultz test. This indicated (Guraya, 1968b) that the droplets contained phospholipids, triglycerides, and cholesterol (esters). Findings essentially similar to those of Guraya (1968b) were reported by Matsukado and Hayashi (1965).

Fienberg and Cohen (1965) found Sudan red- and Schultz-positive lipid in the theca interna but not in granulosa cells of developing follicles. In mature corpora lutea of the menstrual cycle the granulosa lutein cells showed a positive reaction with Sudan red, but were essentially negative with the Schultz reaction; the theca lutein cells, however, displayed strong positive reactions with both the Sudan and Schultz procedures. The luteinization of the theca interna during infancy and childhood has been studied by Kraus and Neubecker (1962).

b. Other Mammals. In the rhesus monkey ovary the granulosa cells of the normal preantral follicle contained heterogeneous lipid bodies consisting of phospholipids; the histochemical reactions were the same as those in corresponding cells in man (Guraya, 1968b), bat, rat, opossum, cat, dog (Guraya and Greenwald, 1964a; Guraya, 1968c), rabbit (Guraya and Greenwald, 1965), guinea pig (Guraya, 1968e), and cattle (Guraya, 1968f). In the hamster ovary (Guraya and Greenwald, 1965), however, the granulosa of normal follicles contained droplets of cholesterol and its esters, as well as triglycerides and phospholipids.

Guraya (1968i) also performed a comparative histochemical study on the lipids in the theca interna cells of bat, guinea pig, hamster, opossum, rat, rabbit, cat, dog, cow, marmoset, rhesus monkey, and man. There were considerable differences between species with respect to cytoplasmic differentiation of the theca cells and their content of histochemicallly demonstrable lipids. Theca interna cells of normal follicles in rhesus monkey and guinea pig ovaries, for example, showed scattered lipid bodies, some of which displayed only phospholipids, whereas other bodies of relatively coarse nature contained both triglycerides and phospholipids; cholesterol and its esters could not be demonstrated (Guraya, 1966b, 1968e). Theca interna cells of the hamster ovary also showed similar lipid bodies (Guraya and Greenwald, 1965; Guraya, 1968i), while the corresponding cells in the rat and rabbit ovary contained lipid bodies with cholesterol and its esters, triglycerides and phospholipids (Guraya, 1968i). In preovulatory rabbit follicles, however, neither cholesterol nor its esters were found (Guraya, 1968d), indicating that the thecal cells (thecal gland cells of Mossman *et al.*, 1964) during this period were metabolically more active, presumably in the production of steroids. After ovulation these cells (theca lutein cells) apparently underwent a change in lipid metabolism and quickly accumulated coarse lipids consisting of triglycerides, cholesterol or its esters, and phospholipids.

The lipids in follicles of the bovine ovary were studied quantitatively by Priedkalns and Weber (1968) and briefly mentioned by Lobel and Levy (1968). Lipids in normal follicular granulosa cells of the pig ovary were demonstrated by Bjersing (1967c), and Guraya (1969b) studied the lipids in developing follicles and corpora lutea of cat and dog ovaries.

The granulosa of estrous follicles in the rabbit (Guraya, 1968d) showed lipid bodies (L_1) consisting of phospholipids and some triglycerides sparsely distributed in the basal portions of the membrana granulosa. Guraya (1968d) suggested that the presence of L_1 bodies in the basal portions near the blood vessels of the theca interna was probably of physiological significance, as similar lipid droplets occurred in theca interna cells. The granulosa layer also displayed some L_2 bodies containing phospholipids. After administration of human chorionic gonadotropin the L_1 bodies increased in number while the L_2 bodies decreased in number and developed triglycerides. Adams *et al.* (1966) found fine droplets positive with Sudan black in peripheral layers of the mature 16-day-old follicle of the guinea pig ovary; these became more pronounced during estrus. Earlier in the cycle, only single small sudanophilic granules were visible in some cells of the cumulus and the granulosal wall. The theca interna of secondary follicles displayed finely sudanophilic cells. Ovarian lipids were also followed throughout the estrous cycle in the rat (Bronzetti *et al.*, 1963) and during normal development and after gonadotropin stimulation in the mouse (Ben-Or, 1963).

c. Other Vertebrates. Guraya has studied the lipids in follicles of amphibian (Guraya, 1968a, 1969c) and snake (Guraya, 1965) ovaries.

3. Enzymes

a. Mammals. The enzyme histochemical studies published since 1963 have been presented in Table I. Because of the tremendous development in the field of enzyme histochemistry this was felt to be the most profitable way to give a systematic account of what has been done since the first edition of this book. The encyclopedic form of course means that the contributions included vary in significance and quality, but the results have proved or may prove important in so many different fields of ovarian biology that it is difficult to exclude single studies as being insignificant; furthermore, a complete account of recent enzyme histochemical studies does not exist, and the table will therefore help to give a survey of such histochemical investigations in different mammals. For a discussion of the biological significance of some of the most interesting results, see Section VII,B of this chapter. Only a few studies on oxidoreductases, transferases, and hydrolases will be briefly mentioned below.

i. Oxidoreductases. Since steroid dehydrogenases are discussed in some detail in Section VII, only some other oxidoreductases are dealt with here. In the rat, Burkl and Slezak-Klemencic (1967b) found that glucose-6-phosphate and succinate dehydrogenase reactions were distinctly positive in the membrana granulosa of large tertiary follicles and still stronger in corpora lutea, theca interna, and interstitial cells; the highest activities were observed during proestrus. Several large follicles in the ovary of a parturient rat showed very strong positive reactions for glucose-6-phosphate dehydrogenase and $NADH_2$- and $NADPH_2$-tetrazolium reductases in the granulosa layer; other follicles, however, were devoid of such reactions (Schultka and

TABLE I

Histochemically Studied Enzymes of the Nonneoplastic Mammalian Ovary at the Light Microscopic Level since 1963[a]

Enzyme[b]	Species	Reference and tissue studied[c]
	1. Oxidoreductases	
Choline dehydrogenase	Rabbit	Hadjiisky (1970) CL
	Man	König (1965a,b)
Cytochrome oxidase	Rat	Schultka and Schmidt (1969)
	Rabbit	Hadjiisky (1970) CL
	Dog	Fowler (1970
	Sheep	Arvy and Mauleon (1964) CL
	Man	König (1965a,b)
Glucose-6-phosphate dehydrogenase	Guinea pig	Zerbian (1966) F,S
	Hamster	Wingate (1970)
	Mouse	Kimura *et al.* (1970)
	Rat	Balogh *et al.* (1966) F,S; Barbanti Silva *et al.* (1966); Brandau (1967) F,S; Brandau and Luh (1969); Bratt *et al.* (1968); Burkl and Slezak-Klemencic (1967b); Georgieva *et al.* (1967); Goldman *et al.* (1965); Kidwell *et al.* (1966); Pupkin *et al.* (1966); Schultka and Schmidt (1969); Turolla and Magrini (1963); Turolla *et al.* (1967, 1968); Wegmann and Lageron (1964); Wiest *et al.* (1968); Woods and Domm (1966)
	Rabbit	Brandau *et al.* (1967) F.S.; Hadjiisky (1970) CL; Okano *et al.* (1966)
	Dog	Fowler (1970)
	Cow	Kenney (1964); Nakama (1969)
	Pig	Mayner (1966) F,S
	Man	Brandau and Lehmann (1970) F,S; Brandau and Luh (1964a, 1965); Brandau *et al.* (1968); Fienberg and Cohen (1965, 1966); Ghosh *et al.* (1965); Hertig and Adams (1967) F; Kern-Bontke (1964) CL; König (1965a,b); Koudstaal *et al.* (1966); Luh and Brandau (1964a,b); Magrini and Moneta (1966) F,S; Mori (1967); Nakayama (1970); Pesonen *et al.* (1968) S; Poliak *et al.* (1968) CL,S; Pryse-Davies (1970) F,S; Scully and Cohen (1964); Taki *et al.* (1966, 1967) *Comment:* The same as for 6-phosphogluconate dehydrogenase.
Glutamate dehydrogenase	Rat	Barbanti Silva *et al.* (1966); Okano *et al.* (1966)
	Rabbit	Brandau *et al.* (1967) F,S; Hadjiisky (1970) CL
	Dog	Fowler (1970)
	Cow	Nakama (1969)

(Continued)

TABLE I (Continued)

Enzyme[b]	Species	Reference and tissue studied[c]
	Man	Kern-Bontke (1964) CL; König (1965a,b); Luh and Brandau (1964a,b); Taki *et al.* (1966) *Comment:* This histochemical reaction is an early and valuable indicator of cell damage; an increased enzyme activity is apparently caused by an increased permeability of mitochondria in damaged cells (see Chayen *et al.*, 1969).
Glyceraldehyde-3-phosphate dehydrogenase	Rat	Brandau (1967) F,S; Brandau and Lehmann (1970) F,S; Brandau and Luh (1969); Georgieva *et al.* (1967); Wegmann and Lageron (1964)
	Rabbit	Brandau *et al.* (1967) F,S; Hadjiisky (1970) CL
	Man	Brandau and Lehmann (1970) F,S; Brandau and Luh (1964a,b, 1965); Brandau *et al.* (1968) F,S; König (1965a,b); Luh and Brandau (1964a,b)
Glycerolphosphate dehydrogenase	Rat	Okano *et al.* (1966); Schultka and Schmidt (1969)
	Rabbit	Brandau *et al.* (1967) F,S; Hadjiisky (1970) CL
	Cow	Nakama (1969)
	Pig	Mayner (1966) F,S
	Man	Brandau and Lehmann (1970) F,S; Brandau *et al.* (1968) F,S; Jones *et al.* (1968b); Kern-Bontke (1964) CL; König (1965a,b); Koudstaal *et al.* (1966); Luh and Brandau (1964a,b); Taki *et al.* (1966)
β-Hydroxybutyrate dehydrogenase	Guinea pig	Zerbian and Goslar (1968)
	Mouse	Zerbian and Goslar (1968)
	Rat	Barbanti Silva *et al.* (1966); Hadjiisky *et al.* (1969b); Okano *et al.* (1966); Schultka and Schmidt (1969); Zerbian and Goslar (1968)
	Rabbit	Hadjiisky (1970) CL
	Cow	Nakama (1969)
	Man	Jones *et al.* (1968b); Kern-Bontke (1964) CL; König (1965a,b); Koudstaal *et al.* (1966); Niemi *et al.* (1965) CL; Novak *et al.* (1965); Pesonen *et al.* (1968) S; Poliak *et al.* (1968) CL,S; Taki *et al.* (1966)
3β-Hydroxysteroid dehydrogenase	Guinea pig	Davies *et al.* (1966); Motta *et al.* (1970); Rubin *et al.* (1963a); Vollrath (1966)
	Hamster	Blaha and Leavitt (1970); Wingate (1970)
	Mouse	Appelgren (1967, 1969a); Ferguson (1965); Hart *et al.* (1966); Kimura *et al.* (1970); Rubin *et al.* (1963b)
	Rat	Barbanti Silva *et al.* (1966); Brandau (1967) F,S; Brandau and Luh (1969); Bratt *et al.* (1968);

(Continued)

TABLE I (Continued)

Enzyme[b]	Species	Reference and tissue studied[c]
		Burkl and Slezak-Klemencic (1967a); Davies *et al.* (1966); Fuhrmann (1963); Goldman and Kohn (1970) S; Goldman *et al.* (1965) Küppers (1967) F,S; Lobel *et al.* (1966); Motta and Bourneva (1970); Motta *et al.* (1970); Okano *et al.* (1966); Prabhu and Weisz (1970); Presl *et al.* (1965) F,S; Pupkin *et al.* (1966); Rubin and Deane (1965); Rubin *et al.* (1963a,b, 1965a,b, 1969); Schlegel *et al.* (1967) F,S; Takikawa and Matsuzawa (1967); Tóth and Szönyi (1966a); Turolla and Magrini (1963); Turolla *et al.* (1966, 1967, 1968); Vollrath (1966); Woods and Domm (1966)
	Squirrel	Seth and Prasad (1967)
	Rabbit	Boucek *et al.* (1967); Davies *et al.* (1966); Fuhrmann (1963); Hadjiisky (1970) CL; Mori and Matsumoto (1970) S; Rubin *et al.* (1963b, 1965b); Wada *et al.* (1969) CL,S
	Cat	Rubin *et al.* (1963b)
	Dog	Fowler and Feldman (1970); Folwer *et al.* (1970); Rubin *et al.* (1963b)
	Ferret	Galil and Deane (1966)
	Cow	Davies *et al.* (1966); Lobel and Levy (1968); Rubin *et al.* (1963b)
	Sheep	Bjersing *et al.* (1970a,b) CL; Deane *et al.* (1966); Rubin *et al.* (1963b)
	Pig	Bergman *et al.* (1966) F; Bjersing (1967a,b)
	Horse	Channing (1969a); Channing and Short (1966) F; Rubin *et al.* (1963b)
	Armadillo	Rubin *et al.* (1963b)
	Monkey	Dennis and Thomas (1967)
	Man	Brandau and Lehmann (1970) F,S; Brandau *et al.* (1968) F,S; Chieffi *et al.* (1965) F,CL; Davies *et al.* (1966); Fienberg and Cohen (1965, 1966); Goldberg *et al.* (1963, 1964); Jirásek *et al.* (1969) S; Jones *et al.* (1968a,b); Karjalainen (1968) CL; Kern-Bontke (1964) CL; König (1965a,b); Koudstaal *et al.* (1966); Magrini and Moneta (1966) F,S; Mori (1969) F; Motta and Takeva (1970); Motta *et al.* (1970); Nakayama (1970); Niemi *et al.* (1965) CL; Novak *et al.* (1965); Pesonen *et al.* (1968) S; Poliak *et al.* (1968) CL,S; Pryse-Davies (1970) F,S; Scully and Cohen (1964); Taki *et al.* (1967); Tóth *et al.* (1970) F,S; Vacek (1967) CL

(Continued)

TABLE I (Continued)

Enzyme[b]	Species	Reference and tissue studied[c]
17β-Hydroxysteroid dehydrogenase	Guinea pig	Davies *et al.* (1966)
	Mouse	Hart *et al.* (1966)
	Rat	Brandau (1967) F,S; Brandau and Luh (1969); Burkl and Slezak-Klemencic (1967a); Davies *et al.* (1966); Fuhrmann (1963); Goldman *et al.* (1965); Tóth and Szönyi (1966a)
	Rabbit	Boucek *et al.* (1967); Davies *et al.* (1966) Fuhrmann (1963)
	Dog	Fowler and Feldman (1970); Fowler *et al.* (1970)
	Cow	Davies *et al.* (1966); Lobel and Levy (1968)
	Pig	Bjersing (1967a,b)
	Man	Brandau and Lehmann (1970) F,S; Brandau *et al.* (1968) F,S; Chieffi *et al.* (1965) F,CL; Davies *et al.* (1966); König (1965a,b); Tóth *et al.* (1970) F,S
20α-Hydroxysteroid dehydrogenase	Guinea pig	Davies *et al.* (1966)
	Rat	Balogh (1964b) F,CL; Balogh *et al.* (1966) F,S; Bratt *et al.* (1968); Burkl and Slezak-Klemencic (1967b); Davies *et al.* (1966); Kidwell *et al.* (1966); Lamprecht *et al.* (1969) CL; Rubin *et al.* (1969); Turolla *et al.* (1966, 1967, 1968, 1969); Wiest *et al.* (1968); Woods and Domm (1966)
	Rabbit	Boucek *et al.* (1967); Davies *et al.* (1966)
	Dog	Fowler and Feldman (1970)
	Cow	Davies *et al.* (1966)
	Man	Brandau *et al.* (1968) F,S; Davies *et al.* (1966); Pupkin *et al.* (1966)
20β-Hydroxysteroid dehydrogenase	Guinea pig	Davies *et al.* (1966)
	Mouse	Baillie *et al.* (1965a) F; Hart *et al.* (1966)
	Rat	Davies *et al.* (1966)
	Rabbit	Boucek *et al.* (1967); Davies *et al.* (1966)
	Cow	Davies *et al.* (1966); Lobel and Levy (1968)
	Man	Davies *et al.* (1966)
Hydroxysteroid dehydrogenases (other than 3β-, 17β-, 20α-, and 20β-hydroxysteroid dehydrogenases)	Guinea pig	Davies *et al.* (1966)
	Mouse	Baillie *et al.* (1965b, 1966b); Hart *et al.* (1966)
	Rat	Brandau (1967) F,S; Davies *et al.* (1966); Fuhrmann (1963); Goldman *et al.* (1965); Tóth and Szönyi (1966a)
	Rabbit	Davies *et al.* (1966); Fuhrmann (1963)
	Dog	Fowler and Feldman (1970); Fowler *et al.* (1970)
	Cow	Davies *et al.* (1966)
	Man	Brandau and Lehmann (1970 F,S; Chieffi *et al.* (1965) F,CL; Davies *et al.* (1966); Ghosh and Sengupta (1969); Koudstaal *et al.* (1966); Tóth *et al.* (1970) F,S

(Continued)

TABLE I (Continued)

Enzyme[b]	Species	Reference and tissue studied[c]
Isocitrate dehydrogenase (NAD)	Rat	Hadjiisky *et al.* (1969a)
	Rabbit	Brandau *et al.* (1967) F,S; Hadjiisky (1970) CL
	Man	Brandau *et al.* (1968) F,S; Kern-Bontke (1964) CL; König (1965a,b)
Isocitrate dehydrogenase (NADP)	Rat	Barbanti Silva *et al.* (1966); Brandau (1967) F,S; Brandau and Luh (1969); Hadjiisky *et al.* (1969a); Okano *et al.* (1966); Schultka and Schmidt (1969); Wegmann and Lageron (1964)
	Rabbit	Brandau *et al.* (1967) F,S; Hadjiisky (1970) CL
	Cow	Kenney (1964); Nakama (1969)
	Man	Brandau and Lehmann (1970) F,S; Brandau and Luh (1964a, 1965); Brandau *et al.* (1968) F,S; König (1965a,b); Luh and Brandau (1964a,b); Mori (1967); Sculley and Cohen (1964); Taki *et al.* (1966)
"Ketoglutarate oxidase" (oxoglutarate dehydrogenase)	Guinea-pig	Zerbian and Goslar (1968)
	Mouse	Zerbian and Goslar (1968)
	Rat	Ferguson (1966); Zerbian and Goslar (1968)
Lactate dehydrogenase	Guinea pig	Zerbian (1966) F,S
	Rat	Brandau and Luh (1969); Bratt *et al.* (1968); Fuhrmann (1963); Georgieva *et al.* (1967); Küppers (1967) F,S; Matsuzawa and Takikawa (1968); Okano *et al.* (1966); Presl *et al.* (1965) F,S; Pupkin *et al.* (1966); Tóth and Szönyi (1966a); Wegmann and Lageron (1964)
	Rabbit	Brandau *et al.* (1967) F,S; Fuhrmann (1963); Hadjiisky (1970) CL
	Dog	Fowler (1970)
	Cow	Nakama (1969)
	Man	Brandau and Lehmann (1970) F,S; Brandau and Luh (1964a,b, 1965); Brandau *et al.* (1968) F,S; Ghosh *et al.* (1965); Kern-Bontke (1964) CL; König (1965a,b); Koudstaal *et al.* (1966); Luh and Brandau (1964a,b); Nakayama (1970); Niemi *et al.* (1965) CL; Pesonen *et al.* (1968) S; Scully and Cohen (1964); Tóth *et al.* (1970) F,S
Lipoamide dehydrogenase	Rat	Balogh (1964a); Hadjiisky *et al.* (1969a)
	Rabbit	Hadjiisky (1970) CL
	Cow	Nakama (1969)
Malate dehydrogenase	Guinea pig	Zerbian (1966) F,S
	Rat	Brandau (1967) F,S; Brandau and Luh (1969); Hadjiisky *et al.* (1969a); Okano *et al.* (1966); Wegmann and Lageron (1964)
	Rabbit	Brandau *et al.* (1967); Hadjiisky (1970) CL

(Continued)

TABLE I (Continued)

Enzyme[b]	Species	Reference and tissue studied[c]
	Dog	Fowler (1970)
	Cow	Nakama (1969)
	Man	Brandau and Lehmann (1970) F,S; Brandau and Luh (1964a,b, 1965); Brandau *et al.* (1968) F,S; Kern-Bontke (1964) CL; Konig (1965a,b); Luh and Brandau (1964a,b); Scully and Cohen (1964); Taki *et al.* (1966)
Monoamine oxidase	Rat	Hadjisky *et al.* (1969b); Nadkarni and Lloyd (1966)
	Rabbit	Hadjiisky (1970) CL
	Dog	Fowler (1970)
	Cow	Nakama (1969)
	Man	König (1965a,b)
$NADH_2$-tetrazolium reductase (NAD-diaphorase)	Guinea pig	Rubin *et al.* (1963a); Zerbian (1966); F,S; Zerbian and Goslar (1968)
	Hamster	Blaha and Leavitt (1970); Wingate (1970)
	Mouse	Appelgren (1967, 1969a); Rubin *et al.* (1963a); Zerbian and Goslar (1968)
	Rat	Barbanti Silva *et al.* (1966); Burkl and Slezak-Klemencic (1967a); Ferguson (1966); Fuhrmann (1963); Hadjiisky *et al.* (1969a); Kuppers (1967) F,S; Rubin *et al.* (1963b); Schultka and Schmidt (1969); Takikawa and Matsuzawa (1967); Zerbian and Goslar (1968)
	Rabbit	Fuhrmann (1963); Hadjiisky (1970) CL; Rubin *et al.* (1963b)
	Cat	Rubin *et al.* (1963b)
	Dog	Fowler (1970); Fowler and Feldman (1970); Fowler *et al.* (1970); Rubin *et al.* (1963b)
	Ferret	Galil and Deane (1966)
	Cow	Lobel and Levy (1968); Rubin *et al.* (1963b)
	Sheep	Rubin *et al.* (1963b)
	Pig	Bjersing (1967a,b); Mayner (1966) F,S
	Horse	Rubin *et al.* (1963b)
	Armadillo	Rubin *et al.* (1963b)
	Monkey	Dennis and Thomas (1967)
	Man	Brandau and Luh (1964a); Hertig and Adams (1967) F; Jones *et al.* (1968a,b); König (1965a,b); Koudstaal *et al.* (1966); Luh and Brandau (1964a,b); Magrini and Moneta (1966) F,S; Nakayama (1970); Niemi *et al.* (1965) CL; Novak *et al.* (1965); Pesonen *et al.* (1968) S; Pryse-Davies (1970) F,S; Scully and Cohen (1964)
$NADPH_2$-tetrazolium reductase	Hamster	Wingate (1970)

(Continued)

TABLE I (Continued)

Enzyme[b]	Species	Reference and tissue studied[c]
(NADP-diaphorase)	Rat	Barbanti Silva *et al.* (1966); Hadjiisky *et al.* (1969a); Schultka and Schmidt (1969); Takikawa and Matsuzawa (1967)
	Rabbit	Hadjiisky (1970) CL
	Dog	Fowler (1970); Fowler and Feldman (1970)
	Cow	Lobel and Levy (1968)
	Sheep	Deane *et al.* (1966)
	Pig	Bjersing (1967a,b); Mayner (1966)
	Man	Brandau and Luh (1964a); Hertig and Adams (1967) F; König (1965a,b); Koudstaal *et al.* (1966); Luh and Brandau (1964a,b); Magrini and Moneta (1966) F,S; Poliak *et al.* (1968) CL,S
Peroxidase	Rabbit	Hadjiisky (1970) CL
	Sheep	Arvy and Mauléon (1964) CL
6-Phosphogluconate dehydrogenase	Rat	Georgieva *et al.* (1967); Wegman and Lageron (1964)
	Rabbit	Hadjiisky (1970) CL
	Man	König (1965a,b) *Comment:* This dehydrogenase uses NADP as co-enzyme and may, thus, be an important source of $NADPH_2$. Since the dehydrogenase is "soluble," 20% polyvinylalcohol in the incubation medium is of value for optimal histochemical results (Chayen *et al.*, 1969).
"Pyruvate oxidase" (Pyruvate dehydrogenase)	Rat	Ferguson (1966)
Secondary alcohol dehydrogenase	Mouse	Appelgren (1967), 1969a)
	Rat	Burkl and Slezak-Klemencic (1967a)
	Man	Ferguson *et al.* (1966); Hardonk (1965); Jones *et al.* (1968b); König (1965a,b); Koudstaal *et al.* (1966); Poliak *et al.* (1968) CL,S; Taki *et al.* (1967)
Succinate dehydrogenase	Guinea pig	Zerbian (1966) F,S
	Rat	Barbanti Silva *et al.* (1966); Brandau (1967) F,S; Brandau and Luh (1969); Bratt *et al.* (1968); Burkl and Slezak-Klemencic (1967b); Chatterjee and Deb (1966) CL; Ghosh *et al.* (1965) Hadjiisky *et al.* (1969a); Küppers (1967) F,S; Lobel *et al.* (1966); Okano *et al.* (1966); Pupkin *et al.* (1966); Schultka and Schmidt (1969); Tóth and Szönyi (1966a); Wegmann and Lageron (1964)
	Rabbit	Brandau *et al.* (1967) F,S; Hadjiisky (1970) CL
	Dog	Fowler (1970)

(Continued)

TABLE I (Continued)

Enzyme[b]	Species	Reference and tissue studied[c]
	Cow	Kenney (1964); Lobel and Levy (1968); Nakama (1969); Priedkalns and Weber (1968)
	Pig	Mayner (1966) F,S
	Man	Brandau and Lehmann (1970) F,S; Brandau and Luh (1964a, 1965); Brandau *et al.* (1968) F,S; Kern-Bontke (1964) CL; König (1965a,b); Koudstaal *et al.* (1966); Luh and Brandau (1964a,b); Nakayama (1970); Pryse-Davies (1970) F,S; Scully and Cohen (1964); Taki *et al.* (1966); Tóth *et al.* (1970) F,S; Vacek (1967) CL
	2. Transferases	
Amylophosphorylase (see phosphorylase)		
γ-Glutamyl transpeptidase (D-glutamyltransferase)	Guinea pig	Jonek and Zieliński (1966); Jonek *et al.* (1965)
	Hamster	Jonek and Zieliński (1966)
	Mouse	Jonek and Zieliński (1966)
	Rat	Jonek and Zieliński (1966); Rutenburg *et al.* (1969)
	Rabbit	Jonek and Zieliński (1966)
	Cat	Jonek and Zieliński (1966)
Phosphorylase	Rat	Georgieva *et al.* (1967)
	Rabbit	Hadjiisky (1970) CL
	Cow	Kenney (1964); Lobel and Levy (1968)
	Man	König (1965a,b)
Ribonuclease	Rat	Daoust (1966) F,S
	3. Hydrolases	
Acetylcholinesterase (see cholinesterase)		
β-Acetylglucosaminidase	Rat	Bulmer (1965b)
	Man	Sato (1969) F,CL
Acid phosphatase	Guinea pig	Adams *et al.* (1966) F,S; Zerbian (1966) F,S
	Mouse	Zawistowska (1965)
	Rat	Banon *et al.* (1964) F,CL; Bulmer (1964a, 1965c); Chatterjee and Deb (1966) CL; Georgieva *et al.* (1967); Hadjiisky *et al.* (1969b); Küppers (1967) F,S; Lobel *et al.* (1966); Tóth (1964) F,S; Tóth and Szönyi (1966a)
	Rabbit	Hadjiisky (1970) CL
	Dog	Kabra *et al.* (1967b)
	Cow	Kenney (1964); Lobel and Levy (1968); Nakama (1969)

(Continued)

TABLE I (Continued)

Enzyme[b]	Species	Reference and tissue studied[c]
	Sheep	Bjersing *et al.* (1970b) CL; Dingle *et al.* (1968) CL
	Pig	Goode *et al.* (1965) CL; Mayner (1966) (1965) CL; Mayner (1966)
	Man	Dokumov and Dashev (1963a,b); Ghosh *et al.* (1965); Hertig and Adams (1967) F; Karjalainen (1968) CL; Kern-Bontke (1964); CL; König (1965a,b); Koudstaal *et al.* (1966); Pryse-Davies (1970) F,S; Taki *et al.* (1966); Tóth *et al.* (1970) F,S; Vacek (1967) CL
Adenosine triphosphatase (see nucleoside phosphatase)		
Alkaline phosphatase	Guinea pig	Adams *et al.* (1966) F,S; Zerbian (1966) F,S
	Mouse	Ben-Or (1963) F,S; Botte (1968); Zawistowska (1965)
	Rat	Bulmer (1964b); Chatterjee and Deb (1966) CL; Georgieva *et al.* (1967); Hadjiisky *et al.* (1969b); Küppers (1967) F,S; Lobel *et al.* (1966); Tóth (1964) F,S; Tóth and Szonyi (1966a); Varma (1970); Varma and Guraya (1968)
	Rabbit	Hadjiisky (1970) CL
	Dog	Kabra *et al.* (1967a) F,S
	Cow	Gropp (1967) F,S; Kenney (1964); Lobel and Levy (1968); Nakama (1969)
	Pig	Goode *et al.* (1965) CL; Mayner (1966) F,S
	Man	Dokumov and Dashev (1963a,b); Ghosh *et al.* (1965); Hertig and Adams (1967) F; Karjalainen (1968) CL; Kern-Bontke (1964) CL; König (1965a,b); Koudstaal *et al.* (1966); Pryse-Davies (1970) F,S; Taki *et al.* (1966); Tóth *et al.* (1970) F,S; Vacek (1967) CL
Aminopeptidase	Guinea pig	Zerbian (1966) F,S
	Mouse	Niemi and Sylvén (1969) F (naphthylamidase)
	Rat	Hadjiisky *et al.* (1969b); Kaneyoshi (1969)
	Rabbit	Hadjiisky (1970) CL
	Cow	Nakama (1969)
	Man	Ghosh *et al.* (1965); Jones *et al.* (1968b); Kern-Bontke (1964) CL; König (1965a,b); Koudstaal *et al.* (1966); Novak *et al.* (1965); Pryse-Davies (1970) F,S; Taki *et al.* (1966)
Aniline hydroxylase	Rat	Grasso *et al.* (1971)
Arylsulfatase (see sulfatase)		
Cholinesterase (including acetylcholinesterase)	Guinea pig	Bulmer (1965a); Jordan (1970) F,S; Zerbian and Goslar (1968)

(Continued)

TABLE I (Continued)

Enzyme[b]	Species	Reference and tissue studied[c]
	Mouse	Zerbian and Goslar (1968)
	Rat	Bulmer (1965a); Hadjiisky *et al.* (1969b); Zerbian and Goslar (1968)
	Rabbit	Bulmer (1965a); Hadjiisky (1970) CL; Jordan (1970) F,S
	Cat	Jacobowitz and Wallach (1967) S
	Monkey	Bulmer (1965a); Jacobowitz and Wallach (1967) S
	Man	Kern-Bontke (1964) CL; Jacobowitz and Wallach (1967) S; König (1965a,b); König and Stier (1963)
Esterase (see nonspecific esterase)		
β-Galactosidase	Guinea pig	Bulmer (1964a)
	Rat	Bulmer (1964a, 1965b); Monis *et al.* (1963) F,S (α-galactosidase)
	Rabbit	Bulmer (1964a)
	Monkey	Bulmer (1964a)
	Man	König (1965a,b)
Glucose-6-phosphatase	Rat	Georgieva *et al.* (1967)
	Rabbit	Nakama (1969)
	Man	König (1965a,b)
α-Glucosidase	Man	König (1965a,b) *Comment:* The reaction was negative in all ovarian structures.
β-Glucuronidase	Guinea pig	Bulmer (1964a)
	Rat	Bulmer (1963, 1964a, 1965b, 1966, 1967a,b); Hayashi (1964) F,CL
	Rabbit	Bulmer (1963, 1964a); Hadjiisky (1970) CL
	Cow	Nakama (1969)
	Monkey	Bulmer (1964a)
	Man	König (1965a,b); Taki *et al.* (1966); Vacek (1967) CL
Inorganic pyrophosphatase[d]	Man	König (1965a,b)
Leucine aminopeptidase (see aminopeptidase)		
Lipase (see nonspecific esterase)		
Nonspecific esterase (including lipase)	Guinea pig	Bulmer (1964a); Zerbian (1966) F,S
	Mouse	Zawistowska (1965)
	Rat	Abe *et al.* (1964) F,CL; Banon *et al.* (1964) F,CL; Bronzetti *et al.* (1963); Bulmer (1964a, 1965c); Burkl and Slezak-Klemencic (1967b); Hadjiisky *et al.* (1969b); Küppers (1967) F,S; Tóth and Szönyi (1966a)
	Rabbit	Abe *et al.* (1964) F,CL; Bulmer (1964a); Hadjiisky (1970) CL

(Continued)

TABLE I (Continued)

Enzyme[b]	Species	Reference and tissue studied[c]
	Cow	Nakama (1969)
	Monkey	Bulmer (1964a)
	Man	Karjalainen (1968) CL; Kern-Bontke (1964) CL; König (1965a,b); König and Stier (1963); Koudstaal *et al.* (1966); Niemi *et al.* (1965) CL; Taki *et al.* (1966); Tóth *et al.* (1970) F,S; Vacek (1967) CL
Nucleoside phosphatase [including nucleoside triphosphatase, (e.g., ATPase-adenosine triphosphatase) and nucleoside diphosphatase]	Guinea pig	Adams *et al.* (1966) F,S; Zerbian (1966) F,S
	Rat	Hadjiisky *et al.* (1969a); Lobel *et al.* (1966)
	Rabbit	Hadjiisky (1970) CL
	Cow	Lobel and Levy (1968); Nakama (1969)
	Pig	Mayner (1966) F,S
	Man	Ghosh *et al.* (1965); Hertig and Adams (1967) F; Kern-Bontke (1964) CL; König (1965a,b); Koudstaal *et al.* (1966)
5′-Nucleotidase	Guinea pig	Adams *et al.* (1966) F,S
	Mouse	Hardonk (1968)
	Rat	Hadjiisky *et al.* (1969b); Hardonk (1968)
	Rabbit	Hadjiisky (1970) CL
	Pig	Mayner (1966) F,S
	Man	Ghosh *et al.* (1965); Hertig and Adams (1967) F; König (1965a,b); Koudstaal *et al.* (1966)
Phosphoamidase	Man	Kern-Bontke (1964) CL; König (1965a,b) *Comment:* The variability of results with the histochemical methods has only recently been overcome (Chayen *et al.* 1969).
Sulfatase	Guinea pig	Wolf *et al.* (1967)
	Mouse	Wolf *et al.* (1967)
	Rat	Hadjiisky *et al.* (1969); Wolf *et al.* (1967)
	Rabbit	Hadjiisky (1970) CL
	Man	König (1965a,b); Wolf *et al.* (1967)
Thiamine pyrophosphatase[d]	Man	König (1965a,b)
Aldolase	Man	König (1965a,b) *Comment:* No enzyme activity was found in any ovarian structure.
Carbonic anhydrase	Man	König (1965a,b)

[a] For enzyme histochemical studies on the ovary before 1963 see essays by Jacoby (1962), Arvy (1963), Velardo and Rosa (1963), König (1965a), and Baillie *et al.* (1966a).

[b] The enzymes have been grouped into the major classes of the IUB classification (see tabulation in Section II,B) and within the classes they have been put in alphabetical order according to their trivial names.

[c] When no specific tissue is indicated, the whole ovary was studied; F, follicle; CL, corpus luteum; S, stroma and interstitial tissue.

[d] The specificity of the histochemical methods for demonstration of carbonic anhydrase, inorganic pyrophosphatase, and thiamine pyrophosphatase activities is open to some doubt.

Schmidt, 1969). No β-hydroxybutyrate dehydrogenase activity was found in the granulosa layer. The reactions for succinate dehydrogenase, glycerolphosphate dehydrogenase, and isocitrate dehydrogenase were always positive in the follicular granulosa cells. In the dog (Fowler, 1970), granulosa cells of developing follicles showed little glucose-6-phosphate dehydrogenase activity; stronger enzyme reactions were observed during estrus. In the cow (Kenney, 1964), very high glucose-6-phosphate dehydrogenase activities were found in granulosa cells of follicles with diameters >3 mm and succinate dehydrogenase activity in different ovarian elements has been quantitatively assessed (Priedkalns and Weber, 1968).

ii. Transferases. Jonek *et al.* (1965) studied γ-glutamyl transpeptidase and found weak or negative reactions in primary and secondary follicles of guinea pig ovaries. Theca cells of large tertiary follicles were usually negative, but in a few instances a strong activity was found, e.g., in the granulosa cells of the corona radiata. Rutenberg *et al.* (1969) also studied γ-glutamyl transpeptidase activity but with a new histochemical method. This method was tested in many tissues at the light microscope level and at the ultrastructural level in rat pancreas. Granulosa cells of maturing rat ovarian follicles showed a positive reaction, but mature follicles were negative. Daoust (1966) used an interesting substrate film method for localizing ribonuclease activity. A positive reaction was mainly found in the follicular fluid and also in granulosa cells before formation of an antrum; after antrum formation the granulosa cells were negative. Weak activities were found in the theca layer, in the stroma and in atretic follicles.

iii. Hydrolases. Sato (1969) studied β-acetylglucosaminidase activity in the human ovary. No reaction was found in the granulosa layer, but the enzyme activity in the theca interna increased as the follicles enlarged.

Banon *et al.* (1964) reported negative acid phosphatase reactions in normal developing and mature follicles of the rat ovary; a moderately positive reaction in granulosa cells with small cytoplasmic granules was found immediately before and after ovulation; granulosa cells of atretic follicles displayed an intense reaction. The esterase activity pattern was similar to that of acid phosphatase. Kabra *et al.* (1967b) reported marked acid phosphatase activity in the granulosa cells, cortical cords, and rete ovarii of the dog. The granulosa layer of large tertiary follicles displayed a marked positive reaction for nonspecific esterase (Burkl and Slezak-Klemencic, 1967b).

b. Other Vertebrates. It is not easy to make direct comparison between histochemical results obtained in nonmammals with those in mammals; and in this chapter no attempt has been made to cover the histochemical literature on nonmammals. However, as examples of mainly enzyme histochemical studies on amphibian and fish ovaries, the following valuable and recent papers may be mentioned: Kessel and Panje (1968), Lambert

(1970), Xavier *et al.* (1970), and Redshaw and Nicholls (1971). Hoar (1965) has stressed the diversity of reproductive processes in fishes, but most other vertebrates, for example, amphibians, have relatively simple ovaries in comparison with mammals.

B. Corpus Luteum

1. Carbohydrates

a. Man. The central granulosa lutein cells of the early corpus luteum of the menstrual cycle were found to contain glycogen; this substance then disappeared but reappeared and increased in granulosa lutein cells from the twenty-fifth day till the time of menstruation (McKay *et al.*, 1961). During the entire menstrual cycle theca lutein cells contained little or no glycogen. In early degenerating granulosa lutein cells a few glycogen granules were sometimes observed but were completely absent in more advanced degeneration. Ghosh and Sengupta (1970) reported a similar distribution of glycogen; their material was, however, less accurately dated. For a few days after ovulation, the granulosa lutein cells were free of glycoprotein, but as the corpus luteum matured a fibrillar network appeared around each cell and successively increased in thickness and staining intensity (McKay *et al.*, 1961). The theca lutein cells, however, were surrounded only by a fine meshwork of faintly PAS-positive fibrils, which remain unchanged throughout the active life of the corpus luteum. By the second week after menstruation, the fibrillar glycoprotein deposits between the granulosa lutein cells had increased considerably in amount. The corpus luteum of pregnancy did not contain the coarse intercellular fibrils that were prominent in the granulosa lutein layer toward the end of the cycle; instead it exhibited curious colloid glycoprotein droplets lying between cells. Larraguibel and Dallenbach-Hellweg (1970) reported PAS-positive inclusions in granulosa lutein cells of the corpus luteum of pregnancy; however, the material was interpreted as lipoprotein and not carbohydrate. Lutein cells in the corpus luteum of pregnancy were free of glycogen (McKay *et al.*, 1961).

b. Other Mammals. Bjersing *et al.* (1970a), studying ovine corpora lutea at accurately timed stages of the estrous cycle, confirmed the findings of McKay *et al.* (1961) that glycogen appeared in lutein cells late in the cycle (see Table II and Fig. 1); they discussed the possibility that an increase in glycogen may be an early sign of luteal regression. Glycogen was also studied in corpora lutea of hysterectomized sheep at about 140 days after estrus (Bjersing *et al.*, 1970b). Lobel and Levy (1968) found very little glycogen or glycoprotein in the luteal cells of mature corpora lutea of cyclic

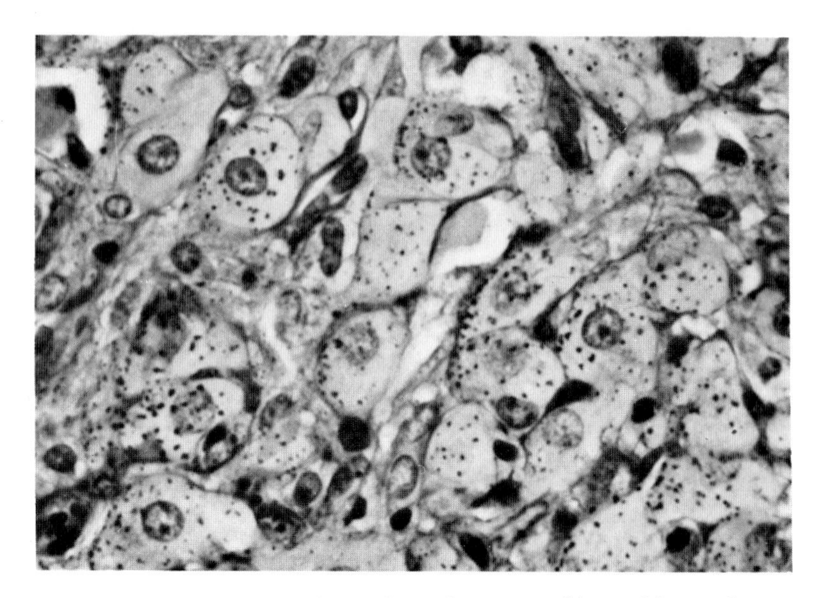

Fig. 1. Paraffin wax section of part of an ovine corpus albicans, 8 hours after onset of estrus, stained with Mayer's hemalun and the PAS procedure. Note the large number of lutein cells. They are smaller than on day 12 of the estrous cycle and almost all of them contain numerous glycogen granules; some still have vesicular nuclei, but others show karyolysis. Glycogen is almost absent from one lutein cell with a vesicular nucleus and from another highly vacuolated cell (central) that is in an advanced state of regression. ×670. (From Bjersing *et al.*, 1970a.)

cows. Between the cells of young corpora lutea of dogs, O'Shea (1966) found pools of acid mucin histochemically similar to liquor folliculi but with greater resistance to neuraminidase and hyaluronidase digestion. The presence of similarly staining granules in the cytoplasm of the luteal cells suggested that the mucin was produced by these cells, although association with formation of connective tissue could not be excluded.

2. Lipids

a. Man. *i. Corpus luteum of the menstrual cycle.* A recent and detailed study of lipids of the human ovary is that by Guraya (1968b). During the transformation of the follicular granulosa cells into lutein cells there was a gradual development of diffuse sudanophilic lipid, interpreted as lipoprotein by Guraya (1968b). He concluded that the so-called luteinization of the granulosa cells was accompanied by this increase of histochemically demonstrable diffuse lipoprotein throughout the cytoplasm. He also thought that the lipoproteins reflected a pronounced increase in agranular endoplasmic reticulum, a structure that is the site of steroidogenically important enzymes and may be a reservoir for both cholesterol and other steroid hormone precursors (see Christensen, 1965; Fawcett, 1965; Bjersing, 1967d). The cytoplasm of granulosa lutein cells also contained L_2 bodies of the same type as found in developing follicles and other discrete lipid bodies, some of which were reported to consist of phospholipids and some to contain triglycerides as well. Niemi *et al.* (1965) reported that theca cells of fully developed mature corpora lutea contained many lipid droplets, while the granulosa cells had fewer. Guraya (1968b) stated that the theca lutein cells displayed the same histochemical reactions as the theca interna cells during follicular development. Sudanophilic lipids, consisting of phospholipids, triglycerides, and cholesterol and its esters were found only in the theca cells of newly ruptured follicles. This finding supplies additional support for the contention of Mossman *et al.* (1964) that the thecal gland cells (the original theca interna cells of the preovulatory follicle) do degenerate after ovulation. The theca lutein cells of more mature corpora lutea may, therefore, represent new paraluteal cells which have differentiated from stromal cells. The regression of the corpus luteum started gradually in 9- to 10-day-old corpora lutea with accumulation of sudanophilic droplets in both granulosa and theca lutein cells. According to Guraya (1968b), the droplets first consisted of phospholipids alone, phospholipids and triglycerides or phospholipids, triglycerides, and cholesterol and its esters. In late stages of corpus luteum regression the droplets contained triglycerides, cholesterol and its esters, and pigments and were of a coarse nature. The theca lutein cells appeared to persist longer than the granulosa lutein cells and were seen in the periphery of the corpus albicans filled with pigmented lipids. Accumulation of similar lipid droplets was reported in other studies on regressing human corpora lutea (Green and Maqueo, 1965; van Lennep and Madden, 1965) and aging rodent corpora lutea (Guraya and Greenwald, 1965; Guraya, 1968d).

Takazawa *et al.* (1964), in an investigation on aging of the ovary, stained a large number of corpora lutea with Sudan IV and performed Schultz's cholesterol test. Matsukado and Hayashi (1965) also studied the lipid histochemistry of human menstrual corpora lutea and paid special attention to granules stainable with Heidenhain's iron hematoxylin.

ii. Corpus luteum of pregnancy. Karjalainen (1968) found comparatively coarse lipid droplets positive with Oil red O in many theca lutein cells up to the eleventh to twelfth gestational week, but thereafter the number of positive cells decreased fairly rapidly. The staining of the granulosa lutein cells was more diffuse, but during the tenth to the eleventh gestational week the number of granulosa cells with large lipid droplets clearly increased. The appearance of large fat droplets in granulosa lutein cells has almost unanimously been interpreted as a sign of fatty degeneration. In the corpus luteum at term it was noticeable that the highly developed cytoplasm lacked sudanophilic lipid droplets (Guraya, 1968h). However, the theca lutein cells and, to some extent, the granulosa lutein cells showed diffuse staining with Sudan black B. Both cell types also displayed lipid granules consisting mainly of phospholipids. Some lipid granules in the granulosa lutein cells also contained triglycerides, but no Schultz-positive substances. The histochemical findings were thought to indicate (Guraya, 1968h) that the corpus luteum cells at term secreted steroid hormones rather than stored their precursors. Intensely refractile inclusions of lipoprotein containing amino acids with SH groups were found in the cytoplasm of granulosa lutein cells by Larraguibel and Dallenbach-Hellweg (1970). They were sparse in corpora lutea of the menstrual cycle. The authors suggested that the inclusions were connected with a nonsteroid hormonal secretion, e.g., relaxin.

b. Other Mammals. In the bovine corpus luteum, general lipid and cholesterol reactivities were assessed quantitatively throughout the estrous cycle in a histochemical investigation by Priedkalns and Weber (1968). The same year, Lobel and Levy (1968) briefly reported on lipids in bovine corpora lutea both of pregnancy and nonpregnancy, and in a biochemical study Coutts and Stansfield (1968) demonstrated cholesterol esterase activity in the corpus luteum of the cow. The latter authors also estimated the size of the available endogenous cholesterol ester pools; this is of interest particularly since cholesterol esters may act as reserves of steroid hormone precursors and also can inhibit cleavage of the side-chain of free cholesterol (Raggatt and Whitehouse, 1966).

In the sheep (Deane *et al.*, 1966; Dingle *et al.*, 1968; Bjersing *et al.*, 1970a) there was an increase in the amount of lipid toward the end of the cycle. Up to day 14, and also on day 15 when the progesterone concentration was still high, lipid was present in only small amounts (see Table II). It was greatly increased in 15-day corpora lutea with low progesterone content (Bjersing *et al.*, 1970a). Large amounts of lipid were present in the corpora albicantia. In all cases, the lipid stainable with Oil red O was completely extracted by acetone. In sections treated with acetone and stained with Sudan black, the lipid droplets could not be demonstrated in the lutein cells, but some background cytoplasmic staining always remained; numerous small sudanophilic granules were also scattered uniformly throughout the tissue. The granules probably corresponded to mitochondria and the diffuse background staining largely to agranular endoplasmic reticulum. As regression proceeded, both in the estrous cycle (Bjersing *et al.*, 1970a) and after hysterectomy (Bjersing *et al.*, 1970b), the background cytoplasmic staining increased in intensity, indicating that more phospholipids were available for staining (Fig. 2) (perhaps partly due to splitting of lipoprotein

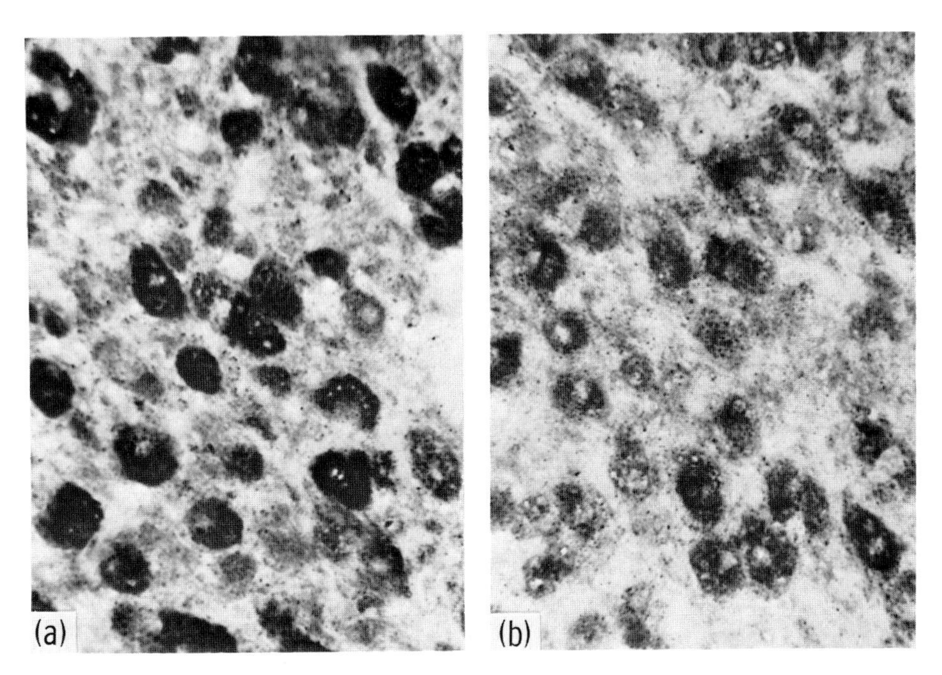

Fig. 2. Alcohol–formalin fixed, frozen sections of corpora lutea of sheep 140 days after hysterectomy; (a) stained with Sudan black, (b) control section extracted with acetone (20 minutes) and stained with Sudan black. The luteal cells are strongly stained but much of the lipid is unaffected by acetone extraction. ×240. (From Bjersing *et al.*, 1970b.)

complexes). Bjersing (1967d) also studied the lipids of the corpus luteum in the pig.

Cats and dogs were subjects of a histochemical investigation of ovarian lipids by Guraya (1969b). The granulosa lutein cells of new corpora lutea (postovulatory follicles) of rabbits displayed more (L_1) lipid bodies consisting of phospholipids and triglycerides than did granulosa cells in estrous follicles (Guraya, 1968d). In the 3- to 4-day-old corpora lutea the cytoplasm was filled with the L_1 bodies, and the granulosa lutein cells also began to show relatively more triglycerides and cholesterol and/or its esters; a significant change in the young corpora lutea was the development of a diffuse, sudanophilic substance, probably lipoprotein, throughout the cytoplasm. This sudanophilic material was thought to be derived from the agranular endoplasmic reticulum (Guraya, 1968d) which develops during the so-called luteinization of granulosa cells (Blanchette, 1966). In the theca interna cells similar diffuse lipoproteins were present even in the estrous follicles as well as in the preovulatory follicles. In 1-day-old corpora lutea, the theca interna cells could still be recognized, but they were shrunken and degenerate. They were filled with coarse lipid bodies containing cholesterol and/or its esters as well as triglycerides and phospholipids. In the 3- to 4-day-old corpora lutea, cells derived from the theca interna could not be distinguished; they had apparently degenerated and disappeared (Guraya, 1968d).

In the hamster ovary, vesicular follicles ovulate during the night of day 4 and give rise to corpora lutea which are fully differentiated on day 1 of the subsequent cycle. Guraya and Greenwald (1965) found that cholesterol and esters appeared in the luteal cells at day 3; after that, regression occurred rapidly and, simultaneously, triglycerides accumulated in the corpora lutea. The authors remarked that it might be of significance that the regression of cyclic corpora lutea at day 3 in the hamster coincided with the onset of atresia in some of the larger developing follicles. In a histochemical study of the corpora lutea of pregnancy, pseudopregnancy, and nonpregnancy in rats, Mor (1966) found that the corpora lutea of early pregnancy could be distinguished from the other corpora lutea by the use of lipid stains. Fomitchev (1969) investigated the distribution of total and unsaturated lipids in the corpus luteum of the rat during pregnancy and lactation; he found a large amount of total lipid and a small amount of unsaturated lipid in the very young lutein cells. By the end of the first week of gestation the sudanophilic granules had diminished in size and increased in number throughout the cytoplasm; the unsaturated lipids were stored with further increase during the first 2 weeks of lactation. The histochemistry of lipids in corpora lutea of alloxan-diabetic rats was studied by Chatterjee and Deb (1966).

In the American opossum there was a striking development of diffuse lipoproteins throughout the cytoplasm during the differentiation (or luteinization) of follicular granulosa cells into granulosa lutein cells (Guraya, 1968c). With regression of the corpus luteum, coarse lipid granules consisting of cholesterol and cholesterol esters, triglycerides, and some phospholipids accumulated in the lutein cells.

3. Enzymes

a. Mammals. The principal histochemical studies of enzymes in corpora lutea since 1963 are assembled in Table I. For a discussion of the biological significance of some of the most interesting results, see Section VII,C. Only a few studies on oxidoreductases, transferases, and hydrolases will be briefly mentioned below.

i. Oxidoreductases. Rubin *et al.* (1963b) studied the Δ^5-3β-hydroxysteroid dehydrogenase reaction in the ovaries of the guinea pig, mouse, rat, rabbit, cat, dog, cow, sheep, horse, and armadillo. The degree of enzyme activity showed considerable differences between species; there were also interesting differences in the location of the enzyme activity within a single species. The authors were unable to obtain positive Δ^5-3β-hydroxysteroid dehydrogenase reactions in corpora lutea of pregnant rabbits. Deane, however, did demonstrate moderate Δ^5-3β-hydroxysteroid dehydrogenase activity in the rabbit corpus luteum in later work (see Galil and Deane, 1966). In the rat, Burkl and Slezak-Klemencic (1967a) found cyclic variations in activity of 3β-hydroxysteroid dehydrogenase and secondary alcohol dehydrogenase in the corpora lutea.

An extensive study on 3β-, 17β-,and 20β-hydroxysteroid dehydrogenase (as well as phosphorylase, acid and alkaline phosphatases, and adenosine triphosphatase) in the cow has been published by Lobel and Levy (1968). In a combined investigation of the endocrine activity, histochemistry, and ultrastructure of ovine corpora lutea (Bjersing *et al.*, 1970a,b) the intensity and distribution of Δ^5-3β-hydroxysteroid dehydrogenase activity was evaluated in normal animals between day 10 of the estrous cycle and the following estrus (see Table II). A similar study was carried out in hysterectomized sheep about 140 days after ovulation (Bjersing *et al.*, 1970b).

The activity of 20α-hydroxysteroid dehydrogenase in the rat ovary has been investigated by several authors. Balogh (1964b) demonstrated a strong reaction in involuting corpora lutea; however, growing and mature ovarian follicles were negative. Turolla *et al.* (1966) found 20α-hydroxysteroid dehydrogenase activity only in corpus luteum cells; in metestrus and diestrus the newly formed corpora lutea totally lacked 20α-hydroxysteroid dehydrogenase activity. Involuting corpora lutea, however, showed a distinctly positive 20α-hydroxysteroid dehydrogenase reaction. Lamprecht *et al.* (1969) performed a combined biochemical and histochemical investigation of 20α-hydroxysteroid dehydrogenase after blockade of pituitary prolactin secretion or after injection of prolactin.

Kern-Bontke (1964) observed that glucose-6-phosphate dehydrogenase showed a much higher activity in theca lutein cells than in granulosa lutein cells of the human ovary. The 3β-hydroxysteroid dehydrogenase levels were reversed.

Arvy and Mauléon (1964) found strong activity of cytochrome oxidase and peroxidase in the corpus luteum of pregnancy. During the estrous cycle only the cytochrome oxidase activity was high.

ii. Transferases. Jonek *et al.* (1965) found a weak activity of γ-glutamyl transpeptidase in the cyclic corpora lutea of guinea pigs but high activity in the corpora lutea of pregnancy.

iii. Hydrolases. In a study of lysosomal function and acid phosphatase in ovine corpora lutea by Dingle *et al.* (1968), biochemical and histochemical

TABLE II

Progesterone Concentration and Histochemical Reactions in Sheep Corpora Lutea and Corpora Albicantia at Different Stages of the Estrous Cycle[a]

Days after onset of estrus	Sheep no.	Progesterone (μg/gm luteal tissue	Glycogen in lutein cells[b]	Steroid dehydrogenase activity[d] Intensity	Steroid dehydrogenase activity[d] Distribution	Lipid reaction[e] with Sudan black	Lipid reaction[e] with Oil red O	Lipid droplets in ~1 μm araldite sections
Corpora lutea								
10	S30	37	±	+++	even	Tr	Tr	no
10	S51	26	±	++	even	+	++	no
12	S31	18	±	++	even	+	++	no
12	S32	39	+	++	even	Tr	Tr	no
12	S33	21	+	++	even	Tr	Tr	no
13	S43	7	±	++	even	Tr	+	no
13	S62	30	+	+++	even	++	+	no
14	S35	27	++	++	even	Tr	Tr	no
14	S53	33	±	+++	even	Tr	Tr	no
14	S54	24	±	+++	even	Tr	Tr	no
14	S55	61	±	+++	even	Tr	+	no
14	S57	37	±	+++	even	+	+	no
14	S65	28	+	++	patchy	+	+	no
14	S66	19	+	+++	even	+	+	no
15	S61	31	++	++	even	Tr	Tr	no
15	S63	6	±/++[c]	+	patchy	+++	+++	+
15	S67	1.8	+++	+	patchy	+++	+++	+
15	S38	0.9	+++	+	patchy	+++	+++	++
Corpora albicantia								
0	S69	0.9	+++	+	even	+++	+++	++
1	S71		none	+	patchy	+++	+++	+++

[a] From Bjersing *et al.* (1970a).

[b] Number of lutein cells containing glycogen: ±, very few; +, about one quarter; ++, about one half; +++, almost all.

[c] This tissue sample was undergoing focal regression.

[d] + → +++, increasing amounts of formazan deposition.

[e] Lipid: Tr, trace; + → +++, increasing lipid content.

results were correlated; the histochemical fragility test of Bitensky (1963) was used. Sato (1969) demonstrated an intense activity of β-acetylglucosaminidase in the granulosa lutein cells of the human corpus luteum of menstruation.

b. Other Vertebrates. See Section IIIA,3,b.

C. Ovarian Stroma and Other Structures

1. Carbohydrates

a. Man. The surface epithelium of the adult ovary displays a well-developed PAS-positive basement membrane in direct contact with the reticular fibers of the ovarian cortex; glycogen granules were found both in the surface epithelium and in the epithelium lining its so-called germinal inclusion cysts (McKay *et al.*, 1961). PAS staining with or without saliva digestion was also used in a histochemical study on the development of the human ovary (Pinkerton *et al.*, 1961). Ghosh and Sengupta (1970) did not detect glycogen in stromal cells of hyperplastic areas, stromal theca cells in areas of thecosis or in intercellular connective tissue. They thought that very little glycogen was stored in the stroma but was instead metabolized as soon as it was formed. Csaba *et al.* (1970) was also unable to find glycogen in the ovarian stroma. Guraya (1966a) detected PAS-positive material in the cytoplasm and Golgi zone of the interstitial gland cells. However, the PAS reaction was positive even after acetylation but not after extraction with pyridine and was therefore interpreted as being due to lipid. Ghosh and Sengupta (1970) obtained a diastase-resistant PAS-positive reaction between the cells of the stroma and interpreted the reactive material as glycoprotein. Stromal cells and stromal theca cells were consistently negative in tests for acid mucopolysaccharides; the connective tissue matrix, however, reacted positively in cases of focal and diffuse stromal hyperplasia (Ghosh and Sengupta, 1970). Csaba *et al.* (1970) reported a marked increase in mucopolysaccharides in the stroma of polycystic ovaries, intercellular areas being filled with acid mucopolysaccharides. Dokumov and Dashev (1963a) studied glycogen, neutral mucopolysaccharides, and acid mucopolysaccharides in the ovaries of normal women and of patients suffering from the Stein–Leventhal syndrome.

b. Other Mammals. O'Shea (1966), in a thorough histochemical study of mucin secretion by subsurface epithelial structures in the canine ovary, found very few cells containing glycogen. The subsurface epithelial structures, partly corresponding to the so-called germinal inclusion cysts, contained intracytoplasm droplets with sialic acid-containing nonsulfated acid mucopolysaccharide. Similar but smaller droplets were sometimes found in the surface epithelium, and on its free surface similar material was frequently seen. In discussing the functional importance of the secretory activity, O'Shea (1966) suggested that the subsurface epithelial structures (perhaps together with the surface epithelium) contributed to the lubricant secretions of the ovarian bursal cavity. The occurrence of carbohydrates in the ovarian stroma and other structures has been described by Ejsmont (1965) and Jirásek *et al.* (1970) in the rat and by Ben-Or (1963) in the mouse; the histochemistry of ovarian macrophages in the rat has been studied by Bulmer (1964c).

2. Lipids

a. Man. Kraus and Neubecker (1962) reported prominent hilus cells in some, but not all, ovaries of infants. Loubet and Loubet (1961) found hilus cells from birth on. However, lipid pigment granules, crystalloids, and hyaline spherules were detected in these cells only after puberty, and hyperplastic changes only after the menopause. The cells often contained sudanophilic inclusions stainable by Schultz's cholesterol method (Loubet and Loubet, 1961). Lipids in follicles and stroma of normal and abnormal ovaries were also studied by Dokumov and Dashev (1963a), Csaba *et al.* (1970), and Ghosh and Sengupta (1970). Fienberg (1963) and Fienberg and Cohen (1965) studied stromal spindle cells (ovarian "fibroblasts") and found that some of these cells formed a Sudan-positive band below the surface of the ovary. Spindle and polygonal stromal theca cells were Sudan- and Schultz-positive, as were the lipid-band cells, but in contrast to lipid-band cells they also showed glucose-6-phosphate dehydrogenase activity and sometimes Δ^5-3β-hydroxysteroid dehydrogenase activity.

The so-called interstitial gland cells of gestation lacked sudanophilic lipid droplets, in contrast to corresponding cells in a variety of other mammalian species (Guraya, 1966a). But the whole cytoplasm stained feebly with Sudan black B; the intensity of the reaction was much stronger in material treated with Zenker, Bouin, and Carnoy solutions instead of the ordinary formaldehyde–calcium. The diffuse lipid component was probably lipoprotein, and the increased reaction probably represented some kind of unmasking. The cytoplasm of fully differentiated interstitial gland cells displayed many sudanophobic vacuoles originating in the Golgi zone and gradually moving toward the plasma membrane; such vacuoles were not found in those interstitial gland cells that were still in the process of differentiation from the theca interna and stromal elements. The interstitial cells also contained scattered heterogeneous bodies (L_2) (Guraya, 1966a) consisting of phospholipids. Guraya (1967a) also studied the lipids of interstitial gland tissue of nonpregnant women. These interstitial gland cells showed the same histochemical reactions as the comparable cells of pregnancy but also displayed lipid droplets containing mainly phospholipids and triglycerides; in interstitial gland cells of relatively large follicles, cholesterol and its esters were also present in the droplets. However, the interstitial cells of nonpregnancy did not show any heterogeneous phospholipid bodies (L_2).

The morphological and histochemical features suggested that the interstitial gland cells of pregnancy were metabolically more active than those of nonpregnancy. After a time, the interstitial cells from small antral and medium-sized follicles began to lose their cytoplasm and lipids and reverted to stromal elements. In contrast, the corresponding cells (thecal gland cells) (Mossman *et al.*, 1964) derived from more mature follicles degenerated and disappeared. The theca interna cells of relatively mature follicles had apparently undergone such advanced differentiation that they had lost their potentiality for reverting to the original stromal elements (Guraya, 1967a). Lipids in ovarian stroma were studied also by Takazawa *et al.* (1964). Guraya (1969a) studied cells with foamy cytoplasm in human ovaries at term. These elderly cells were filled with lipids consisting of pigments, triglycerides, cholesterol or cholesterol esters, and some phospholipids. The cells were apparently so altered because of senescent changes that they could not become active steroid secretors; however, their life might have been prolonged in pregnancy because of strong gonadotropic stimulation.

b. Other Mammals. The ovaries of nonpregnant monkeys and cows (Guraya, 1966b, 1968f,g) were shown to contain interstitial gland tissue developed from the theca interna of degenerating follicles. In cattle and rhesus monkeys, as in women (Guraya, 1967a), the interstitial gland cells remained for a short time. The cells of the primates stored cholesterol and its esters, in contrast to the corresponding elements in cattle (Guraya, 1968f), suggesting a relatively greater metabolic activity in the latter. The interstitial tissue in nonpregnant bats (Guraya, 1967c), rats, cats, and dogs also showed sudanophilic droplets containing cholesterol and its esters in addition to triglycerides and phospholipids (Guraya and Greenwald, 1964a).

The interstitial gland of the rabbit ovary is extensively developed. Guraya and Greenwald (1964b) reported that in the ovaries of estrous rabbits the cells of this glandular tissue were full of lipids consisting of phospholipids, triglycerides, and cholesterol and its esters. As early as 2 hours after injection of human chorionic gonadotropin, mobilization of the lipid droplets could be observed, and in rabbits killed after 11 hours (i.e., about the time of ovulation) there was almost complete depletion of lipid droplets from the cells of the interstitial gland. In the remaining lipid bodies, cholesterol and its esters had disappeared. By 3 days after ovulation the interstitial gland cells were again loaded with lipid droplets of the usual composition. This interesting release of lipid droplets was described in more detail in a later publication by Guraya (1967b). Changes of lipid and lipoprotein lipase activity in nonpregnant rabbit ovaries following administration of human chorionic gonadotropin were also observed by Hanai (1968) in a combined biochemical and histochemical study. Sudan III-positive particles decreased in the interstitial gland cells after administration of the gonadotropin; total cholesterol in the ovaries decreased and phospholipids increased. Triglyceride decreased slightly and nonesterified fatty acids increased in the ovaries as a result of deesterification of triglyceride. The activity of lipoprotein lipase in rabbit ovaries was elevated 48 hours after administration of the gonadotropin. The increase of ovarian phospholipids was probably related to the development of endoplasmic reticulum and mitochondria. It was inferred that the phospholipid were dynamically metabolized following administration of the gonadotropin, and that this played an important role with respect to the permeability of cell membranes. In turn, lipoprotein lipase activity was known to be related to the permeability for lipids. Hanai (1968) suggested that changes in the lipids of nonpregnant rabbit ovaries after administration of human chorionic gonadotropin were mainly influenced by the lipoprotein lipase activity and that these dynamic changes in lipids were closely related to steroidogenesis. Hilliard *et al.* (1963) and Hilliard and Sawyer (1964) found an increased output of 20α-hydroxypregn-4-en-3-one and progesterone after gonadotropic stimulation of rabbits. In this connection, the studies by Nagai *et al.* (1967) on lipid changes in the ovaries of young rats following pituitary (FSH) and chorionic gonadotropic (HCG) hormones are of interest. Fifty hours after subcutaneous injection of FSH, the follicular granulosa cells showed an increase in osmiophilic lipid droplets and smooth endoplasmic reticulum. At the same time there was a relative decrease in total lipids and in cholesterol, mainly its nonesterified portion, whereas the phosphatides were markedly increased.

The interstitial gland cells of the guinea pig ovary contained triglycerides, phospholipids, and cholesterol and its esters (Guraya, 1968e). These lipids, particularly cholesterol and its esters, were reduced in amount during the preovulatory period. This occurred only in relatively young cortical interstitial gland tissue, which in the same period showed an increase in vascularity. However, in comparison with the corresponding gland cells in rabbit, rat, and hamster ovary (Guraya and Greenwald, 1964b), the guinea pig cells stored much less sterol or lipid granules and mobilized much less before ovulation.

Lipids of the interstitial gland cells of the rat ovary were also studied histochemically by Jirásek *et al.* (1970) and by Motta and Bourneva (1970). Ben-Or (1963) followed changes in sudanophilic lipids and cholesterol in the interstitial cells of the developing mouse ovary; she also investigated the effect of gonadotropin treatment.

3. Enzymes

a. Mammals. The principal histochemical studies of enzymes in ovarian stroma and other structures since 1963 are tabulated in Table I. For a discussion of the biological significance of some of the most interesting results, see Section VII,D. Only a few studies on oxidoreductases, transferases, and hydrolases will be briefly mentioned below.

i. Oxidoreductases. Burkl and Slezak-Klemencic (1967a) found cyclic variations in the activity of Δ^5-3β-hydroxysteroid dehydrogenase and secondary alcohol dehydrogenase in interstitial cells of the rat ovary. In pregnant ferrets the degree of Δ^5-3β-hydroxysteroid dehydrogenase activity resembled that seen in the guinea pig; a relatively weak reaction was found in the ovarian interstitial tissue compared with that in the rabbit, another induced ovulator (Galil and Deane, 1966). The enzyme activity of several dehydrogenases in the interstitial tissue of the human ovary was investigated by Brandau and Luh (1965). For the results of an extensive study on these and other enzymes by Koudstaal *et al.* (1966), see Table III.

ii. Transferases. Jonek *et al.* (1965) found a positive histochemical reaction of γ-glutamyl transpeptidase in interstitial cells of the guinea pig ovary.

iii. Hydrolases. Jacobowitz and Wallach (1967) used the histochemical acetylcholinesterase activity to study the distribution of cholinergic ovarian nerve fibers in the cat, monkey, and human. There were fewer cholinergic than adrenergic fibers as demonstrated by a histochemical fluorescence method. Jordan (1970) did not detect any acetylcholinesterase-containing nerves in the ovary of the rabbit or guinea pig; adrenergic nerves were found in the ovaries of both species, e.g., around the follicles.

b. Other Vertebrates. See Section IIIA,3,b.

IV. MISCELLANEOUS SUBSTANCES IN THE OVARY

A. Nucleic Acids

The amount of RNA in the cytoplasm of ovarian cells has been studied histochemically by several workers (e.g., Zawistowska, 1965; Guraya, 1968d, 1971; Csaba *et al.*, 1970; Larraguibel and Dallenbach-Hellweg, 1970). A good idea of the amount of cytoplasmic RNA is also obtained from the numbers of free and membrane-bound ribosomes in ordinary electron micrographs.

Sachs (1968) observed that the nuclei of human granulosa lutein cells of the cycle and in pregnancy were much larger than the nuclei of follicular granulosa cells. He concluded that this difference in size could be explained mainly by an increase in the proteins of the nucleus. However, quantitative histochemical assessment of DNA revealed that an increase in DNA occurred in several of the granulosa lutein cells and that, particularly in pregnancy, some of the cells contained tetraploid nuclei. This endomitotic polyploidism is consistent with the fact that Dübner (1952) could not find any mitoses in human granulosa lutein cells of pregnancy. Sachs (1968) suggested that the simultaneous occurrence of polyploidism and functional edema of the nucleus might be a general principle in the function of hormone-dependent and hormone-producing organs. Recently, Guraya (1971) reported marked decreases in cytoplasmic RNA and nuclear DNA in theca externa cells of the follicles of estrous rabbits during the period of preovulatory swelling, particularly in the area of stigma. He suggested that lysosomal hydrolases, e.g., RNase and DNase might have been set free, and might have brought about the changes in the nucleic acids. Larraguibel and Dallenbach-Hellweg (1970) reported a decrease in nuclear DNA in granulosa lutein cells of women just before and during menstruation.

B. Biogenic Amines

The development of a histochemical procedure for the visualization of catecholamine-containing nerve fibers (Falck, 1962) has made it possible to study the adrenergic innervation of the ovary (Owman and Sjöberg, 1966; Owman *et al.*, 1967; and Volume II, Chapter 2, Section VII). Unsicker (1970a) has studied electron microscopically the innervation of the interstitial gland of the mouse ovary. In all species investigated, some of the adrenergic nerves of the ovary were associated with the vascular system, but nerve terminals were also found to run to various structures of the ovarian parenchyma; these were particularly abundant in the cat. Occasional nerve fibers seen in the theca interna of follicles were judged to be nerves to vessels; no nerve fibers were observed passing through the membrana granulosa (Jacobowitz and Wallach, 1967). Rosengren and Sjöberg (1967) thought that peripheral effects, mediated directly by the ovarian adrenergic innervation, may explain changes in ovarian activity after interference of adrenergic mechanisms with amine-depleting agents or monoamine oxidase inhibitors (Spector, 1961; Hopkins and Pincus, 1963; Coppola *et al.*, 1966). Jacobowitz and Wallach (1967) also studied the cholinergic nerves of the ovary in humans, monkeys, and cats by a histochemical method for acetylcholinesterase. They found fewer cholinergic than adrenergic nerves and suggested that the adrenergic innervation to the ovary had an influence on ovulation.

C. Steroid Hormones

Histochemical Tests for Steroid Hormones

In 1958, Khanolkar *et al.* proposed a new histochemical method for α-ketolic steroids, particularly for corticoids (Clara, 1965). Adams (1965), however, tested the specificity of the method on lipids absorbed on paper.

The results showed that the method was nonspecific *in vitro*; the reaction cannot be recommended for applied histochemical studies.

Dokumov and Dashev (1964, 1965) have published reports of histochemical studies of the ovaries of women suffering from the Stein–Leventhal syndrome. In the latter investigation the method of Khanolkar *et al.* (1958) was used. In the paper from 1964 the so-called Ashbel–Seligman carbonyl reaction (1949) was performed; as Jacoby (1962) and others have remarked this reaction is not by itself specific for ketosteroids.

In two recent papers (Stoward and Adams Smith, 1964; Adams Smith and Stoward, 1967) another procedure was described for the histochemical detection of ketosteroids in the ovary. Sections of formalin-fixed tissue were treated with salicylhydrazide (after pretreatment with methyl hydrazine and sulfobenzaldehyde), then examined for fluorescence before and after treatment with alkali. A greenish-yellow fluorescence was interpreted as indicating the presence of ketosteroids (mostly 3-ketosteroids). It appeared that the alkali quenched the yellow fluorescence of 3-ketosteroids in tissue sections and enhanced the blue fluorescence of other ketosteroids, especially that of 17-ketosteroids. It was pointed out by Adams Smith and Stoward (1967) that the procedure they described did not permit the identification of specific ketosteroids, but that it allowed histological localization of ketosteroids that were known, from biochemical and physiological experiments, to be present in a particular tissue.

As an example of the immunohistochemical demonstration of steroid hormones in the ovary, the excellent study by Woods and Domm (1966) should be mentioned. They identified androgen-producing cells in the gonads of the albino rat and domestic fowl by a fluorescent antibody technique; the limitations of the procedure were discussed. They also studied the same organs with enzyme histochemical methods and summarized the results in a table. In the rat ovary both Δ^5-3β-hydroxysteroid dehydrogenase activity and specific fluorescence were found in stratum granulosum, theca interna and interstitial cells; in corpora lutea the enzyme activity was maximal, while the antibody reaction was negative.

Steroid hormones have also been made visible in ovarian elements by autoradiographic studies using labeled precursors (Appelgren, 1967, 1969b; Stumpf, 1969).

D. Gonadotropins

By an immunofluorescent technique Benitez and Daza (1966) demonstrated HCG in the ovaries of rats that had been given the hormone subcutaneously. Fitko *et al.* (1968) perfused ovine ovaries with luteinizing hormone or prolactin labeled with fluorescein isothiocyanate. The labeled prolactin was almost exclusively localized in active corpora lutea, while the luteinizing hormone was also found in follicular granulosa cells, follicular fluid, interstitial cells, and hilus cells but not in the theca interna. Most recently, Rajaniemi *et al.* (1970) demonstrated labeled pituitary gonadotropins in the ovaries of mature rats by autoradiography. Ovine pituitary gonadotropins (LH and FSH) (supplied by the National Institutes of Health) were labeled with radioactive iodine, using an enzymatic method. Labeled LH produced silver grains in the interstitial gland cells. Labeled FSH was found mainly in the follicular granulosa cells and follicular fluid; some staining was found in the theca layer but none in the interstitial tissue.

E. Relaxin

Zarrow and O'Connor (1966) localized relaxin in corpus luteum cells of pregnant rabbits by the indirect fluorescent antibody technique; the presence or absence of relaxin in ovaries of several other species was discussed. Relaxin was also demonstrated in the ovaries of pregnant mice and rats with an ^{131}I-labeled antibody to relaxin (Zarrow and McClintock, 1966). The results suggested that the ovary may be the primary site of the synthesis of relaxin.

F. Ascorbic Acid

Since the first edition of this book, there have been few histochemical studies of ascorbic acid in the ovary. One reason is apparently the shortcomings of the silver nitrate technique; these have recently been discussed by Pearse (1968). Of distinct value are the elegant autoradiographic investigations by Hammarström (1966). In an effort to elucidate the mechanism of ascorbic acid depletion by tropic hormones, Stansfield and Flint (1967) carried out a biochemical study of the rat corpus luteum. Their results indicated that loss or outflow of ascorbic acid normally occurs by passive diffusion, whereas uptake of ascorbic acid into luteal tissue is an energy-dependent process which is prevented by luteinizing hormone via progesterone.

G. Pigments

Ovarian pigments such as ceroid and lipofuscins have been little studied in recent years (Grecchi *et al.*, 1966; Reichel, 1968). In general, there have been surprisingly few studies of these primary fluorescent substances, probably because of concern about falsely negative or positive results.

However, the use of unfixed cryostat sections eliminates most of the objections to this form of fluorescence microscopy (Chayen and Denby, 1968). The lipid pigments apparently represent a spectrum of different degrees of oxidation of lipid-containing particles; they probably correspond to the residual bodies seen in electron micrographs, which are interpreted as being indigestible products of secondary lysosomes.

H. Reinke Crystalloids

In 1896, Reinke described "crystalloidal" inclusions in human testicular interstitial cells. Sternberg (1949) conclusively demonstrated Reinke crystalloids in both normal and neoplastic hilus cells of the human ovary. They were found to be acidophilic, "not sudanophilic, not doubly refractile"; available histochemical evidence suggested that the rod-shaped crystalloids were protein bodies, and that they might develop from spherical acidophilic structures in the cytoplasm. Recent histochemical evidence indicated that the Reinke crystalloids of ovarian hilus cells consist of globular proteins arranged in a crystalline lattice (Janko and Sandberg, 1970). Reinke crystalloids apparently exist only in man; for further details, including ultrastructural, see Christensen and Gillim (1969), Fawcett *et al.* (1969), and Janko and Sandberg (1970).

V. HISTOCHEMICAL STUDIES ON OVARIAN TUMORS

Volume II, Chapter 4 is especially devoted to ovarian tumors, and histochemical studies on neoplastic lesions of the ovary will therefore be only briefly mentioned here. Moreover, stains for carbohydrates and lipids are almost routinely applied in cases of ovarian tumors, and information on these substances can be easily obtained from standard textbooks of gynecological pathology; for more details, the recent work of Teilum (1971) on special tumors of the ovary, the review by Hughesdon (1966) on ovarian lipoid and theca cell tumors, the comprehensive study by Koudstaal *et al.* (1968), and the paper on mucins in human ovarian neoplasms by Garcia-Bunuel and Monis (1964) should be consulted.

In the main, therefore, only the field of enzyme histochemistry remains to be discussed. Willighagen and Thiery (1968) presented a survey of the literature on previous enzyme histochemical studies of ovarian tumors; their own material consisted of 94 ovarian neoplasms of various kinds, in most of which six hydrolases and five dehydrogenases were investigated. They reported some differences in the enzyme pattern in different types of tumors. To a certain degree, these different patterns were characteristic for the types of tumor, but their tables show that there was considerable overlap and that individual variations existed. Histochemical characteristics are, therefore, generally of limited value for diagnostic purposes. For example, the activity of acid phosphatase was found to be variable and could not be correlated with the type of tumor. On the other hand, mucinous tumors as a group lacked alkaline phosphatase activity, while serous tumors generally showed positive reactions. The most interesting observation in the study was perhaps the detection of so-called enzymatically active stromal cells (EASC) in a wide variety of ovarian neoplasms. This type of cell was named and studied in detail by Scully and Cohen (1964). Willighagen and Thiery (1968) found EASC mainly in mucinous tumors. They assumed that EASC were induced from the original ovarian stroma by proliferating epithelial cells of both primary and secondary ovarian tumors. In another study (Pfleiderer and Teufel, 1968), EASC were also frequent in mucinous ovarian neoplasms. These authors compared 85 ovarian tumors and 29 metastases of ovarian carcinomas with respect to EASC. Thirty-seven percent of the former cases showed EASC, while the metastases contained none. This strengthens the supposition that the EASC originate from the ovarian stroma. Pfleiderer and Teufel (1968) discussed the possibility that the theca of the normal growing follicle results from compression of the ovarian stroma by the growing follicle and that the hormonally active tissue of granulosa cell tumors develops from compression of the ovarian stroma by the tumor. Koudstaal *et al.* (1968) considered EASC to be derived from normal ovarian stroma and to be activated by both benign and malignant neoplasms. However, the endocrine-active cells of granulosa–theca cell tumors were considered to be parts of the tumors. For details on enzyme activities in a thecoma and a fibroma of the human ovary, see Table III. The steroid histochemistry of three experimental granulosa cell tumors was reported by Maeir and Wagner (1968). They remarked that the stromal elements in some tumors displayed more enzyme activity than the masses of tumor cells.

Woodruff *et al.* (1968) briefly mentioned previous reports of endocrine effects of a variety of ovarian tumors, generally believed to be endocrinologically inert. These problems have also been commented on by Scott *et al.* (1967), Fathalla (1968), and Novak *et al.* (1970). Woodruff *et al.* (1968) studied a large number of enzyme histochemical reactions in two cases of Krukenberg tumors with clinically different endocrine patterns. The results were less striking than expected, but enzyme activities suggestive of steroidogenesis were present in the ovarian stromal elements of the tumors rather than in the epithelial cells; 3β-hydroxysteroid dehydrogenase activity was detected in stromal cells but not in the epithelium. Hardonk and Koud-

TABLE III

Enzyme Pattern of Ovarian Follicles, Corpus Luteum of Menstruation, Thecomatosis, Thecoma, and Fibroma of the Human Ovary[a,b]

	Growing follicles		Atretic follicles		Corpus luteum of menstruation		Thecomatosis		Thecoma		Fibroma
	T	G	T	G	T	G	L.C.	H.S.	L.C.	T.S.	
Oil Red O	+/±	−	+	−	+++/++	++/+	+/±	±/−	×/±	++/+	+/−
Sudan Black B	+/±	−	+	−	++/+	+/−	+/±	±/−	+/±	++/+	+/−
Alkaline phosphatase	++	−	++	−	++/+	+/−	++/+	−	++/+	+/−	−
Acid phosphatase	++	+	++	±	++/+	+++	++	±	++	+/−	+/−
ATP-ase	±/−	−	±/−	−	−	−	−	−	−	−	
5-Nucleotidase	±/−	−	−	−	−	±/−	?	++	?	+/−	+/−
α-Naphthyl acetate esterase[c]	++/+	−	++/+	−	+/−	+++	+	−	+	−	−
Naphthol AS-D acetate esterase[c]	±	−	+/−	−	±/−	++/+	±	−	±	−	−
Indoxyl acetate esterase[c]	±	−	±	−	+/−	+++	±	−	±	−	−
NADH-tetrazolium reductase	+++	++	+++/+	+	+++	++	+++	++	+++	+	+/−
Lactic acid dehydrogenase	++	+	++/−	+/−	++	++/+	++	+	++	+/±	±/−
β-Hydroxybutyric acid dehydrogenase	++	+	++/−	−	++/+	+	++	±/−	++	−	−
3β-ol-Hydroxysteroid dehydrogenase											
Dehydroepiandrosterone	±/−	−	±/−	−	+	+++	±/−	−	×/−	−	−
4-Androstene-3β-ol, 17β-ol	+	−	±/−	−	±	+++	×	−	++	−	−
Pregnenolone	±/−	−	+/−	−	±/−	++	±/−	−	±/−	−	−
Dehydroepiandrosterone	+/−	−	+/−	−	±/−	++	±/−	−	×/−	−	−
4-Androstene-3β-ol, 17β-ol	+	−	±/−	−	±	+++	+	−	+	−	−
Second alcohol dehydrogenase	+++	−	+++/+	−	+++	+	+++	−	+++	−	−
Alcohol dehydrogenase	+	±	+/−	−	+	+++	±	+	+	±/−	±/−
NADPH-tetrazolium reductase	+++	++	+++/+	+/−	+++	++	+++	++	+++	+	+
Glucose-6-phosphate dehydrogenase	+++	+	+++/+	±/−	+++	++/+	+++	+	+++	+/±	±/−
Succinic dehydrogenase	++/+	+	++	+/−	++/+	+	++	+	++	+	+/−
α-Glycerophosphate oxidase	±/−	++	±	+/−	±	++/+	+/±	±	+/±	±/+	±/−
Aminopeptidase	++	−	+	−	++/+	−	−	−			−

[a] From Koudstaal *et al.*, (1966.)

[b] T, theca cells; G, granulosa cells; L.C., luteinized cells; H.S., hyperplastic ovarian cortical stroma; T.S., thecal stroma; −, negative; ±, slight; +, moderate; ++, strong; +++, very strong enzyme activity; /, varying activity; ?, not known.

[c] Nonspecific esterases.

staal (1971), however, found a positive 3β-hydroxysteroid dehydrogenase reaction in the epithelial cells of a Krukenberg tumor of a pregnant woman. Pfleiderer *et al.* (1969) studied an extensive series of different primary and secondary ovarian carcinomas for lactate dehydrogenase activity with both histochemical and biochemical methods. Neoplasms with certain histochemical patterns of lactate dehydrogenase appeared to react more favorably to treatment with endoxan and X-rays. Finally, the very detailed enzyme histochemical study on ovarian neoplasia by Jones *et al.* (1967) deserves comment. These authors compared the enzyme histochemistry of presumed steroid- and nonsteroid-producing cells of an arrhenoblastoma with that of cells of normal ovaries, testes, and adrenals. They concluded that a battery of enzymatic histochemical reactions can clarify the histogenesis of the androgen-producing tumors, provided that the reactions are interpreted in relation to the morphological appearance of the neoplastic cells.

VI. HISTOCHEMICAL STUDIES AT THE ULTRASTRUCTURAL LEVEL

The identification of glycogen at the ultrastructural level is more complicated than is generally appreciated (Revel, 1964). Only a few workers have investigated the subcellular distribution of ovarian glycogen in detail. Of considerable interest are the studies by Blanchette (1965, 1966), in which she followed changes in the glycogen of pre- and postovulatory rabbit follicles, both at the light and electron microscope level.

Lipid droplets can be studied in ordinary electron microscopical sections. Since ovarian lipids were recently discussed in an excellent review on the fine structure and function of steroid-secreting cells (Christensen and Gillim, 1969), the large body of literature covered by these authors need not be commented upon here. However, a recent study of the fine structure of ovarian hilus cells (Unsicker, 1970b), with remarks concerning their lipid droplets, pigment granules, and innervation should be mentioned, and also the report of cholesterin crystals in regressing corpora lutea of the human ovary by Carsten and Merker (1965). For a discussion of the appearance and significance of lipids in electron micrographs of endocrine ovarian structures, see also Fawcett *et al.* (1969) and Bjersing (1967d).

To date, little has been published on the subcellular enzymorphology of the ovary. As briefly mentioned in Section II,D, several acceptable methods are available, and many studies will undoubtedly be reported in the near future. Although the physiological significance of some of the results obtained in this way may not be immediately obvious, they may later prove to be important for a composite picture of ovarian biology. The lysosomal enzyme acid phosphatase has so far been the most popular subject for study. Anderson (1970) found acid phosphatase activity in the Golgi complex, multivesicular bodies, and dense spherical bodies in developing oocytes and follicular cells of sexually mature virgin mice, guinea pigs, and rabbits. Acid phosphatase activity has also been reported in corpora lutea of man (van Lennep and Madden, 1965), pig (Belt *et al.*, 1970), and sheep (Bjersing *et al.*, 1970a). Finally, oxidoreductases have been studied by Blanchette (1968); she gave in abstract form a short account of attempts to demonstrate steroid dehydrogenase activity in "luteinized" cells and cell fractions from the ovaries of rats induced to superovulate and pseudopregnant rabbits. In an autoradiographical study at the ultrastructural level, Moricard (1964) localized silver grains, due to the action of ^{35}S, adjacent to the ergastoplasmic sacs in ovarian follicular cells of rats, rabbits, and guinea pigs. The significance of uptake of ^{35}S by the ovary was discussed in detail by Jacoby (1962) and will not be further commented upon here.

VII. COMMENTS ON THE BIOLOGICAL SIGNIFICANCE OF THE HISTOCHEMICAL FINDINGS

A. General Remarks

The most interesting new information about the histochemistry of the ovary has been provided by enzyme histochemical studies. This discussion will therefore focus mainly on the biological significance of the enzyme histochemical results.

In the ovary, as in many other organs, the biological importance of several enzymes is still not fully understood. De Duve (1969) pertinently pointed out that few enzymes have attracted so much attention from different scientific fields as the phosphatases; he found it "remarkable, and to some extent frustrating, that such an enormous amount of work has yielded so little definite information concerning the function of these ubiquitous enzymes." From the point of view of biological significance it seems to be of questionable value to include routinely a large number of such enzymes with unknown function in histoenzymatic studies on the ovary. However, regarding enzymes directly involved in a most important sector of ovarian metabolism, i.e., steroid biosynthesis, histochemical methods have supplied much valuable information. For example, Δ^5-3β-hydroxysteroid dehydrogenase, histochemically the most studied steroid-converting enzyme, is essential in the synthesis of nearly every biologically active ovarian steroid.

Other histochemically studied hydroxysteroid dehydrogenases such as 17β- and 20α-hydroxysteroid dehydrogenases are important in adjusting the potencies of endocrine secretion to meet changing requirements. A variety of other tissues also display hydroxysteroid dehydrogenase activity (Baillie *et al.*, 1966a); the activities in these sites may provide a means for local control of endocrine effect.

1. Validity of the Methods

All of the procedures in use cannot be accepted without reservation, and therefore some remarks on the validity of the histochemical methods are appropriate.

It is necessary to be aware of "false positive" reactions. As discussed by Bjersing (1967a), readily soluble enzymes such as several types of dehydrogenases may pass into the incubation medium and generate reduced pyridine nucleotide. This in turn may act as substrate for tissue-bound diaphorase in a region lacking the particular dehydrogenase. There are other causes of "false positive" reactions; for example, the "nothing dehydrogenase" reaction due to tissue sulfydryl groups or alcohol dehydrogenase, the affinity of formazan for lipoprotein membranes and lipid droplets, and crystal formation of formazan due to physicochemical damage to the tissue (Høyer and Andersen, 1970).

Falsely negative reactions may also occur, for example, inhibition of dehydrogenase activity because of excess of substrate, activator or tetrazolium salt (Chayen *et al.*, 1969). Progesterone may inhibit Δ^5-3β-hydroxysteroid dehydrogenase activity in theca interna and corpus luteum cells when dehydroepiandrosterone is used as substrate (Jones *et al.*, 1968a). Galil and Deane (1966) eventually demonstrated Δ^5-3β-hydroxysteroid dehydrogenase activity in rabbit corpora lutea (in contrast to earlier negative results) after use of redistilled acetone for the washes and media. For further discussion of factors influencing histochemical activity, see Baillie *et al.* (1966a), Bjersing (1967a,b), Chayen *et al.* (1969), Fahimi and Karnovsky (1966), Jones *et al.* (1968a), Lobel and Levy (1968), and Rosenbaum (1964). A critical study on specificity in steroid histochemistry is that by Høyer and Andersen (1970).

2. Oxidoreductases and Hydrolases

Emphasis is placed below on those histochemical studies and results that seem to be of greatest biological significance. Readers particularly interested in enzymes that are not discussed here are directed to the references of Table I. Pertinent and up-to-date information on the biological role of enzymes may also be obtained from Lehninger (1970).

a. Oxidoreductases. The oxidoreductases can to some extent be grouped in functional categories according to their role in metabolism. The following classification is used because of the key position of steroidogenesis in ovarian biology.

i. Steroid-converting enzymes. 3β-, 17β-, 20α-, 20β-, and other hydroxysteroid dehydrogenases. In spite of considerable progress, several questions regarding the enzyme specificity of hydroxysteroid dehydrogenases are still unsettled (see Cheatum and Warren, 1966; Davenport and Mallette, 1966; Davies *et al.,* 1966; Deane and Rubin, 1965; Dorfman and Ungar, 1965; Ewald *et al.,* 1964; Ferguson, 1965; Goldberg *et al.,* 1964; Jones *et al.,* 1968a; Kautsky and Hagerman, 1970; Prabhu and Weisz, 1970). One illustration of this fact is provided by the alternative designations Δ^5-3β-hydroxysteroid dehydrogenase and 3β-hydroxysteroid dehydrogenase. Deane and Rubin (1965), however, remarked that one enzyme activity (Δ^5-3β-hydroxysteroid dehydrogenase) appeared to be restricted to unsaturated steroid substrates, whereas the other (3β-hydroxysteroid dehydrogenase) catalyzed the oxidation of steroids saturated in rings A and B. In conformity with Deane and Rubin (1965), we prefer the precise designation Δ^5-3β-hydroxysteroid dehydrogenase when a Δ^5-3β-hydroxysteroid has been used as substrate in a histochemical reaction. Other facets of enzyme specificity are reflected in the following observations by Prabhu and Weisz (1970). Positive reactions were found with dehydroepiandrosterone and pregnenolone as substrates but negative reactions with their sulfates. With pregnenolone reduced at the 20β- or the 17β-position a negative reaction was obtained. This indicated that alterations of the molecular structure at sites distant from the A- and B-rings can determine whether a Δ^5-3β-hydroxysteroid can serve as substrate for Δ^5-3β-hydroxysteroid dehydrogenase. In contrast to the findings in the rat ovary by Prabhu and Weisz (1970), Ferguson (1965) reported positive reactions in the mouse ovary with 3β-sulfoxy steroids as substrates. Ferguson (1965) also discussed the possibility that the sulfate was removed by a steroid sulfatase. The presence of both alcoholic and phenolic steroid sulfatase activity was demonstrated biochemically in the human ovary by Sandberg and Jenkins (1966). A possible intracellular regulatory role of dehydroepiandrosterone sulfate in ovarian steroid cells was recently discussed by Griffiths *et al.* (1971).

The steroid converting enzymes will be discussed further under subheadings B to D of this Section.

ii. $NADPH_2$-supplying enzymes. Glucose-6-phosphate, isocitrate, malate, and phosphogluconate dehydrogenases.

iii. Other enzymes closely related to steroidogenesis: $NADH_2$- and $NADPH_2$- tetrazolium reductases and secondary alcohol dehydrogenases. Access to cytoplasmic $NADPH_2$ is necessary to several steps in steroid biosynthesis. The activities of the soluble (supernatant) enzymes glucose-6-phosphate, isocitrate, malate, and phosphogluconate dehydrogenases can provide this co-factor in the ovary (Flint and Denton, 1970; Luh and Brandau, 1964a; Lunaas *et al.*, 1968; Nielson and Warren, 1965; Savard *et al.*, 1963; Wiest and Kidwell, 1969). According to biochemical studies the NADP-dependent malate dehydrogenase ("malic enzyme") plays a dominant role in the rat ovary (Flint and Denton, 1970; Lunaas *et al.*, 1968), while glucose-6-phosphate dehydrogenase of the hexose monophosphate pathway is of prime importance in the formation of $NADPH_2$ in the human and bovine ovaries (Nielson and Warren, 1965; Savard *et al.*, 1963). Surprisingly enough, NADP-dependent malate dehydrogenase does not seem to have been studied histochemically. Deane *et al.* (1966) considered $NADPH_2$-tetrazolium reductase as "indicative of the hexose monophosphate pathway" as did Hadjiisky *et al.* (1969a). $NADPH_2$-tetrazolium reductase activity in the ovary might rather reflect the sum of the activities of the enzymes that supply $NADPH_2$ for use in steroid biosynthesis (and other processes).

In a recent paper, Koudstaal and Hardonk (1969) reported biochemical and histochemical studies of the rat liver before and after phenobarbital treatment. They suggested that the NADP-dependent dehydrogenases glucose-6-phosphate and isocitrate dehydrogenases are indirectly involved in the hydroxylation chain of liver microsomes (smooth endoplasmic reticulum) while $NADPH_2$-tetrazolium reductase is directly involved; the $NADPH_2$-tetrazolium reductase activity may be a measure of the hydroxylation activity, and an increase in glucose-6-phosphate and isocitrate dehydrogenase activities may reflect an increased demand for $NADPH_2$ because of intense hydroxylation. In the biosynthesis of ovarian steroids hydroxylations also occur (Christensen and Gillim, 1969). If the relationship between hydroxylation and the dehydrogenase, including reductase, activities in the liver proves true and causal, it may also apply to the ovary since a similar enzymatic hydroxylation system is apparently also present in steroid-producing cells of the ovary (Christensen and Gillim, 1969; Koudstaal and Hardonk, 1969; Grasso *et al.*, 1971). Cytochrome P-450, an important part of the hydroxylation system, has been demonstrated in both the mitochondrial and the microsomal fractions of the bovine corpus luteum (Yohro and Horie, 1967). Grasso *et al.* (1971), with a new histochemical procedure, demonstrated aniline hydroxylase activity in the rat ovary, principally in corpus luteum and theca externa cells of maturing Graafian follicles; a slight but definite activity was also found in granulosa cells of the Graafian follicle and in the interstitial ovarian cells. As remarked by Grasso *et al.* (1971), it is uncertain to what extent steroid hydroxylation enzymes might have contributed to the histochemical reaction with the foreign compound.

Hardonk (1965) demonstrated strong histochemical reactions indicative of secondary alcohol dehydrogenase in steroid-producing cells of human testis, ovary, and adrenal and suggested that this activity is related to the oxidative cleavage of the cholesterol side chain. Other contributions to the histochemistry of alcohol dehydrogenase in the ovary are the studies by, for example, Ferguson *et al.* (1966) and the demonstration by Poliak *et al.* (1968) of marked isopropyl alcohol dehydrogenase activity in hilus cells of the human ovary.

iv. Glycolytic enzymes: glyceraldehyde-3-phosphate and lactate dehydrogenases.

v. Citric acid and glycerol phosphate cycle enzymes: Isocitrate[NAD(P)]-malate(NAD)-, and succinate dehydrogenases and glycerolphosphate dehydrogenase.

vi. Miscellaneous oxidoreductases. As glucose is broken down, energy is freed and acetyl coenzyme A is produced. The histochemically studied enzymes glyceraldehyde-3-phosphate and lactate dehydrogenases are active in that sector of metabolism. Some of the acetyl coenzyme A may then be used in the energy-producing citric acid cycle and some in the synthesis of fatty acids and steroids. The histochemically studied citric acid cycle enzymes are isocitrate[NAD(P)]-, malate(NAD)-, and succinate dehydrogenases. Glycerolphosphate dehydrogenase does not need an exogenous cofactor such as NAD, and, thus, like succinate dehydrogenase, is closely linked to the cytochrome system. There are connections between the glycerolphosphate cycle and the Embden–Meyerhof glycolytic pathway via dihydroacetone phosphate.

b. Hydrolases. In view of the advance in general knowledge of the lysosomes during the past 10 years, the biological importance of lysosomes and lysosomal enzymes such as *N*-acetyl-β-glucosaminidase, acid phosphatase, aminopeptidase, arylsulfatase, β-galactosidase, α-glucosidase, β-glucuronidase, and organophosphate-resistant esterase is of particular interest. Since these problems have recently been covered in a review on reproductive organs by Dott (1969), they need not be discussed here. It should be pointed out, however, as remarked by Fishman *et al.* (1967) regarding acid hydrolases in general and β-glucuronidase in particular, that at least some of the enzymes demonstrated in lysosomes may have nonlysosomal locations as well. The principal nonlysosomal site of β-glucuronidase was shown to be the endoplasmic reticulum (Fishman *et al.*, 1967). Regarding the biological significance of these and other hydrolases as well as transferases and lyases for the organism in general and for ovarian biology see also Barka and Anderson (1963), Chayen *et al.* (1969), Fishman *et al.* (1967), Jacoby (1962), König (1965a), Lehninger (1970), Velardo and Rosa (1963), and others.

Botte (1968) has suggested hydrocortisone 21-phosphate as substrate for phosphatases hydrolyzing steroid phosphates in the ovary and other organs. However, as remarked by Botte and Delrio (1970), the presence of steroid phosphates *in vivo* is uncertain.

B. Follicles

Recent histochemical results have shed new light on the old and important problems of follicular maturation and the steroidogenic function of theca and granulosa cells. Bjersing (1967a), in a previous biochemical study, remarked that only few investigators had demonstrated hydroxysteroid dehydrogenase activity in granulosa cells of healthy follicles with histochemical methods; however, most workers had found positive Δ^5-3β-hydroxysteroid dehydrogenase reactions in theca interna cells (see Baillie *et al.,* 1966a). In contrast to the microsomal enzyme Δ^5-3β-hydroxysteroid dehydrogenase (Christensen and Gillim, 1969), the soluble enzymes 17β- and 20β-hydroxysteroid dehydrogenase are much more difficult to demonstrate histochemically, both in theca and granulosa cells of ovarian follicles, and the results should be interpreted with caution (Bjersing, 1967a,b; Lobel and Levy, 1968). However, follicular granulosa cells of mouse, dog, cow, and pig appeared to display true 17β-hydroxysteroid dehydrogenase activity (Hart *et al.,* 1966; Fowler and Feldman, 1970; Lobel and Levy, 1968; Bjersing, 1967a,b). It is also of interest that greater $NADPH_2$-tetrazolium reductase activity was noted in the granulosa cells than in the theca interna cells of both porcine (Fig. 5) (Bjersing, 1967a) and equine (E. V. YoungLai, M. F. Hay, and L. Bjersing, unpublished) follicles. This enzyme activity probably reflects the sum of activities of $NADPH_2$-dependent enzymes such as 17β-, 20α-, glucose-6-phosphate dehydrogenases and steroid hydroxylating enzymes.

There are now a considerable number of histochemical reports of $\Delta^5 3\beta$-hydroxysteroid dehydrogenase activity in granulosa cells of healthy follicles, indicating that steroid biosynthesis occurs in these cells. Some examples from various species are: hamster (Blaha and Leavitt, 1970; Wingate, 1970), mouse (Ferguson, 1965; Hart *et al.,* 1966; Kimura *et al.,* 1970), rat (Figs. 3 and 4) (Levy *et al.,* 1959; Prabhu and Weisz, 1970; Pupkin *et al.,* 1966; Rubin *et al.,* 1963a; Turolla *et al.,* 1967), squirrel (Seth and Prasad, 1967), dog (Fowler and Feldman, 1970), cow (Lobel and Levy, 1968), pig (Fig. 6) (Bjersing, 1967a,b), and man (Ikonen *et al.,* 1961; Motta and Takeva, 1970).

There may be a number of reasons for the many previous negative histochemical results. In Section VII,A, some causes of falsely negative his-

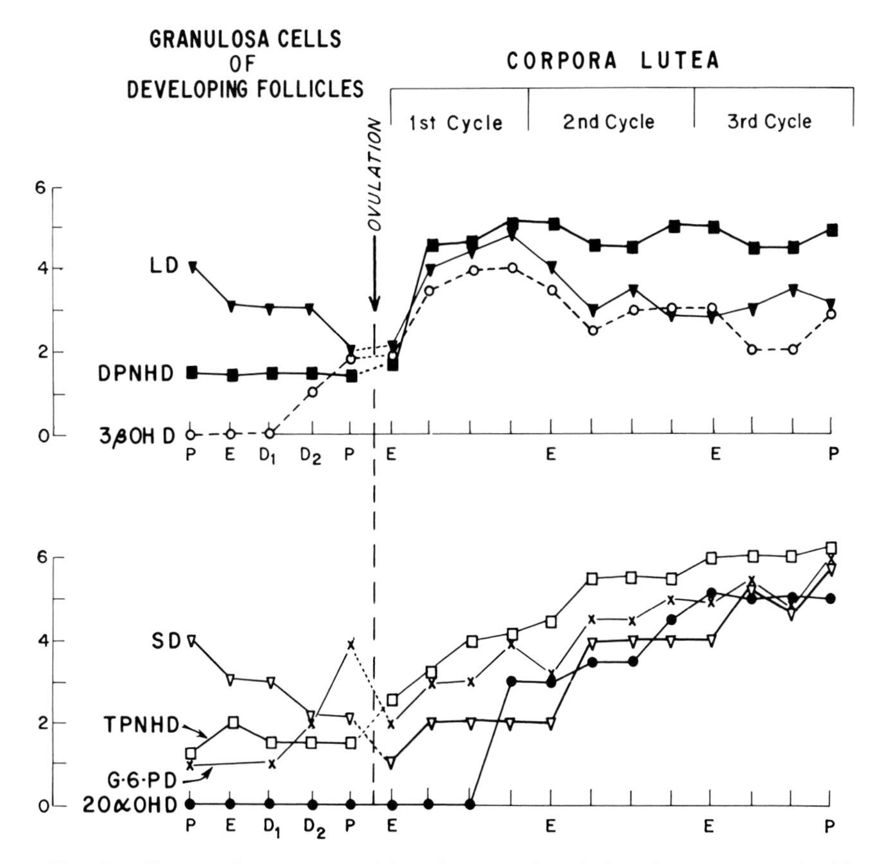

Fig. 3. Changes in enzyme activity of rat ovaries during the estrous cycle. The intensity of the histochemical reactions was evaluated by a single observer in a blind study on a 0–6 plus scale (ordinate). Several factors influencing enzyme reaction velocity vary with enzyme histochemical methods; thus, reported values are only relative to an individual enzyme. Phases of the estrous cycle (P, proestrus; E, estrus; D_1 and D_2, day 1 and day 2 of diestrus) are indicated on the abscissa. DPNHD, diphosphopyridine nucleotide diaphorase; G6PD, glucose-6-phosphate dehydrogenase; LD, lactic dehydrogenase; SD, succinic dehydrogenase; TPNHD, triphosphopyridine nucleotide diaphorase; 3β-OHD, Δ^5-3β-hydroxysteroid dehydrogenase; 20α-OHD, 20α-hydroxysteroid dehydrogenase. (Reproduced by permission from Pupkin *et al.,* 1966.)

tochemical reactions are mentioned. However, the most essential single factor determining a positive or negative result is perhaps the stage of the estrous cycle in which the ovarian specimens were taken. This factor has previously often been overlooked and is still sometimes neglected, and not only in histochemical studies. The importance of knowing the age, sexual maturation, stage of pregnancy or the precise phase of the estrous cycle of the experimental subject cannot be stressed strongly enough. Most of the

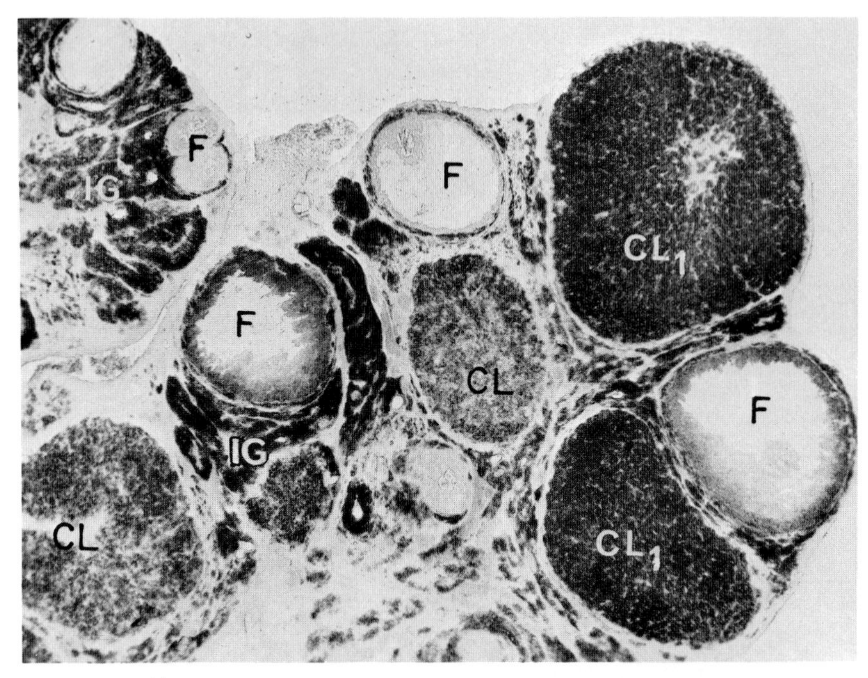

Fig. 4. Δ^5-3β-Hydroxysteroid dehydrogenase in the ovary of a rat at proestrus. Growing follicles show practically no Δ^5-3β-hydroxysteroid dehydrogenase activity, but as they reach the preovulatory stage, enzyme activity becomes clearly distinct. Recently formed lobules of interstitial gland (IG) react more strongly than older ones. First cycle corpora lutea (CL_1) show higher Δ^5-3β-hydroxysteroid dehydrogenase activity than corpora lutea of previous cycles (CL); F, follicle. ×30. (Reproduced by permission from Pupkin *et al.,* 1966.)

positive hydroxysteroid dehydrogenase reactions in the follicular granulosa cells have been obtained in the preovulatory period and in only one or a few large Graafian follicles, presumably those destined to ovulate. This was well illustrated by Pupkin *et al.* (1966) in their histochemical study of rat ovarian specimens from accurately timed stages of the estrous cycle (Figs. 3 and 4). They followed the activities of Δ^5-3β-hydroxysteroid dehydrogenase and

---

Fig. 5. (a), (b), (d); $NADH_2$-tetrazolium reductase activity in the pig ovary; (a) in follicles, ×10; (b) in a large tertiary follicle where the reaction is stronger in the theca interna than in the granulosa cells, ×30; (d) in the corpus luteum 3 to 4 days after ovulation where there is a strong cytoplasmic reaction in the theca luteal (TLC) and moderate one in the granulosa luteal cells (GLC), ×100; (c) $NADPH_2$-tetrazolium reductase activity in a contiguous section of the same follicle as (b), showing a stronger reaction in the granulosa than in the tehca interna cells, ×30. The $NADPH_2$-tetrazolium reductase activity probably reflects the sum of the activities of the enzymes that supply $NADPH_2$. (From Bjersing, 1967a.)

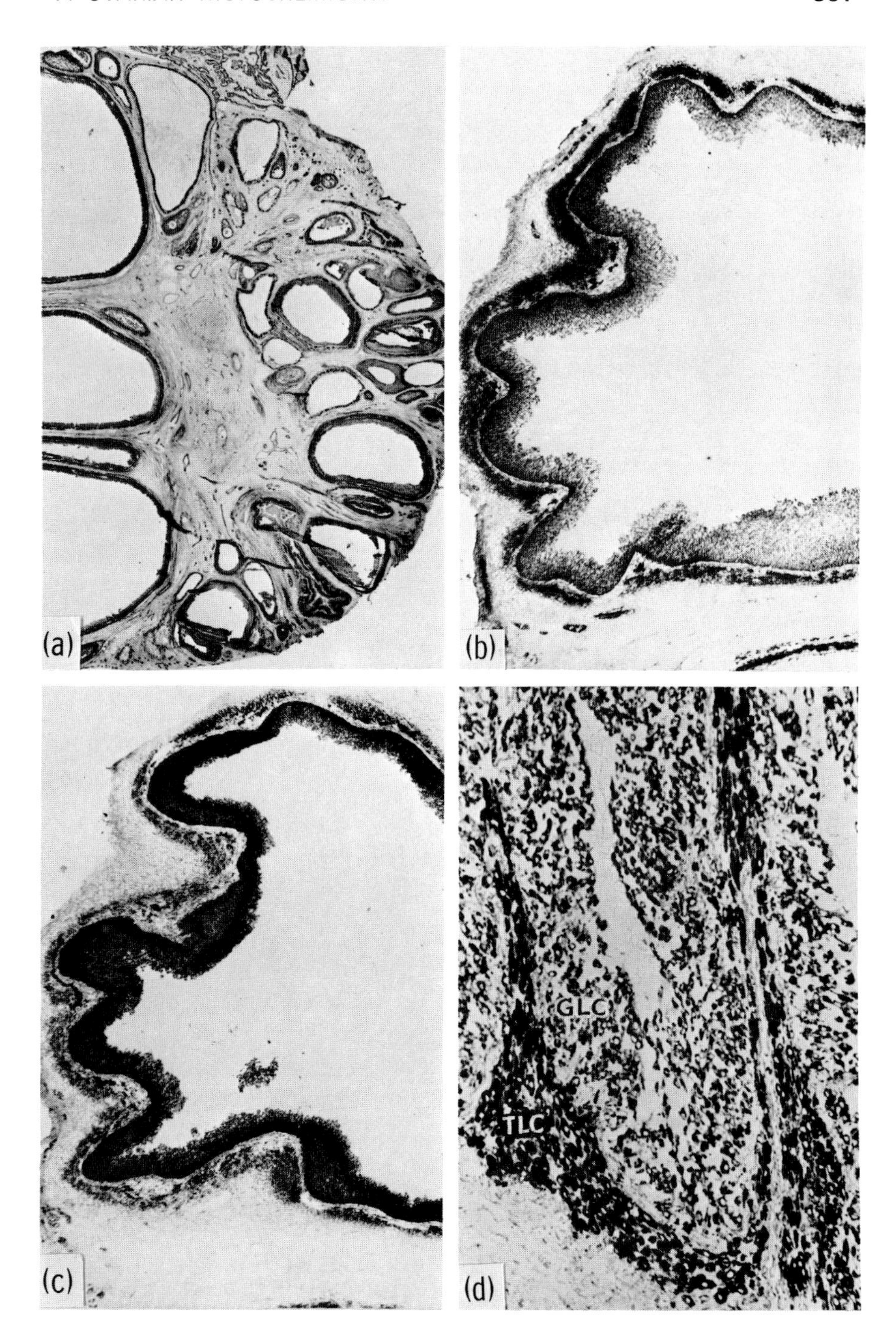
(a)
(b)
(c)
(d)
GLC
TLC

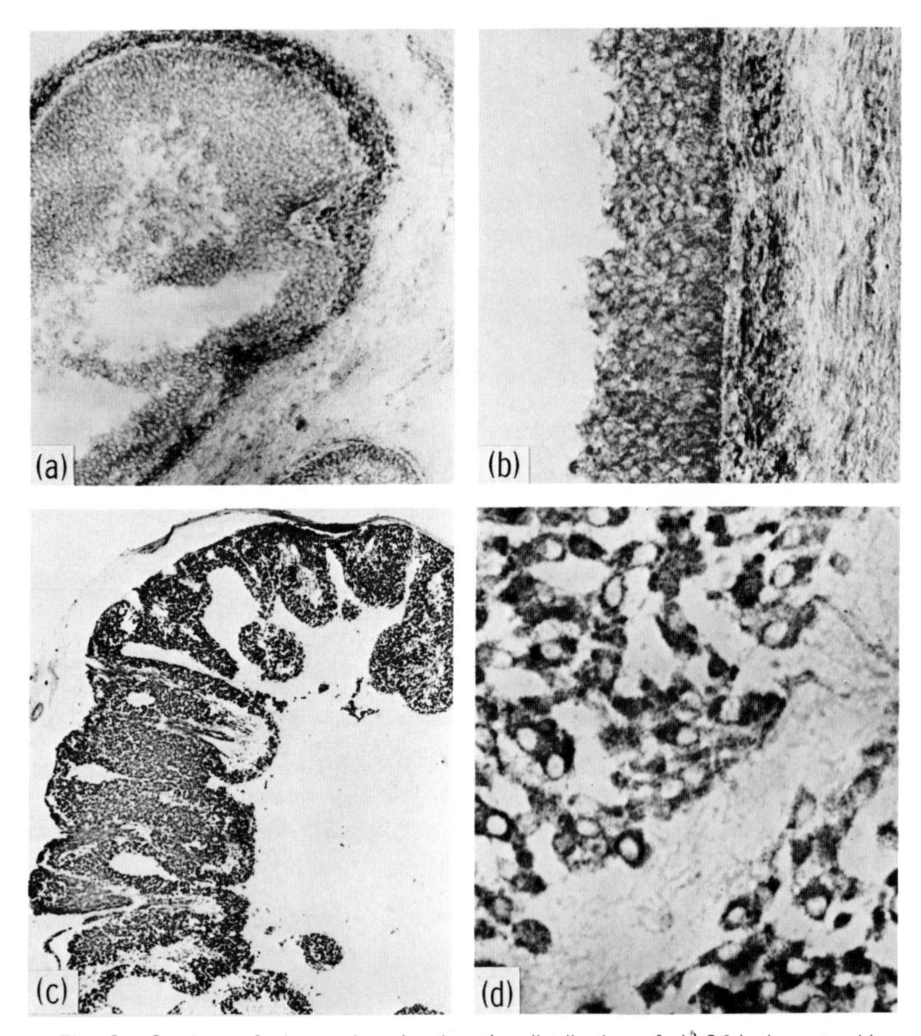

Fig. 6. Sections of pig ovaries showing the distribution of Δ^5-3β-hydroxysteroid dehydrogenase activity. This enzyme is believed to be localized at sites where steroid biosynthesis occurs. (a) ×80, and (b) ×190, large, probably preovulatory follicles showing moderate activity in membrana granulosa and theca interna; (c) corpus luteum 7 to 8 days after ovulation showing a strong reaction in all luteal cells, ×15; (d) active corpus luteum of the cycle showing activity with dehydroepiandrosterone as substrate, ×400. (From Bjersing, 1967a.)

some other enzymes involved in intermediary carbohydrate metabolism throughout the development of the follicles and corpora lutea of the rat ovary. The 3β-hydroxysteroid dehydrogenase activity became demonstrable, and the glucose-6-phosphate dehydrogenase activity increased markedly in the granulosa cells of some Graafian follicles at the time of the presumed "estrogen surge." There were four to six follicles with these enzyme activities in each ovary, and these were supposed to be the preovulatory follicles. The theca interna cells showed strong and uniform activity of both glucose-6-phosphate and 3β-hydroxysteroid dehydrogenase throughout growth and maturation of the follicles. These histochemical results partly confirm the widely held view that follicular granulosa cells have little metabolic activity. More important, however, they clearly demonstrate that marked metabolic changes occur in the granulosa cells of one or more large Graafian follicles during the preovulatory period.

The histochemical observations are of considerable interest in view of the much debated question regarding the respective roles of granulosa and theca cells in steroid hormone secretion by the follicle (Bjersing, 1967b; Bjersing and Carstensen, 1967; Channing, 1969b; Falck, 1959; Ryan *et al.*, 1968; Short, 1964; YoungLai and Short, 1970). They supply further evidence that follicular granulosa cells have the capacity to convert steroid hormones.

In view of recent findings in sheep (Bjersing *et al.*, 1972) that the largest nonatretic follicle (obviously the follicle destined to ovulate in this monovulatory species) is the source of the preovulatory estrogen rise, it seems probable that the follicles with the marked enzyme histochemical reactions in the granulosa cells are the source of the preovulatory estrogen peak. This, however, does not prove that the granulosa cells are the source of, or that they even take part in the preovulatory estrogen secretion; the secretion may partly or wholly depend on the theca cells. It might also be argued that the enzyme changes in the granulosa cells may have been initiated by the preovulatory LH peak and that they are unrelated to the estrogen rise. Prabhu and Weisz (1970), however, recently found that the Δ^5-3β-hydroxysteroid dehydrogenase reactions appeared in the granulosa cells even when the preovulatory LH peak was prevented by sodium pentobarbital. Furthermore, it has now been shown in several species that the estrogen rise precedes the preovulatory LH peak (Bjersing *et al.*, 1972); positive Δ^5-3β-hydroxysteroid dehydrogenase reactions have been demonstrated at the time of the estrogen surge (e.g., Pupkin *et al.*, 1966). It should also be pointed out that structural and histochemical criteria of follicular estrogen production cannot be expected to be as prominent as the corresponding criteria of progesterone production in the corpus luteum. It is often forgotten that the maximum value of the preovulatory estrogen peak in ovarian vein plasma is much less than the highest concentrations of progesterone in the ovarian vein plasma from the side containing an active corpus luteum; in the sheep the estrogen value is about $\frac{1}{1000}$ the progesterone value (Bjersing *et al.*, 1972). Thus, while recent histochemical results can supply no final answer to the question of the respective roles of granulosa and theca cells in follicular estrogen production, they at least suggest that the granulosa cells play a part in the preovulatory estrogen peak (cf., Bjersing, 1967b). Finally, they underline how important it is to perform studies at the right time and on the proper follicles.

Recent work by Turolla *et al.* (1967) has interesting implications regarding both follicular steroidogenesis and maturation. Turolla *et al.* (1967) demonstrated Δ^5-3β-hydroxysteroid dehydrogenase activity in granulosa cells of maturing rat follicles. Furthermore, they studied the effect of steroid treatment on the histochemical activities of ovarian dehydrogenases. The results are of interest from several points of view; for example, for elucidating the mechanism whereby ovulation is prevented in women. A combination of progesterone and estrogen prevented both the maturation of follicles (probably by preventing the tonic release of LH) and the appearance of Δ^5-3β-hydroxysteroid dehydrogenase in the follicular granulosa cells. Ovulation did not occur. Progesterone in large doses did not prevent the maturation of follicles and the appearance of Δ^5-3β-hydroxysteroid dehydrogenase activity in the granulosa cells, but inhibited estrus and ovulation.

The partial inhibition of the Δ^5-3β-hydroxysteroid dehydrogenase activity in the theca interna by means of progesterone or testosterone and its total inhibition with high doses of estradiol or a combination of estradiol and progesterone is of interest in view of recent histochemical observations of strong Δ^5-3β-hydroxysteroid dehydrogenase activity in the granulosa cells but negative reactions in the theca interna cells of horse follicles (E. V. YoungLai, M. F. Hay, and L. Bjersing, unpublished); the $NADH_2$-tetrazolium reductase reaction was positive in both theca and granulosa cells of the equine follicles.

C. Corpus Luteum

It has been demonstrated histochemically that the activities of Δ^5-3β-hydroxysteroid dehydrogenase and several other enzymes in various ovarian structures are influenced by gonadotropins (Levy *et al.*, 1959; Taylor, 1961; Fuhrmann, 1963; Rubin and Deane, 1965; Turolla *et al.*, 1968; Wiest and Kidwell, 1969). 20α-Hydroxysteroid dehydrogenase activity is evident in newly formed rat corpora lutea only in the second half of the estrous cycle (Wiest and Kidwell, 1969). Turolla *et al.* (1967), after administration of estrogen, observed that this enzyme activity was absent in newly formed corpora lutea of the rat. As an explanation they suggested that estrogen inhibited the secretion and/or the release of LH and that estrogen exerted a luteotropic effect through the release of prolactin. Lamprecht *et al.* (1969), on the basis of a combined biochemical and histochemical study in the rat, concluded that prolactin suppressed the synthesis of 20α-hydroxysteroid dehydrogenase in the corpora lutea, and that this constituted part of its luteotropic action. They also suggested that prolactin deprivation over a critical period irreversibly induced 20α-hydroxysteroid dehydrogenase synthesis; thus the ability of the lutein cells to secrete progesterone was reduced. Turolla *et al.* (1969) obtained similar results, and concluded that the production of 20α-hydroxyprogesterone depends on the onset of 20α-hydroxysteroid dehydrogenase activity in the corpus luteum. Wiest and Kidwell (1969), in a paper on the regulation of progesterone secretion by ovarian dehydrogenases, also commented on the limiting effect of prolactin on the development of 20α-hydroxysteroid dehydrogenase. Villee in discussing the paper by Wiest and Kidwell (1969), inferring that LH and prolactin operate at the transcription level, suggested that, for example, LH turned on and prolactin turned off the transcription of specific DNA to RNA.

The so-called luteinization of ovarian granulosa cells has recently been studied in detail in culture experiments, and discussed from several aspects (Channing, 1970). The mechanisms underlying luteinization are still not clearly understood, but recent experimental data suggest that gonadotropic induction of luteinization of the granulosa cells may be mediated by an increase of cyclic AMP (Channing, 1970; Channing and Seymour, 1970). To the many morphological and steroidogenic criteria of luteinization enumerated by Channing (1970), Guraya (1968b) has added another, based on lipid histochemistry. He found that luteinization of granulosa cells was accompanied by a gradual development of diffuse sudanophilic lipid, interpreted as lipoprotein; furthermore, the lipoproteins were thought to reflect a pronounced increase in agranular endoplasmic reticulum, a structure that is the site of steroidogenically important enzymes, for example, Δ^5-3β-hydroxysteroid dehydrogenase, and which may be a reservoir for both cholesterol and other steroid hormone precursors (Christensen, 1965; Fawcett, 1965; Bjersing, 1967d). An increase of this subcellular structure could therefore be a common denominator explaining morphological and steroidogenic criteria of luteinization.

D. Ovarian Stroma and Other Structures

This heading covers several different structures which are not always recognized as distinct entities. Mossman *et al.* (1964) emphasized the importance of distinguishing between interstitial gland tissue (formed from theca interna cells of degenerating follicles) and thecal gland tissue (formed from ripening follicles at or near the time of ovulation). The histochemical findings by Guraya (1968b) support the interesting contention by Mossman *et al.* (1964) that the thecal gland cells degenerate after ovulation and that theca lutein cells of mature corpora lutea represent paraluteal cells which have differentiated anew from surrounding stromal cells.

The ovarian hilus cells displayed histochemical reactions similar to those of testicular interstitial cells, suggesting that the hilus cells produce androgens (Jones *et al.*, 1968b). The 3β-hydroxysteroid dehydrogenase activity was obtained in the hilus cells of the human ovary but more readily with Δ^4-androstene-3β, 17β-diol than with dehydroepiandrosterone as substrate (Novak *et al.*, 1965; Jones *et al.*, 1968a; Poliak *et al.*, 1968). Poliak *et al.* (1968) investigated the effect of HCG on postmenopausal women; the hilus cells showed a marked hyperplasia. The reactions with glucose 6-phosphate dehydrogenase, β-hydroxybutyrate dehydrogenase, and $NADH_2$-tetrazolium reductase were maximal in these cells. Koudstaal *et al.* (1966) reported that hilus cells displayed the same enzyme pattern as the theca cells of growing follicles, but aminopeptidase activity was absent. Signs of endocrine activity are hard to detect in single cells or in small groups of cells, and the observations of enzymatic activity in the hilus cells illustrate the special advantage of histochemistry.

VIII. FUTURE DEVELOPMENTS

It is always dangerous to speculate about the future. It might be of some value, however, to discuss briefly trends in ovarian histochemistry and to mention some probable and desirable developments.

Progress in ovarian histochemistry results from the application of methods used with other tissues, by the development of new methods, and by combinations of existing methods. A general trend shared with other fields of histochemistry is the tendency toward quantitative and ultrastructural methods. The application to the ovary of established histochemical methods may help to solve intricate problems. The action of prostaglandins and the interrelationships between these substances, LH, adenyl cyclase, cyclic AMP, and ovarian steroidogenesis are complex and only partly understood. LH is known to stimulate the production of cyclic AMP and steroids, and cyclic AMP is known to stimulate steroidogenesis in ovarian tissues (Savard *et al.*, 1965; Marsh *et al.*, 1966; Dorrington and Kilpatrick, 1969). Prostaglandins have been shown to stimulate the production of cyclic AMP and steroids in the ovary (Kuehl *et al.*, 1970; Marsh, 1970; Speroff and Ramwell, 1970). Kuehl *et al.* (1970) found that the stimulating effect of LH on production of cyclic AMP was prevented by a prostaglandin inhibitor.* They suggested that prostaglandins are mediators

* For a recent and critical discussion of the functional relation between LH and prostaglandins, see Lindner *et al.* (1974), Supplementary References, Section VIII.

of LH stimulation, and that they stimulate adenyl cyclase to convert ATP into cyclic AMP. Kuehl *et al.* (1970) excluded the possibility that prostaglandins had an inhibitory effect on cyclic AMP phosphodiesterase, the enzyme that degrades cyclic AMP, by adding the potent phosphodiesterase inhibitor theophylline to the incubation mixtures. LH has been shown to mobilize cholesterol esters and release fatty acids in ovarian tissues (Armstrong *et al.*, 1969; Armstrong, 1970); these effects may be related to and stimulate the synthesis of prostaglandins from fatty acids (Armstrong, 1970; Speroff and Ramwell, 1970). Prostaglandins have, in fact, been demonstrated in the ovary (Speroff and Ramwell, 1970). Histochemistry may help to clarify the localizations, activities and biological roles of prostaglandins and adenyl cyclase by application to the ovary of two recently developed methods for demonstration of prostaglandin dehydrogenase activity (Nissen and Andersen, 1968) and adenyl cyclase activity (Reik *et al.*, 1970). The electron microscopic procedure developed by Reik *et al.* (1970) should permit localization of adenyl cyclase at the cellular and subcellular level.

The method for ultrastructural demonstration of peroxidase (Graham and Karnovsky, 1966) is another method with good prospects of giving valuable information about the ovary, and a recent histochemical investigation of endogenous peroxidase in the rat uterus yielded interesting results (Brökelmann and Fawcett, 1969).

The perchloric acid–naphthoquinone method developed by Adams (1961) for cholesterol and other steroids with 3-hydroxy-Δ^5 or 3-hydroxy-$\Delta^{5,7}$ configurations is more sensitive, and because it localizes the steroids more precisely than the Schultz reaction it will probably be preferred in future histochemical studies.

The method for localizing cholesterol at the ultrastructural level by the digitonin reaction (Ökrös, 1968) has apparently not been used on ovarian tissues.

Steroids, including steroid hormones, have also been demonstrated in the ovary by autoradiography (Appelgren, 1967, 1969b; Stumpf, 1969) and by immunohistochemistry (Woods and Domm, 1966). These methods will doubtless be used more in the future. However, the demonstration of steroids in tissues is fraught with difficulty because of the tendency of steroids to move about during various technical procedures. The stringent conditions necessary in autoradiography of diffusible substances even at the light microscopical level have been discussed extensively by Stumpf and Roth (1966).

As regards new histochemical methods, better possibilities for localizing steroid-converting enzymes as well as steroids at the ultrastructural level are urgently needed. The steroids should be followed from their entry into the cell as precursors, throughout biosynthesis up to their release into the extracellular space. The problems of soluble substances are naturally greater at the electron than at the light microscope level. Christensen (1969), in the hope of localizing steroids within cells by autoradiography at the ultrastructural level, has devised an ingenious new method for that purpose. Preliminary work with frozen thin sections of fresh corpus luteum tissue of a pregnant guinea pig was mentioned in a congress report (Christensen, 1970), but so far no full report on the use of the method on adrenals or gonads has been published.

Examples of fruitful combinations of established methods are the investigations by Curri *et al.* (1964) and by Hack and Helmy (1967). Although isolated observations are important in efforts to identify tissue lipids, there is a need for more specific methods and systematic studies. The combination of histochemistry with chromatography of lipids in cryostat sections (Curri *et al.*, 1964; Hack and Helmy, 1967) and the rational approach by Chayen *et al.* (1969) represent distinct progress. The lucid paper of Chayen (1968) is an interesting step toward deeper understanding of tissue lipid metabolism, particularly that of phospholipids. A reinvestigation of ovarian lipid histochemistry with special attention to these methods and to the distribution of bound versus free lipids may give new and interesting results.

For an orientation of the development of ovarian histochemistry since this Chapter was written, Supplementary References have been added.

REFERENCES

Abe, M., Kramer, S. P., and Seligman, A. M. (1964). The histochemical demonstration of pancreatic-like lipase and comparison with the distribution of esterase. *J. Histochem. Cytochem.* **12,** 364–383.

Adams, C. W. M. (1961). A perchloric acid-naphthoquinone method for the histochemical localization of cholesterol. *Nature (London)* **192,** 331–332.

Adams, C. W. M. (1965). Histochemistry of lipids. *In* "Neurohistochemistry" (C. W. M. Adams, ed.), pp. 6–66. Elsevier, Amsterdam.

Adams, E. C., Hertig, A. T., and Foster, S. (1966). Studies on guinea pig oocytes. II. Histochemical observations on some phosphatases and lipid in developing and in atretic oocytes and follicles. *Am. J. Anat.* **119,** 303–340.

Adams Smith, W. N., and Stoward, P. J. (1967). A fluorescence histochemical study of ketosteroids in gonadal and adrenal tissue of the rat. *J. R. Microsc. Soc.* [3] **87,** 47–51.

Anderson, E. (1970). The localization of acid phosphatase activity in developing oocytes and associated follicle cells of mammals. *Anat. Rec.* **166,** 271.

Appelgren, L.-E. (1967). Sites of steroid hormone formation. Autoradiographic studies using labelled precursors. *Acta Physiol. Scand.* **68,** Suppl. 301, 1–108.

Appelgren, L.-E. (1969a). Histochemical demonstration of drug interference with progesterone synthesis. *J. Reprod. Fertil.* **19,** 185–186.

Appelgren, L.-E. (1969b). The distribution of labelled testosterone in mice. *Acta Endocrinol. (Copenhagen)* **62,** 505–512.

Armstrong, D. T. (1970). Reproduction. *Annu. Rev. Physiol.* **32,** 439–470.
Armstrong, D. T., Jackanicz, T. M., and Keyes, P. L. (1969). Regulation of steroidogenesis in the rabbit ovary. *In* "The Gonads" (K. W. McKerns, ed.), pp. 3–25. North-Holland Publ., Amsterdam.
Arvy, L. (1963). "Histo-enzymologie des glandes endocrines," pp. 173–201. Gauthier-Villars, Paris.
Arvy, L., and Mauléon, P. (1964). Evolution des activités enzymatiques histochimiquement décelables dans le corps jaune chez la brebis. I. Cytochromoxydase et peroxydase. *C. R. Soc. Biol.* (*Paris*) **158,** 453–457.
Ashbel, R., and Seligman, A. M. (1949). A new reagent for the histochemical demonstration of active carbonyl groups. A new method for staining ketonic steroids. *Endocrinology* **44,** 565–583.
Baillie, A. H., Calman, K. C., Ferguson, M. M., and Hart, D. McK. (1965a). Histochemical demonstration of 20β-hydroxysteroid dehydrogenase. *J. Endocrinol.* **32,** 337–339.
Baillie, A. H., Ferguson, M. M., Calman, K. C., and Hart, D. McK. (1965b). Histochemical demonstration of 11β-hydroxysteroid dehydrogenase. *J. Endocrinol.* **33,** 119–125.
Baillie, A. H., Ferguson, M. M., and Hart, D. McK. (1966a). "Developments in Steroid Histochemistry." Academic Press, London.
Baillie, A. H., Calman, K. C., Ferguson, M. M., and Hart, D. McK. (1966b). Histochemical utilization of 3α-, 6β-, 11α-, 12α-, 16α-, 16β-, 17α-, 21- and 24-hydroxysteroids. *J. Endocrinol.* **34,** 1–12.
Baker, J. R. (1946). Further remarks on the histochemical recognition of lipine. *Q. J. Microsc. Sci.* (N.S.) **88,** 463–465.
Balogh, K., Jr. (1964a). Dihydrolipoic dehydrogenase activity: a step in formation of acyl-coenzyme A, demonstrated histochemically. *J. Histochem. Cytochem.* **12,** 404–412.
Balogh, K., Jr. (1964b). A histochemical method for the demonstration of 20α-hydroxysteroid dehydrogenase activity in rat ovaries. *J. Histochem. Cytochem.* **12,** 670–673.
Balogh, K., Jr., Kidwell, W. R., and Wiest, W. G. (1966). Histochemical localization of rat ovarian 20α-hydroxysteroid dehydrogenase activity initiated by gonadotrophic hormone administration. *Endocrinology* **78,** 75–81.
Banon, P., Brandes, D., and Frost, J. K. (1964). Lysosomal enzymes in the rat ovary and endometrium during the estrous cycle. *Acta Cytol.* **8,** 416–425.
Barbanti Silva, C., Trentini, G. P., Albertazzi, E., and Botticelli, A. (1966). Modificazioni istomorfologiche ed istoenzimatiche dell'ovaio de dell'utero di ratta adulta consequenti ad epifisectomia. *Riv. Ital. Ginecol.* **50,** 515–570.
Barka, T., and Anderson, P. J. (1963). "Histochemistry. Theory, Practice, and Bibliography." Harper (Hoeber), New York.
Barman, T. E. (1969). "Enzyme Handbook." Springer-Verlag, Berlin and New York.
Belt, W. D., Cavazos, L. F., Anderson, L. L., and Kraeling, R. R. (1970). Fine structure and progesterone levels in the corpus luteum of the pig during pregnancy and after hysterectomy. *Biol. Reprod.* **2,** 98–113.
Benítez, E. V., and Daza, L. G., Jr. (1966). Sitio de acción de la HGC en el ovario de la rata, determinado por un método inmunohistoquímico. *Rev. Invest. Clin.* **18,** 239–244.
Ben-Or, S. (1963). Morphological and functional development of the ovary of the

mouse. I. Morphology and histochemistry of the developing ovary in normal conditions and after FSH treatment. *J. Embryol. Exp. Morphol.* **11,** 1–11.

Bergman, S., Bjersing, L., and Nilsson, O. (1966). Histochemical demonstration of Δ^5-3β-hydroxysteroid dehydrogenase activity in cultivated granulosa cells of the porcine ovary. *Acta Pathol. Microbiol. Scand.* **68,** 461–462.

Bitensky, L. (1963). The reversible activation of lysosomes in normal cells and the effects of pathological conditions. *Lysosomes, Ciba Found. Symp., 1963* pp. 362–375.

Bjersing, L. (1967a). Histochemical demonstration of Δ^5-3β- and 17β-hydroxysteroid dehydrogenase activities in porcine ovary. *Histochemie* **10,** 295–304.

Bjersing, L. (1967b). On the morphology and endocrine function of granulosa cells in ovarian follicles and corpora lutea. Biochemical, histochemical, and ultrastructural studies on the porcine ovary with special reference to steroid hormone synthesis. *Acta Endocrinol.* (*Copenhagen*) *Suppl.* **125,** 1–23.

Bjersing, L. (1967c). On the ultrastructure of follicles and isolated follicular granulosa cells of porcine ovary. *Z. Zellforsch. Mikrosk. Anat.* **82,** 173–186.

Bjersing, L. (1967d). On the ultrastructure of granulosa lutein cells in porcine corpus luteum. With special reference to endoplasmic reticulum and steroid hormone synthesis. *Z. Zellforsch. Mikrosk. Anat.* **82,** 187–211.

Bjersing, L., and Carstensen, H. (1967). Biosynthesis of steroids by granulosa cells of the porcine ovary *in vitro*. *J. Reprod. Fertil.* **14,** 101–111.

Bjersing, L., Hay, M. F., Moor, R. M., Short, R. V., and Deane, H. W. (1970a). Endocrine activity, histochemistry and ultrastructure of ovine corpora lutea. I. Further observations on regression at the end of the oestrous cycle. *Z. Zellforsch. Mikrosk. Anat.* **111,** 437–457.

Bjersing, L., Hay, M. F., Moor, R. M., and Short, R. V. (1970b). Endocrine activity, histochemistry and ultrastructure of ovine corpora lutea. II. Observations on regression following hysterectomy. *Z. Zellforsch. Mikrosk. Anat.* **111,** 458–470.

Bjersing, L., Hay, M. F., Kann, G., Moor, R. M., Naftolin, F., Scaramuzzi, R. J., Short, R. V., and YoungLai, E. V. (1972). Changes in gonadotrophins, ovarian steroids and follicular morphology in sheep at oestrus. *J. Endocrinol.* **52,** 465–479.

Blaha, G. C., and Leavitt, W. W. (1970). The distribution of ovarian Δ^5-3β-hydroxysteroid dehydrogenase activity in the golden hamster during the estrous cycle, pregnancy, and lactation. *Biol. Reprod.* **3,** 362–368.

Blanchette, E. J. (1965). Electron microscopic studies on the formation of glycogen in the granulosa follicle cells of the rabbit. *Anat. Rec.* **151,** 325.

Blanchette, E. J. (1966). Ovarian steroid cells. I. Differentiation of the lutein cell from the granulosa follicle cell during the preovulatory stage and under the influence of exogenous gonadotrophins. *J. Cell Biol.* **31,** 501–516.

Blanchette, E. J. (1968). The fine structural demonstration of steroid dehydrogenase activity in ovarian cells and cell fractions. *Anat. Rec.* **160,** 318.

Botte, V. (1968). The hydrocortisone 21-phosphate as substrate for alkaline phosphatase. Histochemical remarks. *Histochemie* **16,** 195–196.

Botte, V., and Delrio, G. (1970). The activity of mouse uterine acid phosphatase on oestradiol-3- and oestradiol-17β-phosphate. *J. Steroid Biochem.* **1,** 335–339.

Boucek, R. J., Telegdy, G., and Savard, K. (1967). Influence of gonadotrophin on histochemical properties of the rabbit ovary. *Acta Endocrinol.* (*Copenhagen*) **54,** 295–310.

Brandau, H. (1967). Enzymhistochemische Untersuchungen über die Wirkung von Gonadotropinen auf das infantile Rattenovar. *Arch. Gynaek.* **204,** 130–131.
Brandau, H., and Lehmann, V. (1970). Histoenzymatische Untersuchungen an menschlichen Gonaden während der intrauterinen Entwicklung. *Z. Geburtshilfe Gynaekol.* **173,** 233–249.
Brandau, H., and Luh, W. (1964a). Zur Lokalisation der innersekretorischen Funktion des menschlichen Ovars. *Acta Endocrinol.* (*Copenhagen*) **46,** 580–596.
Brandau, H., and Luh, W. (1964b). Enzymologische Untersuchungen am Corpus luteum graviditatis des Menschen. Verteilungs- und Aktivitätsmuster von Enzymen der Embden-Meyerhof-Kette und des Krebszyklus. *Z. Geburtshilfe Gynaekol.* **162,** 304–313.
Brandau, H., and Luh, W. (1965). Die Histotopik von Oxydoreduktasen des Intermediärstoffwechsels im interstitiellen Gewebe des menschlichen Ovars. Ein Beitrag zur funktionellen Morphologie der interstitiellen Zellen. *Arch. Gynaekol.* **200,** 407–420.
Brandau, H., and Luh, W. (1969). Enzymhistochemische Untersuchungen am Ovar der androgensterilisierten Ratte. (Early androgen syndrome.) *Z. Geburtshilfe Gynaekol.* **171,** 206–217.
Brandau, H., Lehmann, U., and Luh, W. (1967). Veränderungen der Enzymverteilungsmuster im Kaninchenovar nach Röntgenbestrahlung. *Z. Geburtshilfe Gynaekol.* **168,** 51–64.
Brandau, H., Lehmann, V., and Kazancigil, A. (1968). Histiogramme von Oxydoreduktasen des Kohlenhydratstoffwechsels und von Steroid-Dehydrogenasen im Ovar des Neugeborenen. *Gynaecologia* **165,** 428–441.
Bratt, H., Pupkin, M., Lloyd, C. W., Weisz, J., and Balogh, K., Jr. (1968). Dehydrogenases in the rat ovary. II. A histochemical study of steroid and carbohydrate metabolizing enzymes in micropolycystic ovaries induced by continuous illumination. *Endocrinology* **83,** 329–335.
Brökelmann, J., and Fawcett, D. W. (1969). The localization of endogenous peroxidase in the rat uterus and its induction by estradiol. *Biol. Reprod.* **1,** 59–71.
Bronzetti, P., Cucchiara, G., Mazza, E., and Milio, G. (1963). Osservazioni istochimiche sulla reattivitá lipasica e sulla presenza di materials sudanofilo nell'ovaio e nell'utero della ratta. *Biol. Lat.* **16,** 349–384.
Bulmer, D. (1959). Dimedone as an aldehyde blocking reagent to facilitate the histochemical demonstration of glycogen. *Stain Technol.* **34,** 95–98.
Bulmer, D. (1963). The histochemical distribution of β-glucuronidase activity in the ovary. *J. Endocrinol.* **26,** 171–172.
Bulmer, D. (1964a). The histochemical distribution of β-galactosidase activity in the ovary and female genital tract. *Anat. Rec.* **149,** 699–706.
Bulmer, D. (1964b). The histochemical distribution of certain ovarian enzymes. *J. Anat.* **98,** 27–36.
Bulmer, D. (1964c). The histochemistry of ovarian macrophages in the rat. *J. Anat.* **98,** 313–319.
Bulmer, D. (1965a). A histochemical study of ovarian cholinesterases. *Acta Anat.* **62,** 254–265.
Bulmer, D. (1965b). Histochemical studies on glycosidases of the rat ovary and placenta. *J. Histochem. Cytochem.* **13,** 396–403.
Bulmer, D. (1965c). Cytochemical studies on the non-specific esterases of the rat ovary. *J. R. Microsc. Soc.* [3] **84,** 189–198.
Bulmer, D. (1966). Comparison of histochemical methods for β-glucuronidase in rat ovary and placenta. *J. Anat.* **100,** 697.

Bulmer, D. (1967a). The distribution of β-glucuronidase activity in rat ovary and placenta with the naphthol AS-BI post-coupling method. *J. R. Microsc. Soc.* [3] **86,** 377–383.
Bulmer, D. (1967b). Simultaneous coupling azo dye methods for β-glucuronidase. *J. R. Microsc. Soc.* [3] **86,** 385–395.
Burkl, W., and Slezak-Klemencic, E. (1967a). Über die Aktivität der 3β-Hydroxysteroid-Dehydrogenase und der "sek.-Alkohol-Dehydrogenase" im ovariellen Zyklus der Ratte. *Acta Histochem.* **28,** 100–111.
Burkl, W., and Slezak-Klemencic, E. (1967b). Über die Aktivität der Glucose-6-phosphatdehydrogenase, der Succinodehydrogenase und der unspezifischen Esterase im ovariellen Zyklus bei der Ratte. *Histochemie* **10,** 353–361.
Carsten, P. M., and Merker, H. J. (1965). Die Darstellung von Cholesterinkrit stallen in elektronenmikroskopischen Bild. *Frank. Z. Pathol.* **74,** 539–543.
Channing, C. P. (1969a). Tissue culture of equine ovarian cell types: culture methods and morphology. *J. Endocrinol.* **43,** 381–390.
Channing, C. P. (1969b). Steroidogenesis and morphology of human ovarian cell types in tissue culture. *J. Endocrinol.* **45,** 297–308.
Channing, C. P. (1970). Influences of the *in vivo* and *in vitro* hormonal environment upon luteinization of granulosa cells in tissue culture. *Recent Prog. Horm. Res.* **26,** 589–622.
Channing, C. P., and Seymour, J. F. (1970). Effects of dibutyryl cyclic-3′,5′-AMP and other agents upon luteinization of porcine granulosa cells in culture. *Endocrinology* **87,** 165–169.
Channing, C. P., and Short, R. V. (1966). The biosynthesis of progesterone in tissue culture by equine granulosa cells. *Excerpta Med. Fndn. Int. Congr. Ser.* **111,** 261.
Chatterjee, A., and Deb, C. (1966). Activity of the corpora lutea in alloxan diabetic rats. A histochemical study. *Acta Anat.* **65,** 299–307.
Chayen, J. (1968). The histochemistry of phospholipids and its significance in the interpretation of the structure of cells. *In* "Cell Structure and its Interpretation" (S. M. McGee-Russell and K. F. A. Ross, eds.), pp. 149–156. Arnold, London.
Chayen, J., and Bitensky, L. (1968). Multiphase chemistry of cell injury. *Biol. Basis Med.* **1,** 337–368.
Chayen, J., and Denby, E. F. (1968). "Biophysical Technique as Applied to Cell Biology." Methuen, London.
Chayen, J., Bitensky, L., Butcher, R. G., and Poulter, L. W. (1969). "A Guide to Practical Histochemistry." Oliver & Boyd, Edinburgh.
Cheatum, S. G., and Warren, J. C. (1966). Purification and properties of 3β-hydroxysteroid dehydrogenase and Δ^5-3-ketosteroid isomerase from bovine corpora lutea. *Biochim. Biophys. Acta* **122,** 1–13.
Chieffi, G., La Torretta, G., Del Bianco, C., and Tramontana, S. (1965). Distribuzione istochimica delle 3β-, 17α- e 17β-idrossisteroide deidrogenasi nell'ovario umano nel ciclo menstruale. *Arch. Ostet. Ginecol.* **70,** 492–500.
Christensen, A. K. (1965). The fine structure of testicular interstitial cells in guinea pigs. *J. Cell Biol.* **26,** 911–935.
Christensen, A. K. (1969). A way to prepare frozen thin sections of fresh tissue for electron microscopy. *In* "Autoradiography of Diffusible Substances" (L. J. Roth and W. E. Stumpf, eds.), pp. 349–362. Academic Press, New York.
Christensen, A. K. (1970). Frozen thin sections of fresh tissue for autoradiography

of diffusible substances. *In* "Microscopie Électronique 1970"(P. Favard, ed.), Vol. 1, pp. 503–504. Société Française de Microscopie Electronique, Paris.

Christensen, A. K., and Gillim, S. W. (1969). The correlation of fine structure and function in steroid-secreting cells, with emphasis on those of the gonads. *In* "The Gonads" (K. W. McKerns, ed.), pp. 415–488. North-Holland Publ., Amsterdam.

Clara, M. (1965). Methoden der Lipidhistochemie. *In* "Handbuch der Histochemie" (W. Graumann and K. Neumann, eds.), Vol. 6, Part 1, pp. 54–373. Fischer, Stuttgart.

Coppola, J. A., Leonardi, R. G., and Lippman, W. (1966). Ovulatory failure in rats after treatment with brain norepinephrine depletors. *Endocrinology* **78,** 225–228.

Coutts, J. R., and Stansfield, D. A. (1968). Cholesteryl esterase and cholesteryl ester pools in corpus luteum. *J. Lipid Res.* **9,** 647–651.

Csaba, I., Major, A., and Bùcs, G. (1970). Die histochemische Untersuchung der Ovarien im Stein-Leventhal-Syndrom. *Z. Geburtshilfe Gynaekol.* **172,** 55–66.

Curri, S. B., Raso, M., and Rossi, C. R. (1964). Die Anwendung der Dünnschichtchromatographie zum Nackweis von Fett- und Lipoidsubstanzen in Gewebsschnitten. *Histochemie* **4,** 113–119.

Daoust, R. (1966). Modified procedure for the histochemical localization of ribonuclease activity by the substrate film method. *J. Histochem. Cytochem.* **14,** 254–259.

Davenport, G. R., and Mallette, L. E. (1966). Some biochemical properties of rabbit ovarian hydroxysteroid dehydrogenase. *Endocrinology* **78,** 672–678.

Davies, J., Davenport, G. R., Norris, J. L., and Rennie, P. I. C. (1966). Histochemical studies of hydroxysteroid dehydrogenase activity in mammalian reproductive tissues. *Endocrinology* **78,** 667–671.

Deane, H. W., and Rubin, B. L. (1965). Identification and control of cells that synthesize steroid hormones in the adrenal glands, gonads and placentae of various mammalian species. *Arch. Anat. Microsc. Morphol. Exp.* **54,** 49–66.

Deane, H. W., Hay, M. F., Moor, R. M., Rowson, L. E. A., and Short, R. V. (1966). The corpus luteum of the sheep: relationships between morphology and function during the oestrous cycle. *Acta Endocrinol.* (*Copenhagen*) **51,** 245–263.

de Duve, C. (1969). Summation of the session on intracellular localization. *Ann. N.Y. Acad. Sci.* **166,** 602–603.

Dennis, P. M., and Thomas, G. H. (1967). Δ^5-3β-Hydroxysteroid dehydrogenase in the ovary of the rhesus monkey. *J. Endocrinol.* **39,** 459–460.

Dingle, J. E., Hay, M. F., and Moor, R. M. (1968). Lysosomal function in the corpus luteum of the sheep. *J. Endocrinol.* **40,** 325–336.

Dokumov, S. I., and Dashev, G. I. (1963a). A histochemical study of the ovaries of women suffering from the Stein-Leventhal syndrome. *Am. J. Obstet. Gynecol.* **86,** 183–187.

Dokumov, S. I., and Dashev, G. (1963b). Quelques recherches histo-chimiques sur les ovaires normaux de femmes en maturité sexuelle. *Endokrynol. Pol.* **14,** 505–512.

Dokumov, S. I., and Dashev, G. I. (1964). Histochemical study of 17-ketosteroids in ovaries of women suffering from Stein-Leventhal syndrome. *Am. J. Obstet. Gynecol.* **88,** 671–675.

Dokumov, S. I., and Dashev, G. I. (1965). Corticosteroid-like substances detected

histochemically in the ovaries of women suffering from the Stein-Leventhal syndrome. *Am. J. Obstet. Gynecol.* **91**, 185–189.
Dorfman, R. I., and Ungar, F. (1965). "Metabolism of Steroid Hormones." Academic Press, New York.
Dorrington, J. H., and Kilpatrick, R. (1969). The synthesis of progestational steroids by the rabbit ovary. *In* "The Gonads" (K. W. McKerns, ed.), pp. 27–54. North-Holland Publ., Amsterdam.
Dott, H. M. (1969). Lysosomes and lysosomal enzymes in the reproductive tract. *In* "Lysosomes in Biology and Pathology" (J. T. Dingle and H. B. Fell, eds.), Vol. 1, pp. 330–360. North-Holland Publ., Amsterdam.
Dübner, R. (1952). Zellkerngrössen in Follikeln und Gelbkörpern menschlicher Eierstöcke. *Z. Mikrosk.-Anat. Forsch.* **58**, 147–195.
Ejsmont, G. (1965). The occurrence of mucopolysaccharides and glycogen in the developing Graafian follicle of the rat. *Folia Histochem. Cytochem.* **3**, 291–300.
Ewald, W., Werbin, H., and Chaikoff, I. L. (1964). Evidence for two substrate-specific Δ^5-3-ketosteroid isomerases in beef adrenal glands, and their separation from 3β-hydroxysteroid dehydrogenase. *Biochim. Biophys. Acta* **81**, 199–201.
Fahimi, H. D., and Karnovsky, M. J. (1966). Cytochemical localization of two glycolytic dehydrogenases in white skeletal muscle. *J. Cell Biol.* **29**, 113–128.
Falck, B. (1959). Site of production of oestrogen in rat ovary as studied in microtransplants. *Acta Physiol. Scand.* **47**, Suppl. 163, 1–101.
Falck, B. (1962). Observations on the possibilities of the cellular localization of monoamines by a fluorescence method. *Acta Physiol. Scand.* **56**, Suppl. 197, 1–25.
Fathalla, M. F. (1968). The role of the ovarian stroma in hormone production by ovarian tumours. *J. Obstet. Gynaecol. Br. Commonw.* **75**, 78–83.
Fawcett, D. W. (1965). Structural and functional variations in the membranes of the cytoplasm. *In* "Intracellular Membranous Structure" (S. Seno and E. V. Cowdry, eds.), pp. 15–40. Chugoku Press, Okayama, Japan.
Fawcett, D. W., Long, J. A., and Jones, A. L. (1969). The ultrastructure of endocrine glands. *Recent Prog. Horm. Res.* **25**, 315–380.
Ferguson, M. M. (1965). 3β-Hydroxysteroid dehydrogenase activity in the mouse ovary. *J. Endocrinol.* **32**, 365–371.
Ferguson, M. M. (1966). Histochemical demonstration of pyruvate oxidase and α-ketoglutarate oxidase. *Histochemie* **6**, 185–186.
Ferguson, M. M., Baillie, A. H., Calman, K. C., and Hart, D. McK. (1966). Histochemical distribution of alcohol dehydrogenases in endocrine tissue. *Nature* (*London*) **210**, 1277–1279.
Fienberg, R. (1963). Thecosis: a study of diffuse stromal thecosis of the ovary and superficial collagenization with follicular cysts (Stein-Leventhal Ovary). *Obstet. Gynecol.* **21**, 687–700.
Fienberg, R., and Cohen, R. B. (1965). A comparative histochemical study of the ovarian stromal lipid band, stromal theca cell, and normal ovarian follicular apparatus. *Am. J. Obstet. Gynecol.* **92**, 958–969.
Fienberg, R., and Cohen, R. B. (1966). Oxidative enzymes in luteinization. A histochemical study with special reference to persistent follicular cysts and the stromal theca cell. *Obstet. Gynecol.* **28**, 406–415.
Fishman, W. H. (1963). Recent studies on β-glucuronidase. *Farmaco* **18**, 397–407.
Fishman, W. H., Goldman, S. S., and DeLellis, R. (1967). Dual localization of β-glucuronidase in endoplasmic reticulum and in lysosomes. *Nature* (*London*) **213**, 457–460.

Fitko, R., Chamski, J., and Sawicka, K. (1968). The localization of pituitary gonadotropins in sheep ovaries. *Endokrynol. Pol.* **19**, 657–669.
Flint, A. P. F., and Denton, R. M. (1970). The role of nicotinamide-adenine dinucleotide phosphate-dependent malate dehydrogenase and isocitrate dehydrogenase in the supply of reduced nicotinamide-adenine dinucleotide phosphate for steroidogenesis in the superovulated rat ovary. *Biochem. J.* **117**, 73–83.
Fomitchev, N. I. (1969). Distribution of lipids at different developmental stages of corpus luteum in rat. *Arch. Anat. Gistol. Embriol.* **57**, 40–44.
Fowler, E. H. (1970). Histochemical demonstration of oxidoreductase activity in the canine ovary. *Gen. Comp. Endocrinol.* **14**, 475–483.
Fowler, E. H., and Feldman, M. K. (1970). Histochemical demonstration of hydroxysteroid dehydrogenase activity in the mature canine ovary. *Gen. Comp. Endocrinol.* **14**, 484–490.
Fowler, E. H., Feldman, M. K., and Wilson, G. P. (1970). Histochemical evaluation of ovarian hydroxysteroid dehydrogenase activity in bitches with mammary neoplasms. *Am. J. Vet. Res.* **31**, 2045–2053.
Fuhrmann, K. (1963). Histochemische Untersuchungen über Hydroxysteroid Dehydrogenase-Aktivität in Nebennieren und Eierstöcken von Ratten und Kaninchen. *Arch. Gynaekol.* **197**, 583–600.
Galil, A. K. A., and Deane, H. W. (1966). Δ^5-3β-Hydroxysteroid dehydrogenase activity in the steroid-hormone producing organs of the ferret (*Mustela putorius furo*). *J. Reprod. Fertil.* **11**, 333–338.
Garcia-Bunuel, R., and Monis, B. (1964). Histochemical observations on mucins in human ovarian neoplasms. *Cancer N.Y.* **17**, 1108–1118.
Georgieva, Y., Hadjiiski, P., and Wegmann, R. (1967). Etudes sur l'histoenzymologie de l'ovaire de la ratte blanche durant l'ontogénèse et le cycle oestral. Le métabolisme glucidique. *Ann. Histochim.* **12**, 335–346.
Geyer, G. (1969). "Ultrahistochemie. Histochemische Arbeitsvorschriften für die Elektronenmikroskopie." Fischer, Jena.
Ghosh, B. K., and Sengupta, K. P. (1969). A histochemical method for demonstration of a new steroid dehydrogenase in the human ovary (17-α-hydroxysteroid dehydrogenase). *Indian J. Pathol. Bacteriol.* **12**, 1–3.
Ghosh, B. K., and Sengupta, K. P. (1970). Glycogen and glycoprotein, acid mucopolysaccarides and lipids: histochemical study of the ovary in functional uterine bleeding. *J. Indian Med. Assoc.* **55**, 122–126.
Ghosh, B. K., Sengupta, K. B., and Aikat, B. K. (1965). Histochemical study of the ovary in functional uterine bleeding. (A preliminary report.) *Indian J. Pathol. Bacteriol.* **8**, 37–47.
Glauert, A. M. (1965). Section staining, cytology, autoradiography and immunochemistry for biological specimens. *In* "Techniques for Electron Microscopy" (D. H. Kay, ed.), 2nd ed., pp. 254–310. Blackwell, Oxford.
Goldberg, B., Jones, G. E. S., and Turner, D. A. (1963). Steroid 3β-ol dehydrogenase activity in human endocrine tissues. *Am. J. Obstet. Gynecol.* **86**, 349–359.
Goldberg, B., Jones, G. E. S., and Borkowf, H. I. (1964). A histochemical study of substrate specificity for the steroid 3β-ol dehydrogenase and isomerase systems in human ovary and testis. *J. Histochem. Cytochem.* **12**, 880–889.
Goldman, A. S., and Kohn, G. (1970). Rat ovarian 3β-hydroxysteroid dehydrogenase: normal developmental appearance in tissue culture. *Proc. Soc. Exp. Biol. Med.* **133**, 475–478.
Goldman, A. S., Yakovac, W. C., and Bongiovanni, A. M. (1965). Persistent effects

of a synthetic androstene derivative on activities of 3β-hydroxysteroid dehydrogenase and glucose-6-phosphate dehydrogenase in rats. *Endocrinology* **77,** 1105–1118.

Gomori, G. (1939). Microtechnical demonstration of phosphatase in tissue sections. *Proc. Soc. Exp. Biol. Med.* **42,** 23–26.

Gomori, G. (1952). "Microscopic Histochemistry. Principles and Practice." Univ. of Chicago Press, Chicago, Illinois.

Goode, L., Warnick, A. C., and Wallace, H. D. (1965). Alkaline and acid phosphatase activity in the endometrium and ovary of swine. *J. Anim. Sci.* **24,** 955–958.

Graham, R. C., and Karnovsky, M. J. (1966). The early stages of absorption of injected horseradish peroxidase in the proximal tubules of mouse kidney: ultrastructural cytochemistry by a new technique. *J. Histochem.* **14,** 291–302.

Grasso, P., Williams, M., Hodgson, R., Wright, M. G., and Gangolli, S. D. (1971). The histochemical distribution of aniline hydroxylase in rat tissues. *Histochem. J.* **3,** 117–126.

Grecchi, R., Mariano, M., Saliba, A. M., and Zezza Neto, L. (1966). Ceroid pigment in ovary of mare. Histochemical study. *Rev. Bras. Biol.* **26,** 361–366.

Green, J. A., and Maqueo, M. (1965). Ultrastructure of the human ovary. I. The luteal cell during the menstrual cycle. *Am. J. Obstet. Gynecol.* **92,** 946–957.

Griffiths, K., Fahmy, D., Henderson, W. J., Turnbull, A. C., and Joslin, C. A. F. (1971). Ovarian tumours and steroid hormone biosynthesis. *In* "Control of Gonadal Steroid Secretion" (D. T. Baird and J. A. Strong, eds.), Pfizer Med. Monogr. No. 6, pp. 281–292. Edinburgh Univ. Press, Edinburgh.

Gropp, A. (1967). Enzymhistochemische und morphologische Befunde zur Entwicklung der Gonaden, insbesondere der Derivate des Gonadenblastems. *Anat. Anz., Suppl.* **120,** 33–40.

Guraya, S. S. (1965). A histochemical study of follicular atresia in the snake ovary. *J. Morphol.* **117,** 151–159.

Guraya, S. S. (1966a). Histochemical analysis of the interstitial gland tissue in the human ovary at the end of pregnancy. *Am. J. Obstet. Gynecol.* **96,** 907–912.

Guraya, S. S. (1966b). A histochemical study of the rhesus monkey ovary. *Acta Morphol. Neerl.-Scand.* **6,** 395–406.

Guraya, S. S. (1967a). Histochemical study of the interstitial gland tissue in the ovaries of nonpregnant women. *Am. J. Obstet. Gynecol.* **98,** 99–106.

Guraya, S. S. (1967b). Cytochemical observations concerning the formation, release, and transport of lipid secretory products in the interstitial (thecal) cells of the rabbit ovary. *Z. Zellforsch. Mikrosk. Anat.* **83,** 187–195.

Guraya, S. S. (1967c). Cytochemical study of interstitial cells in the bat ovary. *Nature* (*London*) **214,** 614–616.

Guraya, S. S. (1968a). Histochemical study of granulosa (follicular) cells in the preovulatory and postovulatory follicles of amphibian ovary. *Gen. Comp. Endocrinol.* **10,** 138–146.

Guraya, S. S. (1968b). Histochemical study of granulosa and theca interna during follicular development, ovulation, and corpus luteum formation and regression in the human ovary. *Am. J. Obstet. Gynecol.* **101,** 448–457.

Guraya, S. S. (1968c). Histochemical observations on the granulosa and theca interna during follicular development and corpus luteum formation and regression in the American opossum. *J. Endocrinol.* **40,** 237–241.

Guraya, S. S. (1968d). A histochemical study of pre-ovulatory and post-ovulatory follicles in the rabbit ovary. *J. Reprod. Fertil.* **15,** 381–387.

Guraya, S. S. (1968e). Histochemical study of the guinea pig ovary. *Acta Biol. Acad. Sci. Hung.* **19**, 279–287.

Guraya, S. S. (1968f). Histochemical study of the interstitial cells in the cattle ovary. *Acta Anat.* **70**, 447–458.

Guraya, S. S. (1968g). Histophysiology and histochemistry of the interstitial gland tissue in the ovaries of non-pregnant marmosets. *Acta Anat.* **70**, 623–640.

Guraya, S. S. (1968h). Histochemical study of human corpus luteum at term. *Am. J. Obstet. Gynecol.* **102**, 219–221.

Guraya, S. S. (1968i). A comparative histochemical study of the theca interna in the mammalian ovary. *Acta Morphol. Neerl.-Scand.* **7**, 51–68.

Guraya, S. S. (1969a). Histochemical study of cells with foamy cytoplasm in the human ovaries at term. *S. Afr. Med. J.* **43**, 1179–1180.

Guraya, S. S. (1969b). Some observations on the histochemical features of developing follicle and corpus luteum in the cat and dog ovary. *Acta Vet. Acad. Sci. Hung.* **19**, 351–362.

Guraya, S. S. (1969c). Histochemical study of follicular atresia in the amphibian ovary. *Acta Biol. Acad. Sci. Hung.* **20**, 43–56.

Guraya, S. S. (1971). Histochemical changes of nucleic acids in the follicular stroma of the rabbit's ovary before and during ovulation. *J. Reprod. Fertil.* **24**, 107–108.

Guraya, S. S., and Greenwald, G. S. (1964a). A comparative histochemical study of interstitial tissue and follicular atresia in the mammalian ovary. *Anat. Rec.* **149**, 411–433.

Guraya, S. S., and Greenwald, G. S. (1964b). Histochemical studies on the interstitial gland in the rabbit ovary. *Am. J. Anat.* **114**, 495–519.

Guraya, S. S., and Greenwald, G. S. (1965). A histochemical study of the hamster ovary. *Am. J. Anat.* **116**, 257–267.

Hack, M. H., and Helmy, F. M. (1967). Correlative lipid histochemistry. *Acta Histochem.* **27**, 74–84.

Hadjiisky, P. (1970). Fonctions histoenzymatiques lors de l'évolution et de la lutéolyse du corps jaune de la lapine. *Histochemie* **22**, 125–135.

Hadjiisky, P., Georgiewa, Y., and Wegmann, R. (1969a). Etudes cyto-enzymatiques de l'ovaire de la ratte durant l'ontogénèse et le cycle oestral. Enzymes du métabolisme énergétique. *Ann. Histochim.* **14**, 19–32.

Hadjiisky, P., Georgiewa, Y., and Wegmann, R. (1969b). Etudes cyto-enzymatiques de l'ovaire de la ratte blanche durant l'ontogénèse et le cycle oestral. Enzymes du métabolisme des lipides et des protides. *Ann. Histochim.* **14**, 199–213.

Hammarström, L. (1966). Autoradiographic studies on the distribution of C^{14}-labelled ascorbic acid and dehydroascorbic acid. *Acta Physiol. Scand.* **70**, Suppl. 289, 3–83.

Hanai, J. (1968). Changes of lipids (especially phospholipids) and lipoprotein lipase activity in non-pregnant rabbit ovaries under the influence of human chorionic gonadotrophin. *Folia Endocrinol. Jpn.* **44**, 655–673.

Hardonk, M. J. (1965). A new method for the histochemical demonstration of steroid producing cells in human tissues. *Histochemie* **5**, 234–243.

Hardonk, M. J. (1968). 5′-Nucleotidase. I. Distribution of 5 -nucleotidase in tissues of rat and mouse. *Histochemie* **12**, 1–17.

Hardonk, M. J., and Koudstaal, J. (1971). Histochemische und enzymhistochemische Befunde bei Ovarialgeschwülsten. *Acta Histochem., Suppl.* **9**, 593–599.

Hart, D. McK., Baillie, A. H., Calman, K. C., and Ferguson, M. M. (1966).

Hydroxysteroid dehydrogenase development in mouse adrenal and gonads. *J. Anat.* **100,** 801–812.
Hayashi, M. (1964). Distribution of β-glucuronidase activity in rat tissues employing the naphthol AS-BI glucuronide hexazonium pararosanilin method. *J. Histochem. Cytochem.* **12,** 659–669.
Hertig, A. T., and Adams, E. C. (1967). Studies on the human oocyte and its follicle. I. Ultrastructural and histochemical observations on the primordial follicle stage. *J. Cell Biol.* **34,** 647–675.
Hilliard, J., and Sawyer, C. H. (1964). Synthesis and release of progestin by rabbit ovary *in vivo. Horm. Steroids; Biochem., Pharmacol., Ther., Proc. Int. Congr., 1st, 1962* Vol. 1, pp. 263–272.
Hilliard, J., Archibald, D., and Sawyer, C. H. (1963). Gonadotropic activation of preovulatory synthesis and release of progestin in the rabbit. *Endocrinology* **72,** 59–66.
Hoar, W. S. (1965). Comparative physiology: hormones and reproduction in fishes. *Annu. Rev. Physiol.* **27,** 51–69.
Hopkins, T. F., and Pincus, G. (1963). Effects of reserpine on gonadotropin-induced ovulation in immature rats. *Endocrinology* **73,** 775–780.
Hotchkiss, R. D. (1948). A microchemical reaction resulting in the staining of polysaccharide structures in fixed tissue preparations. *Arch. Biochem.* **16,** 131–141.
Høyer, P. E., and Andersen, H. (1970). Specificity in steroid histochemistry, with special reference to the use of steroid solvents. Distribution of 11β-hydroxysteroid-dehydrogenase in kidney and thymus from the mouse. *Histochemie* **24,** 292-306.
Hughesdon, P. E. (1966). Ovarian lipoid and theca cell tumors; their origins and interrelations. *Obstet. & Gynecol. Surv.* **21,** 245–288.
Ikonen, M., Niemi, M., Pesonen, S., and Timonen, S. (1961). Histochemical localization of four dehydrogenase systems in human ovary during the menstrual cycle. *Acta Endocrinol. (Copenhagen)* **38,** 293–302.
Jacobowitz, D., and Wallach, E. E. (1967). Histochemical and chemical studies of the autonomic innervation of the ovary. *Endocrinology* **81,** 1132–1139.
Jacoby, F. (1962). Ovarian histochemistry. *In* "The Ovary" (S. Zuckerman, ed.), 1st ed., Vol. 1, pp. 189–245. Academic Press, New York.
Janko, A. B., and Sandberg, E. C. (1970). Histochemical evidence for the protein nature of the Reinke crystalloid. *Obstet. Gynecol.* **35,** 493–503.
Jirásek, J. E., Šulcová, J., Čapková, A., Röhling, S., and Stárka, L. (1969). Histochemical and biochemical investigations of 3β-hydroxy-Δ^5-steroid dehydrogenase in the chorion, adrenals and gonads of human foetuses. *Endokrinologie* **54,** 173–183.
Jirásek, J. E., Presl, J., and Horsky, J. (1970). Analysis of the postnatal development of the rat ovary by histochemical methods. *Acta Histochem.* **37,** 162–169.
Jonek, J., and Zieliński, Z. (1966). Localisation of gamma-glutamyl-transpeptidase in ovaries of some laboratory animals. *Endokrynol. Pol.* **17,** 551–565.
Jonek, J., Zieliński, Z., Baranowski, Z., and Denk, J. (1965). Untersuchungen über die Lokalisation der γ-Glutamyl-Transpeptidase in Ovarium und Uterus von normalen und graviden Meerschweinchen. *Acta Histochem.* **22,** 364–373.
Jones, G. E. S., Goldberg, B., and Woodruff, D. J. (1967). Enzyme histochemistry of a masculinizing arrhenoblastoma. *Obstet. Gynecol.* **29,** 328–343.
Jones, G. E. S., Goldberg, B., and Woodruff, J. D. (1968a). Cell specific steroid

inhibitions in histochemical steroid 3β-ol dehydrogenase activities in man. *Histochemie* **14,** 131–142.

Jones, G. E. S., Goldberg, B., and Woodruff, J. D. (1968b). Histochemistry as a guide for interpretation of cell function. *Am. J. Obstet. Gynecol.* **100,** 76–83.

Jordan, S. M. (1970). Adrenergic and cholinergic innervation of the reproductive tract and ovary in the guinea-pig and rabbit. *J. Physiol.* (*London*) **210,** 115P-117P.

Kabra, S. G., Chaturvedi, R. P., and Ujwal, Z. S. (1967a). A study of alkaline phosphatase and mitochondria in the dog ovary. *Indian J. Med. Res.* **55,** 279–283.

Kabra, S. G., Chaturvedi, R. P., and Ujwal, Z. S. (1967b). A study of topographical distribution of acid phosphatase in dog ovary and its role in follicular development and rupture. *Indian J. Med. Res.* **55,** 1213–1220.

Kaneyoshi, A. (1969). The effect of gonadotropin on the aminopeptidase activities in the ovary of rats. *Folia Endocrinol. Jpn.* **44,** 1269–1271.

Karjalainen, O. (1968). The human corpus luteum of pregnancy with reference to the state of the trophoblastic tissue. A histological, histochemical and biochemical study. *Acta Obstet. Gynecol. Scand.* **47,** 1–72.

Kautsky, M. P., and Hagerman, D. D. (1970). 17β-Estradiol dehydrogenase of ovine ovaries. *J. Biol. Chem.* **245,** 1978–1984.

Kenney, R. M. (1964). "Histochemical and Biochemical Correlates of the Estrous Cycle in the Normal Cyclic Uterus and Ovary of the Cow." Ph.D. Thesis, Cornell University, Ithaca, New York.

Kern-Bontke, E. (1964). Histochemische nachweisbare Fermentaktivität in den Theca- und Granulosaluteinzellen des Corpus luteum. *Histochemie* **4,** 56–64.

Kessel, R. G., and Panje, W. R. (1968). Organization and activity in the pre- and postovulatory follicle of Necturus maculosus. *J. Cell Biol.* **39,** 1–34.

Khanolkar, V. R., Krishnamurthi, A. S., Bagul, C. D., and Sahasrabudhe, M. B. (1958). A new method of histochemical localization of corticoids. *Indian J. Pathol. Bacteriol.* **1,** 84–89.

Kidwell, W. R., Balogh, K., Jr., and Wiest, W. G. (1966). Effects of luteinizing hormones on glucose-6-phosphate and 20α-hydroxysteroid dehydrogenase activities in superovulated rat ovaries. *Endocrinology* **79,** 352–361.

Kimura, T., Yaoi, Y., Kumasaka, T., Kato, K., Nishi, N., and Fujii, K. (1970). Ovarian morphology following HMG treatment with special reference to histological, histochemical and electron microscopic studies. *Bull. Tokyo Med. Dent. Univ.* **17,** 39–51.

König, P. A. (1965a). Histotopochemie von Enzymen im menschlichen Ovarium. *Fortschr. Geburtshilfe Gynaekol.* **23,** 77–203.

König, P. A. (1965b). Über die Lokalisation von Enzymen in den endokrinen Formationen des menschlichen Ovarium. *Gynaecologia* **159,** 107–124.

König, P. A., and Stier, R. (1963). Histotopochemie von Carbonsäure-Esterasen im menschlichen Ovar. *Arch. Gynaekol.* **197,** 561–572.

Korn, E. D., and Weisman, R. A. (1966). I. Loss of lipids during preparation of amoebae for electron microscopy. *Biochim. Biophys. Acta* **116,** 309–316.

Koudstaal, J., and Hardonk, M. J. (1969). Histochemical demonstration of enzymes related to NADPH-dependent hydroxylating systems in rat liver after phenobarbital treatment. *Histochemie* **20,** 68–77.

Koudstaal, J., Bossenbroek, B., and Hardonk, M. J. (1966). An investigation into thecomatosis of the ovary and related tumours by histochemical and enzyme histochemical methods. *Eur. J. Cancer* **2,** 313–323.

Koudstaal, J., Bossenbroek, B., and Hardonk, M. J. (1968). Ovarian tumors investigated by histochemical and enzyme histochemical methods. *Am. J. Obstet. Gynecol.* **102,** 1004–1017.

Kraus, F. T., and Neubecker, R. D. (1962). Luteinization of the ovarian theca in infants and children. *Am. J. Clin. Pathol.* **37,** 389–397.

Kuehl, F. A., Jr., Humes, J. L., Tarnoff, J., Cirillo, V. J., and Ham, E. A. (1970). Prostaglandin receptor site: evidence for an essential role in the action of luteinizing hormone. *Science* **169,** 883–886.

Küppers, S. (1967). Zur Fermenthistochemie des Rattenovariums während der postnatalen Entwicklung. *Acta Histochem.* **27,** 267–284.

Lambert, J. G. D. (1970). The ovary of the guppy *Poecilia reticulata.* The granulosa cells as sites of steroid biosynthesis. *Gen. Comp. Endocrinol.* **15,** 464–476.

Lamprecht, S. A., Lindner, H. R., and Strauss, J. F., III. (1969). Induction of 20α-hydroxysteroid dehydrogenase in rat corpora lutea by pharmacological blockade of pituitary prolactin secretion. *Biochim. Biophys. Acta* **187,** 133–143.

Larraguibel, R., and Dallenbach-Hellweg, G. (1970). Histochemische Untersuchungen zur Frage spezifischer Einschlüsse in den Granulosa-Luteinzellen des Corpus luteum graviditatis. *Arch. Gynaekol.* **208,** 224–234.

Lehninger, A. L. (1970). "Biochemistry. The Molecular Basis of Cell Structure and Function." Worth, New York.

Lev, R., and Spicer, S. S. (1964). Specific staining of sulphate groups with Alcian blue at low pH. *J. Histochem. Cytochem.* **12,** 309.

Levy, H., Deane, H. W., and Rubin, B. L. (1959). Visualization of steroid-3β-ol-dehydrogenase activity in tissues of intact and hypophysectomized rats. *Endocrinology* **65,** 932–943.

Lobel, B. L., and Levy, E. (1968). Enzymic correlates of development, secretory function and regression of follicles and corpora lutea in the bovine ovary. *Acta Endocrinol.* (*Copenhagen*) **59,** Suppl. 132, 1–63.

Lobel, B. L., Shelesnyak, M. C., and Tic, L. (1966). Studies on the mechanism of nidation. XIX. Histochemical changes in the ovaries of pregnant rats following ergocornine. *J. Reprod. Fertil.* **11,** 339–348.

Loubet, R., and Loubet, A. (1961). Les cellules du hile de l'ovaire et leurs rapports avec les autres éléments endocrines de la glande. *Ann. Anat.-Pathol.* **6,** 189–212.

Luh, W. (1967). Wirkung von Gonadotropinen auf Enzymaktivitätsmuster des infantilen Rattenovars. *Arch. Gynaekol.* **204,** 131–133.

Luh, W., and Brandau, H. (1964a). Die Lokalisation von Oxydoreduktasen im normalen menschlichen Ovar. Ein Beitrag zur funktionellen Morphologie des Ovars. *Z. Geburtshilfe Gynaekol.* **162,** 113–132.

Luh, W., and Brandau, H. (1964b). Enzymologische Untersuchungen am Corpus luteum graviditatis des Menschen. Verteilungs- und Aktivitätsmuster der Glucose-6-phosphate- und TPN-gebundenen Isocitrat-Dehydrogenase. *Z. Geburtshilfe Gynaekol.* **162,** 294–303.

Lunaas, T., Baldwin, R. L., and Cupps, P. T. (1968). Ovarian activities of pyridine nucleotide dependent dehydrogenases in the rat during pregnancy and lactation. *Acta Endocrinol.* (*Copenhagen*) **58,** 521–531.

McKay, D. G., Pinkerton, J. H. M., Hertig, A. T., and Danziger, S. (1961). The adult human ovary: a histochemical study. *Obstet. Gynecol.* **18,** 13–39.

Maeir, D. M. and Wagner, L. (1968). Steroid histochemistry of three experimental granulosa cell tumors. *Proc. Int. Cong. Histochem. Cytochem., 3rd, 1968* Summary Reports, pp. 165–166.

Magrini, U., and Moneta, E. (1966). Osservazioni istoenzimatiche in alcuni casi di ovaio ipoplasico. *Ann. Ostet. Ginecol.* **88,** 602–608.

Marsh, J. M. (1970). The stimulatory effect of prostaglandin E_2 on adenyl cyclase in the bovine corpus luteum. *FEBS Lett.* **7,** 283–286.

Marsh, J. M., Butcher, R. W., Savard, K., and Sutherland, E. W. (1966). The stimulatory effect of luteinizing hormone on adenosine 3′,5′-monophosphate accumulation in corpus luteum slices. *J. Biol. Chem.* **241,** 5436–5440.

Matsukado, M., and Hayashi, A. (1965). Lipid histochemistry in human ovary, especially on significance of granules stainable with Heidenhain's iron hematoxylin. Report I. Histochemical observation on human menstrual corpus luteum in reproductive age. *J. Jpn. Obstet. Gynecol.* **12,** 67–80.

Matsuzawa, T., and Takikawa, H. (1968). The alteration of isozymic patterns in lactate dehydrogenase during artificial maturation of rat ovary. *Endocrinol. Jpn.* **15,** 291–297.

Mayner, D. A. (1966). Enzymatic histochemistry of the ovary, testis and epididymis of the fetal pig. *Anat. Rec.* **154,** 384.

Modak, S. P., and Kamat, D. N. (1968). A study of periodic acid Schiff material in the ovarian follicles of tropical bats. *Cytologia* **33,** 34–59.

Monis, B., Tsou, K.-C., and Seligman, A. M. (1963). Development of a histochemical method for α-D-galactosidase and its distribution in the rat. *J. Histochem. Cytochem.* **11,** 653–661.

Mor, A. (1966). Histochemical study of corpora lutea in the rat. Corpora lutea of pregnancy, pseudopregnancy, and menstruation. *Int. J. Fertil.* **11,** 205–209.

Mori, H., and Matsumoto, K. (1970). On the histogenesis of the ovarian interstitial gland in rabbits. I. Primary interstitial gland. *Am. J. Anat.* **129,** 289–306.

Mori, M. (1967). Histochemical evaluation of NADP-dependent glucose-6-phosphate and isocitrate dehydrogenases in steroid producing organs and tumors. *Arch. Histol. Iaponicum* **28,** 45–57.

Mori, T. (1969). Histological and histochemical studies of human anovulatory ovaries correlated with endocrinological analyses. *Acta Obstet. Gynaecol. Jpn.* **16,** 156–164.

Moricard, R. (1964). Essais de localisation ultrastructurale du 35-S dans quelques cellules du tractus genital. Autoradiographies en microscopie électronique. *Ann. Histochim.* **9,** 155–162.

Mossman, H. W., Koering, M. J. and Ferry, D., Jr. (1964). Cyclic changes of interstitial gland tissue of the human ovary. *Am. J. Anat.* **115,** 235–256.

Motta, P. (1965). Ricerche sui corpi di Call ed Exner nell'ovaio dei mammiferi. (Origine, evoluzione e significato.) *Biol. Lat.* **18,** 253–258.

Motta, P. (1967). Sui corpi di Call ed Exner in diverse specie di mammiferi. *Biol. Lat.* **20,** 303–318.

Motta, P., and Bourneva, V. (1970). A histochemical study of the lipid content and the Δ^5-3β-hydroxysteroid dehydrogenase activity of the interstitial cells of the rat ovary during the oestrous cycle and pregnancy. *C. R. Acad. Bulg. Sci.* **23,** 879–882.

Motta, P., and Takeva, Z. (1970). An electron microscopic and histochemical study of Δ^5-3β-hydroxysteroid dehydrogenase in the interstitial cells of human ovary in different stages of foetal and adult life. *C. R. Acad. Bulg. Sci.* **23,** 883–886.

Motta, P., Takeva, Z., and Bourneva, V. (1970). A histochemical study of Δ^5-3β-hydroxysteroid dehydrogenase activity in the interstitial cells of the mammalian ovary. *Experientia* **26,** 1128–1129.

Nadkarni, V. B., and Lloyd, C. W. (1966). Histochemical study of monoamine oxidase (MAO) in rat ovaries. *Fed. Proc., Fed. Am. Soc. Exp. Biol.* **25,** 251.

Nagai, K., Lindlar, K., and Stolpmann, H. J. (1967). Morphologische und chemische Untersuchungen über die Lipoide des hormonal stimulierten Ovars der Ratte. *Z. Zellforsch. Mikrosk. Anat.* **79,** 550–561.

Nakama, S. (1969). Enzyme histochemistry of the bovine ovaries in normal and cystic conditions. *Acta Histochem.* **32,** 157–177.

Nakayama, M. (1970). Histochemical study on localization of dehydrogenase systems in human ovary. *Acta Obstet. Gynaecol. Jpn.* **17,** 283.

Nielson, M. H., and Warren, J. C. (1965). TPNH generation in the human ovary. *Acta Endocrinol.* (*Copenhagen*) **49,** 58–64.

Niemi, M., and Sylvén, B. (1969). The naphthylamidase reaction as a diagnostic tool for the demonstration of cellular injury and autophagy. *Histochemie* **18,** 40–47.

Niemi, M., Ikonen, M., Pesonen, S., Saure, A., and Timonen, S. (1965). Histochemistry of the corpus luteum in human ovary. *Ann. Chir. Gynaecol. Fenn.* **54,** 339–346.

Niles, N. R., Bitensky, L., Chayen, J., Cunningham, G. J., and Brainbridge, M. V. (1964). The value of histochemistry in the analysis of myocardial dysfunction. *Lancet* **1,** 963–965.

Nissen, H. M., and Andersen, H. (1968). On the localization of a prostaglandin-dehydrogenase activity in the kidney. *Histochemie* **14,** 189–200.

Novak, E. R., Goldberg, B., Jones, G. S., and O'Toole, R. V. (1965). Enzyme histochemistry of the menopausal ovary associated with normal and abnormal endometrium. *Am. J. Obstet. Gynecol.* **93,** 669–682.

Novak, E. R., Jones, G. S., and Jones, H. W., Jr. (1970). "Novak's Textbook of Gynecology," 8th ed. Williams & Wilkins, Baltimore, Maryland.

Ogawa, K., and Barnett, R. J. (1965). Electron cytochemical studies of succinic dehydrogenase and dihydronicotinamide-adenine dinucleotide diaphorase activities. *J. Ultrastruct. Res.* **12,** 488–508.

Okano, K., Matsumoto, K., Koizumi, T., Mizushima, T., and Mori, M. (1966). Histochemical alterations of oxidative enzymes in rodent ovaries under non-pregnancy, pregnancy, through lactation and hormone treated conditions. *Acta Histochem.* **25,** 12–39.

Ökrös, I. (1968). Digitonin reaction in electron microscopy. *Histochemie* **13,** 91–96.

O'Shea, J. D. (1966). Histochemical observations on mucin secretion by sub-surface epithelial structures in the canine ovary. *J. Morphol.* **120,** 347–358.

Owman, C., and Sjöberg, N. O. (1966). Adrenergic nerves in the female genital tract of the rabbit. With remarks on cholinesterase-containing structures. *Z. Zellforsch. Mikrosk. Ant.* **74,** 182–197.

Owman, C., Rosengren, E., and Sjöberg, N. O. (1967). Adrenergic innervation of the human female reproductive organs: a histochemical and chemical investigation. *Obstet. Gynecol.* **30,** 763–773.

Pearse, A. G. E. (1968). "Histochemistry: theoretical and Applied." Churchill, London.

Pesonen, S., Ikonen, M., Plrocopé, B.-J., and Saure, A. (1968). Androst-5-ene-3β,17β-diol in ovary of post-menopausal hyperoestrogenic women. *In vitro* metabolism; histological and histochemical studies; clinical data of the patients. *Acta Endocrinol.* (*Copenhagen*) **58,** 364–376.

Pfleiderer, A., Jr., and Teufel, G. (1968). Incidence and histochemical investigation of enzymatically active cells in stroma of ovarian tumors. *Am. J. Obstet. Gynecol.* **102,** 997–1003.

Pfleiderer, A., Jr., Kidess, E., and Jung, G. (1969). Klinik und Pathologie der Lactatdehydrogenase im Ovarialcarcinom. *Z. Krebsforsch.* **72,** 329–340.

Pinkerton, J. H. M., McKay, D. G., Adams, E. C., and Hertig, A. T. (1961). Development of human ovary. A study using histochemical technics. *Obstet. Gynecol.* **18,** 152–181.

Poliak, A., Jones, G. E. S., Goldberg, B., Solomon, D., and Woodruff, J. D. (1968). Effect of human chorionic gonadotropin on postmenopausal women. Ovarian histochemistry and urinary hormone excretion. *Am. J. Obstet. Gynecol.* **101,** 731–739.

Prabhu, V. K. K., and Weisz, J. (1970). Effect of blocking ovulation in the rat by pentobarbital on ovarian 3β-hydroxysteroid dehydrogenases: a histochemical study. *Endocrinology* **87,** 481–485.

Presl, J., Jirásek, J., Horský, J., and Henzl, M. (1965). Observations on steroid-3β-ol dehydrogenase activity in the ovary during early postnatal development in the rat. *J. Endocrinol.* **31,** 293–294.

Priedkalns, J., and Weber, A. F. (1968). The succinic dehydrogenase and lipid content of follicular and luteal cells of the bovine ovary. *Acta Anat.* **71,** 542–564.

Pryse-Davies, J. (1970). Enzyme histochemistry of the developing ovary and endometrium. *Arch. Dis. Child.* **45,** 708.

Pupkin, M., Bratt, H., Weisz, J., Lloyd, C. W., and Balogh, K., Jr. (1966). Dehydrogenases in the rat ovary. I. A histochemical study of Δ^5-3β- and 20α-hydroxysteroid dehydrogenases and enzymes of carbohydrate oxidation during the estrous cycle. *Endocrinology* **79,** 316–327.

Quintarelli, G., Scott, J. E., and Dellovo, M. C. (1964). The chemical and histochemical properties of Alcian blue. II. Dye binding of tissue polyanions. *Histochemie* **4,** 86–98.

Raggatt, P. R., and Whitehouse, M. W. (1966). Substrate and inhibitor specificity of the cholesterol oxidase in bovine adrenal cortex. *Biochem. J.* **101,** 819–830.

Rajaniemi, H., Tuohimaa, P., and Niemi, M. (1970). Enzymic radioiodination of pituitary gonadotrophins for radioautography. *Histochemie* **23,** 342–348.

Redshaw, M. R., and Nicholls, T. J. (1971). Oestrogen biosynthesis by ovarian tissue of the South African clawed toad, *Xenopus laevis* Daudin. *Gen. Comp. Endocrinol.* **16,** 85–96.

Reichel, W. (1968). Lipofuscin pigment accumulation and distribution in five rat organs as a function of age. *J. Gerontol.* **23,** 145–153.

Reik, L., Petzold, G. L., Higgins, J. A., Greengard, P., and Barrnett, R. J. (1970). Hormone-sensitive adenyl cyclase: cytochemical localization in rat liver. *Science* **168,** 382–384.

Reinke, F. (1896). Beiträge zur Histologie des Menschen. I. Über Krystalloidbildungen in den interstitiellen Zellen des menschlichen Hodens. *Arch. Mikrosk. Anat.* **47,** 34–44.

Revel, J. P. (1964). Electron microscopy of glycogen. *J. Histochem. Cytochem.* **12,** 104–114.

Rosenbaum, R. M. (1964). The intracellular localization of enzymes. *In* "Handbuch der Histochemie" (W. Graumann and K. Neumann, eds.), Vol. 7, Part 4, pp. 1–108. Fischer, Stuttgart.

Rosengren, E., and Sjöberg, N. O. (1967). The adrenergic nerve supply to the female reproductive tract of the cat. *Am. J. Anat.* **121,** 271–283.

Rossi, F. (1964). Histochemie der Enzyme bei der Entwicklung. *In* "Handbuch der Histochemie" (W. Graumann and K. Neumann, eds.), Vol. 7, Part 4, pp. 109–282. Fischer, Stuttgart.

Rubin, B. L., and Deane, H. W. (1965). Effects of superovulation on ovarian Δ^5-3β-hydroxysteroid dehydrogenase activity in rats of different ages. *Endocrinology* **76,** 382–389.

Rubin, B. L., Deane, H. W., Hamilton, J. A., and Driks, E. C. (1963a). Changes in Δ^5-3β-hydroxysteroid dehydrogenase activity in the ovaries of maturing rats. *Endocrinology* **72,** 924–930.

Rubin, B. L., Deane, H. W., and Hamilton, J. A. (1963b). Biochemical and histochemical studies on Δ^5-3β-hydroxysteroid dehydrogenase activity in the adrenal glands and ovaries of diverse mammals. *Endocrinology* **73,** 748–763.

Rubin, B. L., Hamilton, J. A., Karlson, T. J., and Tufaro, R. I. (1965a). Effect of the age of the rat at the time of hemicastration on the weight, estimated secretory activity and Δ^5-3β-hydroxysteroid dehydrogenase activity of the remaining ovary. *Endocrinology* **77,** 909–916.

Rubin, B. L., Hilliard, J., Hayward, J. N., and Deane, H. W. (1965b). Acute effects of gonadotrophic hormones on rat and rabbit ovarian Δ^5-3β-hydroxysteroid dehydrogenase activities. *Steroids, Suppl.* **1,** 121–130.

Rubin, B. L., Deane, H. W., and Balogh, K., Jr. (1969). Ovarian steroid biosynthesis and Δ^5-3β- and 20α-hydroxysteroid dehydrogenase activities. *Trans. N.Y. Acad. Sci.* [2] **31,** 787–802.

Rutenburg, A. M., Kim, H., Fischbein, J. W., Hanker, J. S., Wasserkrug, H. L., and Seligman, A. M. (1969). Histochemical and ultrastructural demonstration of γ-glutamyl transpeptidase activity. *J. Histochem. Cytochem.* **17,** 517–526.

Ryan, K. J., Petro, Z., and Kaiser, J. (1968). Steroid formation by isolated and recombined ovarian granulosa and thecal cells. *J. Clin. Endocrinol. Metab.* **28,** 355–358.

Sachs, H. (1968). Quantitative histochemical studies of the granulosa lutein cells of the human ovaries. *Arch. Gynaekol.* **205,** 105–109.

Sandberg, E. C., and Jenkins, R. C. (1966). Aryl and steroid sulfatase in the human ovary. *Biochim. Biophys. Acta* **113,** 190–193.

Sato, K. (1969). Histochemical studies on N-acetyl-β-galactosaminidase in the ovary of sexually mature women. *Tohoku J. Exp. Med.* **98,** 75–79.

Savard, K., Marsh, J. M., and Howell, D. S. (1963). Progesterone biosynthesis in luteal tissue: role of nicotinamide adenine denucleotide phosphate and NADP-linked dehydrogenases. *Endocrinology* **73,** 554–563.

Savard, K., Marsh, J. M., and Rice, B. F. (1965). Gonadotropins and ovarian steroidogenesis. *Recent Prog. Horm. Res.* **21,** 285–365.

Schlegel, R. J., Farías, E., Russo, N. C., Moore, J. R., and Gardner, L. I. (1967). Structural changes in the fetal gonads and gonaducts during maturation of an enzyme, steroid 3β-ol-dehydrogenase, in the gonads, adrenal cortex and placenta of fetal rats. *Endocrinology* **81,** 565–572.

Schultka, R., and Schmidt, R. (1969). Histochemische Untersuchungen an Tube und Ovar weisser Ratten ante et post partum. *Acta Histochem.* **34,** 360–374.

Scott, J. E., Quintarelli, G., and Dellovo, M. C. (1964). The chemical and histochemical properties of Alcian blue. I. The mechanism of Alcian blue staining. *Histochemie* **4,** 73–85.

Scott, J. S., Lumsden, C. E., and Levell, M. J. (1967). Ovarian endocrine activity in association with hormonally inactive neoplasia. *Am. J. Obstet. Gynecol.* **97,** 161–170.

Scully, R. E., and Cohen, R. B. (1964). Oxidative-enzyme activity in normal and pathologic human ovaries. *Obstet. Gynecol.* **24,** 667–681.

Seligman, A. M., Ueno, H., Morizono, Y., Wasserkrug, H. L., Katzoff, L., and Hanker, J. S. (1967). Electron microscopic demonstration of dehydrogenase activity with a new osmiophilic ditetrazolium salt (TC-NBT). *J. Histochem.* **15,** 1–13.

Seth, P., and Prasad, M. R. N. (1967). Seasonal changes in the histochemical localization of Δ^5-3β-hydroxysteroid dehydrogenase in the ovary of the five-striped Indian palm squirrel, *Funambulus pennanti* (Wroughton). *Gen. Comp. Endocrinol.* **9,** 383–390.

Short, R. V. (1964). Ovarian steroid synthesis and secretion *in vivo. Recent Prog. Horm. Res.* **20,** 303–340.

Spector, W. G. (1961). Suppression of fertility in rats by an inhibitor of monoamine oxidase. *J. Reprod. Fertil.* **2,** 362–368.

Speroff, L., and Ramwell, P. W. (1970). Prostaglandin stimulation of *in vitro* progesterone synthesis. *J. Clin. Endocrinol. Metab.* **30,** 345–350.

Spicer, S. S. (1960). A correlative study of the histochemical properties of rodent acid mucopolysaccharides. *J. Histochem. Cytochem.* **8,** 18–33.

Stansfield, D. A., and Flint, A. P. (1967). The entry of ascorbic acid into the corpus luteum *in vivo* and *in vitro* and the effect of luteinizing hormone. *J. Endocrinol.* **39,** 27–35.

Sternberg, W. H. (1949). The morphology, androgenic function, hyperplasia, and tumors of the human ovarian hilus cells. *Am. J. Pathol.* **25,** 493–511.

Stoward, P. J. (1963). Cited in Pearse (1968, p. 347).

Stoward, P. J., and Adams Smith, W. N. (1964). The histochemical demonstration of ketosteroids. *J. Endocrinol.* **30,** 273–274.

Stumpf, W. E. (1969). Nuclear concentration of ^{3}H-estradiol in target tissues. Dry-mount autoradiography of vagina, oviduct, ovary, testis, mammary tumor, liver and adrenal. *Endocrinology* **85,** 31–37.

Stumpf, W. E., and Roth, L. J. (1966). High resolution autoradiography with dry mounted freeze-dried frozen sections. Comparative study of six methods using two diffusible compounds, ^{3}H-estradiol and ^{3}-mesobilirubinogen. *J. Histochem. Cytochem.* **14,** 274–287.

Takamatsu, H. (1939). Histologische und biochemische Studien über die Phosphatase. *Trans. Soc. Pathol. Jpn.* **29,** 492–498.

Takazawa, K., Ito, T., and Matsumoto, S. (1964). Histochemical study on aging of ovary. *J. Jpn. Obstet. Gynecol.* **11,** 105–124.

Taki, I., Hamanaka, N., and Mori, M. (1966). Histochemical observations of enzymatic patterns in human ovaries. *Am. J. Obstet. Gynecol.* **96,** 388–399.

Taki, I., Iijima, H., Uetsuki, M., Hamanaka, N., Morishita, M., and Mori, M. (1967). Histochemical studies of steroid 3β-ol dehydrogenases of the human ovaries. *Am. J. Obstet. Gynecol.* **98,** 107–115.

Takikawa, H., and Matsuzawa, T. (1967). Simplified procedure for the histochemical demonstration of dehydrogenase activity in rat ovaries. *Endocrinol. Jpn.* **14,** 276–278.

Taylor, F. B., Jr. (1961). Histochemical changes in the ovaries of normal and experimentally treated rats. *Acta Endocrinol.* (*Copenhagen*) **36,** 361–374.

Teilum, G. (1971). "Special Tumors of Ovary and Testis and Related Extragonadal Lesions. Comparative Pathology and Histological Identification." Munksgaard, Copenhagen.
Thompson, S. W. (1966). "Selected Histochemical and Histopathological Methods." Thomas, Springfield, Illinois.
Tóth, F. (1964). The effect of synthetic progestogens on the ovaries. *Int. J. Fertil.* **9,** 151–153.
Tóth, F., and Szönyi, I. (1966). Histochemical study of rat organs treated with norsteroid and progesterone, with special reference to hydroxysteroid dehydrogenases. *Acta Morphol. Acad. Sci. Hung.* **14,** 119–135.
Tóth, F., Horn, B., and Csömör, S. (1970). Histochemische und ultrastrukturelle Untersuchungen beim Stein-Leventhal-Syndrom. *Geburtshilfe Gynaekol.* **173,** 103–116.
Turolla, E., and Magrini, U. (1963). Attivitá 3Beta-idrossisteroide deidrogenasica e glucoso-6-fosfato deidrogenasica nell'ovaia di ratta normale. *Folia Endocrinol.* **16,** 464–483.
Turolla, E., Magrini, U., and Gaetani, M. (1966). Histochemistry of ovarian 20α-hydroxysteroid dehydrogenase in the rat during the estrous cycle. *Experientia* **22,** 675–676.
Turolla, E., Magrini, U., Gaetani, M., and Arcari, G. (1967). The effect of steroid treatment on ovarian dehydrogenases in the rat. Histochemical study. *Experientia* **23,** 909–912.
Turolla, E., Gaetani, M., Baldratti, G., and Aguggini, G. (1968). Histochemistry of ovarian 20α-hydroxysteroid dehydrogenase in mature hypophysectomized rats. *Experientia* **24,** 345–347.
Turolla, E., Baldratti, G., Scrascia, E., and Ricevuti, G. (1969). Effect of ergocornine on the luteal 20α-hydroxysteroid dehydrogenase in pseudopregnant and pregnant rats. *Experientia* **25,** 415–416.
Unsicker, K. (1970a). Zur Innervation der interstitiellen Drüse im Ovar der Maus (*Mus musculus* L.). Eine fluoreszenz- und elektronenmikroskopische Studie. *Z. Zellforsch. Mikrosk. Anat.* **109,** 46–54.
Unsicker, K. (1970b). Uber den Feinbau der Hiluszwischenzellen im Ovar des Schweins (*Sus scrofa* L.). Mit Bemerkungen zur Frage ihrer Innervation. *Z. Zellforsch. Mikrosk. Anat.* **109,** 495–516.
Vacek, Z. (1967). Ultrastructure and enzyme histochemistry of the corpus luteum graviditatis and its correlation to the decidual transformation of the endometrium. *Folia Morphol.* (*Prague*) **15,** 375–383.
Vanha-Perttula, T. P., and Hopsu, V. K. (1966). The ovarial enzymes hydrolysing L-leucyl- and DL-alanyl-β-naphthylamide. *Histochemie* **6,** 34–45.
van Lennep, E. W., and Madden, L. M. (1965). Electron microscopic observations on the involution of the human corpus luteum of menstruation. *Z. Zellforsch. Mikrosk. Anat.* **66,** 365–380.
Varma, S. K. (1970). A comparative histochemical study of alkaline phosphatase in the vertebrate ovary. *Acta Morphol. Neerl.-Scand.* **7,** 281–291.
Varma, S. K., and Guraya, S. S. (1968). The localization and functional significance of alkaline phosphatase in the vertebrate ovary. *Experientia* **24,** 398–399.
Velardo, J. T., and Rosa, C. G. (1963). Female genital system. *In* "Handbuch der Histochemie" (W. Graumann and K. Neumann, eds.), Vol. 7, Part 3, pp. 1–168. Fischer, Stuttgart.

Vollrath, L. (1966). Über den Einfluss von Ergocornin auf den histochemischen Fermentnachweis von β-Hydroxysteroiddehydrogenase. *Endokrinologie* **49,** 252–253.

Wada, T., Mori, H., Tanaka, K., Takeyasu, K., and Okano, K. (1969). Activity of 3β-hydroxysteroid dehydrogenase in ovary of pregnant rabbit by biochemical and histochemical methods. *Folia Endocrinol. Jpn.* **45,** 139.

Wegmann, R., and Lageron, A. (1964). Variations enzymatiques au cours du cycle oestral et après traitement hormonal dans l'ovaire de ratte albinos. *Ann. Histochim.* **9,** 81–91.

Wiest, W. G., and Kidwell, W. R. (1969). The regulation of progesterone secretion by ovarian dehydrogenases. *In* "The Gonads" (K. W. McKerns, ed.), pp. 295–325. North-Holland Publ., Amsterdam.

Wiest, W. G., Kidwell, W. R., and Balogh, K., Jr. (1968). Progesterone catabolism in the rat ovary: a regulatory mechanism for progestational potency during pregnancy. *Endocrinology* **82,** 844–859.

Willighagen, R. G. J., and Thiery, M. (1968). Enzyme histochemistry of ovarian tumors. *Am. J. Obstet. Gynecol.* **100,** 393–404.

Wingate, A. L. (1970). A histochemical study of the hamster ovary. *Anat. Rec.* **166,** 399.

Wolf, P. L., Horwitz, J. P., Vazquez, J., Chua, J., Pak, M. S. Y., and Von der Muehll, E. (1967). The indigogenic reaction for histochemical demonstration of sulfatase. *Proc. Soc. Exp. Biol. Med.* **124,** 1207–1209.

Wolman, M. (1964). Histochemistry of lipids in pathology. *In* "Handbuch der Histochemie" (W. Graumann and K. Neumann, eds.), Vol. 5, Part 2. Fischer, Stuttgart.

Wolrab, F., and Cossel, L. (1964). Zur Problematik des elektronenmikro-skopischen Nachweises von Dehydrogenasen in der Zelle. *Z. Mikrosk.-Anat. Forsch.* **71,** 457–477.

Wolrab, F., and Fuchs, U. (1967). Formazan-Stabilität und -Reoxydation in Gewebsstrukturen unter dem Einfluss elektronenmikroskopischer Präparationstechnik. *Histochemie* **11,** 171–174.

Woodruff, J. D., Goldberg, B., and Jones, G. S. (1968). Enzymic histochemical reactions in two Krukenberg tumors associated with clinically different endocrine patterns. *Am. J. Obstet. Gynecol.* **100,** 405–417.

Woods, J. E., and Domm, L. V. (1966). A histochemical identification of the androgen-producing cells in the gonads of the domestic fowl and albino rat. *Gen. Comp. Endocrinol.* **7,** 559–570.

Xavier, F., Zuber-Vogeli, M., and le Quang Trong, Y. (1970). Recherches sur l'activité endocrine de l'ovaire de *Nectophrynoides occidentalis* Angel (Amphibien Anoure vivipare). I. Etude histochimique. *Gen. Comp. Endocrinol.* **15,** 425–431.

Yohro, T., and Horie, S. (1967). Subcellular distribution of P-450 in bovine corpus luteum. *J. Biochem.* (*Tokyo*) **61,** 515–517.

YoungLai, E. V., and Short, R. V. (1970). Pathways of steroid biosynthesis in the intact Graafian follicle of mares in oestrus. *J. Endocrinol.* **47,** 321–331.

Zarrow, M. X., and McClintock, J. A. (1966). Localization of ^{131}I-labelled antibody to relaxin. *J. Endocrinol.* **36,** 377–387.

Zarrow, M. X., and O'Connor, W. B. (1966). Localization of relaxin in the corpus luteum of the rabbit. *Proc. Soc. Exp. Biol. Med.* **121,** 612–614.

Zawistowska, H. (1965). Histochemical studies of immature ovaries of the white mouse. *Folia Histochem. Cytochem.* **3,** 301–326.
Zerbian, K. (1966). Über das histochemische Verhalten einiger Enzyme bei der Follikelatresie im Ovarium des Meerschweinchens. *Acta Histochem.* **23,** 303–321.
Zerbian, K. U., and Goslar, H. G. (1968). Das histochemische Verhalten der Azolesterasen sowie weiterer Oxydoreduktasen bei der Follikelatresie im Ovarium einiger Nager. *Histochemie* **13,** 45–56.

SUPPLEMENTARY REFERENCES

The following list is a selection of relevant reports and books published since this account was first compiled.

Section II

Dubach, U. C., and Schmidt, U., eds. (1971). Recent advances in quantitative histo- and cytochemistry. Methods and applications. *In* "Current Problems in Clinical Biochemistry" (H. Aebi, H. Mattenheimer, and F. W. Schmidt, eds.), Vol. 3. Huber, Bern.
Geyer, G. (1973). "Ultrahistochemie. Histochemische Arbeitsvorschriften für die Elektronenmikroskopie." Fischer, Jena.
Pearse, A. G. E. (1972). "Histochemistry. Theoretical and Applied," 3rd ed., Vol. 2. Livingstone, Edinburgh.

Section III

Arvy, L. (1971). Histoenzymology of the endocrine glands. *In* "International Series of Monographs in Pure and Applied Sciences" (P. Alexander and Z. M. Bacq, eds.), pp. 415–476. Pergamon, Oxford.
Chaffee, V. W. (1974). Localization of ovarian acetylcholinesterase and butyrylcholinesterase in the guinea pig during the reproductive cycle. *Am. J. Vet. Res.* **35,** 91–95.
Dott, H. M. (1973). Lysosomes and lysosomal enzymes in reproduction. *Adv. Reprod. Physiol.* **6,** 231–277.
Fischer, T. V., and Kahn, R. H. (1972). Histochemical studies of rat ovarian follicular cells in vitro. *In Vitro* **72,** 201–205.
Fuchs, A.-R., and Mok, E. (1974). Histochemical study of the effects of prostaglandins $F_{2\alpha}$ and E_2 on the corpus luteum of pregnant rats. *Biol. Reprod.* **10,** 24–38.
Guraya, S. S. (1971). Morphology, histochemistry, and biochemistry of human ovarian compartments and steroid hormone synthesis. *Physiol. Rev.* **51,** 785–807.

Guraya, S. S. (1972). Comparative studies on the histochemical features of ovarian compartments in the rat and golden hamster, with special reference to steroid hormone synthesis. *Acta Anat.* **82,** 284–304.

Hatakeyama, S., Kovacs, K., Yeghiayan, E., and Blascheck, J. A. (1971). Aniline-induced changes in the corpora lutea of rats. *Am. J. Obstet. Gynecol.* **109,** 469–476.

Holzinger, M. (1972). Das Verhalten der unspezifischen Esterasen in den Tertiärfollikeln des Ovars normaler und oestrogenbehandelter Ratten. *Acta Histochem.* **42,** 41–55.

Moskov, M., Takeva, Ts., and Konstantinov, N. (1971). Histochemische Untersuchungen über den Eierstock der weissen Ratte nach einseitiger Kastration. *Anat. Anz., Suppl.* **130,** 195–201.

Nicosia, S. V. (1972). Luteinization of rabbit preovulatory granulosa cells cultured in vitro in presence of follicular oocytes. II. Histochemical and electron microscopic observations. *Fertil. Steril.* **23,** 802–816.

Strauss, J. F., III, and Stambaugh, R. L. (1972). Ovarian dehydrogenase activities during pregnancy in the rabbit. *Proc. Soc. Exp. Biol. Med.* **140,** 1143–1147.

Varute, A. T. (1972). β-Glucuronidase in testes and ovaries of frogs (*Rana tigrina*): a correlation of observations on histochemistry and biochemistry. *Acta Histochem.* **42,** 77–86.

Zamorska, L. (1972). The activity of some intracellular enzymes in the *Corpus luteum spurium* of the domestic pig (*Sus scrofa domestica* L.). *Folia Histochem. Cytochem.* **10,** 293–312.

Section IVB

Bahr, J., Kao, L., and Nalbandov, A. V. (1974). The role of catecholamines and nerves in ovulation. *Biol. Reprod.* **10,** 273–290.

Burden, H. W. (1972). Adrenergic innervation in ovaries of the rat and guinea pig. *Am. J. Anat.* **133,** 455–462.

Fink, G., and Schofield, G. C. (1971). Experimental studies on the innervation of the ovary in cats. *J. Anat.* **109,** 115–126.

Leontiuk, L. A. (1973). The development of the adrenergic structures of the ovaries. *Acta Physiol. Pol.* **24,** 233–241.

Section IVD

Amsterdam, A., Koch, Y., Lieberman, M. E., and Lindner, H. R. (1975). Distribution of binding sites for human chorionic gonadotropin in the preovulatory follicle of the rat. *J. Cell Biol.* **67,** 894–900.

Rajaniemi, H., and Vanha-Perttula, T. (1972). Specific receptor for LH in the ovary: evidence by autoradiography and tissue fractionation. *Endocrinology* **90,** 1–9.

Section V

Brandau, H., and Brandau, L. (1971). Enzymologische Studien am Arrhenoblastom. *Acta Endocrinol.* (*Copenhagen*) **67,** 577–589.

Laffargue, P., Argemi, B., Laffargue, F., and Adechy-Benkoël, L. (1971). Gonadoblastome (Histo-enzymologie et structure fine). *Ann. Anat. Pathol.* **16,** 467–476.

Vallette, Cl., Leymarie, P., Castagnier, M., and Chamlian, A. (1971). Un cas de thécome de l'ovaire (hormonologie, structure fine, histochimie, incubation). *Ann. Endocrinol.* **32,** 608–616.

Section VI

Bara, G., and Anderson, W. A. (1973). Fine structural localization of 3β-hydroxysteroid dehydrogenase in rat corpus luteum. *Histochem. J.* **5,** 437–449.

Cajander, S., and Bjersing, L. (1975). Fine structural demonstration of acid phosphatase in rabbit germinal epithelium prior to induced ovulation. *Cell Tiss. Res.* **164,** 279–289.

Kessel, R. G., and Decker, R. S. (1972). Cytodifferentiation in the *Rana pipiens* oocyte. IV. Ultrastructural localization of thiamine pyrophosphatase and horseradish peroxidase. *Z. Zellforsch. Mikrosk. Anat.* **126,** 1–16.

Long, J. A. (1973). Corpus luteum of pregnancy in the rat—ultrastructural and cytochemical observations. *Biol. Reprod.* **8,** 87–99.

Paavola, L. G. (1974). Cytochemical localization of an acyltransferase in guinea pig corpus luteum during lutein cell differentiation. *J. Cell Biol.* **63,** 255a.

Section VIIB

Blaha, G. C., and Leavitt, W. W. (1974). Ovarian steroid dehydrogenase histochemistry and circulating progesterone in aged golden hamsters during the estrous cycle and pregnancy. *Biol. Reprod.* **11,** 153–161.

Friedrich, F., Breitenecker, G., Salzer, H., and Holzner, J. H. (1974). The progesterone content of the fluid and the activity of the steroid-3β-ol-dehydrogenase within the wall of the ovarian follicles. *Acta Endocrinol.* (*Copenhagen*) **76,** 343–352.

Hadjioloff, A. I., Bourneva, V., and Motta, P. (1973). Histochemical demonstration of Δ^5-3β-OHD activity in the granulosa cells of ovarian follicles of immature and mature mice correlated with some ultrastructural observations. *Z. Zellforsch. Mikrosk. Anat.* **136,** 215–228.

Section VIII

Abro, A., and Kvinnsland, S. (1974). Adenylate cyclase in an estradiol sensitive tissue: a cytochemical study. *Histochemistry* **42,** 333–344.

Böck, P. (1972). Peroxysomen im Ovar der Maus. *Z. Zellforsch. Mikrosk. Anat.* **133,** 131–140.

Christensen, A. K. (1971). Frozen thin sections of fresh tissue for electron microscopy, with a description of pancreas and liver. *J. Cell Biol.* **51,** 772–804.

Daoust, R., and Morais, R. (1972). The histochemical demonstration of nuclease activity with films of soluble ribonucleic acid and polyadenylic acid. *J. Histochem. Cytochem.* **20,** 350–357.

Lindner, H. R., Tsafriri, A., Lieberman, M. E., Zor, U., Koch, Y., Bauminger, S., and Barnea, A. (1974). Gonadotropin action on cultured Graafian follicles: induction of maturation division of the mammalian oocyte and differentiation of the luteal cell. *Recent Prog. Horm. Res.* **30,** 79–138.

8

*Natural and Experimental Modification of Ovarian Development**

Katy Haffen and E. Wolff

I. INTRODUCTION

The normal course of ovarian development, that is, sexual differentiation, is discussed in Volume I, Chapter 3. In the first edition the subject of modifications of ovarian development was considered by Wolff (1962a,b) in Chapters 13 and 14B and also in part in Chapter 15 (Raynaud, 1962). This revised chapter demonstrates the two main lines of research that have been followed to determine the factors controlling sexual differentiation of the gonads in vertebrates. Naturally occurring factors, such as aberrant karyotypes and discrepancies between the genetic constitution and endocrine physiology, are discussed in Section II. The effects of embryonic and adult hormones on the gonadal primordia are considered in Section III.

II. NATURAL MODIFICATIONS

This section is restricted to a summary, based on only a few examples, of the natural modifications of ovarian development that may be found in

* Dedicated Posthumously to Professor Louis Gallien.

birds and mammals. The two that most commonly occur are the cases of gonadal dysgenesis and intersexuality.

A. Gonadal Dysgenesis

This condition has been mainly studied in man (see Volume II, Chapter 5), in whom it has been shown to be linked with heteroploidy of the sex chromosomes. The origin of the nonmosaic anomalies is, in general, attributed to meiotic nondisjunctions (Polani, 1961; Stewart, 1961; Ford, 1963; Turpin and Lejeune, 1965). To explain the mosaic varieties, postzygotic mitotic nondisjunctions have been postulated (Ford, 1963; Turpin and Lejeune, 1965). A third possibility is mentioned below (see Section II,B,1,b,ii) when the problem of intersexes is discussed.

1. Turner's Syndrome

This syndrome, described by Turner in 1938, was first recorded by Morgagni in 1749. Details of the clinical symptoms are discussed in Volume II, Chapter 5. The subjects are phenotypically female and of short stature. The external genitalia remain infantile and secondary sex characters, such as pubic and axillary hair growth, do not appear. The gonads, which are in the ovarian position, are reduced to a sterile ovarian stroma beneath a thin tunica albuginea. Cytogenetic techniques have enabled Ford *et al.* (1959) to investigate the correlation between the gonosomal formula XO and this syndrome. Singh and Carr (1966) reported a detailed anatomical and histological study of human embryos from 35 days to 4 months of age showing the XO karyotype. At 35 to 43 days, the gonadal histology was similar to that of XX individuals of similar age, but it differed considerably in older fetuses (70 to 106 days) in which there were no germ cells but only an increase in connective tissue. The problem therefore arises of the mechanisms that affect the survival of the germ cells. Other gonosomal aberrations of number and of structure have been noted during investigations of the Turner's syndrome phenotype. Many of these examples correspond to the varieties XO/XX, XO/XXX, XO/XX/XXX, XO/XY, or XO/XYY which probably represent chromosomal mosaics (Turpin and Lejeune, 1965).

The XO animal equivalent has only been reported in the mouse (Welshon and Russell, 1959) in which it results in a normal, fertile female.

2. Klinefelter's Syndrome

Testicular dysgenesis occurs in man when the gonosomal formula is XXY, and the affected individuals are sterile males with testicular hypodevelopment. Other chromosome formulas have been noted, XXXY, XYY, XXYY without mosaicism and XY/XXY with mosaicism (Turpin and Lejeune, 1965).

B. Intersexuality

This may be defined as the presence, in a single individual belonging to a gonochoric species, of male and female sex characters or of sex characters intermediate between the male and the female. These intersexes are often designated as hermaphrodites, a more vague term, but one hallowed by usage. The difference between true hermaphrodites and pseudohermaphrodites is that in true hermaphroditism the individuals carry the gonads of both sexes. There may be an ovary on one side and a testis on the other (lateral hermaphroditism), or an ovotestis on one side and an ovary or a testis on the other (unilateral hermaphroditism), or an ovotestis on both sides (bilateral hermaphroditism). The secondary sex characters show characteristics of both sexes. The term pseudohermaphroditism is applied to individuals which have gonads of one sex, but the genital ducts and external morphology of the other sex, and they are termed male or female pseudohermaphrodites according to whether the gonads are testes or ovaries. Pseudohermaphroditism is not discussed further in the present chapter.

The origin of intersexes is linked either to genetic causes or to the intervention of epigenetic factors which affect the differentiation of the gonads and the other sex characters.

1. Intersexuality Linked to Genetic Causes

a. Hereditary Factors. Cases of hereditary gonadal intersexuality have been described in birds. Taber (1964) has initiated work on this subject following the original studies by Riddle and his colleagues (1945). Interspecific hybridization, which in mammals brings about sterility, may result in intersexuality in birds (see Taber, 1964), as in the case of female intersexes in the duck (Gomot *et al.,* 1966; Deray and Gomot, 1969; Ridgon, 1967). Gomot *et al.* (1966) have reported that females produced by crossing a Barbary drake and a Pekin duck layed eggs during the first 2 years and then became intersexes. Females from the reverse cross possessed a rudimentary ovary with macroscopically visible follicles but did not lay. According to these authors, this intersexuality is determined by modifications of the genome, occurring with age, which bring about a deviation of the hormonal secretions responsible for the morphogenetic phenomena resulting in these transformations. In a cross between a Barbary drake and a Khaki Campbell duck, signs of intersexuality were already visible in the embryo (Lutz-Ostertag and Gomot, 1960; Lutz-Ostertag, 1968, 1969). After the seventeenth day of incubation, histological examination of the gonads of female embryos showed a well-developed cortex, but a nonregressed medulla with testicular tubules. The right gland had a compact medulla surmounted by a cortex. These signs of intersexuality were found to different degrees in females between the stage of hatching and the age of 4 years but were less accentuated than in the embryo.

In mammals, there are known cases of intersexuality in which the anomalies are sometimes transmitted hereditarily without the genetic constitution being abnormal. In the pig, cow, and goat such anomalies have been attributed to a recessive autosomal gene (Koch, 1963; Soller *et al.*, 1969). In man, there exists the syndrome of the feminizing testis affecting certain individuals with the genetic constitution XY, in which the presence of characteristic testis is concomitant with female genital ducts and sex characters of the female type (see Polani, 1970).

b. Chromosomal Aberrations. i. *Birds.* Gynandromorphism is a phenomenon best known in, and characteristic of, the invertebrates. Under this term one may group all the anomalies of sexual determination which bring about a modification of the genetic sex in part of the body. Gynandromorphism is rare in the vertebrates, although it is known among the birds. All the accounts of this subject (see Taber, 1964, p. 293) suggest that the following pattern is typical: on the left side there is an ovary or an ovotestis, on the right side a testis, and, externally, female plumage on the left half of the body and male plumage on the right. In certain cases, the presence of ripe spermatozoa and oocytes has been reported.

ii. *Mammals.* As in the case of gonadal dysgenesis, use has been made of cytogenetics to correlate certain cases of gonadal intersexuality with gonosomal aberrations. This has been studied particularly in man, and agreement has been indicated between gonadal intersexuality and gonosomal aberrations of the mosaic types XO/XY, XX/XY, XX/XX, and XX/XXY/XXYYY (see Turpin and Lejeune, 1965). As an example we may cite an intersex with the particularly demonstrative karyotype XX/XY which has been described by Gartler *et al.* (1962). The subject, aged 2 years and 4 months, had on the right an ovotestis with seminiferous tubules and ovarian follicles, and on the left an apparently normal ovary. Studies on the karyotype carried out on tissue cultures of the two gonads showed that the right ovotestis was mainly XY for the testicular portion (ratio of the cells XX/XY: 2/12) and mainly XX for the ovarian portion (ratio of cells XX/XY: 9/2). In mammals, a correspondence between the abnormal chromosomal formula and gonadal intersexuality has been reported by Nes (1966) in the case of a mink (*Mustela vison*) which showed bilateral hermaphroditism characterized by the presence of two ovotestes composed of an ovarian zone occupied by follicles in different stages of development and by a testicular zone without spermatogenesis. Study of the karyotype revealed the presence of a diploid line 2A-XX and of a triploid line 3A-XX/XY. Two other cases, in cattle, have been described by Dunn *et al.* (1968, 1970). The first case was a diploid intersex of the type XX/XY, the second case a diploid-triploid of the type XX/XXY. These animals showed on one side a quasinormal ovary and on the other side an ovotestis composed of a cortical zone with small follicles and a testicular zone with seminiferous tubules and rare germ cells. Study of the karyotype carried out, among others, in the mitotic metaphases of the gonadal tissue revealed that in the XX/XY animal the testicular portion of the ovotestis had 80% XX cells and 20% XY cells. In the second intersex of the XX/XXY type, karyotype study showed that, in the left ovary, the ovarian zone of the ovotestis and the testicular zone of the ovotestis the ratios of XX to XXY cells were respectively: 88/11, 72/25, and 50/50.

To explain the origin of individuals with the XX/XY karyotype, a number of authors have suggested the intervention of a double fertilization combined with anomalies in meiosis (Gartler *et al.*, 1962; Zuelzer *et al.*, 1964; Josso *et al.*, 1965). In these cases, serological tests have shown that the two cell lines are genetically different and that these individuals are chimeras rather than mosaics. The terms "chromosomal mosaicism" and "chromosomal chimerism" are often used without discrimination by authors in referring to individuals in which cell lines of opposed genotype coexist. A distinction between these two terms has been proposed by Chu *et al.* (1964) and also by Tarkowski (1969). A "chromosomal mosaic" is an individual in which cell populations of different genotype result from mutational or postzygotic mitotic errors. A "chromosomal chimera" is an individual in which the cell populations of different genotype result from the incorporation in a single embryo of cells derived from two separate zygotes as in vascular chorionic anastomoses, double fertilization, and experimental fusion of zygotes. Double fertilization has also been invoked to explain the origin of spontaneous animal chimeras associated with gonadal intersexuality (Nes, 1966; Dunn *et al.*, 1968, 1970) or not associated with gonadal intersexuality in male tricolored cats XX/XY (Malouf *et al.*, 1967), 38 XX/57 XXY (Chu *et al.*, 1964), and in a female mouse XX/XX (Russell and Woodiel, 1966).

It should be noted, however, that cases of gonadal intersexuality not linked with gonosomal aberrations but of normal karyotype XX or XY are much more frequent in man (Turpin and Lejeune, 1965) and in animals (McFeely *et al.*, 1967). Polani (1970) has classified the cases of human hermaphroditism reported at random between 1899 and 1969. This author draws attention to the fact that out of 310 true intersexes the karyotype had been examined in 108 cases with the following results: eleven cases of XO/XY mosaics; six cases of XX/XY mosaics; five cases of XX/XXY mosaics; four cases not classified; 59 gonadal intersexes corresponding to karyotypes XX and 21 to karyotype XY.

Although several authors agree in explaining the origin of testicular tissue when a Y chromosome is present, others do not discard the possibility, in the absence of a Y chromosome, of a mosaic occurring unnoticed (Dunn *et al.*, 1970) or of a translocation of chromosome material containing the part determining masculinity onto the X chromosome (Ferguson-Smith, 1966). In fact, the mode of action of the Y chromosome has not yet been elucidated in mammals. One hypothesis has been advanced by Hamerton (1968) which reconciles the classical theory of genetic balance with the reality of the bipotentiality of the gonad at the start of its development; both sexes possess genes guiding ovarian and testicular development but the Y chromosome would carry an additional factor controlling the functioning of the genes for maleness.

2. Intersexuality Linked to Epigenetic Factors

The most famous example of a natural modification of ovarian development under the influence of epigenetic factors is provided by the "natural experiment" of the freemartin which was initially investigated by Tandler and Keller (1911), Keller and Tandler (1916), and Lillie (1916, 1917). These studies were later extended by Lillie's pupils Chapin (1917), Willier (1921), and Bissonette (1924). Stockbreeders have long known that if a cow gives birth to twins, one of which is male, the other often shows sexual abnormalities and is sterile. It was Keller and Tandler, followed by Lillie, who showed that these abnormalities only occur if the twins are of unlike sex. The male develops normally, whereas the female becomes an intersex. These authors also studied the developmental stages in great detail. The female twin of a pair of unlike sex becomes an intersex if vascular anastomoses develop between the placentas of the two embryos. Some substance, presumably hormonal in nature, crosses from the male to the female embryo and modifies its development. The female embryo has no reciprocal effect on the male twin. The figures reported by Lillie (1917) from a study of 92 pairs of twins are particularly impressive. In 29 pairs, both twins were male; in six pairs, one was male and the other a normal female; in 33 pairs, one twin was male and the other an intersex; and in the remaining 24 pairs, both were female. The results of this study are all the more convincing in that Lillie was able to demonstrate vascular connections between the twins in the 33 male intersex pairs, whereas no such connections could be found in the six pairs consisting of a normal male and a normal female. From the positive and negative findings, Lillie (1917) postulated hormonal modifications. By study of the sex chromatin, Moore *et al.* (1957) have shown that freemartins are genetically female.

The changes induced in the female embryo affect both the gonads and the genital tract, but the present discussion will be mainly concerned with the gonadal changes.

a. Changes in the Gonads. The changes induced in the gonads of the female by substance(s) derived from the male vary in severity. They may be classified into three main types (Fig. 1):

i. The gonads remain in the position of the ovaries but are small and atypical in appearance (Fig. 1b). Histologically, Pflüger's cords may sometimes be present, but only in a rudimentary form. In most cases, the germinal epithelium has degenerated without giving rise to cortical cords. In their place there is a thick tunica albuginea similar to that of the testis. Within the gonad, the medullary cords do not have the characteristic structure of the testis, they are sterile, and do not communicate with the rete testis via the straight tubules.

ii. The gonads are larger and inguinal in position (Fig. 1c) but still intraabdominal. The medullary cords have developed into seminiferous cords, with a lining of typical Sertoli cells, but they lack spermatogonia. They are arranged in regular fashion around the straight tubules which link them to the rete testis.

iii. The gonads are the same size as testes and have descended to the inguinal region (Fig. 1d). They lie between the abdominal muscles and the skin, and are surrounded by a dense tunica albuginea. The primary cords have formed typical seminiferous tubules. Germ cells are seldom present and seem to have disappeared after birth. The interstitial tissue is well developed. The seminiferous tubules are arranged around normal urogenital connections which communicate in the usual way with the epididymis. Thus, these are typical but sterile testes. In one particular freemartin studied by Hay (1950), a degree of masculinization was found which has seldom been equaled. The testes, which were 25 mm long at autopsy, were completely normal, and contained healthy spermatogonia, though fewer in number than in the male twin.

b. Changes in the Genital Tract. The related changes in the genital tract are equally of interest.

i. Corresponding to the first type of gonad, the female genital tract is more or less rudimentary, and vestiges of the male tract are found fairly well developed. For instance, remnants of the distal end of the Wolffian system may persist, though they are not organized into an epididymis; incomplete vasa deferentia and small seminal vesicles may occur but there is no prostate. The external genitalia are female in pattern.

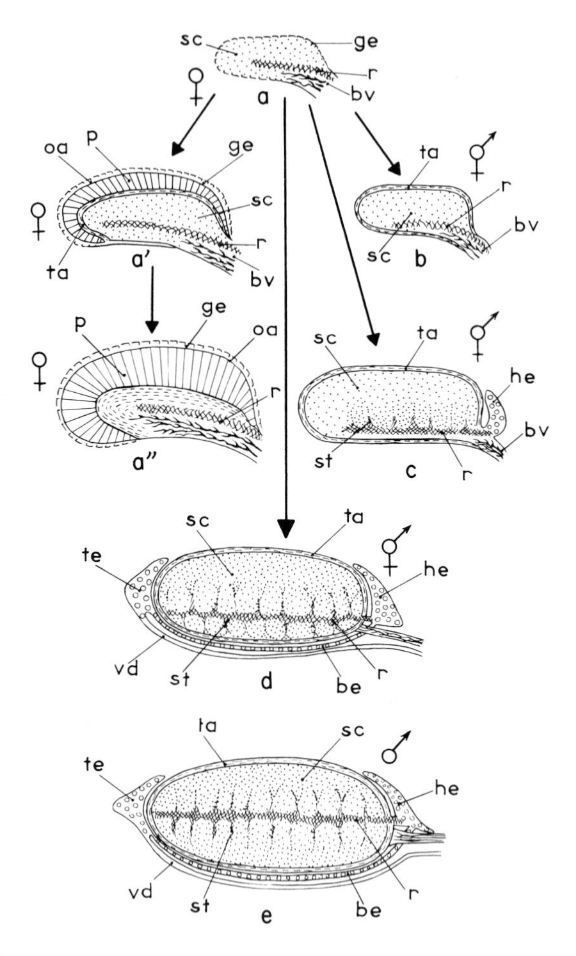

Fig. 1. Diagrammatic reconstructions to show a normal testis (e) and an "indifferent" ovary (a) which in normal conditions differentiates into a normal ovary (shown in the two stages a′ and a″), but under the influence of male sex hormone may form gonads exhibiting three stages of transformation in the male direction (b, c, and d; low, medium, and high degrees of transformation, respectively). bv, blood vessels; ge, germinal epithelium; he, te, be, head, tail, and body of the epididymis, respectively; oa, ovarian albuginea; p. Pflüger's cords; r, rete; sc, medullary cords; st, connections between the rete and the testicular cords; ta, tunica albuginea; vd, vas deferens. (Reproduced with permission from Willier, 1921.)

ii. Corresponding to the second type of gonad, the genital tract is more masculinized. The atrophied uterus persists in the form of two narrow canals, the vagina is rudimentary and the clitoris hypertrophied. Of the male components of the genital tract, the vasa deferentia are usually complete, the epididymis present, the seminal vesicles larger, the gubernaculum testis well developed, and the urogenital sinus intermediate in pattern between male and female.

iii. Corresponding to the third type of gonad, the genital tract is almost completely male. Nothing remains of the uterine horns or the vagina. The vasa deferentia are complete and the epididymis well developed. The urogenital sinus is intermediate in type, with seminal vesicles but seldom with a prostate or bulbourethral glands. The urethra is male and the clitoris is often penile. The external genitalia, however, are never characteristically male. In exceptional cases, the urethra runs through the penis, opening on its ventral surface as in an hypospadiac male (Buyse, 1936).

These findings show that there is a relationship between the degree of intersexuality of the gonads and the extent of the transformation in the genital tract, although the degree of intersexuality in the gonads is always more marked than in the genital tract. The ovaries may be completely changed into testes while the genital ducts, and in particular the external genitalia, are only partly masculinized. These differences are probably due to the fact that vascular anastomoses between the placentas do not always become established at the same stage. Or, of course, the anastomoses may differ in size. As a result, the concentration of the male hormone reached in the freemartin's blood is variable; it may even vary at different developmental stages. Finally, the target organs may vary in their sensitivity to the same dose of male hormone. While the differences between the responses of the gonads, genital tract, and external genitalia in freemartins could be tentatively accounted for on this basis, Witschi (1939) postulated a possible alternative hypothesis. He attributed the masculinization of the genital tract not to the direct action of the hormone derived from the male partner, but to hormones secreted by the gonads of the masculinized female. This possibility has also been suggested by Short *et al.* (1969) and the hypothesis would account for the less intense masculinization of the ducts and the external genitalia as compared with that of the gonads.

A cellular theory has been proposed by a number of authors (Booth *et al.*, 1957; Fechheimer *et al.*, 1963 Goodfellow *et al.*, 1965; Herschler and Fechheimer, 1967) who found a mixture of cells of XX and XY karyotype in the tissues, mainly blood, of cattle twins with vascular anastomoses. Thus, it has been suggested that male and female germ cells coexist in the gonads of the twins. Cells of XX karyotype have, in fact, been found in the testis of a male twin after birth (Ohno *et al.*, 1962; Teplitz *et al.*, 1967). The possibility of an exchange of germ cells between two embryo twins showing vascular anastomoses is plausible in view of the observations of Ohno and Gropp (1965) and of Jost and Prépin (1966) who found germ cells in the circulation of bovine fetuses. A similar situation has been described in the human fetus (Semenova-Tian-Shanskaya, 1969). These observations show that the migration of germ cells in the species studied is not exclusively by the interstitial route. However, it is not possible to attribute the masculinization or the sterility of freemartins to the presence of male germ cells in the developing gonad. We know, in fact, that germ cells play no part in the differentiation of somatic tissues. The problem of the survival and differentiation of germ cells of one sex in the somatic tissues of the other sex is discussed in Volume I, Chapter 3, Section III,2. Jost (1967, 1970) and Jost *et al.* (1972) suggested that the simultaneous inhibition of the presumptive ovaries in the freemartin and of the Müllerian ducts in the male cotwin and the freemartin could be produced by a testicular "antifeminine" substance or hormone inhibiting both the ovary and the Müllerian ducts. The development of male characters in males and in freemartins would be caused by the production of another testicular masculinizing hormone.

Whatever the true explanation, the hormonal theory of intersexuality which was first formulated was strongly supported by the "naturally occurring experiment" of freemartinism, and has provided the impetus for many fruitful researches undertaken on a variety of vertebrates, particularly on amphibians, birds, and mammals.

III. EXPERIMENTAL MODIFICATION OF OVARIAN DEVELOPMENT

Research involving transplantation and parabiosis was first undertaken on amphibians (Burns, 1925; Witschi, 1927; Humphrey, 1929a,b), while the first observations on the effect of hormone adminstration were made on birds (Dantschakoff, 1935; Willier *et al.*, 1935; Wolff and Ginglinger, 1935a,b). Since all the relevant experiments involve gonads which are as yet undifferentiated and similar in structure in the two sexes (see Volume I, Chapter 3), we cannot refer to effects on an ovary but only on a gland which is morphologically bisexual though potentially female (its subsequent development into an ovary being determined by genetic factors). It will be seen later that in the different groups of vertebrates it is sometimes the one sex and sometimes the other which is sensitive to hormones of the opposite sex.

A. Amphibians

1. Experiments on Parabiosis

a. Experiments on *Ambystoma*. The earliest work was carried out by Burns (1925, 1930) on the axolotl (*Ambystoma punctatum*). Two embryos at the tail-bud stage were united on their lateral surfaces. Similar experiments were performed and their range extended by Witschi (1927, 1937) who introduced many variations on this theme, linking together in a chain several embryos of the same or of different species of urodeles and anurans. One of Burns' series (1930) was performed on 57 pairs of parabionts; in 16 pairs both partners were male, in 29 pairs one was male and the other female, and in the remaining 12 pairs both were female. All the heterosexual pairs showed intersexuality. In the majority of cases, the male partner was normal but the female showed variable degrees of masculinization. Similar results were obtained on *A. tigrinum* by Humphrey (1936). The early stages of ovarian differentiation proceeded normally, but later the ovarian cortex underwent partial or complete degeneration. The ovary was reduced to a fine strand of gonadal tissue which was sometimes sterile and consisted only of the medullary primordium. These gonads were able to differentiate into typical testicular structures. They were frequently characterized by alternating sterile regions, nodules of active testicular tissue, and zones showing the structure of an ovotestis. Witschi (1936) made similar observations on parabiotic pairs of *A. tigrinum* and *A. mexicanum*.

The developmental pattern which occurs naturally in freemartins is, thus, closely reproduced by parabiotic axolotls. As in mammals, the hormone derived from the male partner is dominant, and induces both an inhibition of the ovarian cortex and a stimulation of the testicular medulla. In the most severely affected cases, the transformation into a male is complete. Experimental chimeras of *Pleurodeles waltlii* have been produced by Houillon (1964) who associated an anterior part with a posterior part. The adults then grow with two systems of genital glands (Houillon and Charlemagne, 1971). This technique, which bears a resemblance to a graft in parabiosis, has shown that in heterosexual chimeras the testicular development is normal, but that of the ovary is profoundly inhibited. While these observations indicate that the male hormone is dominant, Witschi's (1937) elegant experiments indicate that this may not always be the case. By varying the size and the age of the parabionts, Witschi detected that the opposite effect of feminization of the male gonads may occur, the dominance of one or other sex apparently being correlated with the quantity of hormone entering the blood of the partner. Thus, if a male of the dwarf species *A. jeffersonianum* was united with a female of the giant species *A. tigrinum*, the gonad of the male at first developed into a somewhat atypical ovary, but 2 years later became transformed into a large ovotestis. A similar result was obtained when an *A. tigrinum*, a large salamander characterized by rapid sex differentiation, was joined in parabiosis with an *A. maculatum*, a small species with slow sex differentiation. The ovary of the *A. tigrinum* prevented the development of the testicular medulla in the male gonad of the *A. maculatum*, which grew into small, atypical ovaries.

When the parabionts reached sexual maturity, this inhibition by the female partner was no longer complete. The medulla resumed its development into a testis, and, in some cases, ripe ova and spermatozoa could be seen co-existing in the same gonad.

b. Experiments on Other Amphibians. Witschi (1927, 1930, 1931) and Witschi and McCurdy (1929) carried out a great number of experiments on other species of urodeles, on anurans, and on combinations of various species of urodeles and anurans. Newts in parabiosis yielded results similar to those obtained with *Ambystoma*. The male partner had an inhibitory effect on the female, preventing ovarian development and inducing sterile gonads with a small central cavity and a medulla of variable size. This effect Witschi appropriately enough called "the freemartin effect."

In anurans, parabiosis between individuals of different sex produces results similar to those seen in urodeles. In the frogs, *Rana esculenta, R. sylvatica, R. aurora,* and *Hyla regilla,* the male partner was always dominant and induced transformation of the ovaries into typical testes. These experiments on parabiosis also indicated, however, that some new factor is involved in that the distance separating the genital regions was found to influence the degree of gonadal transformation. In parabiotic frogs, the female gonads were found to be more transformed the nearer they were to the gonads of the male partner. Thus, the right gonad was more severely masculinized than the left, which was further from the male genital regions. Within each gonad, some parts were more affected than others depending on the extent to which they came into the sphere of influence of the male's gonads; the effect could be represented schematically by a series of concentric rings of decreasing activity. In one case, the parabionts were firmly united, and had many viscera in common. Thus, the two adjacent kidneys had fused into a single one, and there was only one gut and one liver. The two gonads nearest each other were linked by a bridge; one was derived from the male partner, the other from the female. The latter gonad was completely masculinized. In contrast, the other female gonad, lying further from the male partner, contained female, intersexual, and male regions. This effect of distance suggested that hormones derived from the male did not travel via the blood vessels, as in freemartins and as seemed to be the case in urodeles, but by diffusion across the intervening tissues. This factor could thus account for the decreasing effect at progressively greater distances.

Huchon (1971) extended the parabiotic union of differently sexed embryos to six anuran species characterized by direct differentiation of their gonads into ovaries and testes, before or at the time of metamorphosis. An intersex effect of the gonads of the genetically female parabiotic twin was observed in *Discoglossus pictus* and *Pelodytes punctatus* but not in *Bombina variegata, Pelobates cultripes, Alytes obstetricans,* and *Limnodynates tasmaniensis*.

2. Transplantation Experiments

Humphrey (1929a,b, 1931a,b, 1933) performed a brilliant series of transplantation experiments between amphibian embryos. Working on embryos of *Ambystoma tigrinum* or *Rana sylvatica* at the tail-bud stage (Harrison's stages 24–32), he removed the somatic primordium of one gonad, corresponding to the intermediate mesoderm of the part of the trunk which lies between the eighth and seventeenth somites (Fig. 2). The donor was allowed to survive, and its remaining gonad served to indicate its genetic sex. The transplanted primordium was grafted into the site of one of the gonadal primordia of a host which, in 50% of cases, proved to be of the opposite sex. With a female donor and a male host, the transplant started its differentiation in the female pattern but then underwent more or less total inversion. The cortex was inhibited and the central cavity of the ovary disappeared, while the medulla became hypertrophied and developed into an ovotestis or an atypical testis. On reaching sexual maturity, these testes or ovotestes contained nodules packed with mature spermatozoa. The complementary experiment yielded similar results. The graft from a male donor prevented differentiation of the host's ovary. This developed into a gonad reduced in

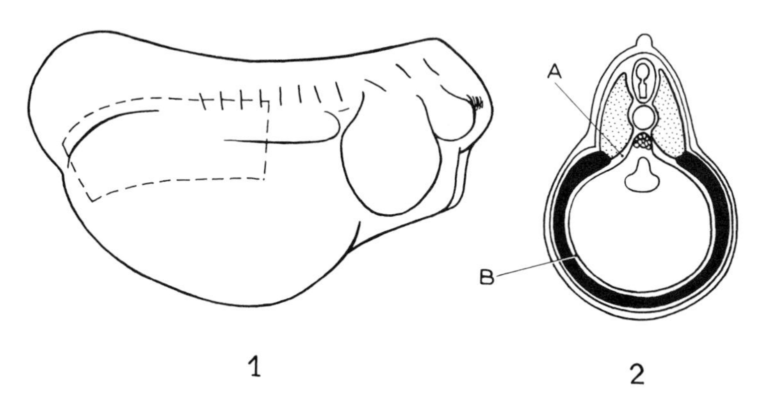

Fig. 2. Diagrams of *Rana sylvatica* embryos of the age used for operation (early tail-bud stage). (1) Lateral aspect; the broken lines indicate the extent of the area of ectoderm and mesoderm removed (or replaced by an orthotopic transplant). (2) Diagrammatic transection to illustrate the relationships of the primordial germ cells to the mesoderm extirpated (or implanted) at operation. The lines A and B represent the longitudinal cuts by which the mesoderm and overlying ectoderm are removed. (Reproduced with permission from Humphrey, 1933.)

size and without a cortex. Testicular structures tended to appear in the hilar region. The cavity of the ovary was lost and solid strands arising from the hilum divided within the substance of the gonad. Lobules of testicular tissue developed between these strands.

This change was reversible if the testis which was inducing the atrophy and sex reversal was removed early enough. Oogonia lying within the germinal epithelium resumed cell division, and formed an ovarian cortex around a central cavity which arose by coalescence of the smaller cavities of the strands of the rete testis. This resumption of normal ovarian development was no longer possible if changes in the gonad had gone as far as the formation of lobules of testicular tissue. In this event, the masculinized gonad evolved into a testis as large as or larger than the normal. The development of such a testis seemed to be stimulated by the removal of the grafted testis at a stage when all oogonia had disappeared from the germinal epithelium.

While the dominance of the male gonad over that of the female is the general rule, it is not invariable. In special circumstances, feminization of the male gonads may be induced, as for example if the male graft in a female host is considerably smaller than a normal testis, or if a male gonad is transplanted in a female of another species whose development proceeds particularly rapidly, as, for example, when a primordial testis of *Ambystoma punctatum* is implanted into the position of the corresponding gonadal primordium of a female *A. jeffersonianum* or *A. mexicanum* (Humphrey, 1935a,b).

Experiments performed since 1942 on the induction of sex reversal by means of transplantation have yielded particularly interesting results. Humphrey (1942a,b, 1945) successfully raised to sexual maturity genetically female axolotls (*A. mexicanum*) in which one gonad had been changed into a testis by a transplanted male gonad either from the same species or from *A. tigrinum*. An exploratory laparotomy in one of these animals 9 months after transplantation showed that the host's gonad had undergone considerable, though incomplete, masculinization. After the grafted testis had remained in the donor for a sufficient time, it was removed. The host female, transformed into a functional male, repeatedly produced ripe sperm. Two and 3 years after the original transplantation, it was mated with normal females. In this way two genetic females were mated with each other. The sex ratio in the F_1 generation of such a mating established the genetic formula of the two sexes in the axolotl. If females were homogametic (XX), such a mating should produce nothing but females. In fact, the proportions found were approximately 25% males and 75% females. This indicates that the male is homogametic (ZZ) and the female heterogametic (WZ). It also shows that in this particular cross (WZ $\times$ ZW), females are produced which are homogametic (WW) as well as heterogametic (WZ), the proportions being: 25% males (ZZ), 50% females (WZ), and 25% females (WW). Humphrey verified this hypothesis by crossing F_1 females with normal males. Some of these matings gave only female offspring (WW $\times$ ZZ $\rightarrow$ WZ). These same homogametic females when mated with sex-reversed females (ZW) gave 50% normal females, and 50% homogametic females (WW). The latter were also characterized by the fact that, on mating with normal males, their offspring were only females (WZ). This observation confirms that in the axolotl the female genotype is WZ and the male ZZ.

This conclusion is confirmed by Humphrey's (1957) experiment on axolotls, in which he transplanted minute fragments of a primordial testis into one gonad of a female host. In the majority of cases, the females' gonads were masculinized by the graft. Occasionally, however, the testis became incorporated into the female gonad. If the graft was taken from a pigmented strain, the ova derived from it were subsequently recognizable from the pigment they contained, and were found to have developed from male gonocytes. Fertilization by spermatozoa from normal males produced offspring that were exclusively male, thus, confirming that it is the male axolotl that is homogametic. The heterogamety of the female sex has also been demonstrated in *Xenopus laevis* by experimental transplantation of male gonads (Mikamo and Witschi, 1963, 1964).

Experimental sex reversal was also carried out on young female embryos of the genotype WW (Humphrey, 1948) using the same technique of transplanting gonadal primordia. As with normal females, the ovarian primordium of these homogametic females proved capable of developing into a testis. This shows that the Z chromosome is not necessary for the formation of typical testicular structures in amphibians.

Experiments on transplantation and parabiosis in batrachians confirm that hormones derived from the gonads pass from one partner to the other, either via the circulation or by diffusion through the tissues. The hormone produced by the male gonads is generally dominant and induces either intersexuality or sex reversal in the gonads of the female partner. If the gonads of the male are small or slow to develop, they may be influenced by the gonads of the female. It would thus appear that dominance depends on the relative quantities of hormones secreted by one or the other gonad. The fact that the male gonad is largely medullary and the female largely cortical in structure has led Witschi to postulate (see Volume I, Chapter 3, Section V) an antagonism between a medullary organizer (which induces male differentiation) and a cortical organizer (responsible for female differentiation). Experiments on transplantation and parabiosis upset the dominance of one or other of these sex factors by inhibiting development of one component of the gonad. It appears that the medullary organizer derived from the male partner is sufficiently powerful to prevent, even at some distance, the development of the female cortex.

3. Polyploidy and Sexual Transformations of the Ovary

Polyploidy has been observed in amphibians in the natural state. The most usual chromosomal aberration is represented by triploidy. Fankhauser and Humphrey (1942) have shown that cold treatment of axolotl eggs produces a certain percentage of polyploid individuals.

Triploidy in *Triturus viridescens* (Fankhauser, 1940) and in *T. alpestris* (Fischberg, 1945) involves the development of rudimentary ovaries. In the female axolotl triploids (apparently of genotype ZZW, ZWW and WWW) obtained by Humphrey and Fankhauser (1946), the development of the ovaries differed little from that of diploid individuals during the beginning of development, but after 3 months the majority of the oocytes degenerated. This partial degeneration was not due to a disequilibrium of the mechanism of genetic determination but was the result of an abnormal synapsis of the chromosomes due to the polyploid state. A small number of oocytes may reach sexual maturity and are capable of being fertilized. Thus, the crossing of partly fertile female triploids with male diploids has produced aneuploid, tetraploid, and sometimes pentaploid offspring (Fankhauser and Humphrey, 1950; Humphrey and Fankhauser, 1956). Similar results have been obtained in *Pleurodeles* by Gallien and Beetschen (1959).

Gallien and Beetschen (1960) have also obtained haploid individuals by applying a thermal shock to the eggs of *Pleurodeles*. These individuals are lethal androhaploids, but it is possible to make them develop by associating them with diploid individuals. When associated with a diploid male, the haploid individual may reach adult age and its gonads develop into testes, but spermatogenesis is interrupted at the stage of primary spermatocytes. When a haploid male is associated with a diploid female the gonads of the two parabionts most often develop in accordance with their genetic sex. However, feminization of the testis does take place in certain cases and the feminized male gonad remains infantile. These experiments show once more that in the urodeles, in which the male sex is dominant (see above), the ovary may, under precise experimental conditions, exert a feminizing action on the differentiation of the male partner (Gallien, 1967). In *Rana micromaculata,* the individuals obtained by thermal shock are gynohaploids; Miyada (1960) has obtained a few male individuals, one juvenile hermaphrodite, and a majority of females with ovaries poorly developed or in involution. This author suggests that the effect of the recessive genes acting in these haploid individuals finds expression in an inhibition of cortical development.

Other cases of intersexuality or of sex change have been obtained by application of thermal shock to amphibian eggs. Thus, in female triploid frogs the ovarian cortex does not develop normally. This results in a complete transformation of the ovaries into testes before metamorphosis in *Rana pipiens* (Humphrey *et al.*, 1950) and in *R. japonica* (Kawamura and Tokunaga, 1952). The same sexual transformations have been observed in female triploid axolotls, *Siredon mexicanum* (Brunst, 1970).

4. The Administration of Sex Hormones

Research into intersexes produced by the effects of sex hormones has resulted in an extraordinary mass of results since 1936. The results, while complex, have shown that amphibians are particularly sensitive to the effects of sex hormones.

a. The Masculinizing Effect of Androgens. Particularly clear-cut results have been obtained with the ranid anurans. Complete masculinization of females has been produced in *Rana temporaria* by Gallien (1937, 1938a,b, 1944), in *R. sylvatica* by Mintz and Witschi (1946) and Mintz (1948), in *R. pipiens* by Witschi and Crown (1937) and Foote (1938), in *R. catesbeiana* by Puckett (1940), in *R. clamitans* by Mintz *et al.* (1945), in *R. agilis* by Vannini (1941), in *R. dalmatina* by Padoa (1947) and in *R. japonica* by Kawamura and Yokota (1959). In this last case, the mating of the sex-reversed female with normal females resulted in 100% female offspring and gave proof of the XX genetic constitution of the female.

b. The Feminizing Effects of Estrogens. Many workers have induced sex reversal of male gonads by the administration of estrogens (estrone, estradiol or their esters). Complete feminization occurs in urodeles: *Ambystoma punctatum* (Burns, 1938); *A. tigrinum* (Ackart and Leavy, 1939); *A. opacum* (Foote, 1940); *Hynobius retardatus* (Hanaoka, 1941a,b); *Pleurodeles waltlii* (Gallien, 1950b,c, 1954); *Triturus helveticus* (Gallien and Collenot, 1960). Similar observations have been reported for the anuran *Xenopus laevis* (Witschi and Allison, 1950; Gallien, 1953, 1956). Some transformed males have been raised to sexual maturity and their eggs fertilized by spermatozoa derived from normal males. The crossing of two males has been brought about in the urodele *Pleurodeles waltlii* (Gallien, 1954) and the anuran *Xenopus laevis* (Gallien, 1955, 1956; Chang and Witschi, 1955). In both instances the transformed males, when mated with normal females, gave rise to exclusively male offspring. These findings not only reaffirm Humphrey's (1945) conclusion that the urodele male is homogametic (ZZ) but also show that the same is true for certain anurans, e.g., *Xenopus* (see also Mikamo and Witschi, 1963, 1964). A very interesting case has been reported by Gallien (1961, 1962, 1965, 1967) for *Pleurodeles* in which a genetic male (ZZ) changed into a functional female with all offspring 100% males (ZZ) can develop functional testes after several years. This observation illustrates the strength of the genetic sex.

Sex reversal induced by the administration of estrogens has been shown to be unstable or temporary in *Rana temporaria* (Gallien, 1938b) and *Pelobates cultripes* (Gallien and Collenot, 1958; Collenot, 1965), and is only partial in *Pelodytes punctatus* (Gallien and Collenot, 1958; Collenot, 1965), *Discoglossus pictus* (Gallien, 1950a) and *Triturus alpestris* (Gallien and Collenot, 1960). The sex-converting efficiency of the sex steroids (androgens and estrogens) is a function therefore of the genetic strength of the individual.

5. Paradoxical Effects of Sex Hormones

Complex results have sometimes been obtained. These may conveniently be grouped together under the heading "paradoxical," because a hormone, instead of producing the expected effect, gives rise to the opposite one.

An example of this effect is provided by Padoa's (1936, 1942) experiments in which sex hormones were administered to amphibians. Padoa reared tadpoles of *Rana esculenta* (belonging to a differentiated strain) in aqueous solutions of estrone and observed that at metamorphosis all the treated larvae possessed testes. A concentration of 60 μg/liter produced 100% females and at 120 μg/liter, intersexes resulted. At concentrations of 250 to 1000 μg/liter all the animals were males, but at higher doses (2000 μg/liter) Gallien (1941) found that estradiol masculinized females although injection of low doses (20 to 80 μg) induced feminization of males. These results demonstrate that a female hormone, believed to be typical of that sex in mammals, can produce an hermaphroditic effect in at least two species of frog, the effect being feminizing or masculinizing, depending on the concentration of hormone.

Several authors, in particular Foote (1940) working on *Ambystoma* and Gallien (1950c) on *Pleurodeles waltlii*, reported that the male hormone, testosterone, induces paradoxical effects. This hormone, though considered one of the most characteristic androgens, is capable of causing complete feminization and this is its only effect in *P. waltlii*, at whatever dose it is given. The paradoxical effect of this hormone seems to have a different explanation from the feminization produced in the same species by estrogens. While female hormones result in stimulation of the cortex and a secondary regression of the medulla, testosterone acts by inhibiting not only the medulla, but the entire region of the mesonephros from which the medulla is derived. The cortex shows signs of partial inhibition, and the fragments of cortex that survive resume development when treatment is discontinued. The gonad is thus feminized by the elimination of the medulla, the male hormone acting by selective destruction.

6. Effects of Antiandrogens and Antiestrogens

Tadpoles of *Rana esculenta* were subjected to treatment for 2.5 months with antiandrogenic or antiestrogenic substances (Chieffi *et al.*, 1974). The antiandrogenic substances (cyproterone and cyproterone acetate) had a strong masculinizing effect in the course of sex differentiation of the gonads. The antiestrogenic substance (I.C.I. 46,474) did not produce any notable effects. Since these antihormones compete with the sex hormones at the target sites in the adult frog, the authors concluded that embryonic sex inductors are not similar to sex hormones.

B. Birds

The effects of sex hormones on the embryonic gonads of birds were studied before experimental work on transplantation and parabiosis was undertaken.

1. Administration of Sex Hormones

In 1935, three groups of workers, Dantschakoff in Lithunia, Willier, Gallagher and Koch in the United States, and Wolff and Ginglinger in France, injected sex hormones into avian embryos.

a. Estrogens. In Wolff and Ginglinger's (1935a,b) study, estradiol benzoate was injected into the allantois of a 5-day embryo. Following the administration of 2–100 μg/embryo, increasing degrees of feminization of male gonads were found. Ovotestes, in which the testicular medulla had partly regressed and the cortex showed various degrees of development, were obtained. Between the two parts of the gonad a zone of lacunae appeared, resembling the spaces seen in the medulla of an ovary (Fig. 3). The most highly feminized gonads were indistinguishable from ovaries. While the gonads showed these intersexual characters, the Müllerian ducts persisted for a variable distance (intersexes of types I to IV; see Wolff and Ginglinger's classification and Fig. 4). The feminization of male gonads was only temporary. Even if sex reversal appeared complete at hatching, the gonad which had been transformed into an ovary reverted to a testis several months later. This change was due to a fresh proliferation of tubules from the germinal epithelium and to regression of the ovarian cortex (Wolff, 1936).

The causes of this regression of the induced ovarian cortex have been studied by Wolff and Haffen (1961). These authors have shown that when the same fragment of intersex male gonad was grafted into several successive hosts, the development of the cortex stopped at a stage close to that

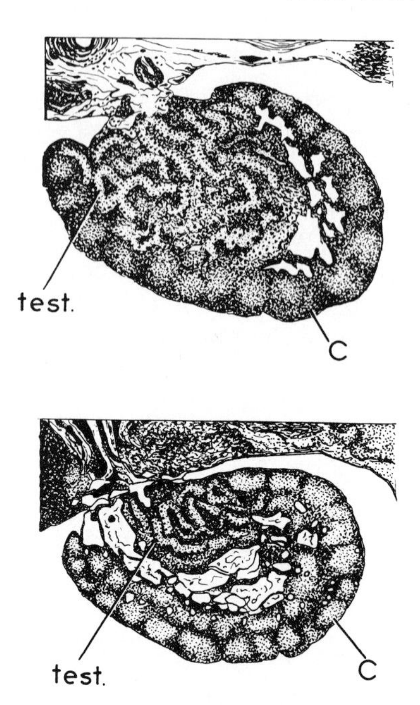

Fig. 3. The left gonads of bird embryos feminized by the administration of female hormones. Two degrees of feminization, illustrating the development of the ovarian cortex, C, and the regression of the testicular medulla, test. Between these two regions medullary lacunae have become increasingly obvious. (From Wolff and Ginglinger, 1935a.)

of hatching. The "oogonia" of the males degenerated after entering premeiosis. However, the feminized structure of the medulla is maintained throughout the whole duration of the experiment and exerts a hormonal action identical to that of a normal female medulla. The substitution of the intersex medulla by a female medulla does not prevent the degeneration of the male "oogonia" (Haffen, 1963, 1969a). When male germ cells are experimentally introduced into a female gonad primordium, Haffen (1968, 1969a) has shown that the male germ cells themselves are involved in the arrest of female differentiation (see Volume I, Chapter 3, Section III,1).

Several estrogens (estrone, estradiol, diethylstilbestrol) have been shown to exert the same effect on male gonads treated before sex differentiation (Wolff, 1939; Wolff and Wolff, 1948b). Lutz-Ostertag (1964) has obtained very well-grown feminizations of embryonic male gonads of quail treated with stilbestrol. Haffen (1965, 1969b), who has studied the development of these intersexes from hatching until 3 months later, has shown that the male "oocytes" may even evolve into large follicles. They are, however,

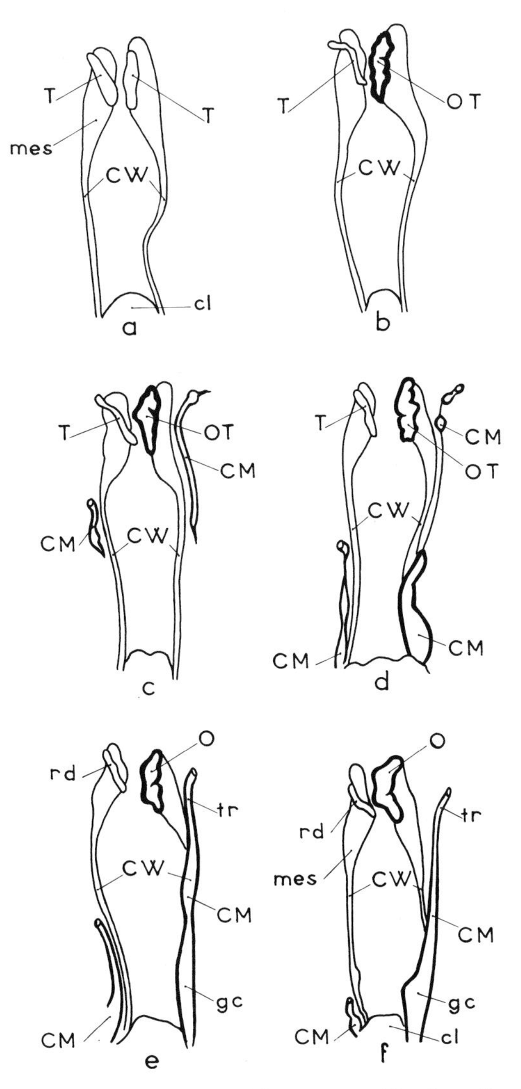

Fig. 4. Diagram of the different types of intersex resulting from the administration of different doses of estrogens to male chick embryos: (a) the normal male; (b), (c), (d), and (e) intersexes types, I, II, III, and IV, respectively; (f) the normal female; cl, cloaca; CM, portions of the Mullerian ducts; CW, Wolffian ducts; gc, shell gland; mes, mesonephros; O, ovary; OT, ovotestis; rd, rudimentary right gonad; T, testis; tr, tube. (From Wolff and Ginglinger, 1935a.)

eliminated by the testicular growth which partly arises in the cortical zone (Fig. 5) and partly proceeds from the medulla (Fig. 6). The testicular tubules of the medulla invade the cortical zone and eventually grow around the follicles by a process similar to that described by Wolff (1936) in the chick. The Sertoli epithelium of the testis tubule comes into direct contact with the follicular epithelium and, thus, the cytoplasm of the oocyte is surrounded by a mixed layer of follicular and Sertoli cells. The further development of the Sertoli epithelium depresses that of the follicular cells which become disrupted (Fig. 6). The cortical zone, deprived of follicles and separated from the medulla by a thick tunica albuginea, may contain tubules of neoformations (Fig. 5). These may be derived from a new proliferation of germinal epithelium, or result from transformation of follicular into Sertoli cells as suggested by Wolff (1936).

b. Androgens. Many different androgenic hormones have been administered to chick embryos. They may be divided into those that have a simple masculinizing action and those which produce both masculinizing and feminizing effects. The masculinizing effect is seen particularly on the

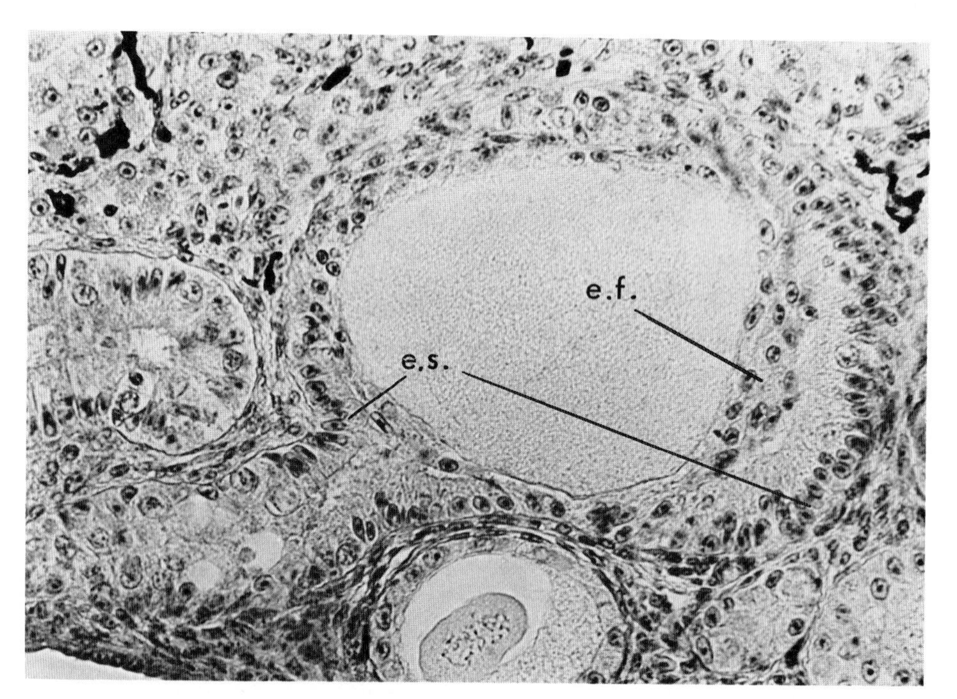

Fig. 5. Detail showing a testicular tubule connected with a follicle in the ovotestis of the quail, 6 weeks after hatching. The development of the Sertoli epithelium (e.s.) finally depresses the follicular cells (e.f.) which become disrupted. (From Haffen, 1965.)

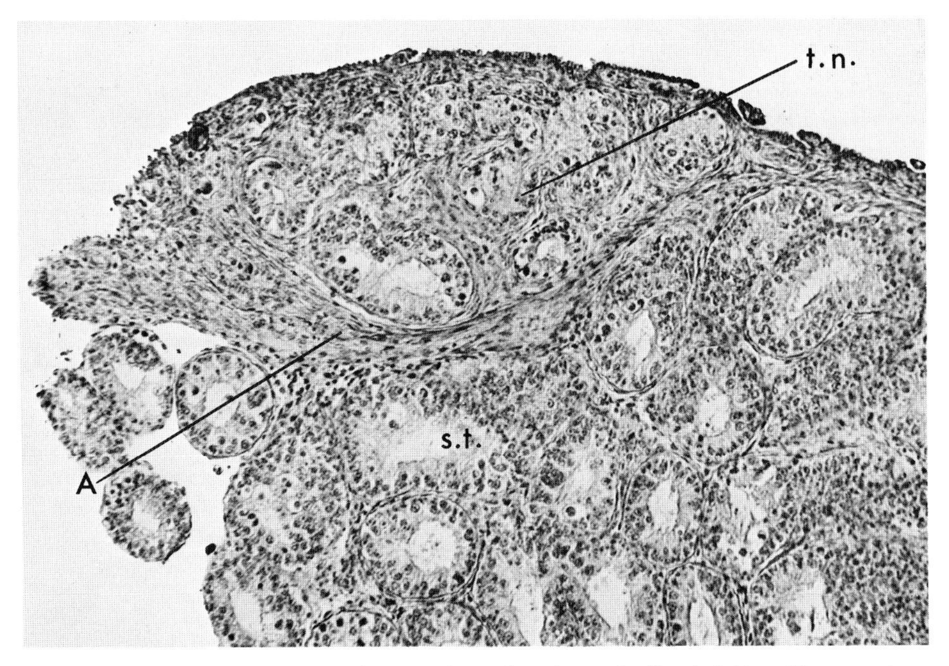

Fig. 6. The development of a transformed male quail after hatching. The gonad reverts to its genetic sex. The medulla contains seminiferous tubules derived from the first proliferation of sex cords. A new proliferation of testicular tubules has invaded the periphery of the cortex. t.n., newly formed tubules; s.t., seminiferous tubules; A, tunica albuginea.

genital duct, while the feminization affects the gonads as well (Wolff *et al.*, 1948). Testosterone and its esters come into the first group. Fairly high doses cause the Müllerian ducts of female embryos to regress (Wolff *et al.*, 1948) or they prevent their development (Stoll, 1948). The hormones have very little effect on the gonads of the female. There is a transient tendency for testicular tubules to form in the hilar region of the medulla, thus, causing retardation in medullary regression rather than stimulation. Androstanediol has effects similar to testosterone but is less toxic and its threshold is lower. The question arises whether these androgens have the same properties as the natural hormones produced by the embryonic male gonad. It is shown in this chapter that they do. The hormone produced by the male chick embryo only acts on the genital ducts and in no way influences the gonads of the female embryo.

A large number of substances have a double action in that they masculinize the genital tract of female embryos and feminize the genital tract of male embryos. The Müllerian ducts, which are stimulated at their cranial end in males, and inhibited at their caudal end in females, come to look alike in the two sexes. The only effect these substances have on the gonads, however, is feminizing. Androsterone, androstanedione, androstenedione, androstenediol, and transhydroandrosterone have the same effect on male gonads as estrogens. (Wolff *et al.*, 1948). Recently, Weniger and Zeis (1973a) showed that after dihydrotestosterone administration, the Mullerian ducts do not regress in the female and are maintained in the male; the left testis is feminized. Induction of female hormone secretion by dihydrotestosterone has been demonstrated (Weniger and Zeis, 1973b, 1974).

c. The Effect of Estrogens on the Right Gonad. Normally, the right gonad does not develop a cortex in normal females and in feminized males because the germinal epithelium over these gonads regresses very early. Even before the fifth day of incubation, it becomes a low undifferentiated epithelium and the right gonad is reduced to a medullary component alone. It follows the pattern of development of the left medulla, forming a testis in the male and remaining rudimentary in the female.

If, on the other hand, estrogen acts on the germinal epithelium before it regresses, the latter gives rise to a cortex (which may not be as large as the normal left cortex) in both females and feminized males (Wolff and Wolff, 1948a; Wolff and Pinot, 1961). This result can be obtained by injecting a highly water-soluble estrogen, such as doisynolic acid, into an embryo at a very early stage (between 2½ and 5 days) of development.

2. The Effect of Substances Other Than Steroid Hormones

a. Chemicals. Salzgeber (1957) has shown that some toxic chemicals, acting on gonads in tissue culture, can selectively inhibit one or the other component of the cortex or medulla. Trypaflavine, for instance, inhibits the medulla both of the testis and of the ovary, changing its cords into lacunae. Colchicine and narcotine destroy the cortex. The destruction of one component of the gonad is never followed by a stimulation of the other and, therefore, sex reversal is not induced. Lutz-Ostertag and Meiniel (1968a,b) studied the action of a pesticide, parathion (*o,o*-diethyl-*o,p*-dinitrophenyl thiophosphate) on 1- to 8-day cultures of gonads from quail embryos (7 to 16 days old) and from chick embryos (9 to 18 days of incubation). An aqueous suspension of parathion in the culture medium caused selective destruction of the gonocytes and the Sertoli cells in testes from quail embryos. In chick testes, the Sertoli cells were not destroyed, only the germ cells were affected. In female gonads, parathion destroyed the cortex, but the medulla was unaffected (see also Lutz-Ostertag and Lutz, 1974).

Melphalan (*p*-dichloroethylamino-L-phenylalanine) is one of the group of alkylating agents used in cancer chemotherapy. Simpson (1969) compared the effect of this substance on 10-day chick embryonic gonads and on fragments of human cancers. After culturing the gonads for 7 days, and depending on the dose, there was an inhibition of growth, a reduction in the number of mitotic figures, and a marked effect on the germ cells. Melphalan destroyed the germ cells but had no effect on the somatic cells at doses that are lethal for tumors. Counts carried out on the surviving germ cells showed that oogonia were more sensitive than spermatogonia.

b. Action of Gonadotropic Hormones. Narbaitz and Adler (1966) cultured male chick gonads of 8 to 10 days' incubation with 8- to 13-day-old ovaries in both the presence and absence of FSH and LH. The gonadotropins were found to have no effect on differentiation of germ cells in the gonads of either sex. The oogonia of the ovarian cortex entered meiotic prophase after 7, 6, or 4 days, depending on whether the ovary had been removed at 8, 10, or 13 days of incubation. Preda *et al.* (1966) cultured the left gonads from 7- to 7½-day-old chick embryos for 7 days in the presence of 10 to 20 I.U. of Coripan (a Rumanian preparation of chorionic gonadotropic hormone). All twenty gonads developed into typical ovaries. The author concluded from this that the chorionic gonadotropin exercized a feminizing effect in culture which is capable of reversing the sexual differentiation of male gonads; the same hormone injected into embryos only gives rise to a condition of intersexuality or testicular agenesis. It is interesting to note that pituitary gonad-stimulating hormones administered to amphibian or reptilian embryos caused various degrees of feminization in male individuals (see Section III, D,4). Manelli and Milano-Crassi (1966) found that when gonads from 6-day-old chick embryos were cultured in contact with hypophyses from embryos of 17 to 18 days' incubation, organ growth and multiplication of germ cells were stimulated.

3. Transplantation Experiments

Minoura (1921) first tried to produce experimental "freemartins" using chick embryos. He grafted fragments of adult testes or ovaries onto the chorioallantoic membrane but the experiments were unsuccessful. Bradley (1941) used coelomic grafts, but his results were not clear cut and did not demonstrate sex reversal. Wolff (1946), using the same technique, was able to graft ovarian or testicular fragments from donor embryos aged 6 to 10 days into embryos aged 72 hours.

a. Ovarian Grafts into Male Embryos. Ovarian grafts turned the male hosts into intersexes by transforming the left gonad to a varying extent. In the most marked cases of intersexuality, some regions of the left gonad were ovarian in structure, while others resembled an ovotestis (Fig. 7). The right gonad was usually slender and atrophied. When intersexuality was less

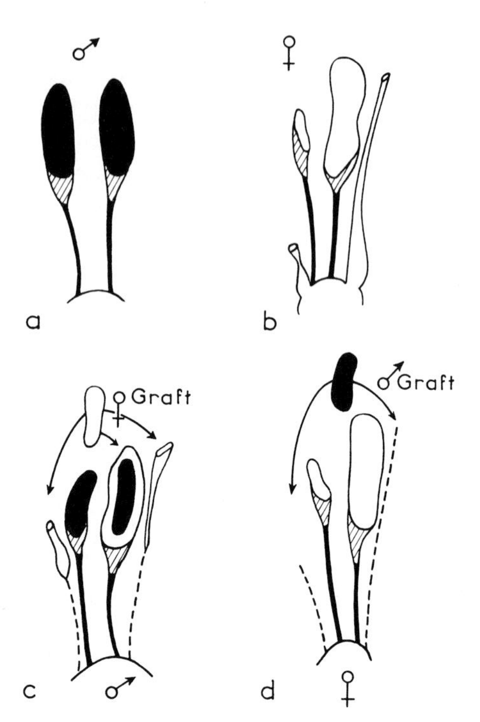

Fig. 7. Diagram showing the results of gonadal grafts to the coelom of chick embryos. (a) normal male; (b) normal female; (c) the effect of an ovarian graft on sex differentiation in a male direction; formation of a left ovotestis and persistence of the cranial segments of the Müllerian ducts; (d) the effect of a testicular graft on the genital tract in a female embryo; regression of the Müllerian ducts. (From Wolff, 1962.)

marked, the left gonad was surrounded by a thin and sometimes fragmentary cortex. The least severe transformations were characterized by testicular tubules which opened abnormally into the coelomic cavity in continuity with the germinal epithelium. These tubules resulted from a late proliferation of the germinal epithelium, like those obtained subsequently by Hampe (1950) with injections of estrogens into embryos at late developmental stages. The most severe feminization obtained by this technique matched the moderate degrees of intersexuality produced by the administration of sex hormones to young embryos. This was also seen in the Müllerian ducts which rarely persisted, and then only at their cranial ends (Fig. 7). The effect of ovarian transplants thus corresponds to that of moderate or low doses of estrogens. These findings were confirmed by Hujbers (1951) who grafted ovaries from chick embryos onto the chorioallantoic membrane. Several ovaries must be grafted onto one embryo to produce a slight degree of intersexuality by this method.

b. Testicular Grafts into Female Embryos. These have no effect whatever on sex differentiation of the ovary. On the contrary, they prevent the Müllerian ducts from developing (Fig. 7). These effects are comparable to those of crystalline male sex hormones such as testosterone.

c. Grafts of the Right Gonad and the Left Medulla. Wolff (1946) showed that an intracoelomic graft of a right gonad into a male embryo exerted the same feminizing effect as did a grafted left gonad. Mintz and Wolff (1954) showed that fragments of ovarian medulla, stripped of cortex, caused a high degree of intersexuality in male gonads. These results indicate that the ovarian medulla produces a female hormone, and that it can, before it atrophies, evoke feminization in male gonads (see also Volume I, Chapter 3, Section IV,B,1 also).

4. Experiments on Parabiosis

The present discussion is divided into a consideration of (1) gonads in parabiosis under tissue culture conditions, and (2) parabiosis of whole embryos.

a. Gonads in Tissue Culture. Wolff and Haffen (1951, 1952a,b,c) perfected a tissue culture technique which allowed gonads taken from duck embryos before the onset of sex differentiation to grow and develop; this has made it possible to culture such gonads in intimate association with each other. The genetic sex of the parabiotic gonads was found by culturing the contralateral gonads of the same embryos in isolation. When two gonads of opposite sex develop in association with each other, the female one grows into a typical ovary while the male one is severely feminized. It develops an ovarian cortex around it, while the medulla remains in an undifferentiated state (Wolff and Haffen, 1952b). The same result is obtained if a male gonad and a right female gonad, containing only medullary tissue, grow together (Wolff and Haffen, 1952c). Feminization of the male gonad also results if it is in parabiosis with the medulla of a left gonad from which the cortex has been removed by tryptic dissociation (Wolff and Haffen, 1959). These experiments confirm that female gonads secrete a feminizing hormone, and that the medulla is probably the tissue responsible (see also Volume I, Chapter 3, Section IV,B,1).

b. Embryos in Parabiosis. Experimental parabiosis between whole chick embryos at very early stages of development has been achieved *in vitro* (Wolff and Simon, 1952). The experiment has not yet been continued, however, beyond the sixth day of incubation, a stage preceding the onset of sex differentiation. However, Lutz and Lutz-Ostertag (1958, 1959) and Ruch (1962, 1968) have observed sex differentiation in spontaneously occurring double-yolked eggs which contain two embryos. Vascular connections arise between the membranes of the two embryos at late stages of development. For this reason, the degree of intersexuality induced is rarely high, but it appears that the gonads of the male embryo may be slightly feminized. The gonads of the female embryo are not affected, but the caudal part of the oviduct is inhibited by hormones produced by the male. These findings agree with some results of hormone administration, of embryonic grafts, and of parabiosis *in vitro.*

5. Castration Experiments

Several authors have investigated the fate of the rudimentary right gonad of the female after removal of the left ovary. Some have worked with adult fowl (Zawadowski, 1926; Domm, 1927; Benoit, 1932; Caridroit and Pezard, 1935), others with very young females (Finlay, 1925; Benoit, 1932). The right gonad in females of the Gallinaceae, as mentioned, is usually reduced to its medullary component. It resumes development after the removal of the left gonad, the cortex of which has a powerful inhibitory effect on the medulla. The same phenomenon occurs on the left side if, in spite of the operation, a few fragments of medulla are preserved; the latter give rise to regenerating testicular tissue. The transformation starts with an increase of interstitial cells; the Leydig cells hypertrophy and divide actively. Next, the medullary cords, reduced to lacunae, become transformed into seminiferous tubules by a thickening of their walls. If gonocytes have not completely disappeared, as for instance in birds castrated while very young, they resume development and give rise to the entire cell series associated with spermatogenesis, even to mature spermatozoa. One may ask why the female germ cells can, under these conditions, accomplish the male gametogenesis, when in the experiments of Haffen (1968, 1969a) the male gonocytes induced to develop precociously in an ovary degenerated. This question has been discussed in Volume I, Chapter 3, Section III,1. The endocrine function of such testes is normal, as indicated by the appearance of male secondary sexual characters, such as the comb, crowing, and the fighting instinct shown by these masculinized hens. The vasa deferentia, which were rudimentary ducts before operation, are also reactivated but they no longer open into the cloaca. Therefore, coitus between these transformed females and normal hens is always infertile.

It is thus evident that the rudimentary medulla of both female gonads is capable of spontaneous evolution into a testis able to secrete male hormones when the inhibitory influence of the ovarian cortex is removed. However, in most cases, secondary female sex characters reappear after a variable time, and the plumage, the oviduct, and the comb return to their original state. It is possible that in some instances this return to normality is associated with the regeneration of cortical tissue. At all events, it indicates that the gonad is once again producing female hormones. Thus, the interstitial tissue of the medulla shows a fluctuating endocrine activity. First, it secretes feminizing hormones in the embryo, and second, it seems to be endowed with the ability to produce spontaneously a male hormone in ovariectomized females. At the same time it cannot be ruled out that it may resume secretion of female sex hormones some months after operation.

Experimental ablation of the left ovary has prompted a number of authors to consider the question of how the left ovary inhibits the development of the right gonad. Research has shown that daily injections of physiological doses of sex hormones (estrogens and androgens) inhibit completely the development of the medullary component of the right gonad of fowl. Simultaneous injection of gonadotropins does not abolish the effect of the steroids (Taber and Salley, 1954; Kornfeld and Nalbandov, 1954; Kornfeld, 1958; Taber *et al.*, 1958). These experiments suggest that it is the estrogens produced by the left ovary that inhibit the development of the right gonad. However, Gardner and Taber (1963) have suggested that nonsteroidal substances elaborated by the left ovary may be influential. Their hypothesis is based on the fact that injection of ovarian extracts lacking steroids or of ovarian grafts from which the estrogen secretion has been inactivated by the liver, more or less completely inhibits the development of the right gonad in the hen.

C. Mammals

Frequent attempts have been made to produce experimental intersexes in mammals that are similar to the naturally occurring freemartins. In general, these experiments have failed to produce changes anywhere but in the genital tract; the case of the young opossum seems to be an exception (Burns, 1950, 1955, 1961). However, parabiosis can only be achieved between embryos at the more advanced stages of development. Administration of hormones has also failed to modify the gonads in eutherian mammals. The problem is whether the hormones, injected either into the mother or directly into the embryo, have been given early enough. At very early stages hormones are toxic and induce abortion. Nevertheless, it does seem that in some experiments they have been injected at a stage when sex differentiation has hardly begun. It is possible that hormones administered to the mother do not cross the placenta soon enough to reach the fetal gonads at the desired time, but the problem is a puzzling one. The administration of hormones induces profound changes in genital ducts and accessory glands, but exerts almost no effect at all on the gonads of higher mammals. Before studying experimental sex reversal in marsupials and the scanty evidence for sex reversal that exists in rodents, the changes in the genital tract in response to sex hormones will be briefly reviewed (see Volume II, Chapter 2).

1. Changes in the Genital Tract

The following summary of the effect of sex hormones on the genital tract and accessory glands of developing mammals is based on the work of Dantschakoff (1936), Greene and Ivy (1937), Raynaud (1937, 1942), Greene *et al.* (1938), Burns (1939a,b, 1942), Jost (1947, 1965, 1967, 1970), and Jost *et al.* (1963).

a. Effects of Male Hormones. The male genital ducts persist in the female embryo after the administration of various androgens. The epididymis and vas deferens may thus appear in the female. At the same time, the urogenital sinus follows the male pattern in that the accessory glands (prostatic buds, seminal vesicles, and coagulating glands) are stimulated. The external genitalia become masculinized but the Müllerian ducts are not suppressed though they often show atrophy at their caudal ends, or fail to unite. The ovary is not affected; if it does undergo any change, this does not involve sex differentiation of the germinal cords.

b. Effects of Female Hormones. The administration of female hormones to male embryos (Greene *et al.*, 1938; Greene, 1942; Raynaud, 1942) results in the persistence of the Müllerian ducts, feminization of the urogenital sinus, and more or less complete atrophy of the male genital ducts and accessory glands. In no case has any degree of intersexuality of the gonads been obtained in experiments on rodents.

Sex hormones can induce the same changes in the genital tract, although often to a greater extent, as are evoked by the natural hormone of the bovine freemartin; but clear-cut changes in the gonads are rare. Since masculinization of the female's gonads is the most frequent and most severe change found in the natural freemartin, it must be questioned whether endogenous hormones differ in their properties from the crystalline hormones (Jost, 1947, 1965, 1967, 1970) or whether experimental conditions, and in particular the method of administration used in the work on rodents, are inadequate.

c. Effects of Antiandrogenic Substances. The studies of Hamada *et al.* (1963), Elger (1966), Jost (1967), and Neumann *et al.* (1970) have shown that antiandrogenic compounds injected into pregnant animals essentially suppressed the masculinizing effects of the fetal testis but did not prevent the differentiation of testis from the undifferentiated gonadal primordium.

2. Experimental Production of Multiple Pregnancies

Jost (1967, 1970), Jost *et al.* (1972), and Vigier *et al.* (1972) reported experiments in which multiple pregnancies were produced in the cow by treatment with gonadotropins. Chromosomal chimerism (XX/XY) was studied in the liver of presumptive freemartin fetuses and male co-twins. The degree of chimerism varied from 0% to 55%, but there was no correlation between the percentage of XY cells and the degree of inhibition of the ovaries or of the Müllerian ducts. The gonads of presumptive freemartin fetuses were studied histologically 39 to 62 days after mating and normal development of ovaries occurred up to day 48. Between days 48 and 52, Müllerian duct inhibition was obvious and ovarian growth ceased. The germinal epithelium was degenerating by day 60. No testicular structures were observed in any of the presumptive freemartin ovaries examined up to day 62 and the authors suggested that seminiferous tubules must develop at approximately 100 days *postcoitum*. Rajakowski and Hafez (1964) also mentioned that freemartin effects in the ovary were noticeable at 60 days in fetuses from multiple pregnancies.

3. Intersexuality in the Gonads of the Opossum

Burns' (1950) experiments on *Didelphis virginiana* provide the only example of a major transformation in the gonads of a mammal, and it is remarkable that these results were obtained in animals treated after birth. Since the young of marsupials are born at a stage that corresponds to a fetal stage in eutherian mammals, hormones may be administered to them directly, before sex differentiation of the gonads is complete. Young opossums are born after a short gestation period of $12\frac{1}{2}$ to 13 days. Burns injected them daily from birth onward with 0.15 to 2 μg estradiol dipropionate for 10 to 30 days. In his first series of experiments, he observed striking changes in the gonads of the male. The germinal epithelium persisted and gave rise to broad but sterile cords which were separated from the testicular medulla by a wide tunica albuginea. In this way a sterile cortex of varying thickness was built up, while the testicular tissue was smaller in size than in a normal testis. The whole structure resembled an ovotestis (Fig. 8). In a subsequent series of experiments (1955) Burns applied the same technique to opossums that had been born a little earlier than usual, after a gestation of 12 days. In these conditions, the gonads of male embryos were almost completely transformed into ovaries. A thick cortex containing oocytes, some of which were enclosed in follicles, developed under the germinal epithelium. The number of oocytes was nevertheless smaller than that in a normal ovary. The medulla was very

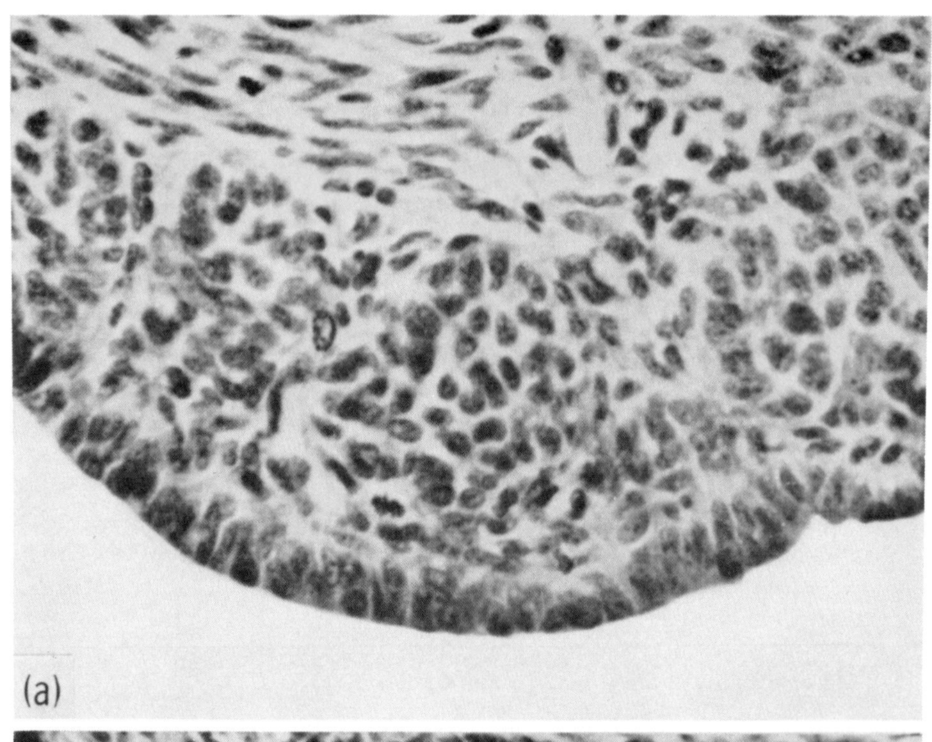
(a)

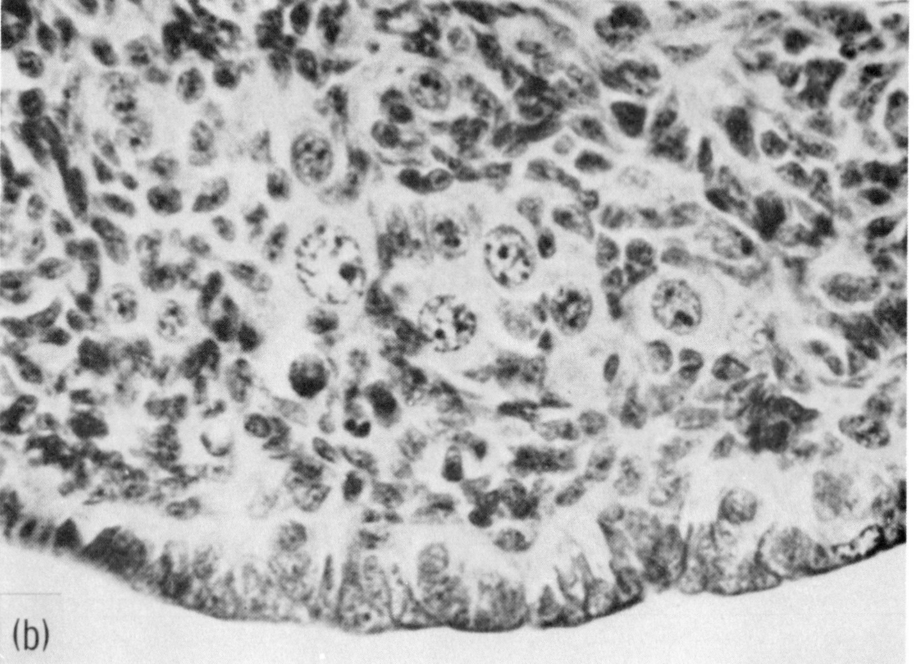
(b)

small, and only few tubules reminiscent of the testicular pattern persisted. More recent results (Burns, 1961) indicate that testes may be transformed into ovaries even more completely, and that the medulla of these gonads contains practically no testicular tubules.

These results indicate that the timely administration of a female hormone to male embryos may stimulate the development of the germinal epithelium at the very moment when the latter is beginning to disappear. Treatment at an earlier stage may cause the gonocytes to persist and divide. The regression of the germinal epithelium usually coincides with the time of birth but if administration of hormones can be accomplished before this time, then female sex hormones can cause sex reversal in the male gonad.

4. Transplantation Experiments and Parabiosis

MacIntyre (1956) and MacIntyre *et al.* (1959) implanted pairs of associated ovaries or testes from 16-day rat fetuses beneath the renal capsule of adult rats castrated 3 weeks before transplantation. The grafts remained in place for 3 weeks. Ovarian differentiation was suppressed in some explants. Other grafts showed ovarian rudiments with degenerate follicles. MacIntyre (1956) also showed that ovarian tissue was transformed into the male type. In seven cases out of twenty-three he observed tubular formations containing oocytes. Holyoke (1956) obtained similar results with 17- to 24-day rabbit fetuses. MacIntyre *et al.* (1960) showed that the tissues had to fuse intimately to obtain an inhibitory effect. These authors grafted the male and female gonads at various distances from one another (1 to 10 mm) and showed that when the two gonads were more than 8 mm apart the male gonad no longer affected the female one. Turner and Asakawa (1964) and Turner (1969) took testes and ovaries from 12½- to 14½-day fetal mice, grafted them beneath the renal capsule of castrated adult male mice, and showed that in ovaries that had been masculinized by the testes the oogonia could develop into spermatocytes.

Parabiosis experiments *in vitro* have been carried out by Holyoke and Beber (1958) and by Salzgeber (1962, 1963). Holyoke and Beber (1958) associated the gonads of fetal rabbits or fetal rats on a medium that was composed of plasma coagulum and 11-day chick embryo extract for which only the anterior part of the embryo, excluding the gonads, was used. In heterosexual combinations, the development of the ovaries was always retarded and no cortical elements were found. In three cases, structures resembling testicular cords were observed in the ovarian medulla. Salzgeber

---

Fig. 8. Detail from a sterile (a) and a fertile (b) ovarian cortex which developed around the gonad of a male opossum treated from birth with female sex hormone. (Reproduced with permission from Burns, 1955.)

(1962, 1963) combined parabiosis experiments *in vitro* with heteroplastic grafting. She associated the ovaries from 12- to 21½-day fetal mice with gonads from 6- to 10-day chick embryos. After culture for 2 to 6 days in Wolff and Haffen's (1952a) medium, the associated explants were grafted into the coelomic cavities of young chick embryos where they developed until the hosts were killed 12 to 19 days later. Culture of the organs before their implantation into the host embryo encouraged the acceptance of the heterograft. In this type of association, the organs fused intimately. When taken at 17 to 19 days of gestation, the ovaries developed follicles but their appearance was altered; the follicular cells were lacking in some cases and flattened in others. This change in the ovaries was even more marked if they were associated with chick testis at an earlier stage in their development. When taken at 13 to 14 days, the ovaries developed in an atypical fashion. They consisted of loose mesenchyme cells and the cortical and medullary zones could not be distinguished. A few isolated oocytes persisted in this undifferentiated tissue. In most of the inhibited ovaries, the tubules of the rete ovarii could be clearly distinguished. They were sinuous hollow tubes with walls composed of a layer of irregularly arranged cells penetrating into the chicken tissue to form a mixed structure with the seminiferous tubules. In some explants, a few cords seen in the vicinity of the rete had a testicular appearance, and these occasionally contained oocytes or degenerating follicles.

These effects have been interpreted as inhibition or, in certain cases, as sexual inversion. In transplantation experiments of genital ridges of 11½-day mouse embryos under the kidney capsule of adult hosts, Ozdzenski (1972) observed structures similar to sterile sex cords in some ovaries but he suggested that these structures are probably caused by degeneration of the oocytes and cannot be considered as evidence of masculinization.

5. Production of Experimental Chimeras in Mice

The problem of genetic and epigenetic factors controlling the sexual differentiation of the gonads of mammals has been considered by the study of experimental chimeras. These studies were carried out by Tarkowski (1961, 1964, 1969, 1970), Mystkowska and Tarkowski (1968, 1970), and Mintz (1964, 1965a,b). We will only briefly describe the technique used by Tarkowski (1961) to produce such chimeras. Eggs taken at the 8-blastomere stage are deprived of their zona pellucida and then placed two by two in a small drop of liquid surrounded by paraffin oil. The oil exerts a pressure on the two eggs which then become joined side by side. They are explanted into culture for 24 to 40 hours. The double blastomeres so obtained show no trace of their dual origin. When implanted into the uterus of a pseudopregnant mouse the sex ratio of the individual chimeras obtained by Tarkowski and Mystkowska between 1961 and 1970 consisted of 48 males, 22 females, and four intersexes. With respect to the gonads, these intersexes could be classified as follows: ovary on one side, ovotestis on the other (two cases), testis on one side, ovotestis on the other (one case), ovotestis on both sides (one case). This predominance of males has also been reported by Mintz (1968) and by McLaren and Bowman (1969). Because of the small proportion of intersexes and the excess of male individuals, Tarkowski has put forward the hypothesis that certain of the phenotypically normal males would correspond to the chimeras XY/XX, in which the male component would be dominant. In 1968, Mystkowska and Tarkowski verified this hypothesis by karyotype studies. They showed that of four males examined, two were chromosomal chimeras. In these two males, respectively 51% and 19% of the cells of the bone marrow were genetically female (XX). Furthermore, the only intersex individual obtained in the course of this experimental study also showed 96% of XX cells. These authors' conclusion is that chromosomal chimerism does not interfere in the development of the male phenotype and that male sexual differentiation occurs more easily than intersexuality. They suggest that these conclusions may be equally valid in the case of spontaneous chromosomal chimeras, in which a quasi-normal male phenotype has been established, as known in man (Zuelzer *et al.*, 1964; Myhre *et al.*, 1965) and also described in a cat (Malouf *et al.*, 1967).

In 1970, Mystkowska and Tarkowski also demonstrated cytogenetically that two of the individuals of normal female phenotype corresponded to chimeric individuals XX/XY and that in these cases, therefore, the chromosomal chimerism did not impede female development. But this does not explain how, in rare cases, an intersexual gonad can be obtained. Tarkowski and Mystkowska do mention, however, that the gonadal tissues of most of the chimeras were not karyotyped; the intersex described in 1970 was an exception. In this ovotestis, XX and XY somatic cells were found and this situation has been demonstrated in the ovotestis of a human intersex (Gartler *et al.*, 1962) and in a bovine intersex (Dunn *et al.*, 1970). It would be very unlikely that the gonads of phenotypically normal but chimeric males and females would be homogeneous, since heterogeneity has been demonstrated in the pelage (Mintz, 1965a,b, 1967; Mystkowska and Tarkowski, 1968), in the external layer of the retina (Tarkowski, 1964; Mystkowska and Tarkowski, 1968), and in a certain number of vascular elements (Mintz and Baker, 1967). Tarkowski (1969) suggests the following explanation as to why the genital ridge sometimes develops into an ovotestis, or into an ovary, or a testis: "It is probably the degree of mixing of the two types of cells in the ridge itself which plays a decisive role—the more thorough the mixing the smaller is the chance of an ovotestis forming"; the development of the female phenotype may perhaps require a predominance of genetically female cells.

Problems concerning the failure of the differentiation of the XX germ cells in the testes of chimeras are discussed in Volume I, Chapter 3, Section III,2.

D. Other Vertebrates

Several studies on the effects of sex hormones on fish and reptiles have been reported.

1. Fish

In teleosts, experiments with sex hormones have shown that sex reversal of the gonads is fairly easy to obtain since many fish have a natural tendency towards intersexuality. Changes in the adult have been induced (e.g., *Xiphiophorus, Lebistes,* and the eel). Among the earliest workers in this field were Padoa (1937), Berkowitz (1937, 1941), and Regnier (1937, 1938).

D'Ancona (1948, 1951, 1959) has defined the conditions under which modifications of sex can be brought about in the eel. In these creatures, the period before sex differentiation is very long, extending over several years. Injections of androgens produce no noticeable change in the gonads. Estradiol, on the other hand, administered to silver eels which are already sexually differentiated, leads to a regression of the seminiferous tubules and stimulates cell division of those gonocytes which have not yet become differentiated. The latter develop into oogonia, enlarge, and enter meiotic prophase. This process first occurs peripherally and spreads progressively towards the center of the gonad. D'Ancona believes that the various parts of the teleostean gonad differ in their sensitivity to hormones. In the absence of corticomedullary differentiation, the response to estrogens represents the onset of differentiation of two potentially different regions. Complete, permanent, functional reversal of the sexual phenotype of the gonads has been obtained by Yamamoto (1953, 1955, 1958, 1959, 1963) in *Oryzias latipes*. The administration of estradiol to the larvae produces 100% females while methyltestosterone results in testicular differentiation of the gonads. The acquired differentiation appears to be stable and sex-reversed individuals can reproduce.

Transplantation experiments of trunk regions containing the gonads of newly hatched fry of *Oryzias latipes* into the anterior chamber of the eyes of adult fish have been performed by Satoh (1973). Only vascularized grafts developed well. A genetic male graft developed into a testis regardless of the sexuality of the host. The gonad of a genetic female fry developed into an ovary if the graft was transplanted into a female fish. On the other hand, the gonad of a genetic female fry transplanted into a male fish failed to develop into an ovary, but formed spermatogenic cells in a gonad of abnormal structure. The possibility that androgens may act as androinducers during natural sex differentiation was suggested by the authors.

Several attempts have been made to modify the gonads of selachians. Thiebold (1954, 1964) observed that crystalline androgens and estrogens exerted a marked effect on the genital ducts of *Scyliorhinus canicula*, but only estrogens acted on the gonads, transforming genetically male gonads into ovotestes. Chieffi (1952a,b, 1953a,b, 1959, 1967) has observed the same effect. Transplantation experiments with the gonads of *S. canicula* have been carried out in the extraembryonic coelom before the stage of sex differentiation by Thiebold (1964). The gonads of opposite sex transplanted side by side do not influence each other during the process of sex differentiation, even if histological connections are seen between them.

2. Reptiles

Two groups of reptiles, the chelonians and the crocodilians, show a prolonged bisexuality in their normal physiology. This disposition has been exploited by workers who have studied the effects of sex hormones on the homo- and heterosexual structures of immature individuals at quite long intervals after hatching (Forbes, 1938a,b, 1939; Godet, 1961; Risley, 1941 on crocodilians; Risley, 1940; Vivien and Stefan, 1958; Vivien, 1959; Stefan, 1958, 1963, on chelonians). Most of these authors observed the persistence and stimulation of the heterosexual territories but the treatments were probably too late for induction of genuine sex inversion. Injections of steroid hormones into reptile embryos before sexual differentiation has given contradictory results, especially with regard to administration of estrogens.

a. Effect of Androgens. The injection of testosterone into the embryo of the chelonian *Chrysemys marginata bellii* at the gastrula stage (Risley, 1940) has produced complete masculinization of the ovaries in five females. Hartley (1945) obtained a slight stimulation of the ovarian medulla with formation of seminiferous cords and cortical inhibition in a female embryo of the ovoviviparous snake *Thamnophis sirtalis*. Injection of testosterone into the embryos of *Agama agama* before sex differentiation does not modify ovarian development (Charnier, 1963), and the female embryos of *Lacerta vivipara* always retain a very distinct ovarian structure (Dufaure, 1966). Testosterone injected into embryonic slowworms, *Anguis fragilis*, induces faint signs of intersexuality in the ovary (Raynaud, 1970).

b. Effect of Estrogens. Dantschakoff (1937) was the first author to obtain significant sexual modifications of the embryos of *L. vivipara* by the injection of estrogens. Raynaud (1965, 1967) showed that estradiol incompletely or completely inhibited the differentiation of the testicular part of the gonad in *L. viridis*. The glands transformed in this way acquired the configuration of an ovotestis or an ovary.

c. Effect of Pituitary Gonadotropins. The administration of hypophysial gonadotropic hormones to the embryos of *L. vivipara* (Dufaure, 1966) provokes feminization of male embryos and the most constant effect is the development of complete Müllerian ducts and their maintenance. When embryos are treated earlier and with greater doses, the gonad develops a cortex and the medulla is feminized. Injection of estrogens with gonadotropins gives different results according to the stage of administration. When administered late, they stimulate the structures which have been induced by the gonadotropins. Raynaud (1967) administered lower doses of gonadotropic hormones than those employed by Dufaure, and found no inversion of the sexual histogenesis of the gonads of male embryos of *L. viridis*. Furthermore, he reported that gonadotropins induce hypertrophy of the cortex of the gonad provoked by the action of the estradiol.

These divergent results remain to be explained. The morphogenetic effect of gonadotropins observed by Dufaure (1966) poses new problems which biochemical evidence may perhaps help to resolve in the future.

IV. CONCLUSIONS

Natural modifications of development of embryonic gonads are known in the case of gonadal dysgenesis and of intersexes. Gonadal dysgenesis is due to heteroploidy of the sex chromosomes and these individuals correspond to aneuploids characterized by the loss of one heterochromosome or by the presence of an excess of heterochromosomes. Such aneuploids have been described in man and the most typical cases correspond to XO individuals (Turner's syndrome) and to XXY individuals (Klinefelter's syndrome). In the first case, the absence of the Y chromosome results in the atrophy or nondevelopment of the gonads. The genital tract is of the immature female type which throws light on castration experiments in mammals. In the second case, the individuals are of the male phenotype characterized by hypodevelopment of the testes. This latter syndrome not only shows the important part played by the Y chromosome in testicular morphogenesis, but also shows that the sole presence of the Y chromosome does not ensure the spermatogenic or endocrine activity of the testis. Correspondence between an abnormal chromosome formula and intersexuality has been demonstrated in other cases. Cytogenetic studies have shown that intersex individuals in reality correspond to mosaics or to chromosomal chimeras. These examples may be compared with experimental chimeras obtained in the mouse by the fusion of two blastomeres where a small proportion of intersexes and an excess of males were obtained. Karyotype analysis has, however, shown that the male individuals also correspond to chimeras of the type XX/XY in which the male component would be dominant. If, in spontaneous intersexes, the presence of a Y chromosome can explain the presence of testicular tissue, the question arises as to how testicular tissue differentiates in the case of a normal XX constitution. Some authors do not reject the possibility of a mosaic passing unnoticed or of a translocation of Y chromosome material containing the part determining maleness on to the X chromosome. Equally, one can invoke a genetic factor by the intervention of a mutant autosomal gene, as in hereditary intersexuality, or an epigenetic factor as in freemartinism.

Experimental modifications of gonad development have been obtained by means of several experimental procedures such as parabiosis between embryos of different sex, grafting gonads of the opposite sex, and injection of sex hormones. The most complete modifications have been obtained in the amphibians where experimental transplantation of heterosexual gonads and hormone injections have resulted in changes of sex such that the transformed individuals have been crossed with normal individuals and have been able to produce offspring. These experiments have had the advantage of showing that in the amphibians, contrary to the birds and mammals, there are two kinds of genetic determination: in *Ambystoma, Pleurodeles,* and *Xenopus* the male sex is homogametic (ZZ), and the female heterogametic (ZW), whereas in the Ranidae the female is homogametic (XX) and the male heterogametic (XY); this is also the case in the teleost fish *Oryzias latipes*.

Experiments involving crossing of transformed individuals of inversed phenotypes have resulted in individuals of new genetic constitution, such as WW males and females in *Ambystoma, Pleurodeles,* and *Xenopus* or YY males and females in *Oryzias latipes*. In fishes and amphibians, the genetic constitution of the primary gonocytes is of little importance for the subsequent orientation of gametogenesis.

In birds, remarkable sex changes have been obtained in male embryos by injection of female hormones. Experiments with gonad grafts and the association of gonads in parabiosis *in vitro* have shown that embryonic gonads produce secretions from the medulla which have the same effects as steroid hormones. Moreover, feminized male gonads are, in their turn, capable of secreting a hormone having the same effects as that which has provoked the transformation. The histochemical and biochemical studies on this subject are reported in Volume I, Chapter 3, Section IV,B and favor the view that they are similar to the sex hormones of the adult. We must, however, remember that the feminization of male gonads produced by steroid hormones or by the embryonic ovary is not maintained after hatching, and eventually the ovarian cortex degenerates and testicular development resumes. The ovarian cortex has been implicated in the causes of this instability of sex inversion in birds. In the fowl, the male germ cells on which female differentiation has been imposed are unable to pass beyond the stage of the meitoic prophase, even when they develop precociously in a genetically female gonad (see Volume I, Chapter 3, Section III,1). It is possible that there is a genetic factor operating whereby differentiation only occurs for female germ cells during the period of the experiment.

Among mammals, which have been the subject of numerous experiments, the injection of sex hormones has not led to the transformation of the gonads. However, the opossum seems to be an exception. In the rodents, only direct parabioses have resulted in inhibition of ovarian development capable, in certain cases, of being interpreted as sexual inversions. The experimental production of chimeras in the mouse seems to indicate that female germ cells made to develop in a male gonad do not proceed to male gametogenesis and degenerate in the same way as do the female germ cells in the masculinized ovary of the freemartin.

Thus, it can be admitted, as suggested by Wolff in 1965, that two types of factors are involved in the differentiation of one of the sexes. One is the hormonal factor which causes stimulation or induction of differentiation, and the other is a genetic factor which controls the receptivity of the tissues competent and incompetent to respond to the inductive agent. One can, thus, attribute the successes, the failures, or the partial successes of experimentation to the activity of one or the other factor. Many authors have so far stressed the more or less efficient role of the hormonal substances utilized in the intersexuality experiments. It is fitting to take into consideration the receptivity of the effector tissue to the inductive action. In a general way, the genetic sex factor exercises its antagonism less in the lower vertebrate groups (fishes, amphibians) than in the higher groups. Thus, in the amphibians sex inversions may be complete and indefinite; in the birds the genetic factor does not hinder the first stages in the female differentiation of the male cortex, but does not countenance its further development. In the mammals it opposes, at least in the majority, the intersexual transformation of the gonads.

REFERENCES

Ackart, R. J., and Leavy, S. (1939). Experimental reversal of sex in salamanders by the injection of estrone. *Proc. Soc. Exp. Biol. Med.* **42,** 720–724.

Benoit, J. (1932). L'inversion sexuelle de la poule, determinée par l'ablation de l'ovaire gauche. *Arch. Zool. Exp. Gen.* **73,** 1–112.

Berkowitz, P. (1937). Effect of estrogenic substances in *Lebistes reticulatus*. *Proc. Soc. Exp. Biol. Med.* **36,** 416–418.

Berkowitz, P. (1941). The effects of estrogenic substances in the fish *Lebistes reticulatus*. *J. Exp. Zool.* **87,** 232–240.

Bissonette, T. H. (1924). The development of the reproductive ducts and canals in the freemartin with comparison of the normal. *Am. J. Anat.* **33,** 267–345.

Booth, P. B., Plant, G., James, J. D., Ikin, E. W., Moores, P., Sanger, R., and Race, R. R. (1957). Blood chimaerism in a pair of twins. *Br. Med. J.* **1,** 1456–1458.

Bradley, E. M. (1941). Sex differentiation of chick and duck gonads as studied in homoplastic and heteroplastic host-graft combinations. *Anat. Rec.* **79,** 507–529.

Brunst, V. V. (1970). Partial sex reversal in adult axolotls developing from cold treated eggs. *J. Exp. Zool.* **175,** 37–67.

Burns, R. K. (1925). The sex of parabiotic twins in Amphibia. *J. Exp. Zool.* **42,** 31–90.

Burns, R. K. (1930). The process of sex transformation in parabiotic *Amblystoma*. I. Transformation from female to male. *J. Exp. Zool.* **55,** 123–170.

Burns, R. K. (1938). The effects of crystalline sex hormones on sex differentiation in *Amblystoma*. I. Estrone. *Anat. Rec.* **71,** 447–467.

Burns, R. K. (1939a). The differentiation of sex in the opossum (*Didelphys virginiana*) and its modification by the male hormone testosterone propionate. *J. Morphol.* **65,** 79–119.

Burns, R. K. (1939b). Sex differentiation during the early pouch stages of the opossum (*Didelphys virginiana*) and a comparison of the anatomical changes induced by male and female sex hormones. *J. Morphol.* **65,** 497–547.

Burns, R. K. (1942). Hormones and experimental modifications of sex in the opossum. *Biol. Symp.* **9,** 125–146.

Burns, R. K. (1950). Sex transformation in the opossum: some new results and a retrospect. *Arch. Anat. Microsc. Morphol. Exp.* **39,** 467–483.

Burns, R. K. (1955). Experimental reversal of sex in the gonads of the opossum *Didelphys virginiana*. *Proc. Natl. Acad. Sci. USA* **41,** 669–676.

Burns, R. K. (1961). Role of hormones in the differentiation of sex. *In* "Sex and Internal Secretions" (W. C. Young, ed.), 3rd ed., Vol. , pp. 76–158. Williams & Wilkins, Baltimore, Maryland.

Buyse, A. (1936). A case of extreme sex modification in an adult bovine freemartin. *Anat. Rec.* **66,** 43–58.

Caridroit, F., and Pezard, J. (1935). Poussée testiculaire autonome à l'intérieur de greffons ovariens autoplastiques chez la poule domestique. *C. R. Acad. Sci. (Paris)* **180,** 2067.

Chang, C. Y., and Witschi, E. (1955). Breeding of sex reversed males of *Xenopus laevis* D. *Proc. Soc. Exp. Biol. Med.* **89,** 150–152.

Chapin, C. L. (1917). A microscopic study of the reproductive system of foetal freemartins. *J. Exp. Zool.* **23,** 453–482.

Charnier, M. (1963). Action de l'hormone mâle (hexa hydro benzoate de testostérone) sur les femelles d'*Agama agama,* Sauriens, *Agamidae*. *C. R. Soc. Biol.* **157,** 1470–1472.

Chieffi, G. (1952a). Sull'organogenesi dell'interrenale e della medulla della gonade in *Torpedo ocellata* e in *Scylliorhinus canicula*. *Pubbl. Stn. Zool. Napoli* **23,** 186–200.

Chieffi, G. (1952b). Azione del testosterone e della diidrofollicolina sul differenziamento sessuale di *Scylliorhinus canicula. Bull. Zool.* **19,** 117–122.

Chieffi, G. (1953a). Azione del testosterone sul differenziamento sessuale di *Scylliorhinus canicula. Ric. Sci.* **23,** 111–117.

Chieffi, G. (1953b). Ulteriori osservazioni sull'azione degli ormoni steroidi sul differenziamento sessuale di *Scylliorhinus canicula. Monit. Zool. Ital.* **82,** Suppl., 446–448.

Chieffi, G. (1959). Sex differentiation and experimental sex reversal in elasmobranch fishes. *Arch. Anat. Microsc. Morphol. Exp.* **48,** 21–36.

Chieffi, G. (1967). The reproductive system of Elasmobranchs: developmental and endocrinological aspects. *Pharmacol. Endocrinol. Immunol.* **37,** 553–580.

Chieffi, G., Iela, L., and Rastogi, R. K. (1974). Effect of cyproterone, cyproterone acetate and ICI 46, 474 on gonadal sex differentiation in *Rana esculenta. Gen. Comp. Endocrinol.* **22,** 532–535.

Chu, E. H. Y., Thuline, H. C., and Norly, D. E. (1964). Triploid-diploid chimerism in a male tortoise-shell cat. *Cytogenetics* **3,** 1–18.

Collenot, A. (1965). Recherches comparatives sur l'inversion sexuelle par les hormones stéroides chez les amphibiens. *Mem. Soc. Zool. Fr.* **33,** 4–141.

d'Ancona, U. (1948). Prime osservazioni sull'azione degli ormoni sessuali sulla gonade dell'anguilla. *Atti Accad. Naz. Lincei, Cl. Sci. Fis., Mat. Nat., Rend.* [8] **5,** 82–87.

d'Ancona, U. (1951). Ulteriori esperienze sull'azione degli ormoni steroidi sulla gonade dell'anguilla. *Atti Accad. Naz. Lincei, Cl. Sci. Fis., Mat. Nat., Rend.* [8] **10,** 284–289.

d'Ancona, U. (1959). Distribution of the sexes and environmental influence in the European eel. *Arch. Anat. Microsc. Morphol. Exp.* **48,** 61–70.

Dantschakoff, V. (1935). Sur l'inversion sexuelle experimentale de l'ebauche testiculaire, chez l'embryon de poulet. *C. R. Acad. Sci.* (*Paris*) **200,** 1983–1985.

Dantschakoff, V. (1936). Réalisation de sexe à volonté par inductions hormonales. I. Inversions du sexe dans un embryon génétiquement mâle. *Bull. Biol. Fr. Belg.* **70,** 241–307.

Dantschakoff, V. (1937). Sur l'action de l'hormone sexuelle femelle chez les reptiles. *C. R. Hebd. Seances Acad. Sci.* **205,** 424–426.

Deray, A., and Gomot, L. (1969). Sur un stade d'ovotestis des gonades droite et gauche de Cane hybride (*Cairina moschata* g × *Anas platyrhynchos* male) en cours d'inversion sexuelle. *C. R. Acad. Sci.* (*Paris*) **268,** 2485–2488.

Domm, L. V. (1927). New experiments on ovariectomy and the problem of sex inversion in the fowl. *J. Exp. Zool.* **48,** 31–119.

Dufaure, J. P. (1966). Recherches descriptives et expérimentales sur les modalités et facteurs du développement de l'appareil génital chez le lézard vivipare (*Lacerta vivipara* Jacquin). *Arch. Anat. Microsc. Morphol. Exp.* **55,** 437–537.

Dunn, H. O., Kenney, M. R., and Leiss, D. H. (1968). XX/XY chimerism in a bovine true hermaphrodite: an insight into understanding of free-martinism. *Cytogenetics* **7,** 390–402.

Dunn, H. O., McEntee, K., and Hansel, W. (1970). Diploid-triploid chimerism in a bovine true hermaphrodite. *Cytogenetics* **9,** 245–259.

Elger, W. (1966). Die Rolle der fetalen Androgene in der Sexualdifferenzierung des Kaninchens und ihre Abgrenzung gegen andere hormonale und somatische Faktoren durch Anwendung eines starken Antiandrogens. *Arch. Anat. Microsc. Morphol. Exp.* **55,** 657–743.

Fankhauser, G. (1940). Sex differentiation in triploid newts (*Triturus viridescens*). *Anat. Rec.* **77,** 227–245.

Fankhauser, G., and Humphrey, R. R. (1942). Induction of triploidy and haploidy in axolotl eggs by cold treatment. *Biol. Bull.* (*Woods Hole, Mass.*) **83,** 367–374.

Fankhauser, G., and Humphrey, R. R. (1950). Chromosome number and development of progeny of triploid axolotl females mated with diploid males. *J. Exp. Zool.* **115,** 207–250.

Fechheimer, N. S., Herschler, M. S., and Gilmore, L. O. (1963). Sex chromosome mosaicism in unlike-sexed cattle twins. *Genet. Today, Proc. Int. Congr., 11th, 1963* Vol. 1, 265.

Ferguson-Smith, M. A. (1966). X-Y chromosomal interchange in the aetiology of true hermaphroditism and of XX Klinefelter's syndrome. *Lancet* **2,** 475–476.

Finlay, G. F. (1925). Studies in sex differentiation in fowls. *Br. J. Exp. Biol.* **2,** 439–468.

Fischberg, M. (1945). Über die Ausbildung des Geschlechts bei triploiden und einem haploiden *Triton alpestris*. *Rev. Suisse Zool.* **52,** 407–414.

Foote, C. L. (1938). Influence of sex hormones on sex differentiation in Amphibia (*Rana pipiens*). *Anat. Rec.* **72,** 120–121.

Foote, C. L. (1940). Response of gonads and gonoducts of *Amblystoma* larvae to treatment with sex hormones. *Proc. Soc. Exp. Biol. Med.* **42,** 519–523.

Forbes, T. R. (1938a). Studies on the reproductive system of the alligator. II. The effects of prolonged injections of oestrone in the immature alligator. *J. Exp. Zool.* **78,** 335–367.

Forbes, T. R. (1938b). Studies on the reproductive system of the alligator. III. The action of testosterone on the accessory sex structures of recently hatched female alligators. *Anat. Rec.* **72,** 87–95.

Forbes, T. R. (1939). Studies on the reproductive system of the alligator. V. The effects of the injections of testosterone propionate in immature alligators. *Anat. Rec.* **75,** 51–57.

Ford, C. E. (1963). The cytogenetics of human intersexuality. *In* "Intersexuality" (C. Overzier, ed.), pp. 86–117. Academic Press, New York.

Ford, C. E., Jones, K. W., Polani, P. E., Almeida, J. C., and Briggs, J. H. (1959). A sex chromosome anomaly in a case of gonadal dysgenesis (Turner's syndrome). *Lancet* **1,** 711–713.

Gallien, L. (1937). Action masculinisante du propionate de testostérone dans la différenciation du sexe chez *Rana temporaria*. *C. R. Hebd. Seances Acad. Sci.* **205,** 375–377.

Gallien, L. (1938a). Action des hormones sexuelles dans la différenciation du sexe chez *Rana temporaria* L. *Bull. Biol. Fr. Belg.* **72,** 270–296.

Gallien, L. (1938b). Action du benzoate de dihydrofolliculine dans la différenciation du sexe chez *Rana temporaria* L. *C. R. Acad. Sci.* (*Paris*) **206,** 282–284.

Gallien, L. (1941). Recherches expérimentales sur l'action amphisexuelle de l'hormone femelle (oestradiol) dans la différenciation du sexe chez *Rana temporaria*, L. *Bull. Biol. Fr. Belg.* **75,** 369–397.

Gallien, L. (1944). Recherches expérimentales sur l'organogénèse sexuelle chez les batraciens anoures. *Bull. Biol. Fr. Belg.* **78,** 258–359.

Gallien, L. (1950a). Action du benzoate d'oestradiol dans la différenciation du sexe chez *Discoglossus pictus*. *C. R. Acad. Sci.* (*Paris*) **230,** 1006–1008.

Gallien, L. (1950b). Inversion du sexe (féminisation) chez l'urodèle *Pleurodeles waltlii*, M., traité par le benzoate d'oestradiol. *C. R. Acad. Sci.* (*Paris*) **231,** 919–920.

Gallien, L. (1950c). Inversion du sexe et effet paradoxal (féminisation) chez l'urodèle *Pleurodeles waltlii*, M., traité par le propionate de testostérone. *C. R. Acad. Sci.* (*Paris*) **231,** 1092–1094.

Gallien, L. (1953). Inversion totale du sexe chez *Xenopus laevis* D. à la suite d'un traitement gynogène par le benzoate d'oestradiol, administré pendant la vie larvaire. *C. R. Acad. Sci.* (*Paris*) **237,** 1565–1566.

Gallien, L. (1954). Inversion experimentale du sexe, sous l'action des hormones sexuelles, chez le triton *Pleurodeles waltlii,* M. Analyse des conséquences génétiques. *Bull. Biol. Fr. Belg.* **88,** 1–51.

Gallien, L. (1955). Descendance unisexuée d'une femelle de *Xenopus laevis,* Daud, ayant subi pendant sa phase larvaire, l'action gynogène du benzoate d'oestradiol. *C. R. Acad. Sci.* (*Paris*) **240,** 913–915.

Gallien, L. (1956). Inversion expérimentale du sexe chez un anoure inférieur, *Xenopus laevis,* D. Analyse des conséquences génétiques. *Bull. Biol. Fr. Belg.* **90,** 163–181.

Gallien, L. (1961). Double conversion du sexe chez le Triton *Pleurodeles waltlii* Michah. *C. R. Acad. Sci.* (*Paris*) **252,** 2768–2770.

Gallien, L. (1962). Evolution chez le triton *Pleurodeles waltlii* des intersexués obtenus après traitement par l'hormone femelle. *Bull. Biol. Fr. Belg.* **96,** 249–280.

Gallien, L. (1965). Genetic control of sexual differentiation in vertebrates. *In* "Organogenesis" (R. L. de Haan and H. Ursprung, eds.), pp. 583–610. Holt, New York.

Gallien, L. (1967). Developments in sexual organogenesis. *Adv. Morphog.* **6,** 259–317.

Gallien, L., and Beetschen, J. C. (1959). Sur la descendance d'individus triploides croisés entre eux ou avec des individus diploides chez le Triton *Pleurodeles waltlii. C. R. Acad. Sci.* (*Paris*) **248,** 3618–3620.

Gallien, L., and Beetschen, J. C. (1960). Différenciation sexuelle et gamétogénèse abortive chez un mâle haploide d'urodèle (*Pleurodeles waltlii*), élevé en parabiose. *C. R. Acad. Sci.* (*Paris*) **251,** 1655–1657.

Gallien, L., and Collenot, A. (1958). Inversion totale (féminisation) du sexe chez *Pelobates cultripes cuv.* et féminisation partielle des gonades (intersexualité) chez *Pelobates punctatus Daud.* à la suite d'un traitement gynogéne par le benzoate d'oestradiol administré pendant la vie larvaire. *C. R. Acad. Sci.* (*Paris*) **247,** 1042–1044.

Gallien, L., and Collenot, A. (1960). Inversion du phénotype sexuel (féminisation) chez *Triturus helveticus Raz.* et *Triturus alpestris Laur.*, á la suite d'un traitement gynogène par le benzoate d'oestradiol administré pendant la vie larvaire. *C. R. Acad. Sci.* (*Paris*) **250,** 926–928.

Gardner, W. A., and Taber, E. (1963). The demonstration of non-steroid gonadal inhibiting substance elaborated by the ovary of the brown leghorn. *Anat. Rec.* **145,** 231.

Gartler, S. M., Waxman, S. H., and Giblett, E. (1962). An XX/XY human hermaphrodite resulting from double fertilization. *Proc. Natl. Acad. Sci. USA* **48,** 332–335.

Godet, R. (1961). Action du propionate de testostérone sur les jeunes crocodiles (*Crocodilus niloticus*) de sexe femelle. *C. R. Soc. Biol.* **155,** 394–395.

Gomot, L., Ardiet, P., and Lutz-Ostertag, Y. (1966). Intersexualité gonadique des hybrides femelles provenant du croisement Canard Pekin x Cane de Barbarie. *C. R. Soc. Biol.* **160,** 1572–1573.

Goodfellow, S. A., Strong, S. J., and Stewart, J. S. S. (1965). Bovine freemartins and true hermaphroditism. *Lancet* **1,** 1040–1041.

Greene, R. R. (1942). Hormonal factors in sex inversion: the effects of sex hormones on embryonic sexual structures of the rat. *Biol. Symp.* **9,** 105–123.

Greene, R. R., and Ivy, A. C. (1937). The experimental production of intersexuality in the female rat with testosterone. *Science* **86,** 200–201.

Greene, R. R., Burrill, M. W., and Ivy, A. C. (1938). Experimental intersexuality: the production of feminized male rats by antenatal treatment with estrogens. *Science* **88,** 130–131.

Haffen, K. (1963). Sur l'évolution en greffes coelomiques, du constituant cortical isolé des gonades femelles et intersexuées de l'embryon de poulet. *C. R. Acad. Sci.* (*Paris*) **256,** 3755–3758.

Haffen, K. (1965). Intersexualité chez le caille (*Coturnix coturnix japonica*). Obtention d'un cas de ponte ovulaire par un mâle génétique. *C. R. Acad. Sci. Ser. 00* **261,** 3876–3879.

Haffen, K. (1968). Sur la greffe prolongée d'ovaires d'embryons de poulet, colonisés experimentalement par des cellules germinales de sexe mâle. *C. R. Acad. Sci. Ser. 00* **267,** 511–513.

Haffen, K. (1969a). Incompatibilité entre détermination chromosomique et stimulation hormonale dans la différenciation sexuelle. *Ann. Embryol. Morphog.* **1,** Suppl., 223–235.

Haffen, K. (1969b). Quelques aspects de l'intersexualité expérimentale chez le poulet (*Gallus gallus*) et la caille (*Coturnix coturnix japonica*). *Bull. Biol. Fr. Belg.* **103,** 401–417.

Hamada, H., Neumann, F., and Junkmann, K. (1963). Intrauterine antimaskuline Beeinflussung von Rattenfeten durch ein stark Gestagen wirksames Steroid. *Acta Endocrinol.* (*Copenhagen*) **44,** 380.

Hamerton, J. L. (1968). Significance of sex chromosome derived heterochromatin in mammals. *Nature* (*London*) **219,** 910–914.

Hampe, A. (1950). Sur la continuation de la première poussée de cordons sexuels mâles, chez l'embryon de poulet, sous l'influence des injections d'hormones femelles. *Arch. Anat. Microsc. Morphol. Exp.* **39,** 35–62.

Hanaoka, L. (1941a). The effect of follicular hormone upon sex differentiation in *Hynobius retardatus*. *J. Fac. Sci., Hokkaido Univ.* **7,** 399–412.

Hanaoka, L. (1941b). The effect of testosterone propionate upon sex differentiation in *Hynobius retardatus*. *J. Fac. Sci., Hokkaido Univ.* **7,** 413–419.

Hartley, R. T. (1945). Effects of sex hormones on the development of the urogenital system in the garter snake. *J. Morphol.* **76,** 115–131.

Hay, D. (1950). Etude d'un freemartin spontané chez les ruminants. *Arch. Anat. Microsc. Morphol. Exp.* **33,** 53–79.

Herschler, M. S., and Fechheimer, N. S. (1967). The role of sex chromosome chimerism in altering sexual development of mammals. *Cytogenetics* **6,** 204–212.

Holyoke, E. A. (1956). The differentiation of embryonic ovaries and testes grafted together in adult hosts in the rabbit. *Anat. Rec.* **124,** 307.

Holyoke, E. A., and Beber, B. A. C. (1958). Cultures of gonads of mammalian embryos. *Science* **128,** 1082.

Houillon, C. (1964). Chimères xénoplastiques entre les urodèles, *Pleurodeles waltlii Michah* et *Triturus alpestris* Laur. *C. R. Acad. Sci.* (*Paris*) **258,** 3901–3903.

Houillon, C., and Charlemagne, J. (1971). Développement d'un double systéme de glandes génitales dans les chimères homoplastiques à corps double chez l'Amphibien Urodèle, *Pleurodeles waltlii Michah. C. R. Acad. Sci. Ser. D* **272,** 1546–1549.

Huchon, D. (1971). Différenciation sexuelle et intersexualité dans la parabiose chez six espèces d'Amphibiens Anoures appartenant aux familles des Discoglossidae, Pelobatidae, Limnodynastidae. *Ann. Embryol. Morphol.* **4,** 237–268.

Hujbers, M. (1951). "Di Involoed van de gonade op de ontwikkeling van het Geslachtsapparat bij het keppenembryo," pp. 1–120. Offco Print, Amsterdam.

Humphrey, R. R. (1929a). Studies on sex reversal in *Amblystoma.* II. Sex differentiation and modification following orthotopic implantation of a gonadic preprimordium. *J. Exp. Zool.* **53,** 171–220.

Humphrey, R. R. (1929b). The early position of the primordial germ cells in urodeles: evidence from experimental studies. *Anat. Rec.* **42,** 301–314.

Humphrey, R. R. (1931a). Studies on sex reversal in *Amblystoma.* III. Transformation of the ovary of *A. tigrinum* into functional testis through the influence of a testis resident in the same animal. *J. Exp. Zool.* **58,** 333–365.

Humphrey, R. R. (1931b). Studies on sex reversal in *Amblystoma.* IV. The development potencies exhibited by the modified (freemartin) ovary of *A. tigrinum* following removal of the testis which had induced its modification. *J. Exp. Zool.* **58,** 367–398.

Humphrey, R. R. (1933). The development and sex differentiation of the gonad in the wood frog (*Rana sylvatica*) following extirpation or orthotopic implantation of the intermediate segment and adjacent mesoderm. *J. Exp. Zool.* **65,** 243–269.

Humphrey, R. R. (1935a). Studies on sex reversal in *Amblystoma.* VII. Reversal of sex type in gonadic preprimordia of *A. punctatum* males implanted in females of more rapidly growing species. *Anat. Rec.* **62,** 223–245.

Humphrey, R. R. (1935b). Sex reversal in *Amblystoma.* VIII. Sex type of gonads developed from gonadic preprimordia of *A. punctatum* implanted in axolotl females. *Proc. Soc. Exp. Biol. Med.* **33,** 102–104.

Humphrey, R. R. (1936). Studies on sex reversal in *Amblystoma.* IX. Reversal of ovaries to testes in parabiotic. *A. tigrinum. J. Exp. Zool.* **73,** 1–21.

Humphrey, R. R. (1942a). Sex of the offspring fathered by two *Amblystoma* females experimentally converted into males. *Anat. Rec.* **82,** 469.

Humphrey, R. R. (1942b). Sex inversion in the Amphibia. *Biol. Symp.* **9,** 81–105.

Humphrey, R. R. (1945). Sex determination in Ambystomid salamanders: a study of the progeny of females experimentally converted into males. *Am. J. Anat.* **76,** 33–66.

Humphrey, R. R. (1948). Reversal of sex in the females of genotype WW in the axolotl (*Siredon* or *Ambystoma mexicanum*) and its bearing upon the role of the Z chromosome in the development of the testis. *J. Exp. Zool.* **109,** 171–185.

Humphrey, R. R. (1957). Male homogamety in the Mexican axolotl: a study of the progeny obtained when germ cells of a genetic male are incorporated in a developing ovary. *J. Exp. Zool.* **134,** 91–101.

Humphrey, R. R., and Fankhauser, G. (1946). The development, structure and functional capacity of the ovaries in triploid ambystomid salamanders. *J. Morphol.* **79,** 467–510.

Humphrey, R. R., and Fankhauser, G. (1956). Structure and functional capacity of the ovaries of higher polyploids (4N), (5N) in the Mexican axolotl (*Siredon* or *Ambystoma mexicanum*). *J. Morphol.* **98,** 161–198.

Humphrey, R. R., Briggs, R., and Fankhauser, G. (1950). Sex differentiation in triploid *Rana pipiens* and the subsequent reversal of females into males. *J. Exp. Zool.* **115,** 399–428.

Josso, N., Grouchy, J., Auvert, J., Nezelof, C., Jayle, M. F., Moullec, J., Frezal, J., Casaubon, A., and Lamy, M. (1965). True hermaphroditism with XX/XY mosaicism, probably due to double fertilization of the ovum. *J. Clin. Endocrinol. Metab.* **25,** 114–126.

Jost, A. (1947). Recherches sur la différenciation sexuelle de l'embryon de lapin. II. Action des androgènes de synthèse sur l'histogénèse génitale. *Arch. Anat. Microsc. Morphol. Exp.* **36,** 241–270.

Jost, A. (1965). Gonadal hormones in the sex differentiation of the mammalian fetus. *In* "Organogenesis" (R. L. de Haan and H. Ursprung, eds.), pp. 611–628. Holt, New York.

Jost, A. (1967). Steroids and sex differentiation of the mammalian foetus. *Proc. Int. Congr. Horm. Steroids, 2nd, 1966* Excerpta Med. Int. Congr. Ser. No. 132, pp. 74–81.

Jost, A. (1970). Hormonal factors in the sex differentiation of the mammalian foetus. *Philos. Trans. R. Soc. London Ser. B* **259,** 119–130.

Jost, A., and Prépin, J. (1966). Données sur la migration des cellules germinales primordiales du foetus de veau. *Arch. Anat. Microsc. Morphol. Exp.* **55,** 161–186.

Jost, A., Chodkiewicz, M., and Mauleon, P. (1963). Intersexualité du foetus de veau produite par des androgènes. Comparaison entre l'hormone foetale responsable du freemartinisme et l'hormone testiculaire adulte. *C. R. Acad. Sci. (Paris)* **256,** 274–276.

Jost, A., Vigier, B., and Prépin, J. (1972). Freemartins in cattle: the first steps of sexual organogenesis. *J. Reprod. Fertil.* **29,** 349–379.

Kawamura, T., and Tokunaga, G. (1952). The sex of triploid frogs *Rana japonica Günther. Proc. Imp. Acad. (Tokyo)* **13,** 121–128.

Kawamura, T., and Yokota, R. (1959). The offspring of sex reversed females of *Rana japonica Günther. J. Sci. Hiroshima Univ.* **18,** 31–38.

Keller, K., and Tandler, J. (1916). Uber das Verhalten der Eihäute bei der Zwillingsträchtigkeit des Rindes. *Monatsschr. Ver. Tieraerz. Oest.* **3,** 513–526.

Koch, W. (1963). Intersexuality in mammals. *In* "Intersexuality" (C. Overzier, ed.), pp. 35–47. Academic Press, New York.

Kornfeld, W. (1958). Endocrine influences upon the growth of the rudimentary gonad of fowl. *Anat. Rec.* **130,** 619–637.

Kornfeld, W., and Nalbandov, A. V. (1954). Endocrine influences on the development of the rudimentary gonad of fowl. *Endocrinology* **55,** 751–761.

Lillie, R. F. (1916). The theory of the freemartin. *Science* **43,** 611–613.

Lillie, F. R. (1917). The freemartin: a study of the action of sex hormones in the foetal life of cattle. *J. Exp. Zool.* **23,** 371–452.

Lutz, H., and Lutz-Ostertag, Y. (1958). Etude d'un freemartin chez les oiseaux. *Arch. Anat. Microsc. Morphol. Exp.* **47,** 205–210.

Lutz, H., and Lutz-Ostertag, Y. (1959). Free-martinisme spontané chez les oiseaux. *Devl. Biol.* **1,** 364–376.

Lutz-Ostertag, Y. (1964). Action du stilboestrol sur les canaux de Müller de l'embryon de caille (*Coturnix coturnix japonica*). *C. R. Acad. Sci. (Paris)* **259,** 879–881.

Lutz-Ostertag, Y. (1968). Intersexualité de l'ovaire de l'hybride de canards (*Cairina moschata* L. male × *Anas platyrhynchos* L. femelle) et activité hormonale. *C. R. Acad. Sci. Ser. D* **267,** 2020–2022.

Lutz-Ostertag, Y. (1969). Hybridation entre le canard Barbarie (*Cairina moschata*

L.) et la cane Khaki campbell (*Anas platyrhynchos* L.); activité des gonades femelles. *Ann. Embryol. Morphog.* **2,** 169–177.

Lutz-Ostertag, Y., and Gomot, L. (1960). Sur l'hybride femelle issu du croisement *Cairina moschata* L. (male) et *Anas platyrhynchos* L. (femelle). *C. R. Soc. Biol.* (*Paris*) **12,** 2282–2283.

Lutz-Ostertag, Y., and Lutz, H. (1974). Sexualité et pesticides. *Année Biol.* **13,** 173–195.

Lutz-Ostertag, Y., and Meiniel, R. (1968a). Action stérilisante du *o*, *o*-diethyl-*o*-paranitro-phenyl triosephosphate sur les testicules de l'embryon de Caille (*Coturnix coturnix japonica*) en culture *in vitro C. R. Acad. Sci. Ser. A* **267,** 96–97.

Lutz-Ostertag, Y., and Meiniel, R. (1968b). Action stérilisante du parathion sur l'ovaire de l'embryon de caille et de poulet en culture *in vitro C. R. Acad. Sci. Ser. D* **267,** 2178–2180.

McFeely, R. A., Hare, W. C. D., and Biggers, J. D. (1967). Chromosome studies in 14 cases of intersex in domestic mammals. *Cytogenetics* **6,** 242–253.

MacIntyre, N. (1956). Effect of the testis on ovarian differentiation in heterosexual embryonic rat gonad transplants. *Anat. Rec.* **124,** 27–47.

MacIntyre, N., Baker, J., and Wykoff, J. W. (1959). Effect of the ovary on testicular differentiation in heterosexual embryonic rat gonad transplants. *Arch. Anat. Microsc. Morphol. Exp.* **48,** 141–154.

MacIntyre, N., Hunter, G. E., and Morgan, A. H. (1960). The partial limits of activity of fetal gonadal inductors in the rat. *Anat. Rec.* **138,** 137–148.

McLaren, A., and Bowman, P. (1969). Mouse chimaeras derived from fusion of embryos differing by nine genetic factors. *Nature* (*London*) **224,** 238–240.

Malouf, N., Benirschke, K., and Hoefnagel, D. (1967). XX/XY chimerism in a tri-colored male cat. *Cytogenetics* **6,** 228–241.

Manelli, H., and Milano-Crassi, E. (1966). Iopfisi e gonadi embrionali di pollo associate in vitro. *Arch. Zool. Ital.* **51,** 855–862.

Mikamo, K., and Witschi, E. (1963). Functional sex reversal in genetic females of *Xenopus laevis* induced by implanted testes. *Genetics* **48,** 1411–1421.

Mikamo, K., and Witschi, E. (1964). Masculinization and breeding of the WW *Xenopus*. *Experientia* **20,** 622–623.

Minoura, T. (1921). A study of testis and ovary grafts on the hen's eggs and their effects on the embryos. *J. Exp. Zool.* **33,** 1–61.

Mintz, B. (1948). Testosterone propionate minimum for induction of male development in anurans; comparative data from other vertebrates. *Proc. Soc. Exp. Biol. Med.* **69,** 358–361.

Mintz, B. (1964). Formation of genetically mosaic mouse embryos and early development of Lethal (t 12/t 12)—Normal mosaics. *J. Exp. Zool.* **157,** 273–292.

Mintz, B. (1965a). Genetic mosaicism in adult mice of quadriparental lineage. *Science* **148,** 1232–1234.

Mintz, B. (1965b). Experimental genetic mosaicism in the mouse. *Preimplantation Stages Pregnancy, Ciba Found. Symp., 1965* pp. 194–207.

Mintz, B. (1967). Gene control of mammalian pigmentary differentiation. 1. Clonal origin of melanocytes. *Proc. Natl. Acad. Sci. USA* **58,** 344–351.

Mintz, B. (1968). Hermaphroditism, sex chromosomal mosaicism and germ cell selection in allophrenic mice. *J. Anim. Sci.* **27,** Suppl. 1, 51–60.

Mintz, B., and Baker, W. W. (1967). Normal mammalian muscle differentiation and gene control of isocitrate dehydrogenase synthesis. *Proc. Natl. Acad. Sci. USA* **58,** 592–598.

Mintz, B., and Witschi, E. (1946). Determination of the threshold dose of testosterone propionate inducing testicular development in genetically female anurans. *Anat. Rec.* **96,** 526–527.

Mintz, B., and Wolff, E. (1954). The development of embryonic chick ovarian medulla and its feminizing action in intra-coelomic grafts. *J. Exp. Zool.* **126,** 511–536.

Mintz, B., Foote, C., and Witschi, E. (1945). Quantitative studies on response of sex characters of differentiated *Rana clamitans* larvae to injected androgens and estrogens. *Endocrinology* **37,** 286–296.

Miyada, S. (1960). Studies on haploid frogs. *J. Sci. Hiroshima Univ.* **19,** 1–55.

Moore, K. L., Graham, M. A., and Barr, M. L. (1957). The sex chromatin of the bovine freemartin. *J. Exp. Zool.* **135,** 101–127.

Morgagni, J. B. (1749). Recherches anatomiques sur le siège et les causes des maladies. "Trad. Franc. de Bestouet XLVIe lettre edit." Victor Rozier, Paris.

Myhre, B. A., Meyer, T., Opitz, J. M., Race, R. R., Sanger, R., and Greenwalt, T. J. (1965). Two populations of erythrocytes associated with XX/XY mosaicism. *Transfusion* (*Philadelphia*) **5,** 501–505.

Mystkowska, E. T., and Tarkowski, A. K. (1968). Observations on CBA-p/CBA = T_6T_6 mouse chimeras. *J. Embryol. Exp. Morphol.* **20,** 33–52.

Mystkowska, E. T., and Tarkowski, A. K. (1970). Behaviour of germ cells and sexual differentiation in late embryonic and early postnatal mouse chimeras. *J. Embryol. Exp. Morphol.* **23,** 395–405.

Narbaitz, R., and Adler, R. (1966). Germ cell differentiation in embryonic gonads cultured in heterosexual pairs. *Experientia* **22,** 677–678.

Nes, N. (1966). Diploid-triploid chimerism in a true hermaphrodite mink (*Mustela vison*). *Heriditas* **56,** 159–170.

Neumann, F., Elger, W., and Steinbeck, H. (1970). Antiandrogens and reproductive development. *Philos. Trans. R. Soc. London, Ser. B* **259,** 179–184.

Ohno, S., and Gropp, A. (1965). Embryological basis for germ cell chimerism in mammals. *Cytogenetics* **4,** 251–261.

Ohno, S., Trujillo, J. M., Stenius, C., Christian, L. C., and Teplitz, R. L. (1962). Possible germ cell chimeras among new born dizygotic twin calves (*Bos taurus*). *Cytogenetics* **1,** 258–265.

Ozdzenski, W. (1972). Differentiation of the genital ridges of mouse embryos in the kidney of adult mice. *Arch. Anat. Microsc. Morphol. Exp.* **61,** 267–278.

Padoa, E. (1936). Effecte paradossale (mascolinizzazione) sulla differenziazione sessuale digirini de *Rana esculenta* trattati con ormone follicolare. *Monit. Zool. Ital.* **47,** 285–289.

Padoa, E. (1937). Differenziazione e inversione sessuale (femminizzazione) di avanotti ditrota (*Salmo iridens*) trattati con ormone femminile. *Monit. Zool. Ital.* **48,** 195–203.

Padoa, E. (1942). Femminizzazione e mascolinizzazione di girmi de *Rana esculenta* in funzione della dose di diidrofollicolina loro somministrata. *Monit. Zool. Ital.* **53,** 210–213.

Padoa, E. (1947). Differente sensibilita al testosterone dei genotipi male et femelle di *Rana dalmatina* Bp. *Arch. Zool. Ital.* **32,** 1–24.

Polani, P. E. (1961). Paternal and maternal non-disjunction in the light of colour vision studies. *Proc. Conf. Hum. Chromosomal Abnorm., 1959* pp. 80–83.

Polani, P. E. (1970). Hormonal and clinical aspects of hermaphroditism and the testicular feminizing syndrome in men. *Philos. Trans. R. Soc. London, Ser. B* **259,** 187–204.
Preda, V., Cracium, O., and Cimpianu, A. P. (1966). L'action de l'hormone gonadotrope choriale sur les cultures de longue durée d'organes embryonnaires. L'action de l'hormone sur la gonade embryonnaire de poulet, cultivée dans le milieu de Wolff et Haffen. *Rev. Roum. Embryol. Cytol., Ser. Cytol.* **3,** 113–115.
Puckett, W. O. (1940). Some effects of crystalline sex hormones on the differentiation of the gonads of an undifferentiated race of *Rana catesbeiana* tadpoles. *J. Exp. Zool.* **84,** 39–52.
Rajakoski, E., and Hafez, E. S. E. (1964). Cytological differentiation of fetal bovine gonads. *Cytogenetics* **3,** 193.
Raynaud, A. (1937). Intersexualité provoquée chez la souris femelle par injection d'hormone mâle à la mère en gestation. *C. R. Soc. Biol.* **126,** 866–868.
Raynaud, A. (1942). Modification expérimentale de la différenciation sexuelle des embryons de souris par action des hormones androgènes et oestrogènes. *Actual. Sci. Ind.* **925** and **926.**
Raynaud, A. (1962). The histogenesis of urogenital and mammary tissues sensitive to oestrogens. *In* "The Ovary" (S. Zuckerman, ed.), 1st ed., Vol. 2, pp. 179–230. Academic Press, New York.
Raynaud, A. (1965). Effets d'une hormone oestrogène sur la différenciation sexuelle de l'embryon de lézard vert (*Lacerta viridis laur*). *C. R. Acad. Sci.* (*Paris*) **260,** 4611–4614.
Raynaud, A. (1967). Effets d'une hormone oestrogène sur le développement de l'appareil génital de l'embryon de lézard vert. *Arch. Anat. Microsc. Morphol. Exp.* **56,** 63–122.
Raynaud, A. (1970). Effet de la testosterone sur l'appareil urogénital de l'embryon d'orvet. *Arch. Anat. Microsc. Morphol. Exp.* **59,** 125–156.
Regnier, M. T. (1937). Action des hormones sexuelles sur l'inversion du sexe chez *Xiphophorus helleri* Heckel. *C. R. Acad. Sci.* (*Paris*) **205,** 1451.
Regnier, M. T. (1938). Contribution à l'étude de la sexualité des cyprinodontes vivipares (*Xiphophorus helleri, Lebistes reticulatus*). *Bull. Biol. Fr. Belg.* **72,** 385–465.
Riddle, O., Hollander, W. F. and Schooley, J. P. (1945). A race of hermaphrodite-producing pigeons. *Anat. Rec.* **92,** 401–429.
Ridgon, R. H. (1967). Gonads in hermaphroditic ducks: a pathological study. *Am. J. Vet. Res.* **28,** 1125–1134.
Risley, P. L. (1940). Intersexual gonads of turtle embryos following injection of male sex hormone. *J. Morphol.* **67,** 439–453.
Risley, P. L. (1941). A comparison of effects of gonadotropic and sex hormones on the urogenital systems of juvenile terrapins. *J. Exp. Zool.* **87,** 477–515.
Ruch, J. V. (1962). Contribution à l'étude des modifications du système génital des embryons issus d'oeufs doubles de poule (*Gallus domesticus*). *Arch. Anat., Histol. Embryol.* **45,** 62–129.
Ruch, J. V. (1968). Modification du système génital observées chez des embryons issus d'oeufs doubles de poule. *Arch. Anat., Histol. Embryol.* **51,** 607–612.
Russell, L. B., and Woodiel, F. N. (1966). A spontaneous mouse chimera formed from separate fertilization of two meiotic products of oogenesis. *Cytogenetics* **5,** 106–119.

Salzgeber, B. (1957). Influence de facteurs teratogènes sur l'évolution des organes sexuées de l'embryon de poulet. 1. Action des facteurs physiques. 11. Action des facteurs chimiques. *Bull. Biol. Fr. Belg.* **91,** 355–438.

Salzgeber, B. (1962). Etude du développement de l'ovaire de souris greffé dans l'embryon de poulet après culture in vitro. *Arch. Anat. Microsc. Morphol. Exp.* **51,** 1–10.

Salzgeber, B. (1963). Modification expérimentale du développement de l'ovaire de souris sous l'influence du testicule embryonnaire de poulet. *J. Embryol. Exp. Morphol.* **11,** 91–105.

Satoh, N. (1973) Sex differentiation of the gonad of fry transplanted into the anterior chamber of the adult eye in the teleost, *Oryzias latipes. J. Embryol. Exp. Morphol.* **30,** 345–358.

Semenova-Tian-Shanskaya, A. G. (1969). Primordial germ cells during migration to gonad anlage in human embryos. *Arch. Anat. Histol. Embryol. Leningrad* **6,** 3–8.

Short, R. V., Smith, J., Mann, T., Evans, E. P., Dickson, J., Fryer, A., and Hamerton, J. L. (1969). Cytogenetic and endocrine studies of a freemartin heifer and its bull co-twin. *Cytogenetics* **8,** 369–388.

Simpson, P. (1969). La sensibilité différentielle d'une tumeur humaine et des tissus somatiques et germinaux des gonades embryonnaires en culture in vitro. *Eur. J. Cancer* **5,** 331–337.

Singh, R. P., and Carr, D. H. (1966). The anatomy and histology of XO human embryos and fetuses. *Anat. Rec.* **155,** 369–375.

Soller, M., Padeh, B., Wysoki, M., and Ayalon, N. (1969). Cytogenetics of Saanen-goats showing abnormal development of the reproductive tract associated with the dominant gene for polledness. *Cytogenetics* **8,** 51–67.

Stefan, Y. (1958). Etude préliminaire de l'action de quelques oestrogènes sur le tractus génital de la jeune tortue d'eau, *Emys leprosa. Ann. Endocrinol.* **19,** 481–506.

Stefan, Y. (1963). Contribution à l'étude expérimentale de l'intersexualité chez un chelonien: *Emys leprosa. Bull. Biol. Fr. Belg.* **97,** 363–467.

Stewart, J. S. (1961). Mechanism of meiotic non-disjunction in man. *Proc. Conf. Hum. Chromosomal Abnorm., 1959* pp. 84–96.

Stoll, R. (1948). Actions de quelques hormones sexuelles sur le développement des canaux de Müller de l'embryon de poulet. *Arch. Anat. Microsc. Morphol. Exp.* **37,** 118–135.

Taber, E. (1964). Intersexuality in birds. *In* "Intersexuality in Vertebrates including Man" (C. N. Armstrong and A. J. Marshall, eds.), pp. 285–310. Academic Press, New York.

Taber, E., and Salley, K. W. (1954). The effects of sex hormones on the development of the right gonad in the female fowl. *Endocrinology* **54,** 415–424.

Taber, E., Claytor, M., Knight, J., Flowers, J., Gambrell, D., and Ayers, C. (1958). Some effects of sex hormones and homologous gonadotropins on the early development of the rudimentary gonad in fowl. *Endocrinology* **63,** 435–448.

Tandler, J., and Keller, K. (1911). Ueber das Verhalten des Chorions bei verschiedengeschlechtlicher Zwillings gravidität des Rindes und über die Morphologie des Genitales der weiblichen Tiere, welche einer solchen gravidität entstammen. *Dtsch. Tieraerztl. Wochenschr.* **19,** 148–149.

Tarkowski, A. K. (1961). Mouse chimaeras developed from fused eggs. *Nature (London)* **190,** 857–860.

Tarkowski, A. K. (1964). The hermaphroditism in chimaeric mice. *J. Embryol. Exp. Morphol.* **12,** 735–757.

Tarkowski, A. K. (1969). Consequences of sex chromosome chimaerism for sexual differentiation in mammals. *Ann. Embryol. Morphog., Suppl.* **1,** 211–222.

Tarkowski, A. K. (1970). Are the genetic factors controlling sexual differentiation of somatic and germinal tissues of a mammalian gonad stable or labile? *Environ. Influences Genet. Expression, 1970* Fogarty Int. Cent. Proc., Vol. 2, pp. 49–60.

Teplitz, R. L., Moony, Y. S., and Basrur, P. K. (1967). Further studies of chimerism in heterosexual cattle twins. *Chromosoma* **22,** 202–209.

Thiebold, J. J. (1954). Etude préliminaire de l'action des hormones sexuelles sur la morphogénèse des voies génitales chez *Scylliorhinus canicula* L. *Bull. Biol. Fr. Belg.* **88,** 130–145.

Thiebold, J. J. (1964). Contribution à l'étude de l'organogénèse uro-génitale et de son déterminisme chez un poisson elasmobranche: la petite roussette *Scyliorhinus canicula* (L). *Bull. Biol. Fr. Belg.* **98,** 253–347.

Turner, C. D. (1969). Experimental reversal of germ cells. *Embryologia* **10,** 206–230.

Turner, C. D., and Asakawa, H. (1964). Experimental reversal of germ cells in ovaries of fetal mice. *Science* **143,** 1344–1345.

Turner, H. H. (1938). A syndrome of infantilism, congenital webbed neck, and cubitus valgus. *Endocrinology* **23,** 566–574.

Turpin, R., and Lejeune, J. (1965). "Les Chromosomes Humains," pp. 217–318. Gauthier-Villars, Paris.

Vannini, E. (1941). Rapida azione masculinizzante del testosterone sulle gonadi di *Rana agilis* in metamorphosi. *Mem. R. Accad. Ital., Cl. Sci. Fis., Mat. Nat.* [7] **2,** 666–676.

Vigier, B., Prépin, J., and Jost, A. (1972). Absence de corrélation entre le chimerisme XX/XY dans le foie et les premiers signes de freemartinisme chez le foetus de veau. *Cytogenetics* **11,** 81–101.

Vivien, J. H. (1959). Réactivité particulière du cortex gonadique et de l'épithélium du canal de Müller à l'action des hormones sexuelles chez le jeune mâle d'*Emys leprosa,* S., traité après l'éclosion. *Arch. Anat. Microsc. Morphol. Exp.* **48,** 297–312.

Vivien, J. H., and Stefan, Y. (1958). Féminisation des gonades de jeunes tortues d'eau, *Emys leprosa,* mâles par action du diethylstilboestrol. *C. R. Soc. Biol.* **152,** 649–652.

Welshon, S., and Russell, L. B. (1959). The Y chromosome as the bearer of male determining factors in the mouse. *Proc. Natl. Acad. Sci. USA* **45,** 560–566.

Weniger, J. P., and Zeis, A. (1973a). Recherches sur la nature chimique de l'hormone testiculaire de l'embryon de poulet. *Ann. Embryol. Morphog.* **6,** 219–228.

Weniger, J. P., and Zeis, A. (1973b). Sécrétion d'une hormone féminisante induite dans le testicule embryonnaire de poulet par la dihydrotestostérone. *Arch. Anat. Micr. Morphol. Exp.* **62,** 173–176.

Weniger, J. P., and Zeis, A. (1974). Sur l'induction de la synthèse d'oestrogènes par les androgènes. *C. R. Acad. Sci. Ser. A* **279,** 85–87.

Willier, B. H. (1921). Structures and homologies of freemartin gonads. *J. Exp. Zool.* **33,** 63–127.

Willier, B. H., Gallagher, T. F., and Koch, F. G. (1935). Sex modification in the chick embryo resulting from injections of male and female hormones. *Proc. Natl. Acad. Sci. USA* **21,** 625–631.

Witschi, E. (1927). Sex reversal in parabiotic twins of the American wood frog. *Biol. Bull.* (*Wood's Hole, Mass.*) **52,** 137–146.

Witschi, E. (1930). Sex development in parabiotic chains of the California newt. *Proc. Soc. Exp. Biol. Med.* **27,** 763–764.

Witschi, E. (1931). Range of the cortex-medulla antagonism in parabiotic twins of *Ranidae* and *Hylidae*. *J. Exp. Zool.* **58,** 113–145.

Witschi, E. (1936). Studies on sex differentiation and sex determination in amphibians. VIII. Experiments on inductive inhibition of sex differentiation in parabiotic twins of salamanders. *Anat. Rec.* **66,** 483–503.

Witschi, E. (1937). Quantitative relationships in the induction of sex differentiation and the problem of sex reversal in parabiotic salamanders. *J. Exp. Zool.* **75,** 313–373.

Witschi, E. (1939). Modification of the development of sex in lower vertebrates and in mammals. *In* "Sex and Internal Secretions" (E. Allen, ed.), 2nd ed., pp. 145–219. Williams & Wilkins, Baltimore, Maryland.

Witschi, E., and Allison, J. (1950). Response of *Xenopus* and *Alytes* to the administration of some steroid hormones. *Anat. Rec.* **108,** 589–590.

Witschi, E., and Crown, E. N. (1937). Hormones and sex determination in fishes and in frogs. *Anat. Rec.* **70,** Suppl., 121–122.

Witschi, E., and McCurdy, H. M. (1929). The freemartin effect in experimental parabiotic twins of *Triturus torosus*. *Proc. Soc. Exp. Biol. Med.* **26,** 655–657.

Wolff, Em., and Pinot, M. (1961). Stimulation du cortex de la gonade droite de l'embryon d'oiseau. *Arch. Anat. Microsc. Morphol. Exp.* **50,** 487–506.

Wolff, Et. (1936). L'évolution après l'éclosion des poulets mâles transformés en intersexués par l'hormone femelle injectée aux jeunes embryons. *Arch. Anat., Histol. Embryol.* **23,** 1–28.

Wolff, Et. (1939). Action du diethylstilboestrol sur les organes génitaux de l'embryon de poulet. *C. R. Acad. Sci.* (*Paris*) **208,** 1532–1533.

Wolff, Et. (1946). Recherches sur l'intersexualité expérimentale produite par la méthode des greffes de gonades à l'embryon de poulet. *Arch. Anat., Histol. Embryol.* **36,** 69–90.

Wolff, Et. (1962a). Experimental modification of ovarian development. *In* "The Ovary" (S. Zuckerman, ed.), 1st ed., Vol. 2, pp. 81–130. Academic Press, New York.

Wolff, Et. (1962b). The effect of ovarian hormones on the development of the urogenital tract and mammary primordia. *In* "The Ovary" (S. Zuckerman, ed.), 1st ed., Vol. 2, pp. 155–178. Academic Press, New York.

Wolff, Et. (1965). Problèmes généraux et problème spécial de la différenciation sexuelle. *Ann. Fac. Sci. Univ. Clermont* **26,** 17–25.

Wolff, Et., and Ginglinger, A. (1935a). Sur la transformation des poulets mâles en intersexués par injection d'hormone femelle—folliculine—aux embryons. *Arch. Anat., Histol. Embryol.* **20,** 219–278.

Wolff, Et., and Ginglinger, A. (1935b). Sur la production expérimentale d'intersexués par l'injection de folliculine à l'embryon de poulet. *C. R. Acad. Sci.* (*Paris*) **200,** 2118–2120.

Wolff, Et., and Haffen, K. (1951). Sur la culture *in vitro* des glandes génitales des embryons d'oiseau: obtention de la différenciation sexuelle normale et de l'intersexualité expérimentale des gonades explantées. *C. R. Acad. Sci.* (*Paris*) **233,** 439–441.

Wolff, Et., and Haffen, K. (1952a). Sur le développement et la différenciation sexuelle des gonades embryonnaires d'oiseau en culture *in vitro*. *J. Exp. Zool.* **119,** 381–399.

Wolff, Et., and Haffen, K. (1952b). Sur l'intersexualité expérimentale des gonades embryonnaires de canard cultivées *in vitro*. *Arch. Anat. Microsc. Morphol. Exp.* **41,** 184–207.

Wolff, Et., and Haffen, K. (1952c). Action feminisante de la gonade droite de l'embryon femelle de canard en culture *in vitro*. *C. R. Soc. Biol.* **146,** 1772–1774.

Wolff, Et., and Haffen, K. (1959). La culture *in vitro* de l'epithelium germinatif isolé des gonades mâles et femelles de l'embryon de canard. *Arch. Anat. Microsc. Morphol. Exp.* **48,** 331–345.

Wolff, Et., and Haffen, K. (1961). Sur la féminisation induite par les gonades mâles intersexuées, chez l'embryon de poulet. *Arch. Anat., Histol. Embryol.* **44,** 273–302.

Wolff, Et., and Simon, D. (1952). L'explantation et la parabiose *in vitro* de blastodermes incubés d'embryons de poulet. L'organisation de la circulation extra-embryonnaire. *C. R. Acad. Sci.* (*Paris*) **241,** 1994–1996.

Wolff, Et., and Wolff, Em. (1948a). Sur l'induction expérimentale de l'ovaire droit chez l'embryon d'oiseau. *C. R. Acad. Sci.* (*Paris*) **226,** 1140–1141.

Wolff, Et., and Wolff, Em. (1948b). Action d'une substance oestrogène artificielle, l'acide *N*-bisdehydrodoisynolique (sel de sodium) racémique sur l'appareil génital mâle et femelle de l'embryon de Poulet. *C. R. Soc. Biol.* **142,** 700–702.

Wolff, Et., Strudel, G., and Wolff, Em. (1948). L'action des hormones androgènes sur la différenciation sexuelle des embryons de poulet. *Arch. Anat., Histol. Embryol.* **31,** 237–310.

Yamamoto, T. (1953). Artificially induced sex reversal in genotypic males of the medaka (*Oryzias latipes*). *J. Exp. Zool.* **123,** 571–594.

Yamamoto, T. (1955). Progeny of artificially induced sex reversals of male genotype (XY) in the medaka (*Oryzias latipes*) with special reference to YY male. *Genetics* **40,** 406–419.

Yamamoto, T. (1958). Artificial induction of functional sex reversal in genotypic females of the medaka (*Oryzias latipes*). *J. Exp. Zool.* **137,** 227–265.

Yamamoto, T. (1959). A further study of induction of functional sex reversal in genotypic males of the medaka (*Oryzias latipes*) and progenies of sex reversal. *Genetics* **44,** 739–757.

Yamamoto, T. (1963). Induction of reversal in sex differentiation of YY zygotes in the medaka, *Oryzias latipes*. *Genetics* **48,** 293–306.

Zawadowski, M. (1926). Bisexual nature of the hen and experimental hermaphroditism in hens. *Trans. Lab. Exp. Biol. Zoo Park Moscow* **2,** 164–179.

Zuelzer, W. W., Beattie, K. M., and Reisman, L. E. (1964). Generalized unbalanced mosaicism attributable to dispermy and probable fertilization of a polar body. *Am. J. Hum. Genet.* **16,** 38–51.

9

The Influence of the Ovaries on Secondary Sexual Characters

H. G. Vevers

I. INTRODUCTION

John Hunter (1780) was one of the first to draw attention to secondary sexual characters when he wrote:

> It is well known, that there are many orders of animals which have the two parts, designed for the purpose of generation, different in the same species; and which are thus divided into male and female: but this is not the only distinguishing mark in many genera of animals, the male being distinguished from the female by various marks.

> A lion is different from a lioness; a cock from a hen, &c; particularly the voice in many animals of the same genus is different: such I shall call secondary properties, which take place only in parts that are neither essential to life nor generation, and which do not take place till towards the age of maturity.

Secondary sexual characters are usually defined as those examples of sexual dimorphism that do not comprise either the primary reproductive organs—the gonads—or the accessory reproductive organs, such as prostate or seminal vesicles, uterus and vagina, and external genitalia. Among the vertebrates, and especially in the birds, secondary sexual characters are often well defined and easily distinguishable from the accessory reproductive organs. In the invertebrates, on the other hand, the distinction between secondary and accessory characters is not always so clear. The most striking dimorphic characters are found in the pattern and coloration of butterflies, fishes, and birds, the differences in size and form of crustacean appendages and mammalian tusks, horns, and antlers, and the assumption at certain times of crests in newts, of a sexual skin in primates, and of enlarged crests and wattles in some birds.

The following pages give a survey of those secondary sexual characters the development of which is influenced or controlled by the ovary. Most of these are female characters, but there are also instances where the ovary affects male secondary sexual characters.

II. INVERTEBRATES

A. Crustaceans

Haemmerli-Boveri (1926) observed that female *Asellus* with degenerated ovaries lost the power to form a brood pouch, and le Roux (1933) found that when *Polymorphus minutus* is parasitic in the body cavity of *Gammarus pulex* there is a reduction in the growth of the host ovary, and absence from the leg oostegites of the bristles which normally conceal the eggs.

The influence of hormones on the development of the gonads and secondary sexual characteristics in crustaceans has been discussed by Carlisle and Knowles (1959) and a critical account of sexual differentiation in Crustacea Malacostraca is given by Carlisle (1960). The development of the gonads in juvenile crustaceans is controlled by the Y organ. Removal of this organ before puberty causes a retardation of gametogenesis, but the operation has no effect after puberty (Arvy *et al.*, 1954). In adults the vas deferens gland is needed for testis development and the hormones produced by this gland can induce the conversion of ovary to testis. The X organ–sinus gland complex produces a hormone which inhibits the ripening of the ovary and prevents yolk formation outside the breeding season. At the approach of the breeding season this ovary-inhibiting hormone decreases in amount, the ovary enlarges, and yolk is formed. With the appearance of the yolk, the ovary starts to produce a hormone which induces the formation of female sexual characters, such as the ovigerous hairs and the brood pouches. The production of these brooding characters is inhibited by ovariectomy.

Carlisle (1953) has shown how the ovary-inhibiting hormone and the ovarian hormone control the attainment of the female condition in the protandrous hermaphrodite prawn *Lysmata seticaudata.* He has also shown (Carlisle, 1959) that the loss of the male condition in the protandrous hermaphrodite prawn *Pandalus borealis* is governed by the vas deferens gland, while the attainment of the female condition is under control similar to that in *Lysmata.*

The ovary-inhibiting hormone is probably a conjugate steroid–protein complex, and Carlisle and Butler (1956) have drawn attention to the similarity between it and the substance produced by queen bees which inhibits the ovaries of worker bees (see also Butler, 1960). Similar ovary-inhibiting substances have been formed in several species of ants (Bier, 1954), and a substance produced by some queen termites has been shown to inhibit the development of supplementary reproductive individuals (Lüscher, 1953, 1956).

In talitrid amphipods the hormone produced by the vas deferens gland induces male differentiation not only in males but also in genetic females. When this hormone is absent, males and females differentiate as females, the first step being the development of the ovary which then secretes an ovarian hormone responsible for the production of the external female characters (Charniaux-Cotton, 1963, 1965, 1967). There is some evidence that in decapod crustaceans the ovaries may influence the development of the female secondary sexual characters (Adiyodi and Adiyodi, 1970).

B. Insects

The female cicada (*Enacanthus*) carries mycetomes which, at the time of egg maturation, form infection tubercles which are passed on to the next generation. In ovariectomized females the mycetomes are still well developed, but they do not form the infection tubercles. It has been suggested that the ovary may control the formation of these tubercles (Buchner, 1925).

Castration of the moth *Lymantria dispar* has given no evidence that the gonads control the sexual dimorphic characters in this species (Oudemans, 1899). Similar results have been obtained following the castration of a number of other insects, and these have been summarized by Hanström (1939). In *Drosophila,* Dantschakoff and Vachkovitchuté (1936) found that male larvae were not affected by an ovarian hormone.

In the caterpillars of the moth *Orgyia* the secondary sexual characters of the wing discs are still not finally determined at the fourth instar. Female wing discs taken at this stage and transplanted into male hosts form male-type wings, but in the reverse experiment (male wing discs growing in a female host) the male type characters are retained (Paul, 1937). This was the first observation which suggested that there might be cases of endocrine control of secondary sexual characters in the Insecta. More recently, Naisse (1966) working on the beetle *Lampyris noctiluca* has made observations which are in some ways parallel to those of Charniaux-Cotton on the Crustacea. *Lampyris* shows striking sexual dimorphism, the main feature being the presence of wings in the male but not in the female; in addition the testicular follicles have apical (mesodermal) tissue not represented in the female. Naisse found that implantation of this apical tissue into female larvae resulted in total masculinization of the larvae, and she has suggested that the apical tissue may produce androgenic hormone. There is as yet no evidence that the ovary produces a hormone influencing sexual differentiation in this insect.

In the butterfly *Cosmotriche* the wing color of male and female castrates carrying ovarian implants resembles that of the female (Prell, 1915), but the evidence is not very convincing as the sexual color dimorphism of this insect is normally variable.

There is, therefore, little evidence that insect gonads have any direct effect on secondary sexual characters, although it is possible that their development may be controlled by an organ not connected with the gonads, perhaps comparable to the vas deferens gland of crustaceans.

C. Other Invertebrates

Giard (1888) found no change in the secondary sexual characters of the turbellarian *Leptoplana* or of the nemertine *Lineus obscurus* suffering parasitic castration, but Vandel (1920, 1921) considered that the development of the copulatory apparatus of planarians was dependent upon the presence of the gonads, especially the testes.

In earthworms, Harms (1912) found that ovariectomy did not affect the development of the clitellum, but that orchidectomy did. Avel (1929), however, was unable to confirm these observations. Visnak (1935) found that crystalline follicular hormone had no effect on the sex characters or reproduction of the oligochaetes *Eisenia foetida* and *Rhynchelmis limosella.*

Linke (1934) noticed that in the winkle *Littorina littorea* destruction of the gonads by parasitic trematodes was accompanied by atrophy of the associated sexual organs, including the penis. In the cephalopods, Sereni (1929) considered that the presence of the gonads was necessary for the full development of the hectocotylus in the male, but Callan (1939) showed that this organ will regenerate after gonadectomy, thus suggesting that secondary sexual characters in cephalopods are not controlled by gonadal hormones.

D. Estrogen Production by Invertebrates

Alcoholic extracts of the protozoan *Colpoda steini* have yielded a substance which produces estrus in mice (Bauer, 1931), and similar estrogenic substances have been recorded in some nemertines, cestodes, and oligochaetes and also in the mollusks *Aplysia, Octopus,* and *Eledone* (Steidle, 1930). Estrogenic substances have also been found in aqueous extracts of the beadlet anemone, *Actinia equina,* of certain spiders and of the wasp, *Vespa crabro* (Schwerdtfeger, 1931), and in the ovaries of the echinoid *Lytechinus variegatus* (Donahue and Jennings, 1937). The ovaries and oviducts of the moth *Attacus atlas* also produce estrogenic substances (Loewe *et al.,* 1922). There is, however, no evidence to show that these substances are related to the chemically known estrogens of mammals, and no work appears to have been done on their effect on the secondary sexual characters of either vertebrates or invertebrates. It should also be remembered that many plants produce substances with similar physiological properties.

III. CYCLOSTOMES AND FISHES

The role of hormones in the reproductive biology of fishes has been the subject of a number of reviews in recent years (Dodd *et al.,* 1960; Ball, 1960; Dodd, 1960; Hoar, 1965). Most of the results so far reported refer to the action of hormones on the development of the gonads and accessory sexual characters, particularly the effects of testis, testicular extracts, or androgens. There is, however, some information on the action of the ovary and of estrogens on secondary sexual characters.

A. Cyclostomes

In the river lamprey, *Lampetra fluviatilis,* modifications of the cloacal region during the breeding season involve swelling of the labial lips and the formation of "ducts" between the coelom and the mesonephric ducts. Knowles (1939) showed that these changes can be induced in immature adults and in ammocoete larvae by the injection of a mammalian pituitary preparation. Treatment of larvae and adults with estrone and testosterone resulted in a cloacal swelling in the adults only. It is possible that the gonadal hormones do not act directly on the cloacal region but may merely stimulate the production of a pituitary hormone. If, as seems likely, the larval pituitary does not produce a hormone affecting the cloacal region, this hypothesis might explain the failure of ammocoetes to respond to injections of male or female gonadal hormones. This isolated instance from the cyclostomes is interesting because these animals show cloacal responses, involving vasodilatation, swelling, and epithelial changes, which are in some respects similar to those found in mammals.

B. Selachians

In this group the best known sexual dimorphic characters are the claspers in the male, and although these might strictly be regarded as accessory sexual organs it is perhaps best to include them here. Brough (1937) has also drawn attention to other morphological differences between the sexes, such as the shape of the lower jaw and the size and arrangement of the teeth.

Dodd (1955) found that injections of testosterone propionate induced small growth increments in the claspers of *Raia radiata,* and Dodd and Goddard (1961) reported that intramuscular implantation of estradiol pellets in female dogfish (*Scyliorhinus caniculus*) resulted in stimulation of the oviduct and the secretion of horny egg-case material.

C. Teleosts and Other Fishes

1. Ovariectomy and Gonad Transplantation

The male bowfin, *Amia calva,* has a brightly marked ocellus on the tail which is lacking in the female. In ovariectomized bowfins, Zahl and Davis (1932) found that a caudal ocellus began to appear 3 to 5 weeks after the operation and became well marked in about 3 months. The ocellus thus produced in the female fish did, however, lack the bright encircling corona which is present in the male only during the breeding season. This suggests that a basic ocellus is potentially present in both genetic sexes but that its development in any form is inhibited by the ovary, and its full development, with corona, is only attained in the male at times of testicular activity.

Ovariectomy of the cichlid *Tilapia macrocephala* caused a decrease in the length of the genital tube and a change in the color of the operculum to that of the immature fish. Treatment of the experimental fish with estradiol benzoate restored the size of the genital tube, but had no effect on opercular coloration (Aronson, 1948).

In the minnow, *Phoxinus laevis,* Kopeč (1927) found that development of the nuptial red coloration depended upon the presence of ovaries or testes. The full development of this secondary sexual character occurs when the erythrophores are fully expanded, and Kopeč suggested that this expansion may be provoked by secretions from the pituitary gland, stimulated by the gonads.

In the poeciliid fish *Gambusia affinis,* Okada and Yamashita (1944b) showed that, although ovariectomy alone produced no visible effect on secondary sexual characters, subsequent implantation of testicular tissue induced the development of male secondary sexual characters, and in particular of the gonopodium from the anal fin. In the medaka, *Oryzias latipes,* the sexes are distinguished by the size and shape of the dorsal and anal fins, by the presence or absence of guanophores, and by the development in the male only of small processes on some of the posterior rays of the anal fin. The latter characters, once formed, are permanent in the male, but the shape of the fins and the distribution of guanophores are seasonal characters, only coming to full development during breeding. Okada and Yamashita (1944a) found that in the medaka ovariectomy had no effect on any of these characters, but that removal of the ovary and its replacement by testis always produced masculinization of the female.

2. The Influence of Natural and Synthetic Estrogens

Early work showed that the fish ovary contains substances which had an estrogenic action on laboratory mammals, e.g., extracts of fish eggs (Sereni *et al.,* 1929; Donahue, 1949) and extracts of the whole ovary (Donahue, 1941).

There appears to have been no work on the effect of fish ovarian extracts on castrated fishes, although estrogens have been found in the ovaries of teleosts (Weisman *et al.,* 1937; Donahue, 1941; Hoar, 1955), of elasmobranchs (Wotiz *et al.,* 1958) and of lungfish (Dean and Chester-Jones, 1959). Most if not all of the work on the effect of ovarian secretions on the secondary sexual characters of fishes has been carried out with mammalian hormones or with synthetic preparations.

The live-bearing cyprinodonts show sex dimorphism in size and color pattern and in the development by functional males of a gonopodium; in some species of *Xiphophorus* the lower lobe of the caudal fin elongates to form the so-called sword. Most of these characters can be regarded as secondary sexual characters, although the gonopodium is more strictly an accessory sexual character.

Many observations have been made on the effect of natural and synthetic estrogens on these characters. In the guppy, *Poecilia* (= *Lebistes*) *reticulata*, males reach a length of about 18 mm, and females may be as long as 30 mm; in addition the males have a bright color pattern and also a typical gonopodium. In a large group of newborn guppies treated with estrogen for periods of 1 to 5 months, none of the fish showed male secondary sexual characters. When estrogen treatment was stopped, male color pattern and gonopodia appeared in some of the fish but, of course, the large feminine size attained by the fish was retained. Adult male guppies treated with estrogen showed no change in their secondary sexual characters (Berkowitz, 1937). In the same species, treatment of immature fishes with pregneninolone, a synthetic progesterone, inhibited all female secondary sexual characters and induced a precocious assumption of male characters (Eversole, 1941). Similar results were obtained by Mohsen (1958) who treated guppies with the same substance during the period from birth to 6 months of age. In all the treated fish the anal fin was transformed into a gonopodium, regardless of the original sex. Guppies treated with progesterone or pregnanediol show no change in secondary sexual characters (Eversole, 1941).

In *Gambusia affinis*, young males treated with estriol or diethylstilbestrol did not develop male secondary sexual characters, but when older males, which already possessed some male characters, were treated with one of these substances they developed subsequently in the female direction and produced mosaics (Okada and Yamashita, 1944b). Hamon (1946) found that estradiol benzoate had comparable effects on the males of *Gambusia*, while females treated with estrogen developed female secondary sexual characters precociously.

In the platy, *Xiphophorus maculatus*, which has secondary sexual characters similar to those of the guppy and mosquito fish (*Gambusia*), Cohen (1946) found that estradiol benzoate inhibited spermatogenesis in genetic males, that this was followed by suppression of the gonopodia, and that estrogen-treated males approached the normal females in body size. Using a strain of the same species in which a sex-linked factor allowed the sexes to be distinguished at birth, Tavolga (1949) confirmed these results using estradiol benzoate but found that treatment of male fish over 19 mm in length with free estradiol stimulated the testes and induced the formation of gonopodia. This difference in action between the free hormone and its ester has apparently not been observed elsewhere, and Tavolga has suggested that the steroids may act indirectly on the ovaries, possibly through the pituitary. She also found that pregneninolone given to male platies induced gonopodium formation, while the same substance administered to females inhibited ovarian development and also induced gonopodia. In addition, pregneninolone has been shown to induce, in genetic female platies, the elongation of the sixth caudal fin ray to form a tiny sword comparable to that of male common swordtails, *Xiphophorus helleri* (Cohen, 1946); this is interesting as swords are not normally found in male platies. Cohen also found that pregneninolone-treated female platies grew fin rays which did not bifurcate as they do in normal males.

In the bitterling, *Rhodeus amarus* (family Cyprinidae), Bretschneider and Duyvené de Wit (1947) found that progesterone caused growth of the ovipositor; they considered that this was due not to direct action by the steroid, but to its stimulation of the pituitary to produce a luteinizing hormone which then induced gonopod development through an ovarian hormone.

In the freshwater goby, *Chaenogobius annularis* (family Gobiidae), the fin-rays of the unpaired fins are longer in the female than in the male. Implants of estrone, estradiol, and testosterone all induced elongation of these structures in adult and immature fish, but the reaction was much more marked in those treated with estrogen (Egami, 1960a). However, in three other goby species (*Tridentiger obscurus, Rhinogobius similis,* and *Pterogobius zonoleucus*) androgen treatment caused an increase in the length of the first dorsal fin-rays but estrogen did not (Egami, 1960b).

Development of the large distal teeth characteristic of male *Oryzias latipes* is completely inhibited when estradiol-17β is added to the diet (Takeuchi, 1968).

In the loach, *Misgurnus anguillicaudatus* (family Cobitidae), estrone inhibits the development of male characters, e.g., the lateral integumental swellings immediately posterior to the dorsal fin (Egami, 1954).

Dean and Chester Jones (1959) showed that extracts of the ovary of the African lungfish, *Protopterus annectens,* contain estriol, estrone, and a trace of estradiol-17β; it is noteworthy that they found much larger amounts of estriol than estradiol, a condition which appears to be reversed in mammals (Burrows, 1949).

In teleosts, therefore, mammalian and synthetic estrogens cause feminization of certain secondary sexual characters, but the amount of experimental evidence is still relatively small, and the observations of Tavolga suggest that the problem may be complicated by the effect of estrogen on other endocrine organs. Progesterone does not influence the development of true secondary sexual characters in teleosts. On the other hand, pregneninolone inhibits the secondary sexual characters of the female and stimulates those of the male. Transplants of skin or of fin components between sexes might provide information on the influence of the ovaries on secondary sexual characters and would not cause the upset of the endocrine environment which probably occurs after gonad implantation or hormone injection.

IV. AMPHIBIANS

A. Anurans

Nuptial pads develop on the prepollex of some frogs and toads during the breeding season, and in *Rana temporaria, R. esculenta,* and *Bufo vulgaris* castration of the male results in a decrease in the size of these pads (Aron, 1926); this has been confirmed by grafting experiments (Welti, 1928). In *Discoglossus* and *Xenopus* the thumb pad area is determined at metamorphosis and a micropad is formed as a result of androgenic treatment, but there is no evidence to suggest that ovarian hormones play any part in the development or inhibition of anuran thumb pads.

In female anurans, in general, the coelomic epithelium is ciliated. Donahue (1934) has shown that injection of estrogen (theelin) into mature male *Rana pipiens* induced ciliated areas in this epithelium.

B. Urodeles

In urodeles the best known secondary sexual characters include the dorsal crest and the silvery tail band developed by European crested newts (*Triturus cristatus*). The crest only develops in castrates of both sexes and in males; its absence from the female is due to the inhibiting action of ovarian hormone (de Beaumont, 1933).

In the newt *Triturus viridescens* the pelvic and abdominal glands of the cloaca normally undergo regression in the adult. In orchidectomized males with ovarian grafts this regression has been found to be somewhat less than that of the same glands in male castrates lacking ovarian tissue (Adams, 1930). This observation may be compared with the findings of Noble and Pope (1929) that, in the salamander *Desmognathus,* the pelvic and abdominal glands regressed after ovariectomy, but that the implantation of testis tissue into such females caused increased activity, and, in some cases, new development in these glands.

The premaxillary and maxillary teeth of *Desmognathus* are elongated, directed forward and monocuspid in the male but short and bicuspid in the female, which also has teeth on the prevomers and on the posterior part of the dentary. Ovariectomized females with testis implants develop male-type teeth (Noble and Pope, 1929). Thus, although the monocuspid male-type of teeth appear to be developed only in the presence of testis, the factors controlling the appearance of the more posterior teeth are not yet known; it is possible that they only develop when ovarian tissue is present, or they might be under genetic control.

V. REPTILES

Secondary sexual characters are not common among the reptiles, but the most striking examples occur among the lizards, where the male may differ from the female in size and coloration, and in the development of dorsal crests, gular folds, and femoral pores. Castration of male *Lacerta agilis* causes the femoral glands to become more like those of the female, fat accumulates and there is a change in coloration to something approaching that of the female (Matthey, 1929). Ovariectomy has little effect on the external features and coloration of the female, but the femoral glands of the castrates appear to be less developed than those of normal females (Regamey, 1935).

In snakes the two sexes are very similar externally. In some, the male is smaller than the female and has fewer ventral and more dorsal scales, but there is no evidence for hormonal control of these differences. Estrogens have been found in the ovaries of the South American snakes *Bothrops jararaca* and *Crotalus terrificus* (Valle and Valle, 1943).

VI. BIRDS

A. Introduction

Secondary sex characters are widespread among birds and during the last 60 years many observations have been made on the factors controlling their production. The greater part of this work has been done on gallinaceous birds, and in particular on their plumage.

In most birds the cortex of the left ovary differentiates into a functional gonad, while the right ovary atrophies. As an exception many diurnal birds of prey have a large right gonad which is, however, usually nonfunctional.

B. Combs and Wattles

The turgidity of the fully developed combs and wattles of the common domestic fowl is due to the production of mucoid tissue in the dermis. In the female the comb and wattles only reach their full development during the period of egg-laying. In ovariectomized hens and in those that are not laying, these structures regress. This suggests that the development of the comb and wattles in the hen is under the control of the ovary, but poulards injected with various female sex hormones do not regenerate the comb and wattles, and normal male birds treated in the same way show a partial inhibition of comb growth. Similarly, Bolton (1953) found that the comb and wattles of immature pullets showed no change after treatment with estradiol dipropionate or progesterone, but that they hypertrophied following injections of testosterone propionate. However, he also noticed that the greatest amount of comb growth occurred when these three hormones were injected together. The key to this apparent paradox may perhaps lie in the nature of the tissue changes induced. Testosterone treatment is followed by hypertrophy of the connective tissue and mucoid element of the comb, whereas the ovarian hormones inhibit connective tissue proliferation but cause an increase in fat deposition in the comb. Régnier (1938) has suggested that estrogens may inhibit comb growth indirectly by a depressing action on the hypophysis.

There is apparently no genetic difference in the response of male and female combs, as a female genotype comb will develop as fully as a male one when the bird is injected with androgen.

C. Bills

In the house sparrow, *Passer domesticus,* males start to develop blue or black bills in October and November and by February the bill is jet-black. This is maintained throughout the breeding season, until in early August a whitish ring appears round the base of the bill and within 2 weeks the whole bill is pale in color. Keck (1934) showed that orchidectomy immediately stopped the deposition of pigment and the bills of the castrates became permanently ivory. Injection of male hormones induced a resumption of melanin production, but progesterone and natural and synthetic estrogens had no effect. This latter finding is significant because hen sparrows also show a change to dark bills during the breeding season, and this coloration can also be induced out of the breeding season when they are injected with androgens. It suggests that the reaction of the bill in normal female sparrows is due to the action of androgenic hormone produced by the ovary. Witschi (1935, 1955) has shown that the seasonal color change of the bill in African weaver finches (*Pyromelana*) is also controlled by androgens, although in these birds the bills of normal females only darken to a pale brown color, whereas after injection of androgens male and female bills turn black.

In the starling (*Sturnus vulgaris*) the bill is orange–yellow in both sexes during the breeding season, but at other times of the year it is black. After orchidectomy and ovariectomy the bills become permanently black. Injection of androgens into such castrates induces yellow pigmentation, but progesterone, deoxycorticosterone, and all estrogens have no effect. Here again, therefore, the ovary appears to be producing sufficient androgen to induce the production of yellow carotenoid pigment (Witschi, 1961). The color changes of the bill of the black-headed gull (*Larus ridibundus*) are comparable and the method of control is similar (van Oordt and Junge, 1933).

Witschi (1961) has described the only case in which the seasonal color of a bill is almost certainly controlled by estrogen. In the masked weaver (*Quelea quelea*) the bills of the males and castrates and of females in eclipse are blood red. During the breeding season this color changes, in the female only, to a pale yellow.

Bill color in the paradise whydah (*Steganura paradisea*) is controlled in yet another way. The bills are pale during the eclipse season and dark during the breeding season, but castrated males show the same seasonal color changes as normal males. This color change is controlled by the luteinizing factor of the pituitary; injection of this factor during the eclipse season is followed by darkening of the bills (Witschi, 1955). This method of control has not yet been confirmed for the female of the species.

D. Plumage

Sexual dimorphism occurs in the plumages of a great number of birds and it is convenient in discussing this phenomenon to denote the dull mottled-brown feathers usually worn by females as the hen-type plumage and the brighter, more ornamental plumages as cock type. The cocks of some species may wear a hen-type plumage either seasonally or throughout life. Conversely there is at least one bird, the red-necked phalarope (*Phalaropus lobatus*), in which the males retain a more or less hen-type plumage throughout life, whereas the females assume a bright cock-type plumage during the breeding season.

1. Plumage Control in Domestic Fowls

In gallinaceous birds the relationship between the gonads and sexual dimorphic plumage characters has been investigated by a number of workers in the United States, France, Britain, and elsewhere. Goodale (1913) showed that ovariectomized Brown Leghorn hens assume a male-type plumage. Other important works of the period are those of Pézard (1918), Zawadowsky (1922), Caridroit (1926), Domm (1924, 1927), Greenwood (1928), and Benoit (1929). Castration of the domestic cock results in no well-defined change in the color and pattern of the plumage, although the feathers of the capon become looser and in some cases longer. These changes, of course, only occur in new generation feathers following a molt or artificial depluming.

After sinistral ovariectomy the plumage of the poulard resembles that of the capon, but later reverts to the female type under the influence of the right gonad, which hypertrophies under these conditions. If the right gonad is removed the plumage becomes permanently capon-like in character (Domm, 1929). The implantation of ovarian tissue into a male or caponized domestic fowl causes the development of a female-type plumage (Caridroit, 1926), whereas the implantation of testis tissue into capon, poulard, or hen produces no plumage changes. From these basic experiments it has been concluded that in general the males, capons, and poulards of fowl races all carry a "neutral" plumage, the development of which is inhibited in normal females by the ovarian secretions.

These early investigations were extended and substantially confirmed by the work of Juhn and Gustavson (1929, 1930) and of Freud *et al.* (1929), who obtained feminization of the plumage of Brown and Golden Leghorn cocks and capons following injections of ovarian extracts. In more recent years these results have been repeatedly confirmed, using injections of pure female sex hormones, and have been extended to other races of the common fowl. In the Leghorn races the feathers studied did not show any barring, but in races of fowl with barred feathers Emmens and Parkes (1940) found that estrogen injections into capons induced the female type of barring. Siller (1956) has described a Brown Leghorn capon with female-type feathers, which had a Sertoli cell tumor, which he considered to have been secreting estrogen.

Skin transplantation has also been used as a method of analyzing plumage characteristics. This method is particularly valuable since it allows the reactivity of feather follicles of one sex to be tested in the hormonal environment of the opposite sex without causing any endocrine disturbance. Skin grafted from one race of fowl to the opposite sex of another race produced feathers showing the race characteristics of the donor, but the sexual type and color of the host sex (Danforth, 1929; Danforth and Foster, 1929).

The apparent completeness of sex reversal in the hen following bilateral ovariectomy and the observation of feminized plumage in the cock with an ovarian implant led Zawadowsky (1928) to postulate an equipotentiality of the two sexes with regard to plumage. Finlay (1925) also regarded the plumage of Brown Leghorn fowls as equipotential in the two sexes. More recent work has shown that, especially in pheasants, the male and female are not equipotential in the development of plumage characteristics, and this is discussed in more detail below. Kopeč and Greenwood (1930), working on the effect of yolk injections on Brown Leghorns, obtained changes in feather color and pattern in the direction of femaleness, but not complete feminization of shape.

Further analysis of the reaction of domestic fowl plumage to female sex hormones has shown that, especially in cocks, feathers growing on different feather tracts have different thresholds of reaction. Juhn *et al.* (1931) found that in Brown Leghorns these regional thresholds were directly related to the growth rate of the feathers. They also found that the threshold of reaction of those parts of the barbs near the feather shaft was lower than that of the outer parts of the barbs (at the edge of the vane), which required a higher dose of estrogen before they reacted by developing female-type coloration.

2. Hen-Feathered Races and Follicle Sensitivity

In Sebright and Campine fowls the plumage of the females is of the typical dull hen-type, but the males show two different types. Some males are cock-feathered with a bright plumage comparable with that of normal races, while others are hen-feathered, having a plumage which is almost identical with that of the female. Hen-feathering of the cocks of these races is conditioned by a single dominant gene, which can be bred into other races of fowl. Castration of the hen-feathered cocks results in a bird with a cock-feathered plumage (Morgan, 1915). Thus, the testes in this case condition the hen-feathering, probably by one of the testicular hormones, for the injection of testosterone into Sebright capons gives a female-type plumage (Deanesly and Parkes, 1937). It seems that in the hen-feathered races the plumage is sensitive not only to estrogens but also to other gonadal steroids. This was confirmed by Danforth (1930) who found that skin from female Campine bantams grafted on to male Leghorns grew female-type feathers, whereas skin from a female Leghorn grafted on to a male Leghorn gave male-type feathers. The feather follicles of hen-feathered races may, therefore, be regarded as having a very low threshold of response.

The sensitivity of the feather follicle of fowls to steroids therefore differs considerably according to race. All were sensitive to natural and synthetic estrogen, but whereas birds of the Phoenix race are quite insensitive to male hormones, some Sebright and Campine cocks are very sensitive.

3. Plumage Control in Pheasants

a. Common and Reeves' Pheasants. In pheasants the injection of ovarian extracts produced plumage changes broadly comparable with those found in domestic races of fowl (Champy, 1935). However, Danforth (1937a,b) noticed that in certain pheasants the female-type plumage of a cock or capon receiving ovarian hormone injections was, in fact, distinct from the normal female plumage of the species. In Reeves' pheasant, *Syrmaticus reevesi,* male skin transplanted to a male host, and female skin transplanted to a female host, produced normal male and female feathers, respectively. But male skin on a female host and female skin growing on a male host produced new type feathers differing from one another, and in many respects intermediate between the normal male and female feather types (Danforth, 1937a). These experiments and similar ones on the common pheasant, *Phasianus colchicus* (Danforth, 1937b), showed clearly that the feather follicles of male and female pheasants are not equipotential and that male skin does not react to female sex hormones in the same way as skin from a female host. The difference in reactivity is probably under the control of genetic factors.

b. Amherst Pheasants. Further experimental work on Amherst pheasants, *Chrysolophus amherstiae,* has shown that it is possible to define more exactly the characters that are influenced in development by natural and synthetic estrogens and has demonstrated that there are a number of sexual differences that are not affected by these hormones (Vevers, 1954). In this work feathers were grown on normal Amherst pheasant cocks receiving injections of estrone, estradiol dipropionate, and diethylstilbestrol. The effects of these injections on final feather size, pigmentation, barbule structure, barring pattern, and growth rate were observed.

Feathers grown on male Amherst pheasants receiving estrogens were smaller than normal male feathers grown from the same follicles, and tended to approach the size of normal female feathers.

Estrogen changed the distribution of the melanin pigment. In the pendent tract the vane of the normal male feathers is white with two dark terminal bars. Males receiving estrogen developed pendent feathers with a spread of dark brown melanin over the whole vane; with large doses bars of pale brown melanin were also deposited. In the back tract the dark iridescent melanin pigmentation of the normal male disappeared in feathers grown on birds receiving estrogen and were replaced by feathers with alternate transverse dark and pale brown bars.

In Amhersts and many other pheasants iridescence is widespread in cock plumages but almost absent from the hens. In noniridescent feathers the barbules are so orientated in relation to the feather vane that only their dorsal edges are apparent to an observer and no interference colors are produced. In iridescent feathers the barbules become twisted during development through an angle of 90 degrees, so that their broad, originally proximal surfaces are presented to the observer. The keratin of these broad surfaces is deposited in the form of fine lamellae, which act as interference films and produce iridescence. Vevers (1954) has shown that in the male Amherst pheasant barbule torsion is lost in feathers grown in the presence of injected estrogen. This loss of barbule torsion was complete in back feathers grown with estrone, estradiol, or stilbestrol injections, but in the pendent tract the barbule torsion in the area of the subterminal black bar was never completely lost, even in the presence of large doses of estrogen. The production of iridescent structure in this particular bar is probably controlled by genetic and not by hormonal factors.

In the back tract of Amherst pheasants there is a definite difference in the angle made by the pale brown transverse bars with the shaft between normal female feathers and those grown on males receiving estrogen injections. This difference is dependent upon the angle at which the barbs join the shaft and is probably determined in the early feather germ. It is a sexual character which is not affected by the hormonal environment.

Estrogens also affect the growth rate of male soma feathers in Amherst pheasants. Normal male feathers on different tracts grow at approximately the same rate, although some grow for longer periods than others and attain a greater final size. Normal female feathers, on the other hand, show considerable differences in growth rate. With injections of estrogens the growth curves of male soma feathers approach those of normal female feathers. In addition, some feathers have a higher threshold of response to hormones than others, and this is correlated not only with faster growth, as found in Leghorns by Juhn *et al.* (1931), but also with a longer growth period, a greater final size, and the larger size of the barbule cells.

These experiments on the plumage of Amherst pheasants show that some sexually dimorphic characters, such as pigmentation, barbule structure, and growth rate, are clearly influenced by estrogens. However, other feather characters, such as the angle of barring and the actual production of certain bars, are not affected by estrogen and appear to be controlled by genetic factors.

4. Plumage Control in Ducks

In the mallard, *Anas platyrhynchos,* the female wears a hen-type plumage throughout life, and the drake carries a similar plumage during the breeding season, but molts after this to a brilliant cock-type which is retained for 8 months until the succeeding nuptial season. These changes are, therefore, the reverse of those occurring in male weaver finches. In some other ducks both sexes wear a hen-type plumage throughout life and there are also species in which the male and female carry a cock-type plumage. The mallard is the only duck in which the method of plumage control has been investigated experimentally. If male or female mallards are castrated their plumages become permanently male in type. The production of the hen-feathering in the drakes only occurs when the testes are fully active. It has therefore been assumed that the hen-feathering of the breeding drake is induced by testis hormone, and Witschi (1961) has suggested that during the 8-month eclipse period a pituitary secretion, probably luteinizing hormone, is responsible for cock-feathering. The position is not, however, entirely clear because Caridroit (1938) succeeded in feminizing the plumage of presumptive mallard drakes by estrogen injections, but not by androgens. Further work on this and on other species of duck is obviously needed.

5. Plumage Control in Other Birds

The widespread sexual dimorphism of other birds has received relatively little attention, considering the numerous orders in the class Aves. In the house sparrow, *Passer domesticus,* the cock has a slate-grey crown, black throat and lores, white cheeks, and red-brown scapulars, whereas the hen has a more uniformly drab grey and olive-brown plumage. Keck (1934) showed that castration of either sex or the injection of gonadal or pituitary hormones has no effect on these plumages. Similar results have been obtained in the bullfinch, *Pyrrhula pyrrhula* (Novikow, 1936) and the chaffinch, *Fringilla coelebs* (Novikow, 1937). It appears, therefore, that in these species, and probably in other related passerines, the sexually dimorphic character of the feathers is under genetic control. This would be in accordance with the clearcut nature of the few cases of bilateral gynandromorphism recorded in this type of bird, e.g., in Gouldian finches, *Poephila gouldiae.* In the Javanese fighting quail, *Turnix pugnax,* the difference between the sexes also appears to be controlled by the genetic constitution (Witschi, 1961).

In some birds the plumage dimorphism is under the control of androgenic hormones. For instance, in the ruff, *Philomachus pugnax*, orchidectomy prevents the development of the typical male plumage (van Oordt and Junge, 1936). Nonetheless, during the breeding season the female, or reeve, assumes a plumage which is rather like an incomplete cock plumage; this may be due to the production of androgen by the female, possibly in the ovary.

Perhaps the most unexpected method of plumage control has been analyzed by Witschi and his associates in Iowa. In weaver finches (*Euplectes*), the hen birds have a typical hen-type plumage throughout life with an annual molt at the end of the breeding season. The cocks, on the other hand, have two molts in the year, wearing a cock-type plumage in the breeding season and hen-type plumage during the eclipse season. Castration of the males has no effect on these seasonal plumages, but ovariectomy causes the female to acquire seasonal plumages similar to those of the male (Witschi, 1935). Injection of estradiol or other estrogens into cocks about to molt to the breeding plumage inhibits the development of the cock-type feathers. Witschi considers that the cock-type plumage of weaver finches is controlled by a pituitary hormone. He found that as little as 1 rat unit of pregnant mares' serum will bring about these effects (Witschi, 1955), and he has postulated that the hormone concerned is probably the luteinizing hormone of the pituitary and that its effect in producing cock-type feathers is inhibited by estrogens (Witschi, 1961). This type of plumage control probably occurs in other related finches, such as the American indigo bunting, *Passerina cyanea*, and the masked weaver or dioch, *Quelea quelea*.

The development of the incubation patch in birds is also influenced by the ovary. This involves the loss of down feathers from a region of the breast and abdomen and a great increase in the thickness of the epidermis and in the water content and vascularization of the integument. Selander and Yang (1966) treated house sparrows (*Passer domesticus*) with estradiol which produced complete defeathering but only small increases in the growth of the integument. The effect of the estradiol was augmented by the addition of prolactin, which greatly increased dermal vascularity and edema. This subject has been recently reviewed by Jones (1971).

VII. MAMMALS

Although secondary sexual characters are relatively common among the mammals there is little experimental evidence to show that their development is influenced by the ovary.

In rabbits, *Oryctolagus cuniculus*, a dewlap is usually present in females but only rarely in males. Ovariectomy causes inhibition of the development of the dewlap or its involution, and injection of estrogens may induce its development in males (Hu and Frazier, 1938).

Periodic swelling of the sexual skin occurs in the females of various primates and appears to be a secondary sexual character associated with the follicular part of the cycle. Zuckerman and Fulton (1934) found that the injection of estrogen into an adolescent ovariectomized chimpanzee caused swelling of the sexual skin. Other anthropoids do not show these marked swellings, although in the orangutan and gorilla there is some activity of the skin in the genital area. Sexual skin changes occur in several catarrhine monkeys, particularly in the baboon (*Papio*), mandrill (*Mandrillus*), and mangabey (*Cercocebus*) (see Eckstein and Zuckerman, 1956). The degree of swelling varies considerably between individuals of the same species (Zuckerman, 1933). In *Papio* the swelling is dependent upon the stage of the menstrual cycle. Thus, Zuckerman and Parkes (1932) have shown that the swelling occurs during the follicular phase of the cycle, with the production of two turgid pink or red areas; during the middle of the cycle these swellings subside and then reappear at the time of the next menstruation. Zuckerman *et al.* (1938) found that injection of estrone into female rhesus monkeys, *Macaca mulatta,* causes maturation of the sexual skin; similar treatment of males also causes comparable changes in the corresponding skin areas.

In rats, mice, and dogs the spontaneous replacement of hair is retarded following the injection of estrogens (Gardner and De Vita, 1940; Emmens, 1942; Forbes, 1942, Hooker and Pfeiffer, 1943; Houssay, 1953; Mulligan, 1943); the pelage grown under these conditions is fine and sparse. It is probable that the adrenal cortex or corticosteroid has to be present for the complete action of estrogens (Baker and Whitaker, 1949), but there may also be a direct action on the skin and hair follicles since small topical applications of estrogen to the skin also inhibit hair growth in dogs (Williams *et al.*, 1946) and in the guinea pig (Whitaker, 1956).

In man, the development of axillary hair may be considered a secondary sexual character (Hamilton, 1958). It does not appear until sexual maturity, and is absent or nearly so in those castrated before puberty. After axillary hair has reached its full growth it still depends on stimulation by gonadal secretions and the hairs may atrophy after ovariectomy (or orchidectomy) and after the menopause. During the second decade of life the total weight of axillary hair is greater in females than it is in males, and it reaches a maximum in the second and third decades. Thereafter, the rate of decrease in weight of axillary hair is higher in females than in males and this appears to be associated with the decline in ovarian secretions. The weight of hair is reduced in young females after ovariectomy. These observations do not, however, take account of any possible changes in responsiveness of the hair follicles with age. It is also probable that genetic factors may influence the development of axillary hair. In Chinese and Japanese the weight of hair produced is considerably lower than that from subjects of Caucasian races, but there may be a genetic difference affecting the responsiveness of the hair follicles to the hormonal environment, perhaps comparable to the differences in feather follicle sensitivity between hen-feathered and cock-feathered races of domestic fowl.

In deer the growth of antlers is initiated at sexual maturity by testosterone. Antlers do not normally occur in female deer, except in the reindeer, *Rangifer tarandus*; in the occasional cases in which they have been observed in other species the ovary has been found to contain testicular tissue (Frazer, 1959).

In the brush-tailed possum, *Trichosurus vulpecula,* estrogen induces the development of the pouch in females (Bolliger and Carrodus, 1940). In the adolescent castrated male of this species estrogen has been shown to induce a transformation of the scrotum into a marsupial pouch (Bolliger and Tow, 1947).

There are many other mammals showing secondary sexual characters, for instance, the horns of some male antelopes, in which the method of control is still unknown.

REFERENCES

Adams, E. A. (1930). Studies on sexual conditions in *Triturus viridescens. J. Exp. Zool.* **55,** 63–79.

Adiyodi, K. G., and Adiyodi, R. G. (1970). Endocrine control of reproduction in decapod Crustacea. *Biol. Rev. Cambridge Philos. Soc.* **45,** 121–165.

Aron, M. (1926). Recherches morphologiques et expérimentales sur de déterminisme des caractères sexuels secondaires mâles chez les Anoures (*Rana esculenta* L. et *Rana temporaria* L.). *Arch. Biol.* **36,** 1–97.

Aronson, L. R. (1948). Problems in the behavior and physiology of a species of African mouth-breeding fish. *Trans. N.Y. Acad. Sci.* [2] **11,** 33–42.

Arvy, L., Echalier, G., and Gabe, M. (1954). Modifications de la gonade de *Carcinides* (*Carcinus*) *moenas* L., (Crustacés décapodes), après ablation dilaterale de l'organe Y. *C.R. Acad. Sci.* (*Paris*) **239,** 1853–1855.

Avel, M. (1929). Recherches expérimentales sur les caractères sexuels somatiques des Lombriciens. *Bull. Biol. Fr. Belg.* **63,** 149–318.

Baker, B. L., and Whitaker, W. L. (1949). Relationship of the adrenal cortex to inhibition of growth of hair by estrogen. *Am. J. Physiol.* **159,** 118–123.

Ball, J. N. (1960). Reproduction in female bony fishes. *Symp. Zool. Soc. London* **1,** 105–135.

Bauer, E. E. (1931). Ueber weibliche Sexualhormone bei einzelligen Tieren. *Naunyn-Schmiedebergs Arch. Exp. Pathol. Pharmakol.* **163,** 602–610.

Benoit, J. (1929). Le déterminisme des caractères sexuels secondaires du coq domestique. Etude physiologique et histo-physiologique. *Arch. Zool. Exp. Gén.* **69,** 217–499.

Berkowitz, P. (1937). Effects of estrogenic substances in *Lebistes reticulatus* (guppy). *Proc. Soc. Exp. Biol. Med.* **36,** 416–418.

Bier, K. (1954). Ueber den Einfluss der Königin auf die Arbeiterinnenfertilität im Ameisenstaat. *Insectes Soc.* **1,** 7–19.

Bolliger, A., and Carrodus, A. (1940). The effect of estrogens on the pouch of the marsupial *Trichosurus vulpecula. J. Proc. R. Soc. N.S.W.* **73,** 218–227.

Bolliger, A., and Tow, A. J. (1947). Late effects of castration and administration of sex hormones on the male *Trichosurus vulpecula. J. Endocrinol.* **5,** 32–41.

Bolton, W. (1953). The effect of sex hormones on comb growth in immature pullets. *J. Endocrinol.* **9,** 440–445.

Bretschneider, L. H., and Duyvené de Wit, J. J. (1947). "Sexual Endocrinology of Non-mammalian Vertebrates." Elsevier, Amsterdam.

Brough, J. (1937). On certain secondary sexual characters in the common dogfish (*Scyliorhinus caniculus*). *Proc. Zool. Soc. London* **107B,** 217–223.

Buchner, P. (1925). Studien an intracellulären Symbionten. *Z. Morphol. Oekol. Tiere* **4,** 88–244.

Burrows, H. (1949). Estrogens. *In* "Hormone Assay" (C. W. Emmens, ed.), p. 391. Academic Press, New York.

Butler, C. G. (1960). Sex determination and caste differentiation in the honey-bee (*Apis mellifera*). *Mem. Soc. Endocrinol.* **7,** 3–8.

Callan, H. G. (1939). The absence of a sex-hormone controlling regeneration of the hectocotylus in *Octopus vulgaris* Lam. *Pubbl. Stn. Zool. Napoli* **18,** 15–19.

Caridroit, F. (1926). Etude histo-physiologique de la transplantation testiculaire et ovarienne chez les Gallinacés. *Bull. Biol. Fr. Belg.* **60,** 135–312.

Caridroit, F. (1938). Recherches expérimentales sur les rapports entre testicules, plumage d'éclipse et mues chez le canard sauvage. *Trav. Stn. Zool. Wimereux* **13,** 47–67.

Carlisle, D. B. (1953). Studies on *Lysmata seticaudata* Risso (Crustacea Decapoda). V. The ovarian inhibiting hormone and the hormonal inhibition of sex-reversal. *Pubbl. Stn Zool. Napoli* **24,** 355–372.

Carlisle, D. B. (1959). On the sexual biology of *Pandalus borealis*: Crustacea Decapoda. II. The termination of the male phase. *J. Mar. Biol. Assoc. U.K.* **38,** 481–491.

Carlisle, D. B. (1960). Sexual differentiation in Crustacea Malacostraca. *Mem. Soc. Endocrinol.* **7,** 9–16.

Carlisle, D. B., and Butler, C. G. (1956). The "Queen-substance" of honeybees and the ovary-inhibiting hormone of crustaceans. *Nature* (*London*) **177,** 276–277.

Carlisle, D. B., and Knowles, F. (1959). "Endocrine control in Crustaceans." Cambridge Univ. Press, London and New York.

Champy, C. (1935). Recherches sur l'action des glandes génitales sur le plumage des oiseaux. *Arch. Anat. Microsc. Morphol. Exp.* **31,** 145–270.

Charniaux-Cotton, H. (1963). Démonstration expérimentale de la sécrétion d'hormone femelle par le testicule inversé en ovaire de *Talitrus saltator* (Crustacé Amphipode). Considerations sur le génétique et l'endocrinologie sexuelles des Crustacés supérieurs. *C.R. Acad. Sci.* (*Paris*) **256,** 4088–4091.

Charniaux-Cotton, H. (1965). Contrôle endocrinien de la différenciation sexuelle chez les crustacés supérieurs. *Arch. Anat. Microsc. Morphol. Exp.* **54,** 405–415.

Charniaux-Cotton, H. (1967). Endocrinologie et génétique de la différenciation sexuelle chez les Invertebrés. *C.R. Soc. Biol.* (*Paris*) **161,** 6–9.

Cohen, H. (1946). Effects of sex hormones on the development of the platyfish, *Platypoecilus maculatus. Zoologica* (*N.Y.*) **31,** 121–128.

Danforth, C. H. (1929). The effect of foreign skin on feather pattern in the common fowl (*Gallus domesticus*). *Wilhelm Roux' Arch. Entwicklungsmech. Org.* **116,** 242–252.

Danforth, C. H. (1930). The nature of racial and sexual dimorphism in the plumage of Campines and Leghorns. *Biol. Gen.* **6,** 99–108.

Danforth, C. H. (1937a). An experimental study of plumage in Reeves' pheasants. *J. Exp. Zool.* **77,** 1–12.

Danforth, C. H. (1937b). Artificial gynandromorphism and plumage in *Phasianus*. *J. Genet.* **34,** 497–506.

Danforth, C. H., and Foster, F. (1929). Skin transplantation as a means of studying genetic and endocrine factors in the fowl. *J. Exp. Zool.* **52,** 443–470.

Dantschakoff, V., and Vachkovitchuté, A. (1936). Sur l'immunité de la Drosophile envers les hormones sexuelles hétérologues. *C.R. Soc. Biol.* (*Paris*) **121,** 755–757.

Dean, F. D., and Chester-Jones, I. (1959). Sex steroids in the lungfish (*Protopterus annectens* Owen). *J. Endocrinol.* **18,** 366–371.

Deanesly, R., and Parkes, A. S. (1937). Multiple activities of androgenic compounds. *Q. J. Exp. Physiol.* **26,** 393–402.

de Beaumont, F. (1933). La différenciation sexuelle dans l'appareil uro-génital du triton et son déterminisme. *Wilhelm Roux' Arch. Entwicklungsmech. Org.* **129,** 120–178.

Dodd, J. M. (1955). The hormones of sex and reproduction and their effects in fish and lower chordates. *Mem. Soc. Endocrinol.* **4,** 166–184.

Dodd, J. M. (1960). Genetic and environmental aspects of sex determination in cold-blooded vertebrates. *Mem. Soc. Endocrinol.* **7,** 17–44.

Dodd, J. M., and Goddard, C. K. (1961). Some effects of oestradiol benzoate on the reproductive ducts of the female dogfish *Scyliorhinus caniculus*. *Proc. Zool. Soc. London* **137,** 325–331.

Dodd, J. M., Evennett, P. J., and Goddard, C. K. (1960). Reproductive endocrinology in cyclostomes and elasmobranchs. *Symp. Zool. Soc. London* **1,** 77–103.

Domm, L. V. (1924). Sex reversal following ovariectomy in the fowl. *Proc. Soc. Exp. Biol. Med.* **22,** 28–35.

Domm, L. V. (1927). New experiments on ovariotomy and the problem of sex inversion in the fowl. *J. Exp. Zool.* **48,** 31–173.

Domm, L. V. (1929). The effects of bilateral ovariotomy in the Brown Leghorn fowl. *Biol. Bull.* (*Woods Hole, Mass.*) **56,** 459–492.

Donahue, J. K. (1934). Sex-limitation of cilia in the body cavity of the frog *Rana pipiens*. *Proc. Soc. Exp. Biol. Med.* **31,** 1166–1168.

Donahue, J. K. (1941). Occurrence of estrogens in the ovaries of the winter flounder. *Endocrinology* **28,** 519–520.

Donahue, J. K. (1949). Determination of natural estrogens in marine eggs by biological and fluorimetric procedures. *Am. J. Physiol.* **159,** 567–568.

Donahue, J. K., and Jennings, E. W. (1937). The occurrence of oestrogenic substances in the ovaries of echinoderms. *Endocrinology* **21,** 690–691.

Eckstein, P., and Zuckerman, S. (1956). Morphology of the reproductive tract. *In* "Marshall's Physiology of Reproduction" (A. S. Parkes, ed.), 3rd ed., Vol. 1, Part 1, Chapter 2. Longmans Green, London and New York.

Egami, N. (1954). Effects of oestrogen on testis of the loach, *Misgurnus anguillicaudatus*. *J. Fac. Sci., Univ. Tokyo, Sect. 4* **7,** 121–130.

Egami, N. (1960a). Stimulative effects of estrogen and androgen on the growth of

the unpaired fins in the gobiid *Chaenogobius annularis*. (Preliminary report.) *Annot. Zool. Jpn.* **33,** 104–109.
Egami, N. (1960b). Comparative morphology of the sex characters in several species of Japanese gobies, with reference to the effects of sex steroids on the characters. *J. Fac. Sci., Univ. Tokyo, Sect. 4* **9,** 67–100.
Emmens, C. W. (1942). The endocrine system and hair growth in the rat. *J. Endocrinol.* **3,** 64–78.
Emmens, C. W., and Parkes, A. S. (1940). The endocrine system and plumage types. IV. Feminization of plumage, with especial reference to henny cocks and eclipse drakes. *J. Genet.* **39,** 503–515.
Eversole, W. J. (1941). The effects of pregneninolone and related steroids on sexual development in fish (*Lebistes reticulatus*). *Endocrinology* **28,** 603–610.
Finlay, G. F. (1925). Studies on sex differentiation in fowls. *Br. J. Exp. Biol.* **2,** 439–468.
Forbes, T. R. (1942). Sex hormones and hair changes in rats. *Endocrinology* **30,** 465–468.
Frazer, J. F. D. (1959). "The Sexual Cycles of Vertebrates." Hutchinson, London.
Freud, J., de Jongh, S. E., and Laqueur, E. (1929). Ueber die Veränderung des Federkleides bei Hühnern durch Menformon. *Proc. K. Ned. Akad. Wet.* **32,** 1054.
Gardner, W. V., and De Vita, J. (1940). Inhibition of hair growth in dogs receiving estrogens. *Yale J. Biol. Med.* **13,** 213–216.
Giard, A. (1888). Sur le castration parasitaire chez les Eukyphotes des genres *Palaemon* et *Hippolyte*. *C.R. Acad. Sci.* (*Paris*) **106,** 502–505.
Goodale, H. D. (1913). Castration in relation to the secondary sexual characters in Brown Leghorns. *Am. Nat.* **47,** 159–169.
Greenwood, A. W. (1928). Studies on the relation of gonadic structure to plumage characterisation in the domestic fowl. IV. Gonad cross-transplantation in Leghorn and Campine. *Proc. R. Soc. London, Ser. B* **103,** 73–81.
Haemmerli-Boveri, V. (1926). Ueber die Determination der sekundären Geschlechtsmerkale (Brutsackbildung) der weiblichen Wasserassel durch das Ovar. *Z. Vergl. Physiol.* **4,** 688–698.
Hamilton, J. B. (1958). Age, sex and genetic factors in the regulation of hair growth in man: a comparison of Caucasian and Japanese populations. *In* "The Biology of Hair Growth" (W. Montagna and R. A. Ellis, eds.), pp. 399–433. Academic Press, New York.
Hamon, M. (1946). Action des hormones sexuelles de synthèse sur la morphologie externe de *Gambusia Holbrooki* Gir. *Bull. Soc. Hist. Nat. Afr. N.* **37,** 122–141.
Hanström, B. (1939). "Hormones in Invertebrates." Oxford Univ. Press (Clarendon), London and New York.
Harms, W. (1912). Ueberpflanzung von Ovarien in eine fremde Art. I. Mitteilung: Versuche an Lumbriciden. *Arch. Entwicklungsmech. Org.* **34,** 90–131.
Hoar, W. S. (1955). Reproduction in teleost fish. *Mem. Soc. Endocrinol.* **4,** 5–24.
Hoar, W. S. (1965). Comparative physiology: hormones and reproduction in fish. *Annu. Rev. Physiol.* **27,** 51–70.
Hooker, C. W., and Pfeiffer, C. A. (1943). Effects of sex hormones upon body growth, skin, hair and sebaceous glands in the rat. *Endocrinology* **32,** 69–76.
Houssay, B. A. (1953). The relationship of the gonads and adrenals to the growth of hair in mice and rats. *Acta Physiol. Lat. Am.* **3,** 232–246.
Hu, C. K., and Frazier, C. N. (1938). Relationship between female sex hormone and dewlap in the rabbit. *Proc. Soc. Exp. Biol. Med.* **38,** 116–119.

Hunter, J. (1780). Account of an extraordinary pheasant. *Philos. Trans. R. Soc. London* **70,** 527–535.
Jones, R. E. (1971). The incubation patch of birds. *Biol. Rev. Cambridge Philos. Soc.* **46,** 315–339.
Juhn, M., and Gustavson, R. G. (1929). Effect of the female hormone upon the sex type of the feathers of Brown Leghorns. *Anat. Rec.* **44,** 204.
Juhn, M., and Gustavson, R. G. (1930). A 48-hour test for the female hormone with capon feathers as indicator. *Proc. Soc. Exp. Biol. Med.* **27,** 747–748.
Juhn, M., Faulkner, G. H., and Gustavson, R. G. (1931). The correlation of rates of growth and hormone threshold in the feathers of the fowl. *J. Exp. Zool.* **58,** 69–106.
Keck, W. N. (1934). The control of secondary sexual characters in the English sparrow, *Passer domesticus* (Linnaeus). *J. Exp. Zool.* **67,** 315–347.
Knowles, F. G. W. (1939). The influence of anterior pituitary and testicular hormones on the sexual maturation of lampreys. *J. Exp. Biol.* **16,** 535–547.
Kopeč, S. (1927). Experiments on the dependence of the nuptial hue in the gonads of fish. *Biol. Gen.* **3,** 259–280.
Kopeč, S., and Greenwood, A. W. (1930). The effect of yolk injections on the plumage of an ovariotomised Brown Leghorn hen. *Wilhelm Roux' Arch. Entwicklungsmech. Org.* **121,** 87–95.
le Roux, M. L. (1933). Recherches sur la sexualité des Gammariens. *Bull. Biol. Fr. Belg., Suppl.* **16,** 1–138.
Linke, O. (1934). Ueber die Beziehungen zwischen Keimdrüse und Soma bei Prosobranchiern. *Verh. Dtsch. Zool. Ges.* **36,** 164–175.
Loewe, S., Randenbusch, W., Voss, H. E., and van Heurn, W. C. (1922). Nachweis des Sexualhormon-Vorkommens bei Schmetterlingen. *Biochem. Z.* **244,** 347–356.
Lüscher, M. (1953). Kann die Determination durch eine monomolekulare Reaktion ausgelöst werden? *Rev. Suisse Zool.* **60,** 524 and 528.
Lüscher, M. (1956). Die Entstehung von Ersatzgeschlechtstieren bei der Termite *Kalotermes flavicollis* Fabr. *Insectes Soc.* **3,** 119–128.
Matthey, R. (1929). La spermatogénèse du lézard *Lacerta muralis* Lin. *Z. Zellforsch. Mikrosk. Anat.* **8,** 671–690.
Mohsen, T. (1958). Masculinizing action of pregneninolone on female gonads in the cyprinodont, *Lebistes reticulatus. Nature* (*London*) **181,** 1074.
Morgan, T. H. (1915). Demonstration of the appearance after castration of cock-feathering in a hen-feathered cockerel. *Proc. Soc. Exp. Biol. Med.* **13,** 31.
Mulligan, R. M. (1943). Hair loss in male dogs fed stilbestrol. *Proc. Soc. Exp. Biol. Med.* **54,** 21–22.
Naisse, J. (1966). Contrôle endocrinien de la différenciation sexuelle chez l'insecte *Lampyris noctiluca* (Coleoptere Malacoderme Lampyride). I. Rôle androgène des testicules. *Arch. Biol.* **77,** 139–201.
Noble, G. K., and Pope, S. H. (1929). The modification of the cloaca and teeth of the adult salamander, *Desmognathus,* by testicular transplants and by castration. *Br. J. Exp. Biol.* **6,** 399–410.
Novikow, B. G. (1936). Die Analyse des Geschlechtsdimorphismus bei den Sperlingsvögsln (Passeres). II. *Biol. Zentralbl.* **56,** 415–428.
Novikow, B. G. (1937). Die Analyse des Geschlechtsdimorphismus bei den Sperlingsvögeln (Passeres). IV. *Acta Zool.* (*Stockholm*) **18,** 447–458.
Okada, Y. K., and Yamashita, H. (1944a). Experimental investigation of the manifestation of secondary sexual characters in fish, using the medaka, *Oryzias*

latipes (Temminck and Schlegel), as material. *J. Fac. Sci., Univ. Tokyo Sec. 4, Zoology* **6,** 383–437.
Okada, Y. K., and Yamashita, H. (1944b). Experimental investigation of the sexual characters of poeciliid fish. *J. Fac. Sci., Univ. Tokyo Sec. 4, Zoology* **6,** 589–633.
Oudemans, J. T. (1899). Falter aus kastrierten Raupen, wie sie aussehen und wie sie sich benehmen. *Zool. Jahrb., Abt. Syst. Vekol. Geogr. Tiere* **12,** 73–88.
Paul, H. (1937). Transplantation und Regeneration der Flügel zur Untersuchung ihrer Formbildung bei einem Schmetterling mit Geschlechts-dimorphismus, *Orgyia antiqua* L. *Wilhelm Roux' Arch. Entwicklungsmech. Org.* **136,** 64–111.
Pézard, A. (1918). Le conditionnement physiologique des caractères sexuels secondaires chez les oiseaux. Du rôle endocrine des glandes génitales. *Bull. Biol. Fr. Belg.* **52,** 1–176.
Prell, H. (1915). Ueber die Beziehungen zwischen den primären und sekundären Sexualcharakteren bei Schmetterlingen. *Zool. Jahrb., Abt. Allg. Zool. Physiol. Tiere* **35,** 183–224.
Regamey, J. (1935). Les caractères sexuels du lézard (*Lacerta agilis* L.). *Rev. Suisse Zool.* **42,** 87–168.
Régnier, V. (1938). Action de la folliculine sur la crête des coqs normaux. *C. R. Soc. Biol.* (*Paris*) **127,** 519–521.
Schwerdtfeger, H. (1931). Beiträge zum Vorkommen und zur Wirkung der weiblichen Sexualhormone. *Naunyn-Schmiedebergs Arch. Exp. Pathol. Pharmakol.* **163,** 486–492.
Selander, R. K., and Yang, S. Y. (1966). The incubation patch of the house sparrow, *Passer domesticus* Linnaeus. *Gen. Comp. Endocrinol.* **6,** 325–333.
Sereni, E. (1929). Correlazioni umorali nei Cefalopodi. *Am. J. Physiol.* **90,** 512.
Sereni, E., Ashbel, R., and Rabinowitz, D. (1929). Azione degli estratti di uova di pesci. *Boll. Soc. Ital. Biol. Sper.* **4,** 746–749.
Siller, W. C. (1956). A Sertoli cell tumour causing feminization in a Brown Leghorn capon. *J. Endocrinol.* **14,** 197–203.
Steidle, H. (1930). Ueber die Verbreitung des weiblichen Sexualhormons. *Naunyn-Schmiedebergs Arch. Exp. Pathol. Pharmakol.* **157,** 89.
Takeuchi, K. (1968). Inhibition of large distal tooth formation in male Medaka, *Oryzias latipes,* by estradiol. *Experientia* **24,** 1061–1062.
Tavolga, M. C. (1949). Differential effects of estradiol, estradiol benzoate and pregneninolone on *Platypoecilus maculatus. Zoologica* (*N.Y.*) **34,** 215–237.
Valle, J. R., and Valle, L. A. R. (1943). Gonadal hormones in snakes. *Science* **97,** 400.
Vandel, A. (1920). Le développement de l'appareil copulateur des planaires est sous la dépendance des glandes génitales. *C.R. Acad. Sci.* (*Paris*) **170,** 249–251.
Vandel, A. (1921). Recherches expérimentales sur les modes de reproduction des planaires triclades paludicoles. *Bull. Biol. Fr. Belg.* **55,** 343–518.
van Oordt, G. J., and Junge, G. C. A. (1933). Die hormonale Wirkung der Gonaden auf Sommer- und Prachtkleid. I. Der Einfluss der Kastration bei männlichen Lachmöwen (*Larus ridibundus* L.). *Wilhelm Roux' Arch. Entwicklungsmech. Org.* **128,** 166–180.
van Oordt, G. J., and Junge, G. C. A. (1936). Die hormonale Wirkung der Gonaden auf Sommer- und Prachtkleid. III. Der Einfluss der Kastration auf männliche Kampfläufer (*Philomachus pugnax*). *Wilhelm Roux' Arch. Entwicklungsmech. Org.* **134,** 112–121.

Vevers, H. G. (1954). The experimental analysis of feather pattern in the Amherst pheasant, *Chrysolophus amherstiae* (Leadbeater). *Trans. Zool. Soc. London* **28,** 305–348.

Visnak, J. (1935). Versuche über die Wirkung des Follikelhormons auf einige Oligochaeten. *Vest. Cesk. Spol. Zool.* **2,** 48–50.

Weisman, A. I., Mishkind, D. I., Kleiner, I. S., and Coates, C. W. (1937). Estrogenic hormones in the ovaries of swordfish. *Endocrinology* **21,** 413–414.

Welti, E. (1928). Evolution des graffes de glandes génitales chez le crapaud (*Bufo vulgaris*). Auto-, homo-, hétérogreffes. *Rev. Suisse Zool.* **35,** 75–200.

Whitaker, W. L. (1956). Local inhibition of hair growth in the guinea pig by the topical application of alpha-estradiol. *Anat. Rec.* **124,** 447–448.

Williams, W. L., Gardner, W. U., and De Vita, J. (1946). Local inhibition of hair growth in dogs by percutaneous application of estrone. *Endocrinology* **38,** 368–375.

Witschi, E. (1935). Seasonal sex characters in birds and their hormonal control. *Wilson Bull.* **47,** 177–188.

Witschi, E. (1955). Vertebrate gonadotrophins. *Mem. Soc. Endocrinol.* **4,** 149–165.

Witschi, E. (1961). Sex and secondary sexual characters. *In* "Biology and Comparative Physiology of Birds" (A. J. Marshall, ed.), Vol. 2, pp. 115–168. Academic Press, New York.

Wotiz, H. H., Botticelli, C., Hisaw, F., Jr., and Ringler, I. (1958). Identification of estradiol-17β from dogfish ova (*Squalus suckleyi*). *J. Biol. Chem.* **231,** 589–592.

Zahl, P. A., and Davis, D. D. (1932). Effects of gonadectomy on the secondary sexual characters in the ganoid fish, *Amia calva* Linnaeus. *J. Exp. Zool.* **63,** 291–304.

Zawadowsky, M. M. (1922). "Das Geschlecht und die Entwicklung der Geschlechtsmerkmale." Gosudarstvennoe Izdatel'stvo Moscow.

Zawadowsky, M. M. (1928). Tissue equipotentiality in male and female birds and mammals. *Trans. Lab. Exp. Biol. Zoo Park Moscow* **4,** 9–67.

Zuckerman, S. (1933). "Functional Affinities of Man, Monkeys and Apes." Kegan Paul, London.

Zuckerman, S., and Fulton, J. F. (1934). The menstrual cycle of the Primates. Part VII. The sexual skin of the chimpanzee. *J. Anat.* **69,** 38–46.

Zuckerman, S., and Parkes, A. S. (1932). The menstrual cycle of the Primates. Part V. The cycle of the baboon. *Proc. Zool. Soc. London* pp. 139–191.

Zuckerman, S., van Wagenen, G., and Gardiner, R. H. (1938). The sexual skin of the rhesus monkey. *Proc. Zool. Soc. London* **108A,** 385–401.

Author Index

Numbers in italics refer to the pages on which the complete references are listed.

C

E

F

G

H

M

S

Subject Index

D

H

I

N

O

P

T

A 7
B 8
C 9
D 0
E 1
F 2
G 3
H 4
I 5
J 6